A Practical Approach to Biofungicide Formulation

Methods and Protocols

Editors

Kamel A. Abd-Elsalam

Plant Pathology Research Institute
Agricultural Research Center
Giza, Egypt

and

Mirza Abid Mehmood

Plant Pathology, Institute of Plant Protection
Muhammad Nawaz Shareef University of Agriculture
Multan, Pakistan

CRC Press
Taylor & Francis Group
Boca Raton London New York

CRC Press is an imprint of the
Taylor & Francis Group, an **informa** business

A SCIENCE PUBLISHERS BOOK

Cover credit: Image reproduced by kind courtesy of Dr. Kamel A. Abd-Elsalam.

First edition published 2026
by CRC Press
2385 NW Executive Center Drive, Suite 320, Boca Raton FL 33431

and by CRC Press
4 Park Square, Milton Park, Abingdon, Oxon, OX14 4RN

CRC Press is an imprint of Taylor & Francis Group, LLC

Library of Congress Cataloging-in-Publication Data (applied for)

ISBN: 978-1-032-98225-0 (hbk)
ISBN: 978-1-032-99230-3 (pbk)
ISBN: 978-1-003-60302-3 (ebk)

DOI: 10.1201/9781003603023

Typeset in Times New Roman
by Prime Publishing Services

Preface

Over the last decade, biofungicides have developed a degree of sophistication in relation to integrated plant disease management, as they possess limited detrimental effects on the environment and beneficial microorganisms, with a minimal chance of resistance development. Their market is growing at a rapid rate due to increased demand for eco-friendly management options. This book provides practical guidelines for the selection of biocontrol agents in biofungicide formulation. Their efficacy will be enhanced by advances in formulation technology, which are critical for strengthening integrated disease management. The compilation gives an extensive update on novel methodologies related to the maintenance, propagation, and formulation of biofungicides.

The use of biocontrol agents necessitates the existence of specialized carriers that transport them to host plants, protecting them against adverse conditions and thus enhancing viability. This book also explores the importance of carriers and stabilizers and their specificity to biocontrol agents. It also provides further details on the utilization of biofungicides and how they combine with precision agriculture. *Trichoderma, Bacillus, Pseudomonas,* and *Streptomyces* play a crucial role as biofungicides, attracting the interest of researchers and farmers. The effectiveness of biofungicides is determined by the shelf life and stabilization of the biocontrol agents, where certain stabilizing compounds are employed to shield them against environmental pressures, guaranteeing extended survival and preservation of viability. Moreover, challenges in the large-scale production of biofungicides and mitigation strategies for advanced formulations that offer sustainable, residue-free, and environmentally safe crop protection are discussed in detail.

In short, this book is an invaluable source of information for students, academicians, and industry professionals interested in selecting suitable biofungicidal agents, carriers, adjuvants, their compatibility, and large-scale production of biofungicides. By highlighting their development and action mechanisms in sustainable disease management, we aim to inspire further research and innovation in this crucial field, contributing to eco-friendly and resilient agricultural practices.

We sincerely thank all the contributing authors for their excellent work, unwavering support, and valuable insights in crafting these thought-provoking and scholarly chapters, which have been instrumental in shaping this compilation. We extend our heartfelt gratitude to each reviewer who generously dedicated their valuable time to complete the review process. Lastly, we thank our friends, family, and colleagues for their continuous support and encouragement throughout this endeavor.

Kamel A. Abd-Elsalam
Agricultural Research Center, Giza, Egypt

Mirza Abid Mehmood
MNS University of Agriculture, Multan, Pakistan

Contents

Preface iii

I. Development and Formulation of Biofungicides

1. **A Comprehensive Guide to Biofungicides' Development** 3
Kamel A. Abd-Elsalam and *Mirza Abid Mehmood*

2. **Selection and Screening of Biofungicidal Agents** 12
Anitha Rajasekaran

3. **Culture Maintenance and Propagation of Biocontrol Agents for Biofungicide Formulation** 34
Muhammad Mazhar Iqbal, Binnian Tian, Mirza Abid Mehmood and *Muhammad Ashfaq*

4. **Biofungicides: Innovative Formulations and Development** 63
Mohammadhassan Gholami-Shabani, Katayoon Nematpour, Masoomeh Shams-Ghahfarokhi, Farnoush Asghari-Paskiabi, Shahrzad Saeidpour, Mehrad Eslamifar, Zeynab Gholami-Shabani, Ali Eslamifar, Fatemehsadat Jamzivar and *Mehdi Razzaghi-Abyaneh*

5. **Carriers, Adjuvants, and Surfactants in the Formulation and Application of Biofungicides** 118
Kamel A. Abd-Elsalam, Ayat F. Hashim, Nada Essam Abo-Rashed, Hassan Almoammar, Rawan K. Hassan and *Toka E. Abdelkhalek*

6. **Development of Biofungicide Formulations: From General Methods to Carrier Selection and Compatibility** 149
Niranjana Rajesh, Gauriparvathy K. B., Dileep Francis and *Sabu Abdulhammed*

7. **Biofungicides: Scale-up, Production, and Emerging Innovations in Sustainable Disease Management** 175
Muhammad Umair Rafiq, Akhtar Hameed, Muhammad Waqar Alam, Abdul Rehman, Rana Binyamin, Muhammad Arslan Khan and *Subhan Ali*

8. Development of a Biofungicide Formulation to Control Wheat Blast Disease: Methods and Protocols **207**
Lamya Muazzeda Medha, Dipali Rani Gupta and *Tofazzal Islam*

9. Formulation of Biofungicides Based on Streptomyces **242**
Khadija Goura, Najwa Seddiqi Kallali, Abdelaaziz Farhaoui and *Rachid Lahlali*

10. *Pseudomonas* Based Biocontrol: Large-scale Production, Formulation, Quality Control and Delivery Methods **267**
Anurag Yadav and *Kusum Yadav*

11. *Trichoderma* Biocontrol: Large-scale Production, Formulation Techniques, Stringent Quality Control Measures, and Efficient Delivery Methods **311**
Saidul Islam, Arghya Banerjee, Bholanath Mondal, Malay Kumar Samanta, Kamel Abd-Elsalam, Maksud Hasan Shah, Sk Naim Aktar and *Mahatab Ali*

Index **351**

About the Editors **353**

I
Development and Formulation of Biofungicides

1

A Comprehensive Guide to Biofungicides' Development

Kamel A. Abd-Elsalam[1,*] and *Mirza Abid Mehmood*[2]

1. Introduction

In recent years, the agricultural landscape has undergone a remarkable transition toward sustainable practices due to the push to reduce chemical inputs and environmental issues. Among the multifaceted sustainable alternatives to traditional crop protection, biofungicides have become a central choice in plant disease management (Palmieri et al. 2022). Biofungicides are derived from natural organisms or substances that offer a potent and eco-friendly means to combat fungal pathogens (Fenta et al. 2024). The successful application of biofungicides relies heavily on effective formulation strategies, and making a practical guide on biofungicide formulation is also essential for researchers, agronomists, and practitioners (Ezeorba et al. 2023). The importance of a solid biofungicide formulation is imperative because, clearly, a formulation that is well designed will deliver and perform better, and enhance the ability of the biocontrol agent to target specific pathogens and reduce impacts on non-target organisms. The formulation includes numerous ingredients that could include active biological agents, carriers (liquid, solid, etc.), surfactants (wetting agents), and stabilizer agents (Katz et al. 2022). Each of these ingredients can have various impacts on total biofungicide performance. Knowledge of formulation principles will aid researchers and practitioners in preparing biofungicides that can lead to improved performance and product development in compliance with regulatory protocols and acceptance in the marketplace (Galindo et al. 2013).

Moreover, the growing global consumption of organic products is resulting in a rapid expansion of biopesticide markets in response to the demand for more

[1] Plant Pathology Research Institute, Agricultural Research Center, Giza 12619, Egypt.
[2] Plant Pathology, Institute of Plant Protection, Muhammad Nawaz Shareef University of Agriculture, Multan, Pakistan.
* Corresponding author: kamelabdelsalam@gmail.com

DOI: 10.1201/9781003603023-2

sustainable agricultural practices. Therefore, this dynamic situation also offers some opportunities and challenges to manufacturers. The purpose of this practical guide is to assist users in navigating through the challenges associated with implementing biopesticides while creating an understanding of current formulations, testing methods, and best practices. Furthermore, by understanding the principles of relevant and practical formulation, users will be able to adopt practices that are likely to engender innovations that will advance the sustainable use of biofungicidal products in modern agriculture (Muhammad et al. 2022).

In conclusion, the formulation of biofungicides is a pivotal construct of modern agricultural practice; it is an area of research that deserves attention and attention to detail. This book aims to underscore the importance of practical experience and formulation strategies, with much of the content highlighting the processes and practical aspects that would allow the user to engage in developing effective, environmentally safe biofungicides. By adopting these practices, we can continue to aid in the more sustainable use of farm inputs and contribute to a healthier planet, more food security, and greater agricultural resilience for the upcoming generations.

1.1 Agriculture and its Challenges

Agriculture is the backbone of many developing economies with crucial contributions to economic development and food security. It contributes in many ways to employment, increased income, and raw material for many sectors. In many developing countries, agriculture employs a large part of the workforce, which reduces poverty and increases economic growth. For example, in much of sub-Saharan Africa, smallholder farmers contribute the bulk of agricultural production, which highlights the importance of agriculture in providing livelihoods and economic stability (Livingston et al. 2011). However, agriculture is now facing many challenges that act as major hindrances to global food security. Some of these challenges include climate change, water scarcity, soil degradation, insect pests and disease outbreaks, population growth, trade and market instability, and technological barriers (Devlet 2021). The population of our global village is increasing at an alarming rate; it is projected to grow from approximately 7.3 billion to 8.5 billion by 2030, surpassing 9.5 billion by 2050, and exceeding 11 billion by 2100 (De Mello-Sampayo et al. 2024). These challenges lead to food insecurity for this growing population.

1.2 Plant Diseases

Plant diseases act as one of the primary threats in this regard, causing 20 to 40 percent crop losses globally (Junaid and Gokce 2024). Losses caused by these diseases result in food insecurity and undermine the global goal of eliminating hunger and malnutrition. The issue becomes more complex with urbanization, climate change, and the development of resistance in fungi against traditional fungicides (Islam et al. 2024). Diseases caused by fungal plant pathogens contribute significantly to these losses.

1.2.1 Synthetic Fungicides

The application of chemical agents, specifically synthetic fungicides, to manage fungal diseases in crops and landscape plants is a core solution for agricultural food production. These fungicides have been continuously used for the last few decades. However, the repeated, often needless, use of fungicides can have negative implications for the environment, non-target organisms, and human health (Yang et al. 2011). The first category of concern with fungicide use is the development of resistance in pathogenic fungi. When fungicides are used repeatedly to control a pathogenic fungus, the pathogen can evolve into a resistant strain, thereby compromising the effectiveness of the fungicides. The producer may then be forced to apply higher doses of the original fungicide or switch to more toxic alternatives (Kim and Hwang 2007). In developing levels of resistance, health implications include that some strains of resistant fungal pathogens can be cryptic in humans as well. For instance, *Botrytis cinerea* has developed resistance to various groups of chemical fungicides, including benzimidazoles, dicarboximides, QoIs, fenhexamid, fluopyram, and iprodione. Some fungicides, such as chlorothalonil and mancozeb, are known to cause problems in soil and water systems, creating issues for ecosystems (Nguyen 2023).

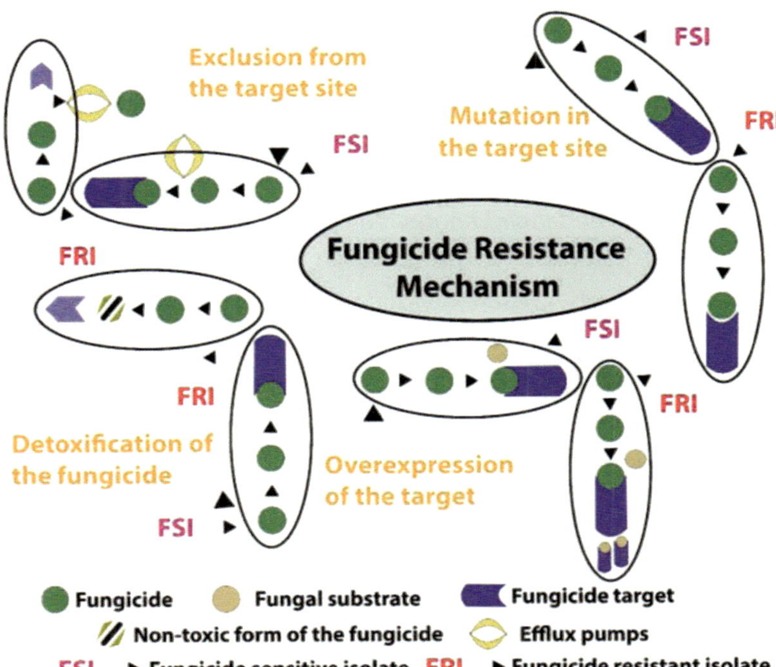

Figure 1: Development of synthetic fungicide resistance among fungal pathogens is multifaceted and is influenced by biological, environmental, and management factors. Proactive resistance management practices are thus required to maintain fungicide efficacy and sustain agricultural practices.

The most common mechanism of fungicide resistance is a change in the target site of fungicides. These changes occur due to mutations that arise at target sites during DNA replication. These mutations result in altered amino acid sequences that change the target sites (Figure 1) (Gisi et al. 2000). Hence, these fungal pathogens escape the effects of fungicides. Sometimes the target sites are overexpressed, which also results in resistance development. This overexpression helps the pathogen survive to some extent, despite the effect of the fungicides. Fungal pathogens can also detoxify chemical fungicide compounds. The fungal body mostly consists of efflux pumps that protect it from toxic compounds and foreign substances. Efflux systems sometimes pump fungicide compounds out of the fungal body (Ma et al. 2005, Jayawardana et al. 2024).

Environmental contamination is another serious issue as these chemicals can leach into soil and water systems and can lead to the accumulation of toxic substances in aquatic environments. These fungicides are also involved in targeting the non-targeted organisms including beneficial organisms such as arbuscular mycorrhizal fungi (AMF) (Calonne et al. 2010). These fungi play a vital role in plant nutrient uptake and soil health. The reduction of AMF populations can compromise soil fertility and plant growth, ultimately jeopardizing crop yields (Hage-Ahmed et al. 2019). In addition, the degradation products of some fungicides can cause more toxic conditions than the parent ones. For example, chlorothalonil fungicide has a breakdown product that is more toxic and persistent in the environment (Figure 2) (Van Scoy and Tjeerdema 2014). An integrated pest management strategy that includes cultural, biological, and mechanical methods, along with judicious use of chemical controls, can prove helpful in overcoming these challenges (Kumar et al. 2019).

1.2.2 Biofungicides and their Importance

Biofungicides are proven to be an efficient and sustainable substitute for synthetic fungicides. Biofungicides, derived from fungi or bacteria that are found in nature, are becoming an essential component of sustainable approaches to agriculture (Rettinassababady and Jeyalakshmi 2014). They offer an environmentally friendly alternative to chemical fungicides, as biofungicides use, exploit, or incorporate biological processes to alleviate or mitigate the effects of plant pathogens. The importance of biofungicides in agriculture is based on their suitability for the intended purpose, strict adherence to environmental safety, and their use in integrated plant disease management (Ezeorba et al. 2023). The principal benefit of using biofungicides is their ability to inhibit or decrease the influence of plant pathogens through various modes of action or mechanisms.

1.2.2.1 Modes of Action

Modes of action include competition for nutrients and space, the production of antimicrobial compounds, or the induction of plant defense responses against invaders (Figure 3) (Asad 2022). For example, microbial-derived biofungicides can prevent pathogen growth because they actively inhibit, compete, or eliminate competing organisms that could inflict disease and reduce disease incidence. Similarly, some

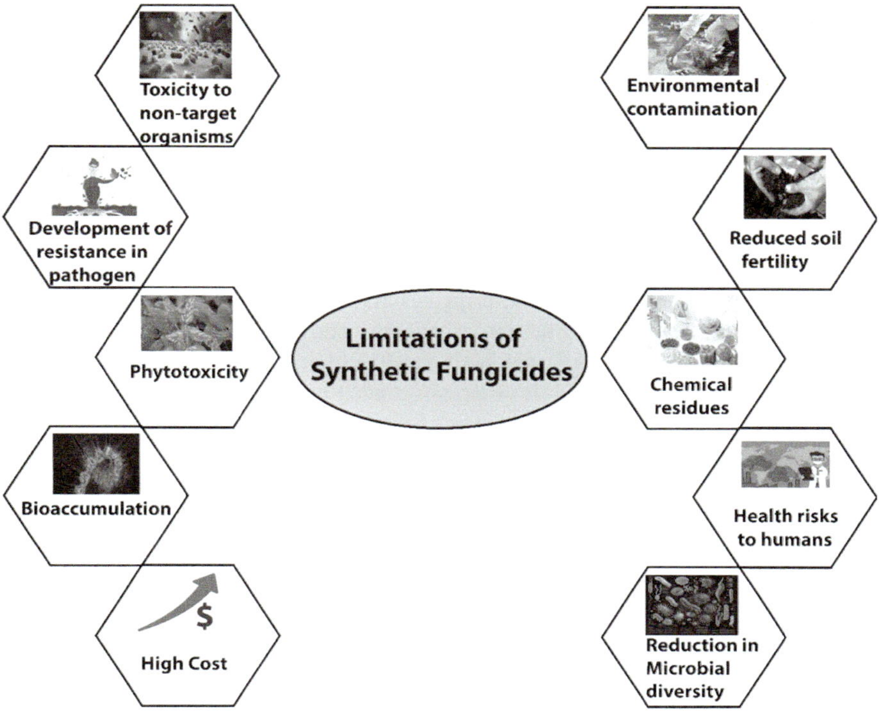

Figure 2: Diagram illustrating the limitations of synthetic fungicides: (although synthetic fungicides are vital to crop protection, their shortfalls call for integrated control practices and alternative methods like biological fungicides, crop rotation, and organic agricultural practices).

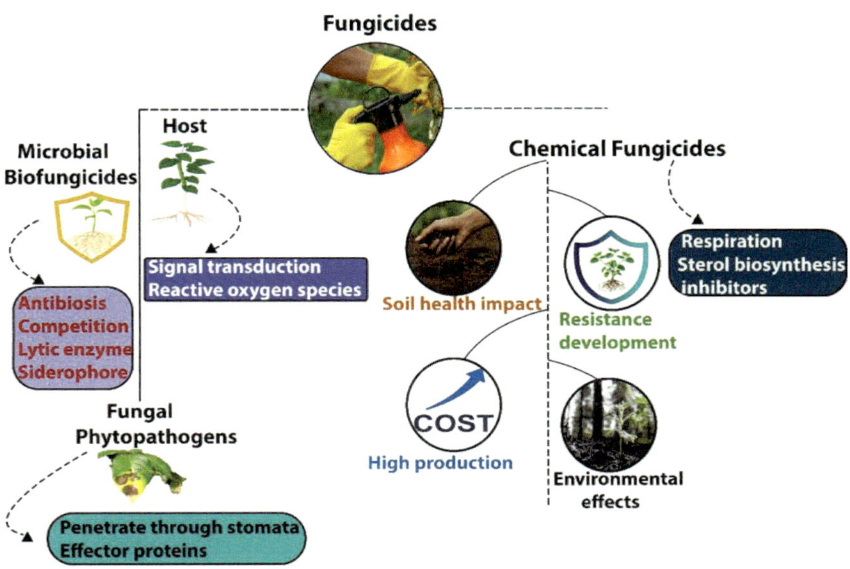

Figure 3: Modes of action of chemical and biofungicides against fungal pathogens along with their defense mechanisms.

biofungicides can also produce specific toxins that can target particular pathogens and enhance the effectiveness of the biofungicide approach (Narayanasamy 2013).

1.2.2.2 Environmental Benefits

Biofungicides are beneficial to the environment in several ways. Unlike synthetic fungicides, which can persist in the environment and harm non-target organisms, biofungicides are generally safer and biodegradable. They limit the probability of contaminating the environment and are less harmful to beneficial insects, soil microorganisms, and aquatic organisms. This notion is consistent with the increasing interest in developing reduced-footprint sustainable agriculture (Rettinassababady and Jeyalakshmi 2014). Biofungicides present numerous human health benefits by offering a safer alternative for both producers and consumers. Economically, biofungicides can be a viable disease management strategy. When applied preemptively, they can be as efficacious as chemical fungicides, especially for root and crown diseases. Some farmers have successfully used preparations like soaking plugs in biofungicide solutions or using biofriendly products as seed or transplant treatments. Furthermore, biofungicide products are often cheaper than conventional fungicides, which generates economic benefits for farmers (Abbey et al. 2019).

1.2.2.3 Integration with Conventional Disease Management strategies

Incorporating biofungicides into crop management programs, including fungicide resistance management programs, will improve overall disease control capabilities. When integrated into crop management systems (i.e., combined with synthetic fungicides), biofungicides can provide reliable levels of disease control and also mitigate the risk of developing resistance to synthetic fungicides due to multiple modes of action (Yin et al. 2023). This integrated approach to disease control provides overall sustainability and effectiveness over the long term for disease management programs and methods. Biofungicides can be used across a wide range of crops and challenges presented by pathogens. For example, in greenhouses, biofungicides can be used as preventive applications in combination with growth media or as seed treatments to manage root and crown diseases, achieving results comparable or superior to chemical fungicides (Keinath and DuBose 2012). Additionally, biofungicides can be used to manage foliar pathogens, as long as they are used as a preventive application (Pertot et al. 2016).

1.2.2.4 Importance of Practical Guide in Biofungicide Formulations

1.2.2.4.1 Role of Formulation in Biofungicide Effectiveness

The formulation and use of a practical guide on biofungicide formulations are required to facilitate effective use and integration of biofungicides into sustainable agricultural practices. Nevertheless, their successful use continues to be challenged by formulation, application, and knowledge gaps. Formulation is essential to the efficacy of biofungicides, as it concerns the manner in which the active biological agent is prepared to achieve viability, stability, and ease of application (Teixidó et al. 2022). Proper formulations protect the microorganism from adverse environmental factors, help sustain the agents during storage, and deliver the agent to the target pathogens. Without an established methodology for formulation, the performance of biofungicides can become unreliable, resulting in unacceptable disease control and diminished user trust. A practical guide would contain methodologies for developing effective formulations that provide biofungicidal agent viability, stability, and efficacy during production and use (Khan and Ahmad 2024).

1.2.2.4.2 Ensuring Product Quality and Standardization

The success of applying biofungicides is just as important as developing them. Chemical fungicides can provide broad-spectrum activity that acts quickly, while the ease of application for biofungicides can vary depending on environmental conditions and timing (Teixidó et al. 2022). Naturally, the timing of application, environmental conditions, and any interactions with other agricultural inputs in the field can influence the performance of a biofungicide. A suitable guide will provide specific application timing and environmental considerations for biofungicide applications based on different crops and growing environments to help improve the biofungicides' functionality and enable more farmers to adopt biofungicides into their routine management practices. The inconsistency in biofungicide production and quality control protocols poses challenges for end-users, as inconsistent products lead to inconsistent results and ultimately impact reliability. Moreover, variability in production and quality control of biofungicides presents significant challenges. Inconsistencies in product quality can lead to unpredictable results, undermining their reliability as a disease management tool. A practical guide would include protocols for quality assessment and standardization. It will enable producers to maintain high-quality products and users to verify the efficacy of biofungicides (Okoro et al. 2024).

1.2.2.4.3 Regulatory and Compliance Aspects

An in-depth understanding of how biofungicides interact with other agricultural inputs and practices is key to incorporating them into pest management plans. Farmers and agricultural practitioners will need assistance in this process, including how to synergistically combine biofungicides with cultural, mechanical, and chemical control methods to formulate effective integrated pest management (IPM) plans (Glare et al. 2022). A guide to Integrated Pest Management with biofungicides would provide useful information not only on synergistic combinations but also on antagonistic interactions. Additionally, the guide could identify ways biofungicide applications can combat resistance development, strengthen resistance management strategies, and help build more effective disease management programs. Furthermore, biofungicide registration is complex and varies not only from state to state but also from country to country. Challenges arise from registration, compliance, and safety requirements for both producers and pesticide users. A reference tool or instruction manual with a complete pathway for regulation of biofungicides to drive compliance and safety would be advantageous by mapping regulatory references and clearly articulating the processes.

References

Abbey, J. A., Percival, D., Abbey, Lord., Asiedu, S. K., Prithiviraj, B., and Schilder, A. (2019). Biofungicides as alternative to synthetic fungicide control of grey mould (*Botrytis cinerea*) – prospects and challenges. *Biocontrol Science and Technology*, 29(3): 207–228.

Asad, S. A. (2022). Mechanisms of action and biocontrol potential of Trichoderma against fungal plant diseases-A review. *Ecological Complexity*, 49: 100978.

Calonne, M., Fontaine, J., Debiane, D., Laruelle, F., Grandmougin-Ferjani, A., and Sahraoui, A. L. (2010). Propiconazole toxicity on the non-target organism, the arbuscular mycorrhizal fungus. *Glomus Irregulare*. Fungicides: 325–346.

De Mello-Sampayo, C., Viana, P., Lopes, A., Carvalho da Silva, R., de Jesus, R., Sarmento, G. et al. (2024). Survey of Antifungal in surface-and groundwater: a portuguese environmental case study. *Sustainability*, 16(2): 594.

Devlet, A. (2021). Modern agriculture and challenges. *Frontiers in Life Sciences and Related Technologies*, 2(1): 21–29.

Ezeorba, T. P. C., Chukwudozie, K. I., Okoye, C. O., Okeke, E. S., Ezugwu, A. L., and Anaduaka, E. G. (2023). Biofungicides: Classification, Applications and Limitations. pp. 12–39. *In:* Biofungicides: Eco-Safety and Future Trends. CRC Press.

Ezeorba, T. P. C., Chukwudozie, K. I., Okoye, C. O., Okeke, E. S., Ezugwu, A. L., and Anaduaka, E. G. (2023). Biofungicides: Classification, Applications and Limitations. pp. 12–39. *In:* Biofungicides: Eco-Safety and Future Trends. CRC Press.

Fenta, L., and Mekonnen, H. (2024). Microbial biofungicides as a substitute for chemical fungicides in the control of phytopathogens: Current perspectives and research directions. *Scientifica*, 2024(1): 5322696.

Galindo, E., Serrano-Carreón, L., Gutiérrez, C. R., Allende, R., Balderas, K., Patiño, M. et al. (2013). The challenges of introducing a new biofungicide to the market: A case study. *Electron. J. Biotech.*, 16(3): 5–5.

Gisi, U., Chin, K. M., Knapova, G., Färber, R. K., Mohr, U., Parisi, S. et al. (2000). Recent developments in elucidating modes of resistance to Phenylamide, DMI and Strobilurin fungicides. *Crop Prot.*, 19: 863–872.

Glare, T., McKinnon, A. C., Nicot, P. C., Pressecq, T., Bardin, M., Dara, S. K. et al. (2022). Improving Integrated Pest Management in Horticulture.

Hage-Ahmed, K., Rosner, K., and Steinkellner, S. (2019). Arbuscular mycorrhizal fungi and their response to pesticides. *Pest Management Science*, 75(3): 583–590.

Islam, T., Danishuddin, Tamanna, N. T., Matin, M. N., Barai, H. R., and Haque, M. A. (2024). Resistance mechanisms of plant pathogenic fungi to fungicide, environmental impacts of fungicides, and sustainable solutions. *Plants*, 13(19): 2737.

Jayawardana, M. A., and Fernando, W. G. D. (2024). The mechanisms of developing fungicide resistance in *Fusarium graminearum* causing Fusarium head blight and fungicide resistance management. *Pathogens*, 13(11): 1012.

Junaid, M. D., and Gokce, A. F. (2024). Global agricultural losses and their causes. *Bulletin of Biological and Allied Sciences Research*, 2024(1): 66–66.

Katz, J. S., Chou, D. K., Christian, T. R., Das, T. K., Patel, M., Singh, S. N. et al. (2022). Emerging challenges and innovations in surfactant-mediated stabilization of biologic formulations. *Journal of Pharmaceutical Sciences*, 111(4): 919–932.

Keinath, A. P., and DuBose, V. B. (2012). Controlling powdery mildew on cucurbit rootstock seedlings in the greenhouse with fungicides and biofungicides. *Crop Protection*, 42: 338–344.

Khan, E., and Ahmad, I. Z. (2024). Production and commercialization of biocontrol products. pp. 165–180. *In:* Biocontrol Agents for Improved Agriculture. Elsevier.

Kim, B. S., and Hwang, B. K. (2007). Microbial fungicides in the control of plant diseases. *Journal of Phytopathology*, 155(11-12): 641–653.

Kumar, N., Kumar, S., Singh, R., and Lal, M. (2019). Integrated disease management is a holistic approach in modern agriculture. *Journal of Pharmacognosy and Phytochemistry*, 8(4S): 98–101.

Livingston, G., Schonberger, S., and Delaney, S. (2011). Sub-Saharan Africa: The state of smallholders in agriculture. *IFAD Conference on New Directions for Smallholder Agriculture*, 24: 25.

Ma, Z., and Michailides, T. J. (2005). Advances in understanding molecular mechanisms of fungicide resistance and molecular detection of resistant genotypes in phytopathogenic fungi. *Crop Protection*, 24(10): 853–863.

Muhammad, M., Abdul Wahab, R., Huyop, F. Z., Rusli, M. H., Syed Yaacob, S. N., and Teo, H. L. (2022). An overview of the potential role of microbial metabolites as greener fungicides for future sustainable plant diseases management. *J. Crop Protect.*, 11(1): 1–27.

Narayanasamy, P. (2013). Mechanisms of Action of Fungal Biological Control Agents. pp. 99–200. *In:* Narayanasamy, P. (Ed.). Biological Management of Diseases of Crops. Springer Netherlands.

Nguyen, K. T. (2023). Effects of the fungicide mixture mancozeb, metalaxyl-M, and chlorothalonil on the degradation of ^ 14} C-MCPA in soil [PhD Thesis, Dissertation, RWTH Aachen University, 2023].

Okoro, C. A., El-Hasan, A., and Voegele, R. T. (2024). Integrating Biological Control Agents for Enhanced Management of Apple Scab (*Venturia inaequalis*): Insights, Risks, Challenges, and Prospects. *Agrochemicals*, 3(2): 118–146.

Palmieri, D., Ianiri, G., Del Grosso, C., Barone, G., De Curtis, F., Castoria, R. et al. (2022). Advances and perspectives in the use of biocontrol agents against fungal plant diseases. *Horticulturae*, 8(7): 577.

Pertot, I., Puopolo, G., Giovannini, O., Angeli, D., Sicher, C., and Perazzolli, M. (2016). Advantages and limitations involved in the use of microbial biofungicides for the control of root and foliar phytopathogens of fruit crops. *Italus Hortus.*, 23(3): 3–12.

Rettinassababady, C., and Jeyalakshmi, C. (2014). Bio-Fungicides: The best alternative for sustainable food security and ecosystem. pp. 401–411. *In:* Kharwar, R. N., Upadhyay, R. S., Dubey, N. K., and Raghuwanshi, R. (Eds.). Microbial Diversity and Biotechnology in Food Security. Springer India.

Teixidó, N., Usall, J., and Torres, R. (2022). Insight into a successful development of biocontrol agents: Production, formulation, packaging, and shelf life as key aspects. *Horticulturae*, 8(4): 305.

Van Scoy, A. R., and Tjeerdema, R. S. (2014). Environmental fate and toxicology of Chlorothalonil. *In*: Whitacre, D. M. (Ed.). Reviews of Environmental Contamination and Toxicology, 232: 89–105. *Springer International Publishing.*

Yang, C., Hamel, C., Vujanovic, V., and Gan, Y. (2011). Fungicide: modes of action and possible impact on nontarget microorganisms. *ISRN Ecology*, 2011, 1–8.

Yin, Y., Miao, J., Shao, W., Liu, X., Zhao, Y., and Ma, Z. (2023). Fungicide Resistance: Progress in Understanding Mechanism, Monitoring, and Management. *Phytopathology*®, 113(4): 707–718.

2

Selection and Screening of Biofungicidal Agents

Anitha Rajasekaran

1. Introduction

Increasing population demands more food production. In the recent decade, more resources have been diverted to increase food production (Tiwari et al. 2017, Kalsoom et al. 2020). A recent report from the United Nations Department of Economic and Social Affairs indicates that the world population is expected to increase from 7.3 billion to 8.5 billion by 2030. The projected population by 2100 would be more than 9.5 billion (Cenobio-Galindo et al. 2024). By 2050, more than 800 million people will not have enough food. Devastating diseases in 2100 will attack crops worldwide. Plant diseases caused by microbes have co-evolved with crops and have been causing significant yield losses. About 20–40% of the world's crop production is lost due to plant diseases (Waldamichael et al. 2022), accounting for 2000 billion dollars per year (Pimentel 2009). Among the plant pathogens, fungi cause a 45 billion dollar crop loss every year. To increase crop productivity and disease control, synthetic fungicides have been used for many years. About 90% of the applied fungicide is lost in the environment and soil, thus increasing the application cost for farmers. Hence, over-application of these substances has caused increased pathogen resistance, resurgence, and extinction. Resistant phytopathogens cause about 60 billion dollars loss every year (Gai and Wang 2024). About 2.5 million tons of pesticides are used every year, accounting for 100 billion dollars in damage annually. Moreover, they are a hazard to living things and the environment (Pelosi et al. 2021, Syromyatnikov et al. 2021, Kumar et al. 2019). They are also non-biodegradable and persist in the environment (Ayilara et al. 2023, Damalas and Eleftherohorinos 2011).

PG and Research Department of Botany, Bharathi Women's College, No-1, Prakasam Salai, Broadway, Chennai, Tamilnadu, India.
Email: anitha.rajasekaran023@gmail.com

DOI: 10.1201/9781003603023-3

The challenge is to overcome diverse plant diseases with sustainable agriculture, striking a balance between maintaining the environment and global food security (Kalogiannidis et al. 2022, Anani et al. 2020). To improve crop productivity, disease-resistant varieties and improved agronomic practices need to be followed (Pawlak and Kołodziejczak 2020, Ansari et al. 2012). In the era of sustainable agriculture, biological control is an effective method that deploys biocontrol agents to reduce pests without affecting the host, enhance growth-promoting factors, and maintain a balanced environment. Biofungicides are eco-friendly and beneficial since fungicide resistance can be avoided. Additionally, they pose less threat to humans and the environment.

Biofungicides can be classified as microbial, botanical, and nanobased fungicides (Figure 1). Microbial fungicides are of microbial origin and are cost-effective, easily accessible, and persist longer than synthetic fungicides. Phytofungicides are phytochemicals of plant origin with diverse modes of action. Nanobiofungicides are target-based, biodegradable, and have good compatibility with the pathogen. Microbial fungicides can be derived from bacteria or fungi. Microbial biofungicides have diverse modes of action, including antagonism, competitive saprophytic ability, antibiosis, hyperparasitism, siderophore production, hydrolytic enzymes, biofilm formation, hydrogen cyanide production, plant growth-promoting substances, and induction of host resistance (Gikas et al. 2022; Fenta et al. 2023).

The advantage of biofungicides is that they are non-toxic and safe for non-target organisms. They can be utilized as spores, living microbes, or mycelia. Application methods include spray drying, chilling, lyophilization, coacervation, extrusion, and electrospraying (de Oliveira Filho et al. 2021). Biofungicides such as *Bacillus, Pseudomonas, Actinomycetes, yeast,* and *Trichoderma* (Ruiu 2020, Samada and Tambunan 2020) are among the very few microbes that have been commercially utilized on a large scale. The demand for biofungicides is increasing as it is possible to produce food without chemical residues (Souza et al. 2018, Abbey et al. 2019, Bazioli et al. 2019). The global biofungicide market is expected to grow and reach $8.5 billion by 2025.

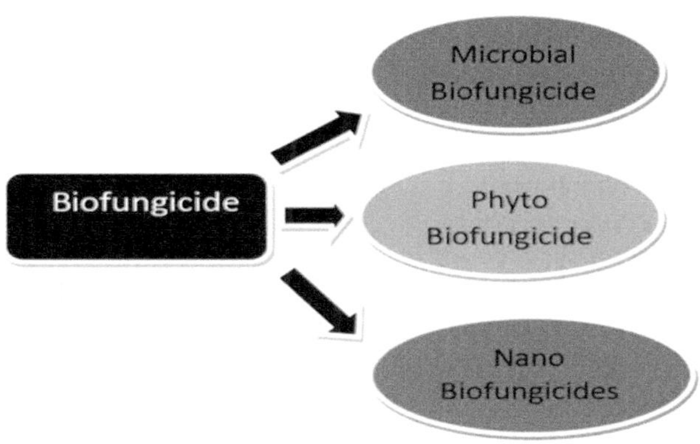

Figure 1: Types of Biofungicides.

Biofungicides must meet the following criteria: the ability to extensively colonize the rhizosphere, high competitive saprophytic ability, broad spectrum of action, compatibility with other bioagents in the soil, and tolerance to desiccation, oxidizing agents, heat, and UV radiation. They should also enhance plant growth, be easy to mass-produce, and offer reliable control (Jeyarajan and Nakkeeran 2000). Therefore, the screening and selection of these agents is an essential step before developing a potent biofungicide. This chapter provides a comprehensive view of the selection and screening methods that could be used to develop a biofungicide.

2. Microbial Biofungicide

Microbial biofungicides are low-cost and less toxic to the environment. They are efficient in controlling several diseases while being safe for non-target organisms (Ayilara et al. 2023). The active components are beneficial in suppressing pathogens without affecting non-target organisms (Kachhawa 2017). Bacteria and fungi are the two main microbial biofungicides that are employed in disease management. The fungi that live in soil and cause decay of several crop plants are inhibited by bacteria such as *Pseudomonas, Yersinia, Bacillus,* and *Trichoderma* (Ruiu 2020, Samada and Tambunan 2020). Microbial biofungicides suppress the growth of pathogens and the production of toxins by competing (Fenta et al. 2024) with the pathogens (Gikas et al. 2022). Microbial biofungicides are being used in pathogen control because of their unique features. Some of the common microbial biofungicides are listed in Table 1 below.

The mechanism that operates in microbial biofungicides includes competition, suppression through siderophores, hydrolytic enzyme production through antagonism and predation, antibiotic production through antibiosis, biofilm formation, and induction of host resistance. The fungicides hamper respiration and sterol biosynthesis (Gigas et al. 2022, Fenta et al. 2023). The application of microbial biofungicides not only suppresses pathogens and induces host resistance, but also reduces greenhouse gas emissions (Ayilara et al. 2023).

2.1 Screening of Microbial Biofungicide Agents

2.1.1 Well Diffusion Method

The well method is similar to the bar method. It is a simple and cost-effective method for screening pure compounds. It may not be useful for screening crude extracts since the activity produced will be due to synergistic effects (Rütten et al. 2022). In this method, 8–10 mm diameter wells are cut in the specific media. The test pathogen spores are spread onto the plate prior to the inoculation of the antagonist. The antagonist bacteria or fungi are grown in broth (PDB for fungi and NB for bacteria) for 48 hours for bacteria and 5 days for fungi. A known aliquot of the broth is added to the wells inoculated with the spores of the pathogen and incubated for 48 hours.

Table 1: Microbial biofungicides effective against phytopathogens.

S. NO	Microbial Biofungicide	Effective against
1	*Streptomyces lydicus*	Powdery mildew, Downy mildew, Botrytis, *Rhizoctonia, Pythium, Phytophthora, Fusarium, Verticillium* Powdery mildew, Downy mildew,
2	*Ulocladium oudemansii* U3	*Botrytis, Sclerotinia*
3	*Bacillus subtillis* GB03	Leaf spots, Powdery mildew, Botrytis, bacterial diseases, *Rhizocotonia, Pythium, Phytophthora*
4	*Bacillus amyloliquefaciens*	*Powdery mildew, Downy mildew, Botrytis, Rhizoctonia, Pythium, others*
5	*Agrobacterium radiobacter K84*	*Agrobacterium tumefaciens*
6	*Streptomyces griseoviridis*	*Botrytis, Rhizoctonia, Pythium, Phytophthora, Alternaria*
7	*Trichoderma harzianum*	*Rhizoctonia, Pythium, Fusarium, Cylindrocladium, Thielaviopis*
8	*Gliocladium catenulatum* JII446	*Botrytis, Rhizoctonia, Pythium, Phytophthora, Fusarium, Verticillium*
9	Gliocladium virens GL-21	*Rhizoctonia, Pythium*
10	Bacteriophages of *Xanthomonas* spp. and *Pseudomonas syringae* pv. tomato	Bacterial spot in pepper and tomatoes and bacterial speck in tomatoes
11	*Pantoea agglomerans* strain E325	Fireblight (*Erwinia amylovora*)
12	*Bacillus pumilus* QST 2808	Rust, powdery mildew, *Cercospora* and brown spot

A clear zone of inhibition will appear around the well. The diameter of the zone is measured, indicating the antagonistic ability. The well method can help in studying antibiosis. This method can also be used to screen botanical fungicides.

2.1.2 Kirby-Bauer Disc Diffusion Method

Sterile discs of 5 mm diameter are made from Whatman No. 1 filter paper. The discs are impregnated with the antagonistic bacterial or fungal broth. After the discs are dried, they are placed in the center of a petri dish containing media that has already been pre-inoculated with the test pathogen. The pathogen, either bacteria or fungi, is spread onto the media. The plates are incubated for 48 hours to observe a zone of inhibition around the disc. The antagonistic ability can be determined by measuring the diameter of the zone of inhibition. The disc method is useful for studying antibiosis. The antibiotic substance can diffuse from the disc through the agar and inhibit the pathogen. This method can also be used to screen botanical fungicides.

2.1.3 Hyperparasitism

Fungal antagonists exhibit hyperparasitism on the pathogen. To study hyperparasitism, a sterile cellophane sheet placed on a sterile slide is inoculated with agar plugs of the antagonist and pathogen. The slides are incubated in a moist chamber and observed after 72 hours. The cellophane sheets are stained with cotton blue and observed under a light microscope (Anitha and Murugesan 2001).

2.1.4 Plate Assay for Chitinase Enzyme Production

Modified Czapek's Dox Agar plates containing colloidal chitin as the sole carbon source with Bromocresol as an indicator can be inoculated with the antagonistic fungi or bacterium at the center. The pH of the medium is maintained at 3.5. The plates are observed for a yellow halo around the culture. As the colloidal chitin is broken down by the action of chitinase, the acidic pH changes to alkaline, resulting in a color change in the medium from blue to yellow (Figure 2) (Merlin and Anitha 2021).

2.1.5 Plate Assay for Detection of Cellulolytic Activity

To detect cellulose activity, the potential fungus or bacterium is inoculated in carboxymethyl cellulose media (KH2PO4, Na2HPO4.7H2O, (NH4)2SO4, MgSO4.7H2O, CaCl2, FeSO4.7H2O, MnSO4, CMC, and agar, pH 7.0) as described by Kumar et al. (2018), and incubated for 24 to 48 hours at 37°C. Gram's iodine is added and the mixture is incubated for 10 minutes. Then, 2M NaCl is added, and a clear halo zone is observed around the culture while the entire plate turns blue (Figure 2).

2.1.6 Organic Volatile Compound Production Assay

Volatile organic compounds produced by the antagonist during metabolism play a vital role in controlling pathogens. Volatile compounds produced by beneficial microbes exert a fungistatic effect and prevent fungal propagules from germinating (Garbeva et al. 2011). A potential biofungicidal agent must possess the property of producing volatile compounds.

A simple modified divided plate method (Yuan et al. 2012) can be used to evaluate the production of volatile organic compounds. A petri dish is divided into two halves, one filled with LB agar and the other side filled with PDA. The LB side is

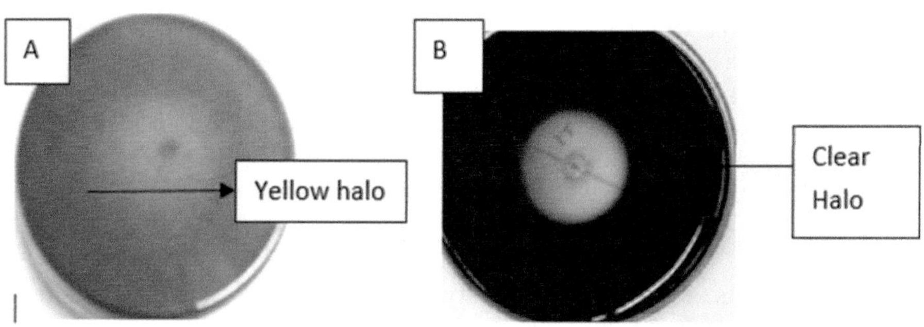

Figure 2: Plate assay for Chitinase (A) and Cellulolytic (B) enzyme detection.

inoculated with a bacterial biofungicidal agent, and the PDA side with the pathogen, then incubated at 37°C. The antifungal effect is observed by a reduction in the dry fungal mass after 7 days. The antifungal effect is exerted by the volatile compounds released by the bacterium. Furthermore, the dry fungal mass can be determined by desiccating the agar portion at 65°C on a pre-weighed filter paper. The difference between the initial and final weight determines the fungal dry mass.

2.2 Evaluation of Biofungicide Agent for Plant Growth-Promoting Traits

A potential biofungicide agent must also be evaluated for its growth-promoting properties. Phosphate solubilizing activity (Wendimu et al. 2023), indole acetic acid production, and volatile organic compounds are properties common to both bacteria and fungi. Hydrogen cyanide and siderophore production are properties attributed to bacteria.

2.2.1 Phosphate Solubilizing Assay

The phosphate-solubilizing ability of the bioagent can be determined using common screening methods described by Gupta et al. (1994). Briefly, PPGM (Pikovskaya's phosphate growth medium, Pikovskaya 1948) with Bromophenol, when inoculated with the culture of the bioagent, produces a yellow halo around the colonies due to the production of organic acids. The National Botanical Research Institute's phosphate growth liquid medium (NBRIP) (Nautiyal et al. 1999) is another widely used method comprising tricalcium phosphate (Berza et al. 2022). The halo produced around the colonies can be used to determine the solubilization index using the formula SI = (CD + HD)/CD, where CD is colony diameter and HD is the halo zone diameter (Edi-Premono et al. 1996).

2.2.2 Indole-3-Acetic Acid Assay

The antagonistic agent can be screened for IAA production according to Sivasankari et al. (2016). Briefly, the bacterium or fungus can be grown in a basic medium with 0.1 g/L tryptophan and incubated at 37ºC at 100 rpm for 48 hours. Upon adding 2 ml of Salkowski's reagent to the culture filtrate, a change in the media color to red can be observed.

2.2.3 Production of Hydrogen Cyanide and Siderophore

Passari et al. (2015) can be used to determine hydrogen cyanide production. The bacterium is streaked onto a basic medium supplemented with glycine (4.4 g/L). After the culture has grown, it is overlaid with 0.5% picric acid and 2% sodium carbonate–moistened Whatman No. 1 filter paper and incubated at 37ºC for seven days. The production of hydrogen cyanide can be observed by a change in color from yellow to orange. The siderophore production by the biofungicidal agent can be assessed following Sheeba et al. (2017). The bacterial agent is inoculated onto a

Chrome Azurol-S medium (CAS) and incubated at 30ºC for 48–72 hours. A change in color from yellow to orange determines the production of siderophore.

3. Phytobiofungicides

Plants are endowed with numerous phytochemicals as a part of their defense mechanisms that can inhibit the growth and development of microbes such as bacteria, fungi, and viruses. This property makes them suitable candidates as antimicrobial agents (Hossain 2024). Botanical fungicide development reduces the hazardous effects of synthetic fungicides. Botanical fungicides are beneficial since they are effective, selective, biodegradable (Yoon et al. 2013, Santra and Banerjee 2020), affordable (Gurjar et al. 2012, Cenobio-Galindo et al. 2024), and reduce the risks caused by synthetic fungicides due to their low residue toxicity (Suteu et al. 2020). Botanicals have an advantage due to multiple modes of action both at the biochemical and molecular level and reduce the pathogen inoculum, delaying the development of resistance (Roger and Philogène 2008). All plant material has antifungal or antimicrobial activity. Plant-based biofungicides benefit the farmers, offer food security, improve quality, and reduce the pathogen inoculum (Tripathi et al. 2020). Moreover, they are safer for pollinating and predatory bees, have less toxicity to non-target organisms, and do not have any allelopathic effect on crop plants (Rana et al. 2024). Some of the important plant-based biofungicides are listed in Table 2. Some of the common methods to screen phytobiofungicides are spore germination assay, poison plate method, and microscopic methods.

Several bioactive compounds from plant extracts such as alkaloids, terpenoids, flavonoids, phenolics, saponins, and tannins are effective in controlling the above fungi and the diseases caused by them. However, there is no defined range of concentration that is used to classify plant extracts as potent antifungal agent. Mesa et al. (2019) reported classification of extracts as either active, slightly active, moderately active, or ineffective (Figure 3).

The phytobiofungicides are known to possess varied effects on the pathogen. Terpenes are known to alter the lipid bilayer's fluidity and membrane permeability. They also destroy the mitochondria and affect electron transport in mitochondria. Alkaloids intercalate with microbial DNA and affect gene replication and transcription. The effectiveness of the biofungicide on plant pathogens can be studied with basic techniques.

3.1 Confocal Laser Scanning Microscopy (CLSM)

CLSM is another recent method used for observation. Briefly, the fungus to be tested is grown in SDA agar for 48 hours. A disc of the culture is rinsed with PBS for 15 minutes. Plant extract is added to the discs and incubated at 37ºC for 24 hours. The extract is then aspirated and removed. Fungal cell viability can be determined by staining with propidium iodide. The disc is then placed on a glass slide, and the ratio of live to dead cells can be determined.

Table 2: List of phytobiofungicide used against plant pathogenic fungi.

S. No.	Plant extract	Plant disease	Causal organism
1	Ethanolic extract *Myristica fragans* Houtt (Cruz et al. 2024a)	Gray mold; anthracnose; trunk canker; chestnut ink; vascular wilting	*Botrytis cinerea; Colletotrichum acutatum; Diplodia corticola; Phytophthora cinnamomic; Fusarium culmorum*
2	Ethanolic *Curcuma longa* L. extract (Cruz et al. 2024b)	Gray mold; anthracnose; trunk canker; chestnut ink; vascular wilting	*Botrytis cinerea; Colletotrichum acutatum; Diplodia corticola; Phytophthora cinnamomic; Fusarium culmorum*
3	Methanolic *Eryngium campestre* L. extract (Al-Askar et al. 2023)	Vascular wilting; damping-off	*Fusarium oxysporum; Rhizoctonia solani*
4	Polyphenol extracts: Mesquite (*Prosopis glandulosa* Torr), Cedar (*Cedrus* Trew), and Oak (*Quercus* L.)	Black mold; vascular wilting; damping-off	*Alternaria alternata; Fusarium oxysporum; Rhizoctonia solani*
5	Methanolic *Cinnamomum camphora* (L.) J. Presl extract (Sobhy et al. 2023)	Black mold; vascular wilting	*Alternaria alternata; Fusarium solani; Fusarium oxysporum*
6	Mixture of ethanol and ethyl acetate: *Lantana hirta* L., *Argemone ochroleuca* L. and *Adenophyllum porophyllum* (Cav.) extract	Crown rot; anthracnose's; gummosis	*Pestalotiopsis clavispora; Colletotrichum gloeosporioides; Lasiodiplodia pseudotheobromae*
7	Orange (Citrus × sinensis L.) peel polyphenolic extract (Hernández et al. 2020)	Brown rot; gray mold; black mold	*Monilinia fructicola; Botrytis cinerea; Alternaria alternata*
8	Ethanolic extract of the fruit of *Sapindus saponaria* L. (Gasca et al. 2020)	Anthracnose	*Colletotrichum spp*
9	Pomegranate (*Punica granatum* L.) peel aqueous extract (Elkhetabi et al. 2020)	Brown rot	*Monilinia laxa; M. fructigena*
10	Hexane extract of *Stevia rebaudiana* Bertoni leaves (Ramírez et al. 2020)	Vascular wilting	*Fusarium oxysporum*
11	Propolis ethanolic extract of Brazilian Red (*Sesuvium portulacastrum* L.) (Dudoit et al. 2020)	Anthracnose	*Colletotrichum musae*
12	Methanol *Pernettya prostrata* (Cav.) and Rubus roseus Schott (Ordóñez et al. 2023)	Banana bacterial wilt; late blight; scab diseases	*Ralstonia solanacearu; Phytophthora infestans; Neopestalotiopsis javaensis*

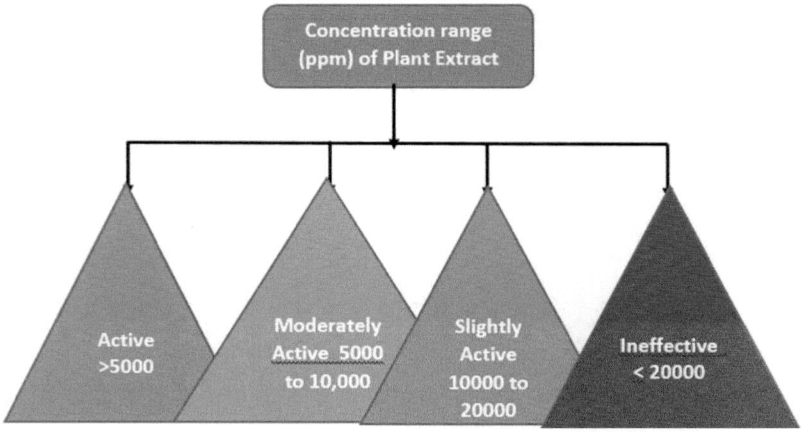

Figure 3: Classification of Effective Concentration of Plant Extract.

3.2 Chromatographic Technique

Chromatography is a simple technique to study biological properties and detect chemical constituents of plant extracts (Choma 2014). Bioautography is a rapid procedure to detect new antimicrobial compounds from plant sources. It allows direct localization of antimicrobial activity on the chromatographic plate (Navarro et al. 1998). Moreover, it saves time by avoiding the isolation of inactive compounds. This method combines activity determination and localization-directed isolation of active constituents (Shahverdi et al. 2007). There are three variants of TLC (Rios et al. 1988).

3.2.1 Thin Layer Chromatography—Direct Bioautography (TLC-DB)

Using TLC, the plant extract is separated and dipped in a microbial suspension grown in nutrient broth. The plates are incubated in a moist chamber. The culture grows on the TLC plate except in the zones of the antimicrobial compound, forming an inhibition zone. The inhibition zone is visualized by spraying tetrazolium salt. Live cells produce dehydrogenase enzymes and convert tetrazolium salt into a purple formazan (Choma and Grzelak 2011, Marston 2011). The background appears purple against the creamy areas which are zones of inhibition (Figure 4).

A modification of the above method involves spraying a spore suspension of 1 x 10^5 CFU in a medium onto a TLC plate with separated compounds. The plate is incubated in the dark at 35°C in a moist chamber and sprayed with 2 mg/ml p-iodonitrotetrazolium violet and left overnight. White bands around the separated compounds indicate the absence of reduction of INT to a colored formazan, suggesting that the microorganism has been inhibited by the plant compound (Begue and Klein 1972).

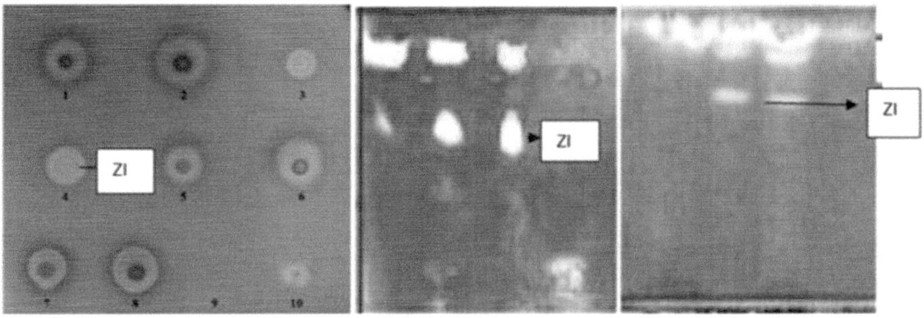

Figure 4: TLC Direct Bioautography. ZI- Zone of inhibition (Image adapted from Choma et al. 2015, Suleimana et al. 2009).

3.2.2 TLC Contact Bioautography

In this method, the antimicrobial compound is scraped out of the TLC plate with a solvent and, using either the well method or disc methods, inoculated in the agar medium with the test pathogen.

3.2.3 TLC Immersion Bioautography

TLC immersion bioautography is a hybrid method of both contact and direct bioautography. A slight modification of the earlier method is immersion TLC bioautography. The phytochemical constituents are separated on a TLC plate, and the media pre-seeded with the microorganism is overlaid over the plate. This method is a hybrid of contact and direct bioautography (Islam et al. 2002).

3.3 Detached Leaf Assay

The screening of plants that could be used as phytobiofungicides can be done by a simple detached leaf assay. Briefly, the leaves of plants screened for activity are collected based on uniform leaf size. The selected leaves are washed with distilled water and surface sterilized using 5% sodium hypochlorite, washed thoroughly, and blotted dry with sterile filter paper, then left in a moist chamber under sterile conditions. A 5 mm plug of fungal pathogen grown on PDA is inoculated onto the leaves and incubated in a growth chamber at 25 ± 2°C with 10 hours of light for 5 days (Theer et al. 2021). The antifungal activity of the plants can be determined by observing the leaf tissue yellowing, discoloration, and dissociation. The damaged leaf area percentage can be estimated using the formula: Width of damaged area/Length of damaged area × Width of total leaf area/Length of total leaf area × 100.

4. Nanobiofungicides

Nanobiofungicides are next-generation fungicides that have high efficacy, durability, and are eco-friendly (Rana et al. 2020). Nanobiofungicides are metallic nanoparticles that are synthesized by a biogenic process using plant extracts, microorganisms, or enzymes (Adeyemi et al. 2022, Malik et al. 2023). Nanoparticles are unique in their physical and chemical properties and range from 1 to 100 nm in size. The small size of nanoparticles increases the surface area of interaction with the pathogens and hence is very effective (Hossain 2024). They can be used in agriculture for the control of pests and diseases. Biological synthesis can be done with microorganisms and plant extracts, which is referred to as green synthesis. When compared to physical and chemical methods of synthesis, the biological method is eco-friendly and cost-effective. Many soil microorganisms can reduce heavy metals like Uranium (VI), Technetium (VII), Chromium (VI), Cobalt (III), and Manganese (IV) (Ghosh et al. 2021). Microbial synthesis occurs through intracellular and extracellular mechanisms through reduction. Silver nanoparticles, synthesized extracellularly by *Amycolatopsis tucumanensis*, showed potential in controlling *Acidovorax avenae subsp. avenae*, the causal agent of Red stripe disease (Guerrero et al. 2022).

The nanoparticles promote plant growth and offer protection against pathogens (Ghosh et al. 2021, Duhan et al. 2017), hence they serve as effective biofungicides. Nanocarriers can precisely and effectively deliver the bioactive molecules. Nanocarriers are designed with a variety of organic and inorganic materials with definite shape, pore size, and surface properties that can protect the bioactive molecules. Fungicidal compounds are well protected from environmental degradation, with increased solubility to produce sustainable release over an extended period, reducing the requirement for frequent application and thereby reducing environmental and human hazards (Maluin et al. 2020, Chandrakala et al. 2022).

Previously, nanoparticles have successfully controlled *Colletotrichum, Fusarium* spp., *Alternaria, Bipolaris*, etc. (Kaushal 2018, Fatima et al. 2021). Essential oils can also be encapsulated with a nanostructured lipid carrier, and they are effective against *Fusarium oxysporum* (Shanmugapriya et al. 2024). Chitosan and Cu-chitosan nanoparticles have uniform size and stability, attributing to a higher fungicidal property against *Alternaria alternata, Macrophomina phaseolina,* and *Rhizoctonia.* The list of nanoparticles used in biofungicide formulations are given in Table 3.

Silver nanoparticles have gained importance in agriculture due to their green synthesis through eco-friendly methods using plant extracts, bacteria, fungi, or yeast to convert silver ions to nanoparticles. Bharose et al. (2024) reported bacteria-mediated green synthesis of silver nanoparticles that are effective against *Aspergillus flavus*. Silver nanoparticles are reported to be effective at 75 ppm and inhibit the mycelium and spore germination of *Pestatiopsis microspora, Alternaria alternate, Botryosphaeria dothidea,* and *Diaporthe actinidae*. Silver nanoparticles increase membrane permeability, resulting in cellular leakage (Alharbi et al. 2022). Concentrations of 150 ppm have strong antifungal activity against *P. digitatum, A. citri,* and *A. alternata* (Abdelmalek and Salaheldin 2016). Copper nanoparticles embedded in chitosan inhibit the growth of *F. oxysporum* and *Neoscytalidium dimidiatum* (Zahrani et al. 2023, Mosa and El-Abeid 2023). Nevertheless, these particles are reported to be toxic at high doses of 3200 ppm on cucumber and pansy plants. Nickel nanoparticles of 60–80 nm are effective against *Fusarium oxysporum*

Table 3: Nanoparticles used in biofungicide formulation.

S. NO	Nanocarrier	Bioactive compound	Effective against
Lipid-based Nanocarriers			
1.	Nanoemulsion	Clove oil	*Neoscytalidium* Blight Disease of *Carum carvi* L. (Hashem et al. 2023)
2.	Nanoemulsion	Eugenol oil	*Fusarium oxysporum* in cottonseeds (Abd-Elsalam et al. 2015)
3.	Nanoemulsion	Peppermint oil	Early blight disease (Pandey et al. 2020)
4.	Nanoemulsion	Clove, black seed, lemon, and orange oils	*Botrytis cinerea* (Ziedan et al. 2022)
Polymeric Nanoparticles			
5.	Chitosan	Thymol	*Botrytis cinerea* (Zhao et al. 2023)
6.	Polymeric micelles	Geranylorcinol compounds	*Botrytis cinerea* (Taborga et al. 2015)
Inorganic Nanoparticles			
7.	Silica nanoparticle	Silica	*Fusarium oxysporum* and *Aspergillus niger* (Goswami et al. 2022)
8.	Silver nanoparticle	Silver	*Alternaria alternata, Alternaria citri and Penicillium digitatum Fusarium avenaceum* and *Fusarium equiseti* (Matras et al. 2022)
9.	Magnetite nanoparticle	Iron oxide	*Fusarium oxysporum f.* sp. *Lycopersici, Aspergillus niger* and *Mucor piriformis* (Ashraf et al. 2022)
10.	Copper nanoparticles	Copper	*Fusarium solani, Neofusicoccum* sp., and *Fusarium oxysporum* (Pariona et al. 2019)

(Sharma et al. 2022). Zinc and copper oxide nanoparticles control *Phytophthora infestans* in greenhouse conditions (Alharethi et al. 2024).

5. Bionanofungicide

A bionanofungicide is composed of a metallic core of silver and copper and a biological portion synthesized using a microbe, plant, or enzyme, which is called a bio-corona (Romero and Davila Costa 2025). The biological portion is composed of proteins, lipids, carbohydrates, and nucleic acids found as a bilayer around a metallic core. The bio-corona molecule is classified into a hard corona consisting of tightly bound molecules and a soft corona, which has loosely bound molecules that can be exchanged (Figure 5) (García-Álvarez and Vallet-Regí 2021). The composition of the bio-corona depends on the biological material used for synthesis. *Macrophomina phaseolina* synthesized silver nanoparticles composed of 46 different proteins, with 60% being hydrolytic enzymes and 21% oxidoreductases (Spagnoletti et al.

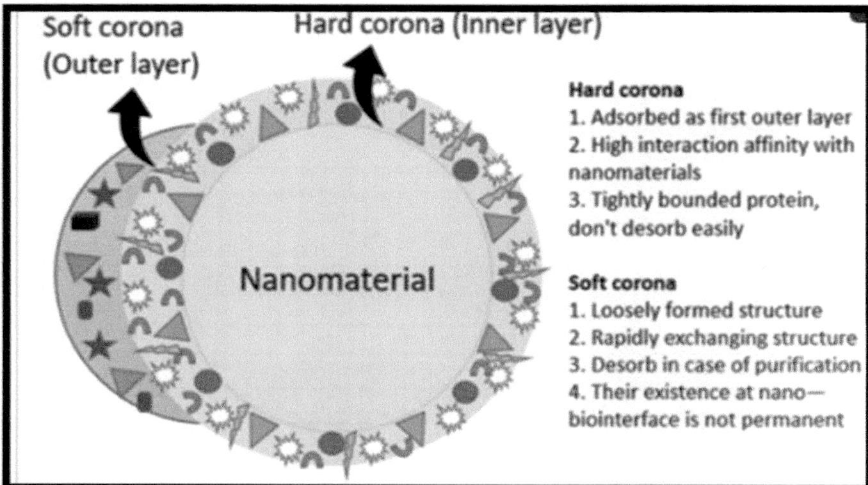

Figure 5: Biocorona molecule depicting two prominent layers soft and hard corona regions of a nanomaterial. Image adapted from Akhter et al. (2021).

2021). Bionanofungicides are more advantageous than nanobiofungicides due to stability, effectiveness, delivery mechanisms, and a reduced negative impact on the environment (Xu et al. 2021, Yu et al. 2023). Molecular modeling and docking studies help in predicting the interaction of proteins in the bio-corona with nanoparticles (Paterlini et al. 2021). Hence, metabolomics and proteomics are excellent approaches to elucidate the structure of the bio-corona. Further, other biomolecules such as lipoproteins, glycolipids, and terpenes as components of the bio-corona need further investigation.

5.1 Mechanism of Action of Bionanofungicides

5.1.1 Surface Interaction

The ability of nanobiofungicides and bionanofungicides to cross the hydrophobic cuticle barrier depends on several factors. Due to their small size, these can effectively traverse the leaf surface. Adhesion and retention on the foliar surface for a longer time have better chances of pathogen control (Gao et al. 2022).

5.1.2 Penetration and Uptake

Once the nano-bio and bio-nanofungicide adhere to the cuticle, penetration is essential to reach the pathogen. The size of the particles enables them to be absorbed through stomata, hydathodes, necrosis spots, and trichomes (Jia et al. 2024). The cuticle forms a larger surface area which creates a non-polar pathway of entry. This is a major advantage when compared to nano-biofungicides (Paterlini et al. 2021, Spagnoletti et al. 2021).

5.1.3 Release and Efficacy

Upon penetration, the nanobiofungicide releases the active component into the leaf tissue to control the pathogen. Organic molecules in the bio-corona offer sustained action on phytopathogens (Fincheira et al. 2023). Recently, Li et al. (2024) introduced a dual-functioning delivery system consisting of fungicide and immune response molecules. Despite such advancements, controlled release of nanobiofungicides in the field and the efficient dosage needs further investigation.

5.2 Screening of Nanobiofungicide Agents

5.2.1 Resazurin Assay

Resazurin is one of the versatile colorimetric methods to measure the metabolic activity in living cells (Sen et al. 2017), which can be used to study the efficacy of nanobiofungicidal agents. Briefly, the principle of the assay is based on the reduction of resazurin (blue-colored dye) to resorufin (pink-colored dye) by metabolically active cells. The amount of resorufin produced is directly proportional to the metabolic activity, which indicates the efficacy of nanobased biofungicides. A modified method, in which the test fungi—either spores or mycelium—are allowed to grow in the presence of a nanobased fungicide, is incubated at room temperature. Further, an optimal concentration of resazurin solution is added and incubated for 3–4 hours. A change in color from blue to pink or purple indicates the poor efficacy of the nanobiofungicide. Lack of color change is an indication of dead cells, which proves the efficacy of the agent.

5.2.2 Superficial Immersion Technique

To screen nanobased biofungicides, the superficial immersion technique can be used. PDA is poured into petri plates, and 1 ml of nanobiofungicide synthesized with plant or microbial sources is spread superficially over the media and air-dried in a laminar flow for 6 hours. Nanomaterial forms a thin film over the media and are in direct contact with the mycelium of the pathogen. The agar plug of the pathogen is inoculated in the center, and the radial mycelial growth is used to determine the inhibition.

6. Virtual Selection of Biofungicide

In silico analysis can be used to select biofungicide compounds with precision and accuracy. Initially, bioactive compounds similar in structure and properties to chemical fungicides have to be selected from the natural product database. Using tools such as Fungi Pad makes screening for fungicide likeness properties much easier. Further identification and selection of the binding pocket can be done with PrankWeb and AutoDock Vina to predict the binding energy. Finally, Discovery Studio is used to evaluate the interaction pattern (Yongpisanphop 2024). *In silico* analysis does not require any sophisticated infrastructure or laboratory. It saves time as well and works out very effective.

7. Conclusion

Improper use of synthetic pesticides has led to the contamination of soil, water, and the environment. Biofungicides offer a broader perspective in integrated pest management. Microbial biofungicides use different microbes, and each one requires different storage conditions, in-depth research, ecological studies, and mass production. Storage and transit may be a huge barrier for production and marketing, and the shelf life may be affected in open environmental conditions. Bionanofungicides and nanobiofungicides are more promising than chemical fungicides and are an eco-friendly alternative for disease management. Using nanotechnology, it is now possible to enhance stability, bioavailability, and steady release of bioactive compounds, enhancing antifungal activity with reduced residues. Although ecotoxicological effects of nanoparticles on plants and soil microorganisms have been well explored, concerns over the biological effects of these nanoparticles, lack of knowledge of their mode of action, and phytotoxicity prevail. It must be mentioned that biofungicides, although safe and eco-friendly to humans and the environment, are generally more expensive compared to synthetic fungicides. Toxicity, product safety, and efficiency need to be studied. Registration of products before commercialization of biofungicides is cumbersome. Moreover, there are no separate guidelines for registration for biopesticides, and similar criteria need to be followed as for chemical fungicides. Environmental risk assessment is also required and is very expensive. Complete substitution of chemical fungicides is possible in the near future with the advent of technology.

References

Abbey, J. A., Percival, D., Abbey, L., Asiedu, S. K., Prithiviraj, B., and Schilder, A. (2019). Biofungicides as alternative to synthetic fungicide control of grey mould (*Botrytis cinerea*) – Prospects and challenges. *Biocontrol Sci. Technol.*, 29: 241–262. http://doi.org.10. 1080 /09583157.2018.1548574.

Abd-Elsalam, K. A., and Khokhlov, A. R. (2015). Eugenol oil nanoemulsion: antifungal activity against *Fusarium oxysporum* f. sp. *vasinfectum* and phytotoxicity on cottonseeds. *Appl. Nanosci.*, 5: 255–265. https://doi.org/10.1007/s13204-014-0398-y.

Abdelmalek, G. A., and Salaheldin, T. A. (2016). Silver nanoparticles as a potent fungicide for citrus phytopathogenic fungi. *J. Nanomedicine Res.*, 3,00065, https:// doi.org/10.15406/ JNMR.2016.03.00065.

Adeyemi, J. O., Oriola, A. O., Onwudiwe, D. C., and Oyedeji, A. O. (2022). Plant extracts mediated metal-based nanoparticles: synthesis and biological applications. *Biomolecules*, 12(5): 627. https://doi.org/10.3390/biom12050627.

Al-Askar, A. A., Bashir, S., Mohamed, A. E., Sharaf, O. A., Nabil, R., Su, Y. et al. (2023). Antimicrobial efficacy and HPLC analysis of polyphenolic compounds in a whole-plant extract of *Eryngium campestre*. *Separations*, 10: 362.

Alharbi, N. S., Alsubhi, N. S., and Felimban, A. I. (2022). Green synthesis of silver nanoparticles using medicinal plants: characterization and application, *J. Radiat. Res. Appl. Sci.*, 15: 1687–8507. https:// doi.org/10.1016/j.jrras.2022.06.012.

AlHarethi, A. A., Abdullah, Q. Y., AlJobory, H. J., Anam, A. R. M., Arafa, R. A., and Farroh, K. Y. (2024). Zinc oxide and copper oxide nanoparticles as a potential solution for controlling *Phytophthora infestans*, the late blight disease of potatoes. *Discov. Nano*, 19(1): 105. https://doi.org/10.1186/ S11671-024-04040-6.

Anani, O. A., Mishra, R. R., Mishra, P., Enuneku, A. A., Anani, G. A., and Adetunji, C.O. (2020). Effects of toxicant from pesticides on food security: current developments. pp. 313–321. *In*: Mishra, P.,

Mishra, R. R., and Adetunji, C. O. (Eds.). Innovations in Food Technology, Springer Nature Pvt. Ltd., Singapore. https://doi.org/10.1007/978-981-15-6121-4_22.

Anitha,R., and Murugesan, K. (2001). Mechanism of action of *Gliocladium virens* on *Alternaria helianthi. Indian Phytopathology*, 54(4): 449–452.

Ansari, M., Ahmad, N., and Hasan, F. (2012). Potential of biopesticides in sustainable agriculture. pp. 529–595. *In:* Malik, A., and Grohmann, E. (Eds.). *Environmental Protection Strategies for Sustainable Development. Strategies for Sustainability*. Springer, Dordrecht. https://doi.org/10.1007/978-94-007-1591-2_17.

Ashraf, H., Batool, T., Anjum, T., Illyas, A., Li, G., Naseem, S. et al. (2022). Antifungal potential of green synthesized magnetite nanoparticles black coffee-magnetite nanoparticles against wilt infection by ameliorating enzymatic activity and gene expression in *Solanum lycopersicum* L. *Frontiers in Microbiology*, 13: 754292. https://doi.org/10.3389/fmicb.2022.754292.

Ayilara, M. S., Adeleke, B. S., Akinola, S. A., Fayose, C. A., Adeyemi, U. T., Gbadegesin, L. A. et al. (2023). Biopesticides as a proising alternative to synthetic pesticides: a case for microbial pesticides, phytopesticides, and nanobiopesticides. *Frontiers in Microbiology*, 14: 1–16. https://doi:10.3389/fmicb.2023.1040901.

Bazioli, J. M., Belinato, J. R., Costa, J. H., Akiyama, D. Y., de Moraes Pontes, J. G., Kupper, K. C. et al. (2019). Biological Control of citrus Postharvest Phytopathogens. *Toxins*, 11: 460. http://doiI: 10.3390/toxins11080460.

Begue, W. J., and Klein, R. M. (1972). The use of tetrazolium salts in bioautographic procedure. *J. Chromato*, 88: 182–184. http://doi:10.1016/s0021-9673(00)92965-0.

Berza, B., Sekar, J., Vaiyapuri, P., Pagano, M. C., Assefa, F. et al. (2022).Evaluation of inorganic phosphate solubilizing efficiency and multiple plant growth promoting properties of endophytic bacteria isolated from root nodules *Erythrina brucei. BMC Microbiol.*, 22: 276. https://doi.org/10.1186/s12866-022-02688-7.

Bharose, A. A., Hajare, S. T., Gajera, H. P., Soni, M., Prajapati, K. K., Singh, S. C. et al. (2024).Bacteria mediated green synthesis of silver nanoparticles and their antifungal potentials against *Aspergillus flavus*. *PLoS One*, 19: 1–16. https://doi.org/10.1371/JOURNAL.PONE.0297870.

Cenobio-Galindo, A. d. J., Hernández-Fuentes, A. D., González-Lemus, U., Zaldívar-Ortega, A. K., González-Montiel, L., Madariaga-Navarrete, A. et al. (2024). Biofungicides Based on Plant Extracts: On the Road to Organic Farming. *International Journal of Molecular Sciences*, 25(13): 6879. https://doi.org/10.3390/ijms25136879.

Chandrakala, V., Aruna, V., and Angajala, G. (2022). Review on metal nanoparticles as nanocarriers: current challenges and perspectives in drug delivery systems. *Emergent Materials*, 5(6): 1593–1615. https://doi.org/10.1007/s42247-021-00335-x.

Choma, I. M., and Wioleta, J. (2014). Effects-directed biological detection: Bioautography. pp. 279–312. *In:* Colin, F. Poole (Eds.). *Instrumental Thin-Layer Chromatography*. Elsevier: Amsterdam, The Netherlands. https://doi.org/10.1016/B978-0-12-417223-4.00011-X.

Choma, I., and Grzelak, E. M. (2011). Bioautography detection in thin-layer chromatography. *J. Chromatogr. A.*, 1218: 2684–2691. http://10.1016/j.chroma.2010.12.069.

Choma, I. M., and Wioleta, J. (2015). TLC-direct bioautography as a high throughput method for detection of antimicrobials in plants *Chromatography*, 2(2): 225–238. https://doi.org/10.3390/chromatography2020225.

Cruz, A., Sánchez-Hernández, E., Teixeira, A., Oliveira, R., Cunha, A., Martín-Ramos, P. et al (2024a). Phytoconstituents and ergosterol biosynthesis-targeting antimicrobial activity of nutmeg (*Myristica fragans* Houtt.) against Phytopathogens. *Molecules*, 29: 471.

Cruz, A., Sánchez-Hernández, E., Teixeira, A., Martín-Ramos, P., Cunha, A., Oliveira, R. et al. (2024b). Antifungal and Antioomycete Activities of a *Curcuma longa* L. hydroethanolic extract rich in *Bisabolene sesquiterpenoids*. *Horticulturae*, 10: 124.

Damalas, C. A., and Eleftherohorinos, I. G. (2011). Pesticide exposure, safety issues, and risk assessment indicators, *International Journal of Environmental Research and Public Health*, 8 (5): 1402–1419. http://doi:10.3390/ijerph8051402.

de Oliveira Filho, J. G., Silva, G. D. C., Cipriano, L., Gomes, M., and Egea, M. B. (2021). Control of postharvest fungal diseases in fruits using external application of RNAi. *Journal of Food Science.*, 86(8): 3341–3348. http:// doi: 10.1111/1750-3841.15816.

Devi, H., Boda, M., Shah, M., Parveen, S., and Wani, A. (2019). Green synthesis of iron oxide nanoparticles using *Platanus orientalis* leaf extract for antifungal activity. *Green Processing and Synthesis*, 8(1): 38–45. https://doi.org/10.1515/gps-2017-0145.

Dudoit, A., Mertz, C., Chillet, M., Cardinault, N., and Brat, P. (2020). Antifungal activity of Brazilian red propolis extract and isolation of bioactive fractions by thin-layer chromatography-bioautography. *Food Chem.*, 1: 127060.

Duhan, J. S., Kumar, R., Kumar, N., Kaur, P., Nehra, K., and Duhan, S. (2017). Nanotechnology: the new perspective in precision agriculture. *Biotechnol Rep.*, 15: 11–23. http://doi:10.1016/j.btre.2017.03.002.

Edi-Premono, M., Moawad, A. M., and Vleck, P. L. G. (1996). Effect of phosphate solubilizing *Pseudomonas Putida* on the growth of maize and its survival in the rhizosphere. *Indones. J. Crop Sci.*, 11: 13–23.

Elkhetabi, A., Lahlali, R., Askarne, L., Ezrari, S., El Ghadaroui, L., Tahiri, A. et al. (2020). Efficacy assessment of pomegranate peel aqueous extract for brown rot (*Monilinia* spp.) disease control. *Physiol. Mol. Plant Pathol.*, 110: 101482.

Fatima, F., Hashim, A., and Anees, S. (2021). Efficacy of nanoparticles as nanofertilizer production: A review. *Environmental Science and Pollution Research*, 28(2): 1292–1303. http://doi:10.1007/s11356-020-11218-9.

Fenta, L., Mekonnen, H., and Kabtimer, N. (2023). The exploitation of microbial antagonists against postharvest plant pathogens. *Microorganisms*, 11(4): 1044, http://doi.org/10.3390/microorganisms11041044.

Fenta, L., and Mekonnen, H. (2024) .Microbial biofungicides as a substitute for chemical fungicides in the control of phytopathogens: Current perspectives and research directions. *Scientifica* Cairo). 28;5322696. http://doi:10.1155/2024/5322696.

Fincheira, P., Hoffmann, N., Tortella, G., Ruiz, A., Cornejo, P., Diez, M. C. et al. (2023). Eco-efficient systems based on nanocarriers for the controlled release of fertilizers and pesticides: Toward smart agriculture. *Nanomaterials*, 13(13): 1978. https://doi.org/10.3390/nano13131978.

Gai, Y., and Wang, H. (2024). Plant disease: a growing threat to global food security. *Agronomy*, 14(8): 1615. https://doi.org/10.3390/agronomy14081615.

Gao, F., Cui, B., Wang, Chong., Li, X., Li, B., Zhan, S. et al. (2022). Nano-EMB-SP improves the solubility, foliar affinity, photostability and bioactivity of emamectin benzoate. *Pest Manag. Sci.*, 78: 3717–3724. https://doi.org/10.1002/PS.7015.

Garbeva, P., Hol W. H. G., Termorshuizen, A. J., Kowalchuk, G. A., and de Boer, W. (2011). Fungistasis and general soil biostasis - a new synthesis. *Soil Biol. Biochem.*, 43: 469–477. http://doi:10.1016/j.soilbio.2010.11.020.

García-Álvarez, R., and Vallet-Regí, M. (2021). Hard and Soft Protein Corona of Nanomaterials: Analysis and Relevance. *Nanomaterials*, 11(4): 888. https://doi.org/10.3390/nano11040888.

Gasca, C. A., Dassoler, M., Brand, G. D., de Medeiros Nóbrega, Y. K., Gomes, S. M., Jamal, C. M. et al. (2020). Chemical composition and antifungal effect of ethanol extract from *Sapindus saponaria* L. fruit against banana anthracnose. *Sci. Hortic.*, 259: 108842.

Ghosh, S., Ahmad, R., Zeyaullah, M., and Khare, S. K. (2021). Microbial Nano-Factories: Synthesis and Biomedical Applications. *Front Chem.*, 16(9): 626834. http://doi: 10.3389/fchem.2021.626834.

Gikas, G.D., Parlakidis, P., Mavropoulos, T., and Vryzas, Z. (2022). Particularities of fungicides and factors affecting their fate and removal efficacy: a review. *Sustainability*, 14(7): 4056. https://doi.org/10.3390/su14074056.

Goswami, P., and Mathur, J. (2022). Application of agro-waste-mediated silica nanoparticles to sustainable agriculture. *Bioresources and Bioprocessing*, 9(1): 9. https://doi.org/10.1186/s40643-022-00496-5.

Guerrero, D. S., Bertani, R. P., Ledesma, A., Frías, M. de los A., Romero, C. M., Dávila Costa, J. S. et al. (2022). Silver nanoparticles synthesized by the heavy metal resistant strain *Amycolatopsis tucumanensis* and its application in controlling red strip https://doi.org/10.1016/J.HELIYON.2022.E09472.

Gupta, R., Singal, R., Shankar, A., Kuhad, R. C., and Saxena, R. K. (1994). A modified plate assay for screening phosphate solubilizing microorganisms. *J. Gen. Appl. Microbiol.*, 40: 255–260. https://doi.org/10.2323/jgam.40.255.

Gurjar, M. S., Ali, S., Akhtar, M., and Singh, K. S. (2012). Efficacy of plant extracts in plant disease management. *Agric. Sci.*, 3(3): 425–433. http://doi:10.4236/as.2012.33050.

Hashem, A. H., Abdelaziz, A. M., Hassanin, M. M. H., Al-Askar, A. A., AbdElgawad, H., and Attia, M. S. (2023). Potential impacts of clove essential oil nanoemulsion as bio fungicides against *Neoscytalidium* blight disease of *Carum carvi* L. *Agronomy*, 13(4): 1114. https://doi.org/10.3390/agronomy13041114.

Hernández, A., Ruiz-Moyano, S., Galván, A. I., Merchán, A. V., Nevado, F. P., Aranda, E. et al. (2020). Anti-fungal activity of phenolic sweet orange peel extract for controlling fungi responsible for post-harvest fruit decay. *Fungal Biol.*, 2: 143–152.

Hernández-Ceja, A., Loeza-Lara, P. D., Espinosa-García, F. J., García-Rodríguez, Y. M., Medina-Medrano, J. R., Gutiérrez-Hernández, G. F. et al. (2021). *In vitro* antifungal activity of plant extracts on pathogenic fungi of blueberry (*Vaccinium* sp.). *Plants (Basel, Switzerland)*, 10(5): 852. https://doi.org/10.3390/plants10050852.

Hossain, T. J. (2024). Methods for screening and evaluation of antimicrobial activity: A review of protocols, advantages, and limitations. *European Journal of Microbiology and Immunology*, 14(2): 97–115. https://doi.org/10.1556/1886.2024.00035.

Islam, N., Parveen, S. A., Nakazawa, N., Marston, A., and Hostettmann, K. (2002).Bioautography with the fungus *Valsa ceratosperma* in the search for antimycotic agents. *Pharm Biol.*, 41: 637–640. http://doi:10.1080/1388020039050262.

Jeyarajan, R., and Nakkeeran, S. (2000). Exploitation of microorganisms and viruses as biocontrol agents for crop disease management. pp. 95–116. *In:* Upadhyay, R. K., Mukerji, K. G., and Chamola, B. P. (eds.). *Biocontrol Potential and its Exploitation in Sustainable Agriculture*. Springer, Boston, MA. https://doi.org/10.1007/978-1-4615-4209-4_8.

Jia, Y., Kang, L., Wu, Y., Zhou, C., Cai, R., Zhang, H. et al. (2024). Nano selenium foliar intervention-induced resistance of cucumber to *Botrytis cinerea* by activating jasmonic acid biosynthesis and regulating phenolic acid and cucurbitacin. *Pest Manag. Sci.*, 80: 554–568. https://doi.org/10.1002/PS.7784.

Kachhawa, D. (2017). Microorganisms as a biopesticides. *Journal of Entomology and Zoology Studies.*, 5: 468–473.

Kalsoom, M., Ur. Rehman, F., Shafique, T., Junaid, S., Khalid, N., Adnan, M. et al. (2020). Biological importance of microbes in agriculture, food and pharmaceutical industry: A review. *Innovare Journal of Life Sciences*, 8(6): 1–4. https://doi.org/10.22159/ijls.2020.v8i6.39845.

Kalogiannidis, S., Kalfas, D., Chatzitheodoridis, F., and Papaevangelou, O. (2022). Role of crop-protection technologies in sustainable agricultural productivity and management, *Land*, 11(10): 1680. https://doi.org/10.3390/land11101680.

Kaushal, M. (2018). Role of microbes in plant protection using intersection of nanotechnology and biology. pp. 111–135. *In*: Abd-Elsalam, K. A., and Prasad, R. (Eds.). *Nanobiotechnology applications in plant protection.* Salmon, Springer International Publishing. https://doi.org/10.1007/978-3-319-91161-8_5.

Kraszewska, J., Wzorek, W., Sztando, E., and Raczyńska, C. J. A. (2005). Aktywność antagonistyczna bakterii fermentacji mlekowej z gatunku *Lactobacillus plantarum. Acta Sci. Pol., Technol. Aliment.*, 4(1): 39–45.

Kumar, B., Alam, A., Agrawal, K., Prasad, H., and Verma, P. B. (2018). Production, purifcation and characterization of an acid/alkali and thermo tolerant cellulase from *Schizophyllum commune* NAIMCC-F03379 and its application in hydrolysis of lignocellulosic wastes. *AMB Express.*, 8: 173. http://doi.10.1186/s13568-018-0696-y.

Kumar, A., Zacharia, S., Maurya, A. K., and John, V. (2019). Effect of fungicides and neem oil on the Rhizoctonia root rot of soybean (*Glycine max* L.). *International Journal of Current Microbiology and Applied Sciences*, 8(1): 368– 372. http://doi: 10.20546/ijcmas.2019.801.037.

Li, R., Jia, X., Xie, H., Wang, X., Wen, J., Yu, G. et al. (2024). Vali@mPEG-TK-COS: A dual function nano-pesticide for fungicidal action and plant immune modulation. *Mod. Agric.*, 2: 10. e70001. https://doi.org/10.1002/MODA.70001.

Malik, A. Q., Mir, T. ul G., Kumar, D., Mir, I. A., Rashid, A., Ayoub, M. et al. (2023). A review on the green synthesis of nanoparticles, their biological applications, and photocatalytic efficiency against environmental toxins. *Environ. Sci. Pollut. Res. Int.*, 30: 69796–69823. https://doi.org/10.1007/S11356-023 27437-9.

Maluin, F. N., Hussein, M. Z., and Idris, A. S. (2020). An overview of the oil palm industry: Challenges and some emerging opportunities for nanotechnology development. *Agronomy*, 10(3): 356. https://doi.org/10.3390/agronomy10030356.

Marston, A. (2011). Thin-layer chromatography with biological detection in phytochemistry. *J. Chromatogr. A.*, 1218: 2676–2683. http://doi:10.1016/j.chroma.2010.12.068.

Matras, E., Gorczyca, A., Przemieniecki, S., and Ocwieja, M. (2022). Surface properties-dependent antifungal activity of silver nanoparticles. *Scientific Reports*. 12: 18046. 10.1038/s41598-022-22659-2.

Merlin, T. M., and Anitha, R. (2021). Antagonistic Action of *Verticillium glaucum* Bonord. on Fungal PlantPathogen. *J. Mycol. Pl. Pathol.*, 51(1): 59–65.

Mesa, A. M., Marin, P. A., Ocampo, O., Calle, J., and Monsalve Fonnegra, Z. I. (2019). Fungicidas a partir de extractos vegetales: Una alternativa en el manejo integrado de hongos fitopatógenos. *Rev. Investig. Agropecu.*, 45: 23–30.

Mosa, M. A., and El-Abeid, S. E. (2023). Chitosan-loaded copper oxide nanoparticles: A promising antifungal nanocomposite against Fusarium wilt disease of tomato plants. *Sustainability*, 15(19): 14295. https://doi.org/10.3390/su151914295.

Navarro, V., Rojas, G., Delgado, G., and Lozoya, X. (1998). Antimicrobial compounds detected in *Bocconia arborea* extracts by direct bioautographic method. *Arch. Med. Res.*, 29: 191–194.

Nautiyal, C. S. (1999). An efficient microbiological growth medium for screening phosphate solubilizing microorganisms. *FEMS Microbiol. Lett.*, 170: 265–270. http://doi: 10.1111/j.1574-6968.1999.tb13383.x.

Ordóñez, Y. F., Ruano, J., Avila, P., Berutti, L., Guerrero, P. C., Ordóñez, P. E. et al. (2023). *In vitro* antimicrobial activity of plant species against the phytopathogens *Ralstonia solanacearum*, *Phytophthora infestans*, and *Neopestalotiopsis javaensis*. *Agriculture*, 13: 2029.

Pandey, S., Giri, V. P., Tripathi, A., Kumari, M., Narayan, S., Bhattacharya, A. et al. (2020). Early blight disease management by herbal nanoemulsion in *Solanum lycopersicum* with bio-protective manner. *Industrial Crops and Products*, 150: 112421. https://doi.org/10.1016/j.indcrop.2020.112421.

Pariona, N., Mtz-Enriquez, A. I., Sánchez-Rangel, D., Carrión, G., Paraguay-Delgado, F., and Rosas-Saito, G. (2019). Green-synthesized copper nanoparticles as a potential antifungal against plant pathogens. *RSC Advances*, 9(33): 18835–18843. https://doi.org/10.1039/c9ra03110c.

Passari, A. K., Mishra, V. K., Gupta, V. K., Yadav, M. K., Saikia, R., and Singh, B. P. K. (2015). *In vitro* and *in vivo* plant growth-promoting activities and DNA fingerprinting of antagonistic endophytic Actinomycetes associates with medicinal plants. *Plos one.*, 10(9):e0139468. http://doi:10.1371/journal.pone.0139468.

Paterlini, P., Rodríguez, C., Ledesma, A., Pereyra, J., Dávila Costa, J. S., Álvarez, A. et al. (2021). Characterization of biosynthesized silver nanoparticles from *Streptomyces aqueous* extract and evaluation of surface-capping proteins involved in the process. *Nano-Structures & Nano-Objects*, 26: 100755. https://doi.org/10.1016/J.NANOSO.2021.100755.

Pawlak, K., and Kołodziejczak, M. (2020). The role of agriculture in ensuring food security in developing countries: considerations in the context of the problem of sustainable food production. *Sustainability*, 12(13): 5488 https://doi.org/10.3390/su12135488.

Pelosi, C., Bertrand, C., Daniele, G., Coeurdassier, M., Benoit, P., Nélieu, S. et al.(2021). Residues of currently used pesticides in soils and earthworms: a silent threat? *Agriculture, Ecosystems and Environment*, 305, D 107167. https://doi.org/10.1016/j.agee.2020.107167.

Picman, A. K., Schneider, E. F., and Gershenzon, J. (1990). Antifungal activities of sunflower terpenoids, *Biochem. Sys. Ecol.*, 18: 325–328. https://doi.org/10.1016/0305-1978(90)90005-Z.

Pikovskaya, R. I. (1948). Mobilization of phosphorus in soil in connection with vital activity of some microbial species. *Mikrobiologiya.*, 17: 362–370.

Pimentel, D. (2009). Pesticide and pest control. *In:* Peshin, P., and Dhawan, A. K. (Eds.). Integrated pest management: innovation-development process. Springer, Dordrecht, Netherlands, 83–87.

Ramírez, P. G., Ramírez, D. G., Mejía, E. Z., Ocampo, S. A., Díaz, C. N., Martínez, R. I. R. et al. (2020). Extracts of Stevia rebaudiana against Fusarium oxysporum associated with tomato cultivation. *Sci. Hortic.*, 259: 108683.

Ramírez, G. E., Contreras-Oliva, A., Salinas-Ruiz, J., Hernández-Ramírez, G., Spinoso-Castillo, J. L., Colmenares Cuevas, S. I. et al. (2023). Plant extracts control *in vitro* growth of disease-causing fungi in Chayote. *Plante*, 12: 1800.

Rana, A., Krishna, Y., and Sheeja, J. (2020). A comprehensive review on green synthesis of nature-inspired metal nanoparticles: Mechanism, application and toxicity. *Journal of Cleaner Production*, 272(5): 122880. http://doi:10.1016/j.jclepro.2020.122880.

Rana, A. K., Kaur, K., and Vyas, P. (2024). Biofungicides and plant growth promoters: Advantages and opportunities in entrepreneurship. *Entrep. Microorg.*, 1: 259–277.

Rios, J. L., Recio, M. C., and Villar, A. (1988). A Screening methods for natural products with antimicrobial activity: A review of the literature. *J. Ethnopharmacol.*, 23: 127–149. http://doi:10.1016/0378-8741(88)90001-3.

Romero, C. M., and Dávila Costa, J. S. (2025). Nano-biofungicides and bio-nanofungicides: state of the art of innovative tools for controlling resistant phytopathogens. *Preprints*. https://doi.org/10.20944/preprints202502.2287.v1.

Roger, R. C., and Philogène, B. J. R. (2008). Past and current prospects for the use of botanicals and plant allelochemicals in integrated pest management. *Pharmaceut. Biol.*, 46(1–2): 41–52. http://doi:10.1080/13880200701729794.

Ruiu., L. (2020). Plant-growth-promoting bacteria (PGPB) against insects and other agricultural pests, *Agronomy*, 10(6): 861. https://doi.org/10.3390/agronomy10060861.

Rütten, A., Kirchner, T., and Musiol-Kroll, E.M. (2022). Overview on Strategies and Assays for Antibiotic Discovery. *Pharmaceuticals*, 15(10): 1302. https://doi.org/10.3390/ph15101302.

Saeed Al-Zahrani, S., and Mohammed Al-Garni, S. (2023). Antifungal potentiality of mycogenic silver nanoparticles capped with chitosan produced by endophytic *Amesia atrobrunnea*. *Saudi Journal of Biological Sciences*, 30(9): 103746. https://doi.org/10.1016/j.sjbs.2023.103746.

Salas-Gómez, A. L., Espinoza Ahumada, C. A., Castillo Godina, R. G., Ascacio-Valdés, J. A., Rodríguez-Herrera, R., Segura Martínez, M. T. d. J. et al. (2023). Antifungal *in vitro* activity of *Phoradendron* sp. extracts on fungal isolates from tomato crop. *Plants*, 12(3): 672. https://doi.org/10.3390/plants12030672.

Samada, L. H., and Tambunan, U. S. F. (2020). Biopesticides as promising alternatives to chemical pesticides: a review of their current and future status. *Online Journal of Biological Sciences*, 20(2): 66–76. https://doi.org/10.3844/ojbsci.2020.66.76.

Santra, H. K., and Banerjee, D. (2020). Natural products as fungicide and their role in crop protection. *Natural Products in Sustainable Agric.*, 131–219. http://doi:10.1007/978-981-15-3024-1_9.

Sen Karaman, D., Manner, S., Fallarero, A., and Rosenholm, J. (2017). Current approaches for exploration of nanoparticles as antibacterial agents. pp 61–86. *In:* Ranjith, N. K (Eds.). *Antimicrobial Agent*. Intech, Crotia. http://dx.doi.org/10.5772/68138.

Shahverdi, A. R., Abdolpour, F., Monsef-Esfahani, H. R., and Farsam, H. (2007). A TLC bioautographic assay for the detection of nitrofurantoin resistance reversal compound. *J. Chromato.* B., 850: 528–530. http://doi:10.1016/j.jchromb.2006.11.011.

Sharma, S., Kumari, P., Thakur, P., Brar, G. S., Bouqellah, N. A., and Hesham, A. E. (2022). Synthesis and characterization of $Ni_{0.5}Al_{0.5}Fe_2O_4$ nanoparticles for potent antifungal activity against dry rot of ginger (Fusarium oxysporum). *Scientific Reports*, 12(1): 20092. https://doi.org/10.1038/s41598-022-22620-3.

Shanmugapriya, S., Jeya S. S., Golla, D. G., Lakshmanan, A., Tarunendu, S., Karthikeyan, G. et al. (2024). Encapsulation of essential oil to prepare environment friendly nanobio-fungicide against *Fusarium oxysporum* f. sp. *lycopersici*: An experimental and molecular dynamics approach, Colloids and Surfaces A: Physicochemical and Engineering Aspects, (681): 132681. http://doi:10.1016/j.colsurfa.2023.132681.

Sheeba, J., Dhamotharan, R., and Baskar, K. (2017). Isolation, screening and characterization of plant growth promoting bacteria and their antifungal effect on *Rhizoctonia solani* (J.G. Kühn 1858). *Adv. Plants Agric. Res.*, 17(7): 369–375.

Sivasankari, B. (2016). Indole-3-acetic acid production by the bacterial strains isolated from vermicomposts in the presence and absence of tryptophan. *Int. J. Innov. Res. Sci. Eng. Technol.*, 5: 8698–8706. http://doi:10.15680/IJIRSET.2016.0505296.

Sobhy, S., Al-Askar, A. A., Bakhiet, E. K., Elsharkawy, M. M., Arishi, A. A., Behiry, S. I. et al. (2023). Phytochemical characterization and antifungal efficacy of camphor (*Cinnamomum camphora* L.) Extract against Phytopathogenic Fungi. *Separations*, 10: 189.

Souza, J. R. B., Kupper, K. C., and Augusto, F. (2018). *In vivo* investigation of the volatile metabolome of antiphytopathogenic yeast strains active against *Penicillium digitatum* using comprehensive two-dimensional gas chromatography and multivariate data analysis. *Microchem. J.*, 141: 204–209. http://doi:10.1016/j.microc.2018.05.047.

Spagnoletti, F. N., Kronberg, F., Spedalieri, C., Munarriz, E., and Giacometti, R. (2021). Protein corona on biogenic silver nanoparticles provides higher stability and protects cells from toxicity in comparison to chemical nanoparticles. *Journal of Environmental Management*, 297: 113434. https://doi.org/10.1016/j.jenvman.2021.113434.

Suleimana, M. M., McGaw, L. J., Naidoo, V., and Eloff, J. N. (2009). Detection of antimicrobial compounds by bioautography of different extracts of leaves of selected South African tree species. *Afr. J. Tradit. Complement Altern. Med.* 15, 7(1): 64–78. http://doi:10.4314/ajtcam.v7i1.57269.

Suteu, D., Rusu, L., Zaharia, C., Badeanu, M., and Daraban, G. M .(2020). Challenge of utilization vegetal extracts as natural plant protection products. *Appl. Sci.*, 10(24): 1–21. https://doi.org/10.3390/app10248913.

Syromyatnikov, M., Savinkova, O., and Popov, V. (2021). Toxicity of new fungicides for eukaryotic microorganisms isolated from the gut of the food-important vegetable pollinator *Bombus terrestris* L. *Proceedings of the Voronezh State University of Engineering Technologies*, 82(4): 54–59. http://doi:10.20914/2310-1202-2020-4-54-59.

Tabassum, B., Khan, A., Tariq, M., Ramzan, M., Khan, M. S. I., Shahid, N., and Aaliya, K. (2017) .Bottlenecks in commercialization and future prospects of PGPR. *Appl. Soil Ecol.*, 121: 102–17. http://doi:10.1016/j.apsoil.2017.09.030.

Taborga, L., Díaz, K., Olea, A. F., Reyes-Bravo, P., Flores, M. E., Peña-Cortés, H. et al. (2015). Effect of polymer micelles on antifungal activity of geranylorcinol compounds against *Botrytis cinerea. Journal of Agricultural and Food Chemistry*, 63(31): 6890–6896. https://doi.org/10.1021/acs.jafc.5b01920.

Theer, K. M., Alheeti, A. A., and Alfalahi, A. O. (2021). A novel and Rapid Method for Detecting Bio-fungicide Properties in Plant Species. *IOP Conf. Ser.: Earth Environ. Sci.*, 761: 012024. http://doi:10.1088/1755-1315/761/1/012024.

Tiwari, S., Vaish, B., and Singh, P. (2017). Population and global food security: issues related to climate change. pp. 40–63. *In:* Singh, R., Singh, A., and Srivastava, V. (Eds.). *Environmental Issues Surrounding Human Overpopulation*. Hershey, PA:. IGI Global. https://doi.org/10.4018/978-1-5225-1683-5.ch003.

Tripathi, Y. N., Divyanshu, K., Kumar, S., Jaiswal, L. K., Khan, A., Birla, H. et al. (2020). Biopesticides: current status and future prospects in India. *Bioeconomy Sustain. Dev.*, 1: 79–109.

Waldamichael, F. G., Debelee, T., Schwenker, F., Ayano, Y. M., and Kebede, S. R. (2022). Machine learning in cereal crops disease detection: a review, *Algorithms*, 15(3): 75. https://doi.org/10.3390/a15030075.

Wendimu, A., Yoseph, T., and Ayalew, T. (2023). Ditching phosphatic fertilizers for phosphate-solubilizing biofertilizers: A step towards sustainable agriculture and environmental health. *Sustainability*. 15(2): 1713. https://doi.org/10.3390/su15021713.

Won, S. J., Kwon, J. H., Kim, D. H., and Ahn, Y. S. (2019) The effect of *Bacillus licheniformis* MH48 on control of foliar fungal diseases and growth promotion of *Camellia oleifera* seedlings in the coastal reclaimed land of Korea. *Pathogens,* 9; 8(1): 6. http://doi: 10.3390/pathogens8010006.

Xu, L., Zhu, Z., and Sun, D. W. (2021). Bioinspired nanomodification strategies: Moving from chemical-based agrosystems to sustainable agriculture. *ACS nano*, 15(8): 12655–12686. https://doi.org/10.1021/acsnano.1c03948.

Yongpisanphop, J. (2024). Virtual screening of potential biofungicide candidate for sustainable fungal disease control. *E3S Web of Conferences*, 536: 02003.

Yoon, M. Y., Cha, B., and Kim, J. C. (2013). Recent trends in studies on botanical fungicides in agriculture. *Plant Pathol. J.*, 29(1): 1–9. https://doi.org/10.5423/PPJ.RW.05.2012.0072.

Yu, Y., Dai, W., and Luan, Y. (2023). Bio- and eco-corona related to plants: Understanding the formation and biological effects of plant protein coatings on nanoparticles. *Environmental pollution (Barking, Essex :* 1987), 317: 120784. https://doi.org/10.1016/j.envpol.2022.120784.

Yuan, J., Shen, Q., and Huang, Q. (2012). Antifungal activity of Bacillus amyloliquefaciens NJN-6 volatile compounds against *Fusarium oxysporum* f. sp. *cubense*. *Appl. Environ. Microbiol.*, 78(16): 5942–5944. http://doi:10.1128/AEM.01357-12.

Zhao, X., Song, P., Hou, D., Li, Z., and Hu, Z. (2021). Antifungal activity, identification and biosynthetic potential analysis of fungi against *Rhizoctonia cerealis*. *Ann. Microbiol.*, 71, 41. https://doi.org/10.1186/s13213-021-01654-4.

Zhao, X., Zhang, Y., Chen, L., Ma, Z., and Zhang, B. (2023). Chitosan-thymol nanoparticle with pH responsiveness as a potential intelligent botanical fungicide against *Botrytis cinerea*. *Pesticide Biochemistry and Physiology*, 195: 105571. https://doi.org/10.1016/j.pestbp.2023.105571.

Ziedan, El Sayed., Saad, M. M., El-Kafrawy, A. A., Sahab, A. F., and Mossa, A. T. (2022). Evaluation of essential oils nanoemulsions formulations on *Botrytis cinerea* growth, pathology and grey mould incidence on cucumber fruits. *Bulletin of the National Research Centre*., 46. 1s0.1186/s42269-022-00765-5.

3

Culture Maintenance and Propagation of Biocontrol Agents for Biofungicide Formulation

Muhammad Mazhar Iqbal,[1] *Binnian Tian,*[2,*]
Mirza Abid Mehmood[1] and *Muhammad Ashfaq*[1]

1. Introduction

Plant diseases pose a major challenge to crop production, which ultimately threatens agricultural and horticultural systems worldwide. Crops are vulnerable to a wide range of pathogens, including bacteria, fungi, viruses, and nematodes, which collectively contribute to considerable economic losses and threaten global food security. It is estimated that approximately 25% of crops globally are destroyed by plant diseases annually (Lugtenberg 2015). Crop losses mainly occur in the staple crops like rice, wheat, and maize. These devastations lead to food shortages as well as unstable economies.

The management of plant diseases poses a significant cost in agricultural production. Traditionally, methods of managing plant diseases have relied on the creation of resistant crop varieties, hygiene to prevent the spread of contaminated soil or seeds, and fungicides for the control of pathogenic fungi (Jahagirdar et al. 2003). However, growing environmental concerns over residues of fungicides in food and their ecological impacts have led to the deregistration of many chemical fungicides. These chemical fungicides lead to environmental pollution, pose a threat to human and animal health, harm beneficial microbes, and can develop

[1] Plant Pathology, Institute of Plant Protection, MNS University of Agriculture, Multan.
[2] College of Plant Protection, Southwest University, Chongqing, China.
* Corresponding author: binniantian@swu.edu.cn

DOI: 10.1201/9781003603023-4

resistance in fungal strains (Fenta et al. 2024). This shift has increased interest in biological control as an alternative and sustainable plant disease management strategy, which does not have such types of threats (O'Brien 2017).

Biological control involves the suppression of plant diseases through the application of biocontrol agents (BCAs) such as fungi, bacteria, viruses, or mixtures of these organisms. BCAs are applied directly to plants or soil surfaces to inhibit the establishment of pathogens. Such agents prevent pathogens from colonizing the plant or soil. These agents work through mechanisms that include competition for essential nutrients, synthesis of antimicrobial substances, activation of plant defense mechanisms, and parasitism through a mechanism that kills pathogenic organisms (Köhl et al. 2019). BCAs are specific in their mode of action, which makes them harmless to non-target organisms. In addition, BCAs can potentially regulate disease development between 79%–98%, depending on the type of BCA used (Alexander and Stewart 2001).

Considerable progress has been made in the last few decades in the isolation and commercial application of effective BCAs for the control of fungal and bacterial diseases in field crops. Currently, a wide variety of BCAs is in commercial application worldwide. For instance, several species of *Trichoderma*, including *T. atroviride*, *T. harzianum*, *T. hamatum*, and *T. viride* are proved to be effective in the control of plant diseases (Alfiky and Weisskopf 2021, Zin and Badaluddin 2020). In addition, several bacterial species, including *Bacillus*, *Streptomyces*, *Pseudomonas*, and *Burkholderia* spp., have been used extensively, which demonstrates their ability to reduce reliance on chemical inputs (LeBlanc 2022).

The most important step in the formulation of biofungicides is the propagation of BCAs, based on their effectiveness in managing plant diseases. Successful mass production of BCAs ensures the viability and effectiveness of their propagules, which serve as active ingredients in biofungicides. These propagules are propagated using optimized methods to maintain their biological efficacy, scalability, and economic viability for commercial production (Khan and Ahmad 2024). Among these techniques, Solid-State Fermentation (SSF) and Submerged Fermentation (SmF) are the most significantly applied. However, challenges associated with new BCA development and commercialization include restricted research and collective knowledge about microbial formulations, patents and industrial issues, along with intricate registration processes, especially in Europe (Frederiks and Wesseler 2019).

2. Common Biocontrol Microorganisms

2.1 Fungal Biocontrol Agents

Fungal biocontrol agents are crucial in integrated and sustainable agricultural systems. They neutralize the pathogenic activities of plant pathogens and pests. These fungi act as natural alternatives to chemical pesticides that function through mycoparasitism, resource competition, production of antimicrobial compounds, and the triggering of induced systemic resistance. Species that belong to this group include *Trichoderma*, *Gliocladium*, and *Penicillium*. The management of plant pathogens by using fungal antagonists is on the rise with time, where they have greater reproductive potential, specificity to target pests, and short generation intervals (Thambugala et al. 2020). Some other important fungal biocontrol agents are mentioned in Table 1.

Table 1: Fungal pathogens and their biocontrol agents along with their mode of action.

Fungal Biocontrol Agent	**Pathogen**	**Disease**	**Host crop**	**Mode of action**	**References**
Trichoderma asperellum	*Sclerotium cepivorum*	White rot	Onion	ISR	(Rivera et al. 2020)
Trichoderma asperellum T8a	*Colletotrichum gloeosporiodes*	Powdery mildew	Mango	Mycoparasitism, Antibiosis	(de los et al. 2013)
Trichoderma hamatum	*Sclerotinia sclerotiorum*	Whit mold	Cabbage	Competition, Mycoparasitism	(Jones et al. 2014)
Trichoderma asperellum			Beans	Competition, ISR	(Geraldine et al. 2013)
Coniothyrium minitans			Oilseed rape	Mycoparasitism	(Li et al. 2006)
			Sunflower		(McQuilken et al. 1997)
Trichoderma viride	*Sclerotium rolfsii*	Stem rot	Groundnut		(Karthikeyan et al. 2006)
Clonostachys rosea	*Fusarium graminearum*	Head blight	Cereals	Mycoparasitism, Competition	(Gimeno 2020)
Clonostachys rosea	*Botrytis cinerea*	Gray mold	Strawberry		(Cota et al. 2008)
Trichoderma hamatum	*Pythium aphanidermatum*	Damping off	Cucumber		(Al-shemmary et al. 2018)
Ampelomyces quisqualis	*Erysiphe necator*	Powdery mildew	Grapes	Hyperparasitism	(Angeli et al. 2009)
Gliocladium virens	*Pythium ultimum*	Damping off	Vegetable crops	Antibiosis, Competition	(Lumsden et al. 1989)
Beauveria bassiana	*Fusarium oxysporum* f. Sp. *Lycopersici*	Fusarium wilt	Tomato		(Culebro-Ricaldi et al. 2017)
Aspergillus niger	*Fusarium oxysporum* f. Sp. *Melongenae*	Wilt	Brinjal		(Patil et al. 2017)
Verticillium lecanii	*Puccinia recondita*	Rust	Wheat	Hyperparasitism	(Spencer and Atkey 1981)
Trichoderma reesei	*Fusarium oxysporum* f. Sp. *Cubense*	Wilt	Banana	Competition	(Damodaran et al. 2020)

2.2 Bacterial Biocontrol Agents

Bacterial biocontrol agents refer to beneficial bacteria that can suppress plant pathogens while having the potential to offer sustainability in agricultural systems. These agents play a significant role in integrated pest management because they reduce dependency on synthetic pesticides. They are soil bacteria commonly found in the rhizosphere, where they colonize plant roots and establish mutual relationships with their host plants. They inhibit pathogens by different mechanisms, such as competing for space and nutrients, producing antimicrobial compounds, and triggering plant immune responses (Compant et al. 2005). These biocontrol agents are particularly valued for their specificity, eco-friendliness and compatibility with other pest management practices. *Pseudomonas* spp., *Streptomyces* spp. and *Bacillus* spp. are the most common and commercially available bacterial biocontrol agents worldwide (Bonaterra et al. 2022). Their efficacy greatly depends on factors like microbial agent, targeted pathogens, environmental conditions, etc. (Schaechter 2009). Some other important bacterial biocontrol agents are mentioned in Table 2.

2.3 Viral Biocontrol Agents

Virus-based biocontrol agents have a unique and critical role in managing plant diseases. Their natural antagonistic properties and specificity against plant pathogens make them unique biocontrol agents. Agents like bacteriophages and mycoviruses are safe alternatives to chemical pesticides. These agents are pathogen-specific in their mode of action and are harmless to non-target organisms. Bacteriophages are viruses that specifically attack and kill bacteria. Considerable research has been carried out on bacteriophages in plant disease control related to bacterial pathogens. One of the most significant uses is in controlling black rot of cabbage, which is a bacterial disease caused by *Xanthomonas campestris* pv. *campestris* on seedlings (Orynbayev et al. 2020), bacterial blight disease in rice caused by *X. oryzae* (Ranjani et al. 2018), and bacterial spot of pepper caused by *X. euvesicatoria* (Gašić et al. 2018). Phage applications reduce populations of pathogens on plant surfaces, thus preventing infection and disease spread with no negative effect on beneficial microbial communities.

Mycoviruses, viruses infecting fungal pathogens, are another promising avenue for the control of plant diseases. These viruses can reduce the virulence, reproduction, and growth of their fungal hosts. For example, *Cryphonectria hypovirus* 1 (CHV1) has been effectively utilized to control chestnut blight caused by *Cryphonectria parasitica* (Akıllı et al. 2013, Zamora et al. 2014). The introduction of CHV1 into fungal populations suppresses the virulence of the pathogen. Cross-protection strategies, in which a mild strain of a virus is introduced in the host to protect plants from more severe strains, are also effective in managing viral diseases of crops such as Citrus tristeza virus (CTV) and Zucchini yellow mosaic virus (ZYMV) (Zhang et al. 2016, Zhou et al. 2012).

Table 2: Bacterial pathogens and their biocontrol agents along with their mode of action.

Biocontrol Agent	**Pathogen**	**Disease**	**Host crop**	**Mode of action**	**References**
Bacillus atrophaeus FA12	*Xanthomonas oryzae pv. oryzae*	Leaf blight	Rice	Antibiosis	(Rajer et al. 2022)
Bacillus subtilis MJMP2				ISR	(Cheng et al. 2016)
Pseudomonas fluorescens VSMKU3054	*Ralstonia solanacearum*	Wilt	Tomato	ISR, Antibiosis	(Suresh et al. 2022)
Bacillus thurinzgiensis				Antibiosis	(Hyakumachi et al. 2013)
Bacillus velezensis FZB42	*Xanthomonas campestris pv. campestris*	Black rot	cabbage	Antibiosis, competition	(Mácha et al. 2021)
Bacillus cereus AR156	*Pseudomonas syringae pv. tomato*	Bacterial speck	*Arabidopsis thaliana*	Antibiosis	(Wang et al. 2018)
Bacillus velezensis	*Xanthomonas citri* subsp. *citri*	Canker	Citrus		(Rabbee et al. 2019)
Pantoea agglomerans	*Neofusicoccum parvum*	Dieback	Grapevine		(Haidar et al. 2021)
Bacillus velezensis strain JIN4	*Pseudomonas syringae pv. actinidiae*	Bacterial canker	Kiwifruit		(Zhao et al. 2024)
Paenibacillus macerans, Paenibacillus polymyxa	*Ralstonia solanacearum*	Wilt	Potato, Tomato	Root colonization, antibiosis	(Li et al. 2010)

3. Mechanisms of Action

Biocontrol agents work through specific mechanisms, as mentioned in Figure 1. The most important of these mechanisms include parasitism, competition, antibiosis, or induction of resistance in their respective host plants.

3.1 Parasitism

Parasitism is an interaction in which microbial biocontrol agents attack or parasitize the pathogens, leading to their death for the purpose of survival (Palmieri et al. 2022). Hyperparasitism is the most common phenomenon in this scenario where beneficial organisms attack and suppress plant disease-causing agents such as fungi, bacteria, or nematodes. *Trichoderma* species are the most commonly used fungal BCAs for the management of soil-borne pathogens including *Rhizoctonia solani*, *Sclerotinia sclerotiorum*, and *Phytophthora* spp. (Guzmán et al. 2023). *Trichoderma*

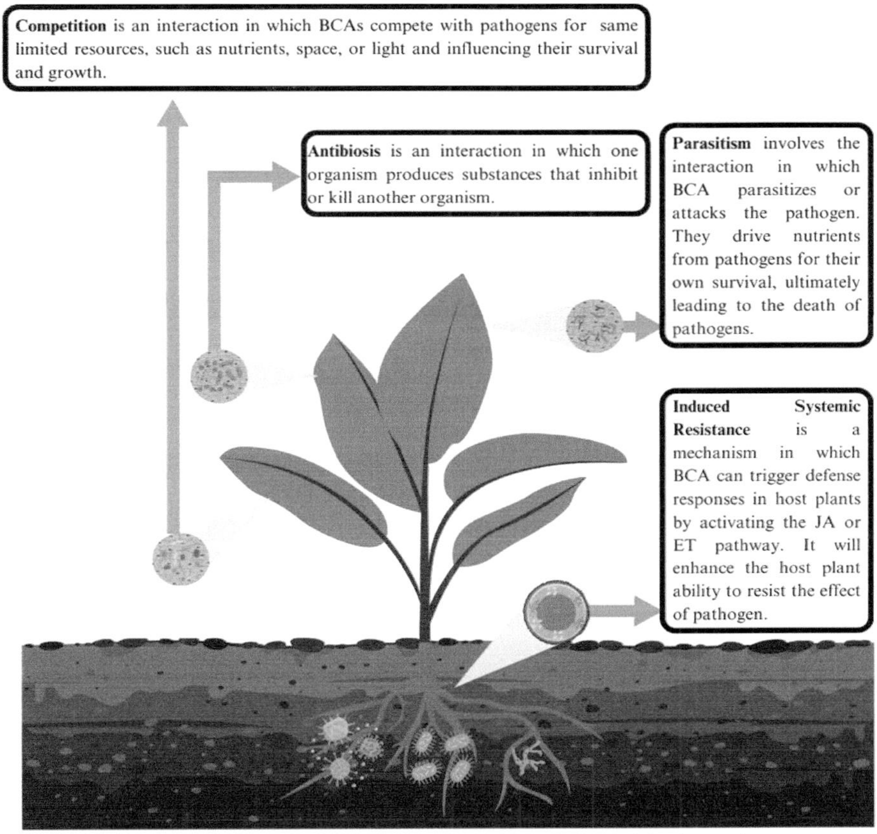

Figure 1: Mechanisms of action of Biocontrol Agents (BCAs) against plant pathogens.

uses the mechanism of mycoparasitism, where it physically interacts with the hyphae of fungal pathogens by coiling around them and then penetrates their cell walls using hydrolytic enzymes such as chitinases, glucanases, and proteases. Once it enters the pathogen, it degrades the pathogen's cellular components, ultimately leading to its death (Viterbo and Horwitz 2014).

Ampelomyces quisqualis is another important mycoparasite and is used worldwide for its ability to control powdery mildew disease on different host plants. *Ampelomyces* penetrates the conidia and mycelia of powdery mildew, consuming their contents to fulfill its nutritional requirements and reducing their ability to reproduce (Caffi et al. 2022). In some cases, multiple mycoparasites may attack a single fungal pathogen. For example, *Gliocladium virens*, *Acrodontium crateriforme*, *A. quisqualis*, *Acremonium alternatum* and *Cladosporium oxysporum* are some representative cases of fungi that have the ability to kill the powdery mildew pathogens (Heydari and Pessarakli 2010).

Some bacteria are also reported as parasites of fungi. Different strains of *Streptomyces* spp. are hyperparasites of different phytopathogenic fungi including *Botrytis cinerea, Alternaria brassicicola, Mycocentrospora acerina, Fusarium oxysporum, S. sclerotiorum, Phomopsis sclerotioides, Rhizoctonia solani* and *Pythium ultimum* (Tapio and Pohto-Lahdenperä 1991). Trunk et al. (2018) reported that these bacteria inject some antifungal effectors in hyphal cells, leading to their death.

3.2 Competition

Competition between microbial communities for nutrients and space is of crucial importance as these resources are generally limited (Ghoul and Mitri 2016). In the rhizosphere area of soil, beneficial microbes like *Pseudomonas fluorescens* and *Bacillus subtilis* establish themselves to compete with pathogenic organisms. These bacteria primarily produce siderophores, iron-chelating molecules (Sivasakthi et al. 2014). Iron is considered an important nutritional component for soil-borne microbes, including pathogens (Colombo et al. 2014). The siderophores produced by beneficial microbes bind with iron molecules present in the soil, making them unavailable for pathogens. For example, *Pseudomonas fluorescens* produces fluorescent siderophores that limit this critical resource for fungal pathogens like *Fusarium oxysporum*, ultimately inhibiting their growth (Sahu et al. 2018).

Some BCAs like *Bacillus subtilis* form biofilms on plant roots, creating a physical barrier. It prevents the attachment and invasion of pathogens (Fessia et al. 2022). This spatial exclusion reduces the likelihood of pathogen infection. Some other BCAs like *Trichoderma* metabolize organic compounds in the soil more efficiently than many plant pathogens. This creates insufficiency of these resources for the pathogens. This type of competition is observed in case of *Rhizoctonia solani* (Ghazanfar et al. 2018).

3.3 Antibiosis

The word "Antibiosis" is derived from two Greek words, ἀντι and βίος, which collectively mean "against life". The term is generally considered an important property of BCA in which they produce some volatile or diffusible antibiotic compounds to kill or inhibit the growth of another microbe (Palmieri et al. 2022). It provides a target suppression of pathogens without harming the environment.

Production of lipopeptides by *Bacillus* spp. *is* one of the most well-documented examples of antibiosis (Ali et al. 2022). It produces compounds such as surfactin, iturin, and fengycin that disrupt the cell membrane of fungal pathogens and stimulate leakage of cellular contents (Nakkeeran et al. 2019). As a result, cell death occurs. For example, iturin produced by *Bacillus subtilis* effectively controls *Rhizoctonia solani* and *Botrytis cinerea* (Kong et al. 2010). Similarly, *Pseudomonas fluorescens* produces phenazine and pyrrolnitrin, which inhibit the growth of fungal pathogens such as *Fusarium oxysporum* and *Phytophthora infestans* (Morrison et al. 2017, Saranraj et al. 2023). These compounds disrupt critical metabolic processes in pathogens. *P. fluorescens* also emits hydrogen cyanide (HCN), which is basically a volatile compound that is toxic to many soil-borne pathogens (Mehmood et al. 2023).

Additionally, beneficial microbes also produce extracellular enzymes that degrade the cell wall of pathogens. These enzymes include chitinases, glucanases, and proteases that break down the structural components of fungal pathogens, effectively neutralizing them. For example, in the case of *Trichoderma* spp., secretion of chitinases degrades the chitin in the cell wall of pathogenic fungi, ultimately leading the cell to death (Sivan and Chet 1989).

3.4 Induced Systemic Resistance (ISR)

Induced Systemic Resistance (ISR) is a defense mechanism of plants activated by certain beneficial microorganisms, by which the host plant enhances its ability to resist a wide range of pathogens (Yu et al. 2022). It primes the plant to respond more effectively to subsequent pathogen attacks and relies on the jasmonic acid (JA) and ethylene signaling pathways (Van Loon et al. 1998). These BCAs enable plants to induce physiological and biochemical changes, such as increased production of pathogenesis-related (PR) proteins, secondary metabolites (i.e., phenolics), and defensive enzymes including peroxidases and polyphenol oxidases (Choudhary and Johri 2009, Nawrocka and Małolepsza 2013).

Some of the first reports to induce ISR include the use of *Pseudomonas* spp. for managing *Fusarium* wilt on carnation (Van Peer et al. 1991) and the use of growth-promoting rhizobacteria for managing *Colletotrichum orbiculare* in cucumber (Wei et al. 1991).

Beneficial microorganisms including *Pseudomonas fluorescens* and *Bacillus subtilis* are some of well-known ISR inducers. These microbes colonize plant roots and release microbe-associated molecular patterns (MAMPs), pathogen-associated molecular patterns (PAMPs) or elicitor molecules (Pieterse et al. 2014). These molecules interact with plant receptors and trigger a cascade of immune responses. For example, *Pseudomonas fluorescens* produces siderophores, lipopeptides, and other signaling molecules that activate ISR (Sahu et al. 2018). This enhanced resistance protects host plants against soil-borne pathogens like *Fusarium oxysporum* and foliar pathogens like *Botrytis cinerea* (David et al. 2018).

Trichoderma spp. are some other widely used BCAs that produce secondary metabolites and enzymes (Sharma et al. 2023). As a result, stimulation of defense-related genes in plants occurs. When *Trichoderma* colonizes the plant roots, it primes the plant to produce antimicrobial compounds including phytoalexins and pathogenesis-related (PR) proteins in case of pathogen invasion (Pacheco et al. 2022). This mechanism is most common against pathogens like *Pythium* and *Phytophthora* (Hadar et al. 1984, Promwee et al. 2017).

4. Maintenance of Biocontrol Cultures

4.1 Short-term Storage Methods

Short-term storage involves the preservation of fungal culture for up to one year. This type of storage can be achieved by periodic serial transfer of the majority of fungal cultures. But this method is labor-intensive and time-consuming and includes a number of drawbacks. Cultures need to be regularly examined for drying and mite or other microbial contamination. Furthermore, the morphology and physiology of cultured fungi may alter over time. Specifically, after multiple transfers, their capacity to sporulate or infect a host may be lost. These drawbacks make the method generally unsuitable for long-term (greater than a year) cultural preservation (Nakasone et al. 2004).

Screw-capped test tubes or petri plates containing the agar medium of choice are used in this method. Fungi are inoculated on culture medium in sterile petri dishes. The plates are then sealed with parafilm, plastic wrap, or airtight lids to minimize contamination and dehydration. These plates are stored at refrigeration temperatures (approximately 4°C) to slow down fungal growth and metabolic activity. This ensures preservation of the culture for a few weeks to months. Regular monitoring and subculturing on fresh media plates are required for culture maintenance (Humber 1997).

4.2 Long-term Storage Techniques

Different techniques are used for long-term preservation, including sclerotization, lyophilization, silica gel, oil overlay, immersion in distilled water, organic substrata (wood, cereal grains, agar strips, insect or plant tissue, etc.), and liquid nitrogen (Lohnoo et al. 2018).

In sclerotization, the hardened, dormant structures called sclerotia are utilized for storage. They are first cultivated on an appropriate nutrient medium such as Potato Dextrose Agar. The collected sclerotia are dried either in a sterile environment or using desiccators to reduce moisture content. This is crucial for preventing decay or microbial growth. Once dried, the sclerotia are stored in sealed containers with silica gel to maintain low humidity. These containers are then stored at –20°C for long-term preservation. These sclerotia can be revived by placing them on fresh nutrient media under optimal growth conditions.

Lyophilization, which is another name for freeze-drying, is the best method for the long-term storage of fungal cultures (Nakasone et al. 2004). This procedure involves harvesting spores or mycelial fragments from a culture plate that are suspended in a sterile cryoprotectant solution like skim milk or sucrose. It protects the cellular structures during freezing and drying. The suspension is rapidly frozen at –80°C or in liquid nitrogen to prevent the formation of ice crystals (Jadhavet et al. 2015). Then they are transferred to a freeze-dryer under vacuum conditions where water sublimates directly from solid (ice) to gas. In this way, cultures of fungi are preserved in a dry state. They are then sealed and placed into sterile vials after preventing contamination; these sealed vials are kept in refrigeration at 4°C (Nakasone et al. 2004).

Oil overlay is an effective method for preserving cultures at room temperature for up to 32 years. This method is effective for mycelial or nonsporulating cultures (Nakasone et al. 2004). It involves growing fungal cultures on agar slants in sterile test tubes or vials. Mineral oil is first sterilized by autoclaving at 15 lb pressure for 2 hours, and entrapped moisture is removed by heating the liquid in a dry oven for 1–2 hours at 170°C (Martin 1963). Once the culture has fully developed, sterile mineral oil is carefully poured over the surface of the agar. The quantity of oil should be such that it completely submerges the fungal growth. This oil creates an anaerobic environment, which limits oxygen availability and reduces the metabolic activity of fungi. The tubes are then sealed and stored at room temperature. To revive the culture, a small portion of the fungal growth under the oil is removed using a sterile needle and transferred to fresh growth media (Martin 1963).

Several species of fungi including *Colletotrichum gloeosporioides*, *Verticillium dahlia, Fusarium oxysporum* and *Phytophthora palmivora* (Pimentel et al. 1980) and bacteria including *Agrobacterium tumefaciens* and *Pseudomonas* spp. have been stored in distilled water for up to 20 years without losing their pathogenicity (Iacobellis and DeVay 1986). Glass vials are first half-filled with sterile distilled water. Bacterial cells or agar culture disks are submerged in these glass vials. The vials are then sealed and placed at room temperature or in refrigerators (Castellani 1964).

Screw-cap tubes are partially filled with silica and dry-heated at 180°C for 90 minutes. They are kept in tightly sealed containers. A 10% (v:v) solution of dry powdered skim milk in distilled water is first chilled at 4°C and then used to suspend the spores. Additionally, the silica gel is immersed in an ice-water bath and chilled at 4°C. The spore suspension is then added to the silica gel and left in the bath for up to 30 minutes. Tubes are first stored at room temperature with loose caps for 1–2 weeks. Then viability is checked, caps are tightened if cultures are viable, and stored in a refrigerator.

Some fungi also have ability to be preserved in dry sterile sand or soil. This is one of the cost-effective methods and has successfully been used for the preservation of fungi including *Rhizoctonia* (Sneh et al. 1991), *Fusarium* (Rizk and Youssef 1985), and *Septoria* (Shearer et al. 1974). Two-thirds of 60 ml glass bottles are filled with loam soil or sand and sterilized at 120°C for 20 minutes. The bottles are cooled and sterilized again under the same conditions. Mycelial growth or spore suspension is created using sterile distilled water. Five ml of this fungal suspension is poured into each of these bottles. The bottles are kept at room temperature for 2–14 days for growth and then stored in a refrigerator at 4°C (Nakasone et al. 2004). Various types of organic substrata, including wood chips, straw, cereal grains, filter paper, agar strips, and plant and insect tissues, are also used for the preservation of fungal cultures (Gehlot et al. 2012).

5. Genetic Stability of Biocontrol Agents

Genetic stability is a critical factor for the successful development and application of BCAs. It ensures that the beneficial traits of the microorganisms, such as their ability to suppress pathogens, their colonization in plants, and their mode of action, remain consistent over time. On the other hand, genetic instability can lead to the loss of efficacy. These biofungicides become less effective or completely non-functional in nature. This genetic instability can arise from genetic drift, spontaneous mutations, or recombination, which can reduce their efficacy. These changes may occur during prolonged cultivation, environmental stress, horizontal gene transfer, or adaptation to laboratory conditions (Bonaterra et al. 2012).

Maintenance of genetic stability starts with proper preservation techniques. For example, cryopreservation at ultra-low temperatures (i.e., –80°C) not only minimizes metabolic activity but also prevents the occurrence of genetic changes. Lyophilization or freeze-drying with stabilizers like skim milk is another important method that effectively maintains the genetic stability of microbial strains. For instance, *Pseudomonas fluorescens* is often preserved using these methods to retain its pathogen-suppressing abilities (Bisutti et al. 2015). Molecular monitoring plays a vital role in assessing genetic stability. Techniques such as genome sequencing and molecular markers like RAPD and SSR allow researchers to detect mutations or genetic drift (Abdel-Mawgood, 2012). Advances in genetic engineering techniques, such as CRISPR/Cas9, have enabled precise modifications to enhance stability and efficacy. Engineered strains of *Pseudomonas putida* improved pathogen suppression while maintaining genetic stability (Glandorf et al. 2001).

Protective additives such as polyethylene glycol (PEG) and antioxidants like ascorbic acid are also added to shield microbial DNA from oxidative stress (Larena et al. 2007, Guijarro et al. 2007). Encapsulation of biocontrol agents in materials like alginate beads or polymers provides an additional layer of protection (Singh and Paithankar 2023). These methods have been successfully applied to *Trichoderma* spp. and *Bacillus subtilis* to maintain their effectiveness in biofungicide formulations (Riseh et al. 2022).

Future research focuses on the integration of molecular tools with field-based assessments to enhance the stability and effectiveness of BCAs. Developing next-generation formulations with genetically stable strains that have extended shelf lives advances the efficacy and reliability of biofungicides. Maintaining the genetic stability of BCAs is essential for their consistent performance in managing plant diseases, ensuring sustainable agricultural practices, and reducing the reliance on chemical pesticides.

6. Mass Propagation Techniques

6.1 Solid-State Fermentation (SSF)

Solid-State Fermentation (SSF) is one of the most widely used mass production methods for microbial-based biocontrol agents (BCAs), especially fungal propagules like conidia and spores. These propagules act as infective units in applications of biocontrol agents. It involves the growth of microorganisms on solid substrates in the absence of free water, but it should contain enough moisture to support microbial metabolism (Manpreet et al. 2005). This technique is superior for the in vitro production of fungal biocontrol agents since it closely mimics natural fungal habitats, and high concentrations of conidia and spores can be produced without losses in quality or infectivity (Kumar et al. 2021). Additionally, the focus of SSF system improvement addresses parameters such as substrate composition, moisture levels, aeration, and temperature control. These aspects significantly impact the conidial yield and quality, and consequently achieve the formation of infective and viable fungal propagules (Méndez et al. 2018). Apart from its utilization in biocontrol, SSF is also applied for the manufacture of fine chemicals such as value-added enzymes, bioethanol, and biofuels. For instance, *Trichoderma* spp. is a broad fungal genus used in SSF for enzyme production, such as cellulases, hemicellulases, xylanases, amylases, and chitinases (Verma et al. 2007). These enzymes have diverse industrial applications like biomass degradation, biofuel production, etc. (Coradi et al. 2013, Kar et al. 2013). This propagation technique includes advantages such as being cost-effective and environmentally sustainable due to the use of agro-industrial residues as substrates, absence of foams, easy control of contamination, mild agitation, and better aeration due to the presence of pores that ensure microbial respiration (Sala et al. 2019).

6.2 Submerged Fermentation (SmF)

Submerged fermentation (SmF) constitutes one of the most significant aspects of the mass production of biocontrol agents, which are microbial in origin. SmF is a process in which microorganisms are grown in a liquid nutrient medium, which guarantees constant agitation and aeration to provide optimal growth conditions for the fungal culture (Krishna 2005). The SmF process is based on nutrient-rich liquid media, synthetic or semi-synthetic. It supports microbial growth and the synthesis of metabolites. These media are frequently supplemented with specific inducers or precursors to improve the synthesis of targeted compounds (Sinha et al. 2022). Moreover, the liquid nature of SmF renders it ideal for processes like filtration and centrifugation, which are very convenient for the effective recovery of microbial products. SmF has potential for propagules that exhibit unique characteristics like submerged conidia and chlamydospores. Papavizas et al. (1984) and Harman et al. (1991) worked on the submerged propagules of Trichoderma in liquid culture systems. The studies established the possibility of submerged propagules with a controlled liquid environment. However, some limitations with SmF do exist, such as lowered desiccation tolerance in the submerged propagules. The submerged conidia and chlamydospores generally have reduced stability under drying and storage conditions. In downstream processing, some of these issues were addressed through the incorporation of stabilizing agents and protective additives by Sriram et al. (2011) and Jin and Custis (2011). However, this does increase production costs and complexity.

7. Stabilization of Biocontrol Agents

7.1 Use of Carriers and Stabilizers

Carriers are essential components in the formulation of biocontrol agents like biofungicides, ensuring their stability and efficacy during storage and providing safe field application. Both organic and inorganic carriers are now being used worldwide. Organic carriers include starch, cellulose, lignin, humic acids, polysaccharides, skimmed milk, etc., and inorganic carriers include silica, vermiculite, zeolite, clay, etc. (Berninger et al. 2017, Palmieri et al. 2022). The selection of a carrier agent depends on factors such as the type of biocontrol agent, target pathogen, mechanism of application, and environmental conditions. Both organic and inorganic carriers have unique advantages, and their combination in certain formulations can offer a synergistic effect (Wang et al. 2024).

Stabilizers are essential in maintaining the viability and efficacy of biofungicides during their storage and application. Most of these stabilizers include polyethylene glycol (PEG) (Larena et al. 2007), ascorbic acid (Guijarro et al. 2007), and tocopherols (Jampílek et al. 2019), with each offering distinct advantages. PEG acts as a cryo- and osmo-protectant. It shields microbial cells during dehydration and freeze-drying and prevents them from membrane disruption and protein denaturation (Mujib et al. 2022). Its hygroscopic nature helps maintain moisture balance in formulations, which reduces desiccation-related stresses.

Ascorbic acid, also known as vitamin C, act as a powerful antioxidant (Iqbal et al. 2004). It neutralizes reactive oxygen species (ROS), protecting microbial cells from oxidative damage (Peng et al. 2005). Tocopherols, a group of vitamin E compounds, target lipid peroxidation in microbial cell membranes (Colombo 2010). They maintain the integrity of membranes and enhance cell survival. They are particularly effective in oil-based formulations, preventing lipid oxidation. These stabilizers enhance the shelf life, storage stability, and field performance of biofungicides.

8. Packaging and Storage

8.1 Packaging Materials for Different Formulations

The selection of appropriate packaging materials, which ensure suitable atmospheric conditions and storage temperatures, plays a pivotal role in achieving successful packaging. Low-barrier packaging materials are more permeable to oxygen and water, which can lead to partial rehydration of formulated cells. This initiates respiration and other degradation processes. On the other hand, high-barrier packaging, such as specialized bags, flasks, or vials, is designed to provide superior resistance to oxygen and moisture penetration (Costa et al. 2002). Polyethylene flasks are widely used worldwide for their excellent resistance to water vapor and gases and their high durability against fats, oils, and chemicals. Some polyethylene flasks are coated with EVOH (ethylene vinyl alcohol copolymer), which enhances their ability to act as an effective barrier against gases (Teixidó et al. 2022). Similarly, bags composed of multi-layered barrier materials offer a versatile solution for the packaging of biofungicides. These bags often include different layers made up of polyester, polyethylene, or polyamide. Each of these contributes to varying degrees of oxygen permeability and ensures protection against environmental degradation (Teixidó et al. 2022).

8.2 Shelf Life

Shelf life is a critical factor for the successful formulation of biocontrol agents (BCA). No doubt that biopesticides typically have a shorter shelf life as compared to synthetic pesticides, but maintenance of their quality and efficacy during storage is essential. Shelf life is basically the duration in which a product retains its desired quality under appropriate storage conditions (Alamprese et al. 2017). Factors such as oxygen, moisture, light, elevated temperatures, microbial contamination, etc. can degrade the effectiveness of the product if not stored properly (Morgan et al. 2006). Therefore, storage conditions and packaging play a crucial role in protecting BCAs. A well-formulated biocontrol product has the ability to retain its genetic stability, cell viability, colonization ability and their specific mechanisms of action (Teixidó et al. 2011). These factors directly influence the product's efficacy against target pathogens. Furthermore, the shelf life of a BCA also depends upon the integration of some protective compounds such as glycine betaine (Zhang et al. 2017). These additives shield the microbial cells from desiccation, oxidative stress, and thermal fluctuations, prolonging their shelf life (Guan et al. 2012).

9. Field Application of Biofungicides

9.1 Methods of Application

The use of biofungicides in agricultural crops includes various methods that deliver beneficial BCAs effectively to crops, aiming at the management of fungal diseases in a sustainable way. The most important of these methods include seed coating, foliar spray, and soil incorporation, each with unique mechanisms, benefits, and considerations.

9.1.1 Seed Coating

It involves the application of a protective layer that contains biofungicides directly on the seed surface before planting. This technique ensures the presence of beneficial microbes at the earliest stages of plant development. It offers protection against soil-borne pathogens. The process typically includes mixing microbial inoculants with binders and sometimes fillers to adhere biofungicide to the seed surface (Rocha et al. 2019). For example, coating seeds with *Trichoderma* spp. not only suppresses diseases like *Fusarium* damping off on maize, but also improves the seedling vigor and overall health of plant (Coninck et al. 2020). However, factors such as seed size, shape, and the presence of inhibitory compounds can influence the effectiveness of seed coatings. Additionally, the survival of the BCAs on the seed surface depends on storage conditions and the duration between coating and planting (Rocha et al. 2019).

9.1.1.1 Film Coating

It is the most widely used seed coating method in which a thin layer of biofungicide is applied to seeds using a polymer-based solution. The biofungicide adheres firmly to the surface of the seed, providing protection against seed-borne and soil-borne pathogens that mostly occur at early stages of crop plants (Boiu-Sicuia et al. 2020). For example, in case of *T. harzianum, P. fluorescens,* and *B. subtilis*, film coating is mostly preferred (Zhang et al. 2022).

9.1.1.2 Pelleting

In pelleting, seeds are coated with multiple layers of biofungicide, inert materials, and adhesives to create a uniform or rounded shape. This technique provides effective seed handling, improves viability, and provides prolonged protection against seed- and soil-borne pathogens. The biofungicide is embedded within the pellet, which ensures gradual release and better adherence to the seed (Sriwati et al. 2019). This method is particularly beneficial for small or irregularly shaped seeds. *B. bassiana* and *G. virens* are mostly applied by this method (Naz et al. 2022).

9.1.1.3 Slurry Coating

In this method, seeds are coated with a liquid suspension of biofungicides. This method ensures even distribution of the biofungicides on the surface of the seed and provides immediate protection against seed- and soil-borne pathogens. Slurry coating is a simple, cost-effective, and widely used seed treatment method in large-scale agriculture. *T. viride* and *P. putida* are mostly applied by this method (Estévez Geffriaud 2021).

9.1.2 Foliar Spray

It includes the application of biofungicides directly on the surface of aerial parts of host plants, such as leaves and stems. This method mostly targets foliar pathogens and can be used as both preventive and curative. Such types of biofungicides are formulated in liquid forms, including emulsifiable concentrates and wettable powders. These are further diluted with water and sprayed directly on the surface of crop plants (Oros and Naár 2017). For example, foliar application of *Bacillus subtilis* has been effective in managing powdery mildew in various crops (Basamma and Kulkarni 2017). Factors like coverage uniformity, environmental conditions, and the timing of application relative to pathogen life cycles affect the success rate of BCA (He et al. 2021).

9.1.3 Soil Incorporation

It involves the introduction of biofungicides into the soil environment that mostly targets soil-borne pathogens. This can be achieved through methods like soil drenching, in which a liquid biofungicide solution is applied to the soil and is allowed to percolate to the root zone, or by mixing granular formulations into the soil particles during or before planting (O'Callaghan et al. 2022). For example, incorporation of *Gliocladium virens* into the soil reduces the incidence of damping-off disease caused by Pythium spp. (Jensen and Wolffhechel 1995). Soil incorporation methods also play a role in establishing beneficial microbial populations in the rhizosphere. However, the efficacy of this method can be influenced by factors like soil properties, including pH, organic matter content, and microbial diversity. Proper soil preparation and moisture management are some of the basic requirements for the survival and proper activity of the introduced biofungicides (O'Callaghan et al. 2022).

9.2 Application Timing and Dosage Optimization

Equally important for effective disease management in crops is the timing and dosage of biofungicide applications. Proper timing and optimal dosages of biofungicides can be significantly effective in controlling pathogens. For instance, an experiment on *Zymoseptoria tritici* in wheat showed that only a single fungicide treatment was effective when the application timing and dosage were optimal. However, three-application programs generally were not optimal as they exerted extensive selection pressure for resistance. In contrast, a two-application program was effective in managing the disease with reasonable flexibility in dose and timing (Van et al. 2016).

Environmental factors including, temperature, humidity, rainfall, UV radiations and soil pH affect the timing of biofungicide application. Treatments are best applied in favorable weather to avoid being washed off by the rain or degraded by sunlight. For instance, foliar applications must be made at times when rainfall and intense sunlight, which can reduce efficacy, are not expected. The timing of application during a day can also affect their efficacy. Biofungicides applied during the periods of the day characterized by lower UV radiation, early morning or late afternoon, exhibited enhanced performance (Courshee 1967). Adjuvants are added to improve the adhesion of the BCAs to plant surfaces, increase penetration, and protect active ingredients from degradation by environmental factors. Some adjuvants, when used, also prevent wash-off during rainfall, maintain coverage during growth, and limit photodecomposition (Butu et al. 2022).

10. Environmental Factors Affecting Efficacy

The efficacy of biofungicides is highly influenced by environmental factors because these factors determine the survival, colonization, and activity of the BCAs. The interaction between BCAs, host plants, and pathogens depends upon these factors. So, these factors ultimately influence the success of disease management. Temperature, humidity, soil pH, UV radiation, soil type, etc., determine the biofungicide's efficacy (Hegde et al. 2023). Zehra et al. (2017) stated that among the four species of *Trichoderma* tested, *T. harzianum* showed the highest inhibition of *Fusarium oxysporum* (68.5%) and *Alternaria alternata* (52.93%) by using dual culture technique. Optimum growth and sporulation was observed at 25–40°C for *T. harzianum* and *T. viride*, and 25–35°C for *T. asperellum* and *T. hamatum*. A pH range of 4.6–7.6 favored all species. NaCl concentrations up to 1000 μM had minimal impact on mycelial growth across all isolates. Ahlem et al. (2012) used different strains of *Bacillus* to manage *Botrytis cinerea* at different temperatures and pH levels. Strain B24 was most effective at 25°C, achieving higher inhibition across temperatures compared to strain I18, which showed lower efficacy. pH also impacted the efficacy; acidic pH led to less than 50% inhibition, while optimal inhibition of the pathogen varied from neutral to alkaline pH (7.5–10). Strain B24 achieved maximum inhibition (62.27%) at pH 7.5.

Soil properties such as pH, nutrient content and texture significantly influence the efficacy of biofungicides. Duffy et al. (1997) used *Trichoderma koningii* for the management of take-all of wheat in Western Australia, across various field sites. In eight silt loams from the Pacific Northwest, *T. koningii* reduced disease severity and increased plant growth. But protection levels were correlated with soil supportiveness for biocontrol, not conduciveness to disease. Biocontrol activity was positively correlated with iron, nitrate-nitrogen, boron, copper, magnesium, and clay, but negatively with soil pH and phosphorus. A regression model ($R^2 = 0.992$) identified nitrate-nitrogen, pH, copper, and magnesium as key factors. Adhikari et al. (2010) conducted growth room experiments to evaluate the effects of soil type, biofungicides, soil compaction and *Plasmodiophora brassicae* pathotypes on clubroot in *Brassica* spp. They used four soil types (peat, mineral, sand, soil-less mix) and two biofungicides (*Bacillus subtilis*, *Clonostachys rosea*). Effectiveness of biofungicide varies with the soil type, such as *C. rosea is* generally more effective on peat and mineral soils, and *B. subtilis* is more effective on sandy soil than on other types. On the other hand, soil bulk density had a greater impact on clubroot severity than soil type, organic matter, or pathotype. Higher severity was observed in soil-less mixtures, and lower severity was observed in reduced-density peat and mineral soils. Soil conditions significantly influenced biofungicide efficacy.

UV radiations are one of the most critical environmental factors for foliar-applied biofungicides. They can damage and reduce the chances of survival of microbial cells on plant surfaces. For example, spores of *B. bassiana* and *Metarhizium anisopliae* are particularly vulnerable to UV radiations. Both of these microbes are entomopathogenes but they also have antifungal properties (Culebro et al. 2017, Ravindran et al. 2014). Exposure of *Metarhizium* spp. to UV-B light for about four hours can significantly reduce the conidial viability (Couceiro et al. 2021). Addition of natural substances such as humic acid into biofungicide formulations has been shown to provide over 90% protection for conidia of *B. bassiana* against UV-B radiation at laboratory stage (Kaiser et al. 2019). Additionally, encapsulation fungal BCAs with the polymers or alginate beads also offer protection against UV radiations (Kaiser et al. 2019).

11. Interactions with Native Microbiota

Soils microbiota plays a very important role in determining the effectiveness of biofungicides on fungal plant pathogens. Some fungal BCAs, such as *Trichoderma* spp., *B. bassiana*, and *M. anisopliae*, are widely used to suppress pathogens; however, their performance is largely determined by the interactions between the applied fungi and indigenous microbial community in soil (Han et al. 2023). Soil microbiota composition and activity may directly or indirectly affect the growth, establishment, and mechanisms of BCAs (De Corato 2020). Beneficial microorganisms in soil may synergize with BCAs by producing secondary metabolites or enzymes that enhance pathogen (Mülner et al. 2019). For instance, bacteria like *P. fluorescens* or *B. subtilis* may produce antifungal compounds that can enhance the fungal BCAs to inhibit pathogens (Dimkić et al. 2022). Moreover, microbial interactions can enhance the persistence of BCAs in soil, leading to long-term disease control.

However, competition for resources, space, and nutrients present in native soil can limit the efficacy of BCAs. BCAs compete with saprophytic and pathogenic fungi and bacteria for organic substrates. In soils with high microbial diversity, BCAs face strong competition, reducing their population density and biocontrol potential. BCAs may establish more effectively and suppress pathogens more efficiently in soils with low microbial activity or diversity (Raaijmakers et al. 2009). For example, some bacteria produce antifungal compounds that inhibit the growth of BCAs. Studies have shown that *T. harzianum*, a widely used fungal BCA, performs poorly in soils with high populations of antifungal chemical producing microbes (Dutta et al. 2022). Suppressive soil, soil having ability to naturally inhibit the growth and activity of plant pathogens due to presence of antagonistic microbes, contains huge quantity of native microbiota (Alabouvette et al. 1996).

One notable example of fungal BCA is *T. koningii*. It suppresses the take-all disease of wheat caused by *G. graminis* (Zhao et al. 2023). Its efficacy varied significantly with the soil, depending on the native microbial community and soil nutrients (Simon and Sivasithamparam 1988). Similarly, *M. anisopliae* was less effective in soils which have high densities of fungistatic bacteria (Jaronski et al. 2007). Soil management practices such as reduced tillage, adding organic matter and inoculating soils with compatible microbial consortia enhance the efficacy of biofungicides (Saikai et al. 2023). Additionally, pre-assessment of soil microbiota can help to identify the conducive environments for BCAs which enable tailored biocontrol strategies (Peddle et al. 2025).

12. Commercial Development of Biocontrol Agents

Commercial development of biocontrol agents requires some important steps as mentioned in Figure 2. The most important of these steps are the development of stable and viable formulations, patent application, and then the registration of the active ingredient (Palmieri et al. 2022). The first and most important of these steps occurs at the laboratory scale, and the second step at the industrial scale. Finally, the formulation is screened by regulatory authorities for its registration.

12.1 Development of Stable and Viable Formulation

This step plays a key role in microbe-based pesticide development. The main purpose of this step is the stabilization of microbial cells to maintain their viability from storage until they are used. Microbial cells undergo a lot of stress throughout the formulation process, which has a significant impact on microbial viability. There are various formulation methods available today, and the method with the least influence on viability can be selected for each microbe (Chaudhary et al. 2020). Additionally, protectants are typically added to the growth medium, which reduce the detrimental effects of harsh conditions on microbial cells. Treatments that take place before formulation, like the adoption of suitable growing conditions, might improve stability throughout storage. Furthermore, appropriate packaging is highly helpful in maintaining formulation stability.

Stabilization in the formulation process also limits microbial contamination under acceptable levels. Physical conditions of the products should also remain unaltered;

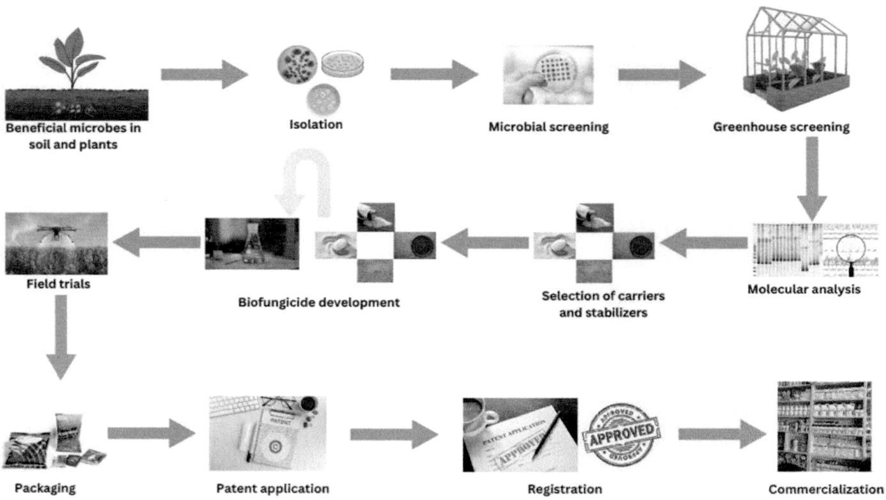

Figure 2: Important steps required for the development of new biofungicide from screening of BCA to its registration and commercialization.

for example, clump formation and aggregation of particles should be prevented in both solid and liquid formulations (Behle and Birthisel 2023). Application of biopesticides in the field, as well as enhancing their persistence in the environment, are some other important functions of this step. Different types of chemical adjuvants are commercially available that enable BCAs to dissolve quickly in water, distribute uniformly on the surfaces of plants, and provide protection against abiotic stresses (Butu et al. 2022).

A variety of biopesticide formulations are available, which are mostly classified as solid or liquid depending on the inert carrier combined with the active ingredients. Depending on the nature of the products, they are subdivided into four types: liquid, granular, slurry, or powder. The inert matter for solid formulations is divided into inorganic (such as silica, vermiculite, zeolite, and clay) and organic (such as cellulose, polysaccharides, humic acids, starch, lignin, and skim milk) (Palmieri et al. 2022). Liquid formulations are composed of suspensions of microbes that are amended with inert substances, which improve stabilization, surfactant properties, stickiness, and dispersal abilities (Sriram et al. 2011).

12.2 Patent Application

A detailed patent application is filed to secure intellectual property rights, covering the microbial strain, formulation process, and application methods. This involves thorough documentation, which includes the novelty of the formulation, experimental evidence, and potential applications (Montesinos 2003). Patent applications are filed with national or international patent offices, depending on the intended market scope. A detailed disclosure of the biofungicide's components, including mode of action and unique features of the formulation, is essential to strengthen the application (Palmieri et al. 2022).

The international treaties include the European Patent Agreement of 1973, which made applications valid in 18 countries of Europe; the Treaty of the Union of Paris of 1883, which has been subscribed by 140 countries; and the Patent Cooperation Treaty of 1970, which is valid for 102 countries. All countries that are members of the World Intellectual Property Organization (WIPO) have signed the Budapest Treaty. It governs the patents of pesticides based on microorganisms and outlines the requirements needed for a pure culture of the microorganism's strain (Montesinos 2003).

12.3 Registration

Only a few biopesticides are registered in response to the large number of patent applications for agricultural use. The registration process depends upon specific requirements, which vary from country to country. These pesticides are first evaluated before commercialization, and studies are conducted to demonstrate their safety for humans, non-target organisms, and the environment. Sometimes, field trials are also required to validate efficacy under real-world conditions. Companies are only allowed to market the product if they meet these requirements. The regulatory bodies such as the EPA in the U.S., EFSA in the EU, or local agricultural authorities are responsible for the evaluation of this data (Frederiks and Wesseler 2019). The registration process may also include product labeling, which contains the necessary information, including usage instructions, safety precautions, and disposal guidelines. Periodic reviews and compliance with Good Agricultural Practices (GAPs) ensure the biofungicide meets regulatory standards throughout its lifecycle.

13. Future Prospects and Innovations

The beneficial microbes, especially those having antifungal properties, are safe and ecofriendly in the management of plant diseases. The success of effective culture maintenance and mass propagation of biofungicides depends on the viability, genetic stability, and efficacy of BCAs. Advances in culture maintenance methods have significantly improved the survival, viability, and activity of BCAs. Techniques such as SmF and SSF have been optimized to yield high-quality BCAs. Despite these benefits, there are some challenges, such as shelf-life limitations, environmental variability, and compatibility with integrated pest management strategies, which act as major hindrances in large-scale adaptation of these biofungicides.

Future perspectives focus on refining production technologies to enhance the scalability and affordability of BCAs worldwide. There is a need for innovations in carrier formulations, such as polymer-based and nanoparticle systems that can improve the stability and delivery of BCAs under field conditions. Moreover, molecular and genome editing tools hold immense potential for enhancing the efficiency and adaptability of BCAs. Integration of bioinformatics and metagenomics can further improve interactions between BCAs, pathogens, and the soil microbiome.

Furthermore, global collaborations between researchers, industries, and policymakers also play an important role in addressing regulatory hurdles and promoting the adoption of BCAs. Training programs for farmers and extension workers can facilitate the effective implementation of biocontrol technologies in diverse agricultural contexts. By addressing these challenges and leveraging technological advancements, BCAs can play a pivotal role in sustainable agriculture, ensuring food security while minimizing environmental impact.

References

Abdel-Mawgood, A. L. (2012). DNA based techniques for studying genetic diversity. *Genetic Diversity in Microorganisms*, 95–122.

Adhikari, K. K. (2010). Effect of temperature, biofungicides and fungicides on clubroot of selected *Brassica* crops [PhD Thesis, University of Guelph]. https://atrium.lib.uoguelph.ca/handle/10214/19550.

Ahlem, H., Mohammed, E., Badoc, A., and Ahmed, L. (2012). Effect of pH, temperature and water activity on the inhibition of *Botrytis cinerea* by *Bacillus amyloliquefaciens* isolates. *Afr. J. Biotechnol.*, 11(9): 2210–2217.

Akıllı, S., Ulubaş Serçe, Ç., Katırcıoğlu, Y. Z., Maden, S., and Rigling, D. (2013). Characterization of hypovirulent isolates of the chestnut blight fungus, *Cryphonectria parasitica* from the Marmara and Black Sea regions of Turkey. *Eur. J. Plant Pathol.*, 135: 323–334.

Alabouvette, C., Hoeper, H., Lemanceau, P., and Steinberg, C. (1996). Soil suppressiveness to diseases induced by soilborne plant pathogens. *Soil biochem.*, 9(1): 371–413.

Alamprese, C., Cappa, C., Ratti, S., Limbo, S., Signorelli, M., Fessas, D. et al. (2017). Shelf life extension of whole-wheat breadsticks: Formulation and packaging strategies. *Food Chem.*, 230: 532–539.

Alexander, B. J. R., and Stewart, A. (2001). Glasshouse screening for biological control agents of *Phytophthora cactorum* on apple (*Malus domestica*). N.Z. J. *Crop. Hortic. Sci.*, 29(3), 159–169.

Alfiky, A., and Weisskopf, L. (2021). Deciphering *Trichoderma*–plant–pathogen interactions for better development of biocontrol applications. *J. Fungi.*, 7(1): 61.

Ali, N., Pang, Z., Wang, F., Xu, B., and El-Seedi, H. R. (2022). Lipopeptide biosurfactants from *Bacillus* spp.: Types, Production, Biological Activities, and Applications in Food. *J. of Food Qual.*, 2022, 1–19.

Al-shemmary, M. K., El-Zaawely, A. A., Al-Taie, A. H., and Elsheery, N. I. (2018). Antifungal activity of some plant extracts and *Trichoderma* spp. Against cucumber damping off caused by *Pythium aphanidermatum*. *Journal of Biodiversity and Environmental Sciences*, 12(6): 195–203.

Angeli, D., Pellegrini, E., and Pertot, I. (2009). Occurrence of *Erysiphe necator* chasmothecia and their natural parasitism by *Ampelomyces quisqualis*. *Phytopathology*, 99(6): 704–710.

Basamma, R. H., and Kulkarni, S. (2017). Growth promoting ability and Bioefficacy of *Bacillus subtilis* against powdery mildew and early blight of Tomato through Foliar Spray in pot culture studies. *J. Pharmac. Phytochem.*, 6(2): 244–246.

Behle, R., and Birthisel, T. (2023). Formulations of entomopathogens as bioinsecticides. pp. 407–429. *In:* Mass production of beneficial organisms. Elsevier.

Berninger, T., Mitter, B., and Preininger, C. (2017). Zeolite-based, dry formulations for conservation and practical application of *Paraburkholderia phytofirmans* PsJN. *J. Appl. Microbiol.*, 122(4): 974–986.

Bisutti, I. L., Hirt, K., and Stephan, D. (2015). Influence of different growth conditions on the survival and the efficacy of freeze-dried *Pseudomonas fluorescens* strain Pf153. *Biocontrol Sci. Technol.*, 25(11): 1269–1284.

Boiu-Sicuia, O.-A., Constantinescu, F., Ursan, M.-D., and Cornea, C. P. (2020). Microbial inoculants applied as seed treatments and their effect on common wheat *Triticum aestivum* L. *Annals of the University of Craiova-Agriculture, Montanology, Cadastre Series*, 49(2): 38–43.

Bonaterra, A., Badosa, E., Cabrefiga, J., Francés, J., and Montesinos, E. (2012). Prospects and limitations of microbial pesticides for control of bacterial and fungal pomefruit tree diseases. *Trees*, 26(1): 215–226.

Bonaterra, A., Badosa, E., Daranas, N., Francés, J., Roselló, G., and Montesinos, E. (2022). Bacteria as biological control agents of plant diseases. *Microorganisms*, 10(9): 1759.

Butu, M., Rodino, S., and Butu, A. (2022). Biopesticide formulations-current challenges and future perspectives. pp. 19–29. *In:* Biopesticides. Elsevier.

Caffi, T., Legler, S. E., Furiosi, M., Salotti, I., and Rossi, V. (2022). Application of *Ampelomyces quisqualis* in the integrated Management of Grapevine Powdery Mildew. *Microbial Biocontrol.* Agents: Developing Effective Biopesticides.

Castellani, A. (1964). The "Water cultivation" of pathogenic fungi. *Ann. Soc. Belges Med. Trop. Parasitol. Mycol.*, 44: 217–219.

Chaudhary, T., Dixit, M., Gera, R., Shukla, A. K., Prakash, A., Gupta, G. et al. (2020). Techniques for improving formulations of bioinoculants. *3 Biotech*, 10: 1–9.

Cheng, J., Jaiswal, K. S., Yang, S. H., and Suh, J. W. (2016). Endophytic *Bacillus subtilis* MJMP2 from Kimchi inhibits *Xanthomonas oryzae pv. oryzae*, the pathogen of Rice bacterial blight disease. *J. Appl. Biol. Chem.*, 59(2): 149–154.

Choudhary, D. K., and Johri, B. N. (2009). Interactions of *Bacillus* spp. And plants–with special reference to induced systemic resistance (ISR). *Microbiol. Res.*, 164(5): 493–513.

Colombo, C., Palumbo, G., He, J.-Z., Pinton, R., and Cesco, S. (2014). Review on iron availability in soil: Interaction of Fe minerals, plants, and microbes. *Journal of Soils and Sediments*, 14: 538–548.

Colombo, M. L. (2010). An update on vitamin E, tocopherol and tocotrienol—Perspectives. *Molecules*, 15(4): 2103–2113.

Compant, S., Duffy, B., Nowak, J., Clément, C., and Barka, E. A. (2005). Use of Plant growth-promoting bacteria for biocontrol of plant diseases: principles, mechanisms of action, and future prospects. *Appl. Environ. Microbiol.*, 71(9): 4951–4959.

Coninck, E., Scauflaire, J., Gollier, M., Liénard, C., Foucart, G., Manssens, G. et al. (2020). *Trichoderma atroviride* as a promising biocontrol agent in seed coating for reducing *Fusarium* damping-off on maize. *J. Appl. Microbiol.*, 129(3): 637–651.

Coradi, G. V., da Visitação, V. L., de Lima, E. A., Saito, L. Y. T., Palmieri, D. A., Takita, M. A. et al. (2013). Comparing submerged and solid-state fermentation of agro-industrial residues for the production and characterization of lipase by *Trichoderma harzianum*. *Ann. Microbiol.*, 63: 533–540.

Costa, E., Usall, J., Teixidó, N., Torres, R., and Vinas, I. (2002). Effect of package and storage conditions on viability and efficacy of the freeze-dried biocontrol agent *Pantoea agglomerans* strain CPA-2. *J. Appl. Microbiol.*, 92(5): 873–878.

Cota, L. V., Maffia, L. A., Mizubuti, E. S., Macedo, P. E., and Antunes, R. F. (2008). Biological control of strawberry gray mold by *Clonostachys rosea* under field conditions. *Biocontrol.*, 46(3): 515–522.

Couceiro, J. da C., Fatoretto, M. B., Demétrio, C. G. B., Meyling, N. V., and Delalibera Jr, Í. (2021). UV-B radiation tolerance and temperature-dependent activity within the entomopathogenic fungal genus *Metarhizium* in Brazil. *Front. Fungal Boil.*, 2: 645737.

Courshee, R. J. (1967). Application and use of foliar fungicides. *Fungicides, an Advanced Treatise*, 1: 239–286.

Culebro-Ricaldi, J. M., Ruíz-Valdiviezo, V. M., Rodríguez-Mendiola, M. A., Ávila-Miranda, M. E., Miceli, F. G., Cruz-Rodríguez, R. I. et al. (2017). Antifungal properties of *Beauveria bassiana* strains against *Fusarium oxysporum* f. Sp. *Lycopersici* race 3 in tomato crop. *J. Environ. Biol.*, 38(5): 821.

Damodaran, T., Rajan, S., Muthukumar, M., Gopal, R., Yadav, K., Kumar, S. et al. (2020). Biological management of banana Fusarium wilt caused by *Fusarium oxysporum* f. sp. *Cubense* tropical race 4 using antagonistic fungal isolate CSR-T-3 (*Trichoderma reesei*). Front. microbiol., 11: 595845.

David, B. V., Chandrasehar, G., and Selvam, P. N. (2018). *Pseudomonas fluorescens*: A plant-growth-promoting rhizobacterium (PGPR) with potential role in biocontrol of pests of crops. pp. 221–243. *In:* Crop improvement through microbial biotechnology. Elsevier.

De Corato, U. (2020). Soil microbiota manipulation and its role in suppressing soil-borne plant pathogens in organic farming systems under the light of microbiome-assisted strategies. *Chem. Biol. Technol. Agric.*, 7(1): 17.

de los Santos-Villalobos, S., Guzmán-Ortiz, D. A., Gómez-Lim, M. A., Délano-Frier, J. P., de-Folter, S., Sánchez-García, P. et al. (2013). Potential use of *Trichoderma asperellum* (Samuels, Liechfeldt et Nirenberg) T8a as a biological control agent against anthracnose in mango (*Mangifera indica* L.). *Biocontrol.*, 64(1): 37–44.

Dimkić, I., Janakiev, T., Petrović, M., Degrassi, G., and Fira, D. (2022). Plant-associated *Bacillus* and *Pseudomonas* antimicrobial activities in plant disease suppression via biological control mechanisms-A review. *Physiol. Mol. Plant Pathol.*, 117: 101754.

Duffy, B. K., Ownley, B. H., and Weller, D. M. (1997). Soil Chemical and Physical Properties Associated with Suppression of Take-all of Wheat by *Trichoderma koningii*. *Phytopathology*, 87(11): 1118–1124.

Dutta, P., Deb, L., and Pandey, A. K. (2022). *Trichoderma*-from lab bench to field application: Looking back over 50 years. *Front. Agron.*, 4: 932839.

Estévez Geffriaud, V. (2021). Seed treatments for the protection of crops.

Fenta, L., and Mekonnen, H. (2024). Microbial biofungicides as a substitute for chemical fungicides in the control of phytopathogens: current perspectives and research directions. *Scientifica*, 2024(1): 5322696.

Fessia, A., Barra, P., Barros, G., and Nesci, A. (2022). Could Bacillus biofilms enhance the effectivity of biocontrol strategies in the phyllosphere? *J. Appl. Microbiol.*, 133(4): 2148–2166.

Frederiks, C., and Wesseler, J. H. (2019). A comparison of the EU and US regulatory frameworks for the active substance registration of microbial biological control agents. *Pest Manag. Sci.*, 75(1): 87–103.

Gašić, K., Kuzmanović, N., Ivanović, M., Prokić, A., Šević, M., and Obradović, A. (2018). Complete genome of the *Xanthomonas euvesicatoria* specific bacteriophage KΦ1, its survival and potential in control of pepper bacterial spot. *Fronti. Microbiol.*, 9: 2021.

Gehlot, P., Kaur, S., and Sharma, K. (2012). Preservation of Anamorphic Fungi. Phytotechnol.: Emerging Trends, 115.

Geraldine, A. M., Lopes, F. A. C., Carvalho, D. D. C., Barbosa, E. T., Rodrigues, A. R., Brandão, R. S. et al. (2013). Cell wall-degrading enzymes and parasitism of sclerotia are key factors on field biocontrol of white mold by *Trichoderma* spp. *Biocontrol.*, 67(3): 308–316.

Ghazanfar, M. U., Raza, M., Raza, W., and Qamar, M. I. (2018). Trichoderma as potential biocontrol agent, its exploitation in agriculture: A review. *Plant Protection*, 2(3).

Ghoul, M., and Mitri, S. (2016). The ecology and evolution of microbial competition. *Trends Microbiol.*, 24(10): 833–845.

Gimeno, A. (2020). New tools to advance the biological control of fusarium graminearum by the antagonist *Clonostachys rosea* [PhD Thesis, University of Zurich].

Glandorf, D. C., Verheggen, P., Jansen, T., Jorritsma, J. W., Smit, E., Leeflang, P. et al. (2001). Effect of genetically modified *Pseudomonas putida* WCS358r on the fungal rhizosphere microflora of field-grown wheat. *Appl. Environ. Microbiol.*, 67(8): 3371–3378.

Guan, N., Li, J., Shin, H. D., Du, G., Chen, J., and Liu, L. (2017). Microbial response to environmental stresses: from fundamental mechanisms to practical applications. *Applied Microbiology and Biotechnology*, 101: 3991–4008.

Guijarro, B., Melgarejo, P., and De Cal, A. (2007). Effect of stabilizers on the shelf-life of *Penicillium frequentans* conidia and their efficacy as a biological agent against peach brown rot. *Int. J. Food Microbiol.*, 113(2): 117–124.

Guzmán-Guzmán, P., Kumar, A., de Los Santos-Villalobos, S., Parra-Cota, F. I., Orozco-Mosqueda, M. del C., Fadiji, A. E. et al. (2023). *Trichoderma* species: Our best fungal allies in the biocontrol of plant diseases—A review. *Plants*, 12(3): 432.

Hadar, Y., Harman, G. E., and Taylor, A. G. (1984). Evaluation of *Trichoderma koningii* and *T. harzianum* from New York soils for biological control of seed rot caused by *Pythium* spp. *Phytopathology*, 74(1): 106–110.

Haidar, R., Amira, Y., Roudet, J., Marc, F., and Patrice, R. (2021). Application methods and modes of action of *Pantoea agglomerans* and *Paenibacillus* sp. to control the grapevine trunk disease-pathogen, *Neofusicoccum parvum*. *Oeno one*, 55(3): 1–16.

Han, V.-C., Michael, P. J., Swift, B., and Bennett, S. J. (2023). Biological control of *Sclerotinia sclerotiorum*: Modes of action of biocontrol agents, soil organic amendments, and soil microbiome manipulation. *Biocontrol.*, 105346.

Harman, G. E., Jin, X., Stasz, T. E., Peruzzotti, G., Leopold, A. C., and Taylor, A. G. (1991). Production of conidial biomass of *Trichoderma harzianum* for biological control. *Biocontrol.*, 1(1): 23–28.

He, D. C., He, M. H., Amalin, D. M., Liu, W., Alvindia, D. G., and Zhan, J. (2021). Biological control of plant diseases: An evolutionary and eco-economic consideration. *Pathogens*, 10(10): 1311.

Hegde, G. M., Dobhal, A., Vijaykumar, K. N., and Jahagirdar, S. (2023). Advances in formulations and efficacy of mycopesticides for plant disease management and sustainable yields. pp. 373–408. *In:* Singh, I., Rajpal, V. R., and Navi, S. S. (Eds.). *Fungal Resources for Sustainable Economy*. Springer Nature Singapore.

Heydari, A., and Pessarakli, M. (2010). A review on biological control of fungal plant pathogens using microbial antagonists. *J. Biol. Sci.*, 10(4): 273–290.

Humber, R. A. (1997). Fungi: Preservation of cultures. In Manual of techniques in insect pathology (pp. 269–279). Elsevier.

Hyakumachi, M., Nishimura, M., Arakawa, T., Asano, S., Yoshida, S., Tsushima, S. et al. (2013). *Bacillus thuringiensis* suppresses bacterial wilt disease caused by *Ralstonia solanacearum* with systemic induction of defense-related gene expression in tomato. *Microbes and Environments*, 28(1): 128–134.

Iacobellis, N. S., and DeVay, J. E. (1986). Long-Term Storage of Plant-Pathogenic Bacteria in Sterile Distilled Water. *Appl. Environ. Microbiol.*, 52(2): 388–389.

Iqbal, K., Khan, A., and Khattak, M. (2004). Biological significance of ascorbic acid (vitamin C) in human health-a review. *Pakistan Journal of Nutrition*, 3(1): 5–13.

Jadhav, T. R., and Moon, R. S. (2015). Review on lyophilization technique. *World Journal of Pharmacy and Pharmaceutical Sciences*, 4(5): 1906–1928.

Jahagirdar, S., Ravikumar, M. R., and Siddaramaiah, A. L. (2003). Traditional methods in the management of plant diseases–A Review. *Agric. Rev.*, 24(2): 142–146.

Jampílek, J., Kráľová, K., Campos, E. V. R., and Fraceto, L. F. (2019). Bio-based nanoemulsion formulations applicable in agriculture, medicine, and food industry. pp. 33–84. *In:* Prasad, R., Kumar, V., Kumar, M., and Choudhary, D. (Eds.). *Nanobiotechnology in Bioformulations*. Springer International Publishing.

Jaronski, S. T., Fuller-Schaeffer, C., Jung, K., Majumdar, A., and Boetel, M. (2007). Challenges in using *Metarhizium anisopliae* for biocontrol of sugarbeet root maggot, *Tetanops myopaeformis*. *IOBC WPRS Bulletin*, 30(7): 119.

Jensen, D. F., and Wolffhechel, H. (1995). The use of fungi, particularly *Trichoderma* spp. And *Gliocladium* spp., to control root rot and damping-off diseases. *Biocontrol Agents: Benefits and Risks*, 177–189.

Jin, X., and Custis, D. (2011). Microencapsulating aerial conidia of *Trichoderma harzianum* through spray drying at elevated temperatures. *Biocontrol.*, 56(2): 202–208.

Jones, E. E., Rabeendran, N., and Stewart, A. (2014). Biocontrol of *Sclerotinia sclerotiorum* infection of cabbage by *Coniothyrium minitans* and *Trichoderma* spp. *Biocontrol Sci. Technol.*, 24(12): 1363–1382.

Kaiser, D., Bacher, S., Mène-Saffrané, L., and Grabenweger, G. (2019). Efficiency of natural substances to protect *Beauveria bassiana* conidia from UV radiation. *Pest Manag. Sci.*, 75(2): 556–563.

Kar, S., Sona Gauri, S., Das, A., Jana, A., Maity, C., Mandal, A. et al. (2013). Process optimization of xylanase production using cheap solid substrate by *Trichoderma reesei* SAF3 and study on the alteration of behavioral properties of enzyme obtained from SSF and SmF. *Bioprocess Biosyst. Eng.*, 36(1): 57–68.

Karthikeyan, V., Sankaralingam, A., and Nakkeeran, S. (2006). Biological control of groundnut stem rot caused by *Sclerotium rolfsii* (Sacc.). Arch. Phytopathol. *Pflanzenschutz.*, 39(3): 239–246.

Khan, E., and Ahmad, I. Z. (2024). Production and commercialization of biocontrol products. In *Biocontrol Agents for Improved Agriculture* (pp. 165–180). Elsevier.

Köhl, J., Kolnaar, R., and Ravensberg, W. J. (2019). Mode of action of microbial biological control agents against plant diseases: Relevance beyond efficacy. *Front. Plant Sci.*, 10: 845.

Kong, H. G., Kim, J. C., Choi, G. J., Lee, K. Y., Kim, H. J., Hwang, E. C. et al. (2010). Production of surfactin and iturin by *Bacillus licheniformis* N1 responsible for plant disease control activity. *Plant Pathol. J.*, 26(2): 170–177.

Krishna, C. (2005). Solid-state fermentation systems—An overview. *Crit. Rev. Biotechnol.*, 25(1–2): 1–30.

Kumar, V., Ahluwalia, V., Saran, S., Kumar, J., Patel, A. K., and Singhania, R. R. (2021). Recent developments on solid-state fermentation for production of microbial secondary metabolites: Challenges and solutions. *Bioresour. Technol.*, 323: 124566.

Larena, I., De Cal, A., and Melgarejo, P. (2007). Effects of stabilizers on shelf-life of *Epicoccum nigrum* formulations and their relationship with biocontrol of postharvest brown rot by *Monilinia* of peaches. *J. Appl. Microbiol.*, 102(2): 570–582.

LeBlanc, N. (2022). Bacteria in the genus Streptomyces are effective biological control agents for management of fungal plant pathogens: A meta-analysis. *BioControl.*, 67(1): 111–121.

Lee, R. F., and Keremane, M. L. (2013). Mild strain cross protection of tristeza: A review of research to protect against decline on sour orange in Florida. *Front. Microbiol.*, 4: 259.

Li, B., Su, T., Yu, R., Tao, Z., Wu, Z., Algam, S. A. et al. (2010). Inhibitory activity of *Paenibacillus macerans* and *Paenibacillus polymyxa* against *Ralstonia solanacearum*. *Afr. J. Microbiol. Res.*, 4(19): 2048–2054.

Li, G. Q., Huang, H. C., Miao, H. J., Erickson, R. S., Jiang, D. H., and Xiao, Y. N. (2006). Biological control of sclerotinia diseases of rapeseed by aerial applications of the mycoparasite *Coniothyrium minitans*. *Eur. J. Plant Pathol.*, 114: 345–355.

Li, Y., Feng, X., Wang, X., Zheng, L., and Liu, H. (2020). Inhibitory effects of *Bacillus licheniformis* BL06 on *Phytophthora capsici* in pepper by multiple modes of action. *Biocontrol.*, 144: 104210.

Lohnoo, T., Yingyong, W., Jongruja, N., and Krajaejun, T. (2018). Comparison of storage methods to preserve the pathogenic oomycete *Pythium insidiosum*. Southeast Asian *J. Trop. Med. Public Health*, 49(3): 421–427.

Lugtenberg, B. (2015). Introduction to plant-microbe interactions. pp. 1–2. *In:* Lugtenberg, B. (Ed.). Principles of Plant-Microbe Interactions. Springer International Publishing.

Lumsden, R. D., and Locke, J. C. (1989). Biological control of damping-off caused by *Pythium ultimum* and *Rhizoctonia solani* with *Gliocladium virens* in soilless mix. *Phytopathol.*, 79(3): 361–366.

Mácha, H., Marešová, H., Juříková, T., Švecová, M., Benada, O., Škríba, A. et al. (2021). Killing effect of *Bacillus velezensis* FZB42 on a *Xanthomonas campestris pv. campestris* (Xcc) strain newly isolated from cabbage *Brassica oleracea* convar. *capitata* (L.): A metabolomic study. *Microorganisms*, 9(7): 1410.

Manpreet, S., Sawraj, S., Sachin, D., Pankaj, S., and Banerjee, U. C. (2005). Influence of process parameters on the production of metabolites in solid-state fermentation. *Malays. J. Microbiol.*, 2(1): 1–9.

Martin, S. M. (Ed.). (1963). Culture Collections: Perspectives and Problems. University of Toronto Press. https://doi.org/10.3138/9781487582715.

McQuilken, M. P., Budge, S. P., and Whipps, J. M. (1997). Biological control of *Sclerotinia sclerotiorum* by film-coating *Coniothyrium minitans* on to sunflower seed and sclerotia. *Plant Pathol.*, 46(6): 919–929.

Mehmood, N., Saeed, M., Zafarullah, S., Hyder, S., Rizvi, Z. F., Gondal, A. S. et al. (2023). Multifaceted Impacts of Plant-Beneficial *Pseudomonas* spp. In Managing Various Plant Diseases and Crop Yield Improvement. *ACS Omega*, 8(25): 22296–22315.

Méndez-González, F., Loera, O., and Favela-Torres, E. (2018). Conidia production of *Metarhizium anisopliae* in bags and packed column bioreactors. *Current Biotechnology*, 7(1): 65–69.

Montesinos, E. (2003). Development, registration and commercialization of microbial pesticides for plant protection. *Int. J. Microbiol.*, 6: 245–252.

Morgan, C. A., Herman, N., White, P. A., and Vesey, G. (2006). Preservation of micro-organisms by drying; a review. *Journal of Microbiological Methods*, 66(2): 183–193.

Morrison, C. K., Arseneault, T., Novinscak, A., and Filion, M. (2017). Phenazine-1-Carboxylic Acid Production by *Pseudomonas fluorescens* LBUM636 Alters *Phytophthora infestans* Growth and Late Blight Development. *Phytopathol.*, 107(3): 273–279.

Mujib, A., Fatima, S., and Malik, M. Q. (2022). Cryo-derived plants through embryogenesis showed same levels of vinblastine and vincristine (anticancer) in *Catharanthus roseus* and had normal genome size. *Sci. Rep.*, 12(1): 16635.

Mülner, P., Bergna, A., Wagner, P., Sarajlić, D., Gstöttenmayr, B., Dietel, K. et al. (2019). Microbiota Associated with Sclerotia of Soilborne Fungal Pathogens – A Novel Source of Biocontrol Agents Producing Bioactive Volatiles. *Phytobiomes J.*, 3(2): 125–136.

Nakasone, K. K., Peterson, S. W., and Jong, S.-C. (2004). Preservation and distribution of fungal cultures. *Biodiversity of Fungi*, 37–47.

Nakkeeran, S., Vinodkumar, S., Senthilraja, C., and Renukadevi, P. (2019). Antimicrobial peptides of Bacillus species: Biosynthesis, mode of action and their role in plant disease management. pp. 487–514. *In: Microbial antagonists:* Their role in biological control of plant diseases. Today & Tomorrow's Printers and Publishers New Delhi.

Nawrocka, J., and Małolepsza, U. (2013). Diversity in plant systemic resistance induced by *Trichoderma*. *Biocontrol.*, 67(2): 149–156.

Naz, R., Asif, T., Mubeen, S., and Khushhal, S. (2022). Seed application with microbial inoculants for enhanced plant growth. In *Sustainable Horticulture* (pp. 333–368). Elsevier.

O'Brien, P. A. (2017). Biological control of plant diseases. Australas. *Plant Pathol.*, 46(4): 293–304.

O'Callaghan, M., Ballard, R. A., and Wright, D. (2022). Soil microbial inoculants for sustainable agriculture: Limitations and opportunities. *Soil Use and Management*, 38(3): 1340–1369.

Oros, G., and Naár, Z. (2017). Mycofungicide: *Trichoderma* based preparation for foliar applications. *Am. J. Plant Sci.*, 8(2): 113–125.

Orynbayev, A. T., Dzhalilov, F. S. U., and Ignatov, A. N. (2020). Improved efficacy of formulated bacteriophage in control of black rot caused by *Xanthomonas campestris* pv. *Campestris* on cabbage seedlings. *Arch. Phytopathol. Pflanzenschutz.*, 53(7–8): 379–394.

Pacheco-Trejo, J., Aquino-Torres, E., Reyes-Santamaría, M. I., Islas-Pelcastre, M., Pérez-Ríos, S. R., Madariaga-Navarrete, A. et al. (2022). Plant defensive responses triggered by *trichoderma* spp. As tools to face stressful conditions. *Hortic.*, *8*(12): 1181.

Palmieri, D., Ianiri, G., Del Grosso, C., Barone, G., De Curtis, F., Castoria, R. et al. (2022a). Advances and perspectives in the use of biocontrol agents against fungal plant diseases. *Hortic.*, 8(7): 577.

Papavizas, G. C., Dunn, M. T., Lewis, J. A., and Beagle-Ristaino, J. (1984). Liquid fermentation technology for experimental production of biocontrol fungi. *Phytopathol.*, 74(10): 1171–1175.

Patil, V. M., Patole, K. R., Paprikar, M. S., and Rajput, J. C. (2017). Biological control of brinjal wilt caused by *Fusarium oxysporum f. Sp. Melongenae* using soluble powder formulation of *Aspergillus niger*. *Int. J. Adv. Res. Biol. Sci.*, 4(11): 66–71.

Peddle, S. D., Hodgson, R. J., Borrett, R. J., Brachmann, S., Davies, T. C., Erickson, T. E. et al. (2025). Practical applications of soil microbiota to improve ecosystem restoration: Current knowledge and future directions. *Biol. Rev.*, 1–18.

Peng, Y., Kwok, K. H. H., Yang, P.-H., Ng, S. S., Liu, J., Wong, O. G. et al. (2005). Ascorbic acid inhibits ROS production, NF-κB activation and prevents ethanol-induced growth retardation and microencephaly. *Neuropharmacology*, 48(3): 426–434.

Pieterse, C. M. J., Zamioudis, C., Berendsen, R. L., Weller, D. M., Van Wees, S. C. M., and Bakker, P. A. H. M. (2014). Induced Systemic Resistance by Beneficial Microbes. *Annu. Rev. Phytopathol.*, 52(1): 347–375.

Pimentel, C. P. V., Pitta, G. P. B., and Figueiredo, M. B. (1980). Preservation of the pathogenicity of some fungi kept in distilled water.

Promwee, A., Yenjit, P., Issarakraisila, M., Intana, W., and Chamswarng, C. (2017). Efficacy of indigenous *Trichoderma harzianum* in controlling Phytophthora leaf fall (*Phytophthora palmivora*) in Thai rubber trees. *Journal of Plant Diseases and Protection*, 124: 41–50.

Raaijmakers, J. M., Paulitz, T. C., Steinberg, C., Alabouvette, C., and Moënne-Loccoz, Y. (2009). The rhizosphere: A playground and battlefield for soilborne pathogens and beneficial microorganisms. *Plant and Soil*, 321(1–2): 341–361.

Rabbee, M. F., Ali, M. S., and Baek, K. H. (2019). Endophyte *Bacillus velezensis* isolated from Citrus spp. Controls streptomycin-resistant *Xanthomonas citri subsp. citri* that causes citrus bacterial canker. *J. Agron.*, 9(8): 470.

Rajer, F. U., Samma, M. K., Ali, Q., Rajar, W. A., Wu, H., Raza, W. et al. (2022). *Bacillus* spp.-mediated growth promotion of rice seedlings and suppression of bacterial blight disease under greenhouse conditions. *J. Pathog.*, 11(11): 1251.

Ranjani, P., Gowthami, Y., Gnanamanickam, S. S., and Palani, P. (2018). Bacteriophages: A new weapon for the control of bacterial blight disease in rice caused by *Xanthomonas oryzae*. *Microbiology and Biotechnology Letters*, 46(4): 346–359.

Ravindran, K., Chitra, S., Wilson, A., and Sivaramakrishnan, S. (2014). Evaluation of Antifungal Activity of *Metarhizium anisopliae* against plant phytopathogenic fungi. pp. 251–255. *In:* Kharwar, R. N., Upadhyay, R. S., Dubey, N. K., and Raghuwanshi, R. (Eds.). *Microbial Diversity and Biotechnology in Food Security*. Springer India.

Riseh, R. S., Hassanisaadi, M., Vatankhah, M., Soroush, F., and Varma, R. S. (2022). Nano/microencapsulation of plant biocontrol agents by chitosan, alginate, and other important biopolymers as a novel strategy for alleviating plant biotic stresses. *Int. J. Biol. Macromol.*, 222: 1589–1604.

Rivera-Méndez, W., Obregón, M., Morán-Diez, M. E., Hermosa, R., and Monte, E. (2020). *Trichoderma asperellum* biocontrol activity and induction of systemic defenses against *Sclerotium cepivorum* in onion plants under tropical climate conditions. *Biocontrol.*, 141: 104145.

Rizk, R. H., and Youssef, B. A. (1985). Some methods for preservation of *Fusarium oxysporum* f. sp. *Vasinfectum.*, 53–57.

Rocha, I., Ma, Y., Souza-Alonso, P., Vosátka, M., Freitas, H., and Oliveira, R. S. (2019). Seed coating: A tool for deliviring beneficial microbes to agricultural crops. *Front. Plant Sci.*, 10: 1357.

Sahu, B., Singh, J., Shankar, G., and Pradhan, A. (2018). Pseudomonas fluorescens PGPR bacteria as well as biocontrol agent: A review. *Int. J. Chem. Stud.*, 6(2): 01–07.

Saikai, K. K., Oduori, C., Situma, E., Njoroge, S., Murunde, R., Kimenju, J. W. et al. (2023). Biocontrol-based strategies for improving soil health and managing plant-parasitic nematodes in coffee production. *Front. Plant Sci.*, 14: 1196171.

Sala, A., Barrena, R., Artola, A., and Sánchez, A. (2019). Current developments in the production of fungal biological control agents by solid-state fermentation using organic solid waste. *Critical Reviews in Environmental Science and Technology*, 49(8): 655–694.

Saranraj, P., Sayyed, R. Z., Kokila, M., Salomi, V., Sivasakthivelan, P., Manigandan, M. et al. (2023). Evolving Concepts of Biocontrol of Phytopathogens by Endophytic *Pseudomonas fluorescence*. pp. 365–388. *In:* Mawar, R., Sayyed, R. Z., Sharma, S. K., and Sattiraju, K. S. (Eds.). Plant Growth Promoting Microorganisms of Arid Region. Springer Nature Singapore.

Schaechter, M. (2009). Encyclopedia of microbiology. Academic Press.

Sharma, V., Chandoliya, R., and Sharma, S. (2023). 7 Role of microbial oxidative enzymes in biological control of plant pathogens. pp. 129–154. *In:* Gupta, A., and Kumar, P. (Eds.). *Microbial Oxidative Enzymes*. De Gruyter.

Shearer, B. L., Zeyen, R. J., and Ooka, J. J. (1974). Of *Septoria* species isolated from cereals. *Phytopathol.*, 64: 163–167.

Shemshura, O., Alimzhanova, M., Ismailova, E., Molzhigitova, A., Daugaliyeva, S., and Sadanov, A. (2020). Antagonistic activity and mechanism of a novel *Bacillus amyloliquefaciens* MB40 strain against fire blight. *J. Plant Pathol.*, 102(3): 825–833.

Simon, A., and Sivasithamparam, K. (1988). Interactions among *Gaeumannomyces graminis* var. *Tritici* , *Trichoderma koningii* , and soil bacteria. *Can. J. Microbiol.*, 34(7): 871–876.

Singh, G., and Paithankar, I. (2023). Encapsulation of Biofertilizers, Biopesticides and Biocontrol Agents. pp. 121–150. *In:* Singh, N. K., Chattopadhyay, A., and Lichtfouse, E. (Eds.), Sustainable Agriculture Reviews 60 (Vol. 60,). Springer Nature Switzerland.

Sinha, R., Anand, S., Singh, D., Tripathi, S., Shukla, P., Singh, S. et al. (2022). Bioprocessing strategies for microbial production and purification of immunosuppressants: An insight for process intensification. *Chemical Engineering and Processing-Process Intensification*, 172: 108797.

Sivan, A., and Chet, I. (1989). Degradation of Fungal Cell Walls by Lytic Enzymes of *Trichoderma harzianum*. *Microbiol.*, 135(3): 675–682.

Sivasakthi, S., Usharani, G., and Saranraj, Pjaj. (2014). Biocontrol potentiality of plant growth promoting bacteria (PGPR)-*Pseudomonas fluorescens* and *Bacillus subtilis*: A review. *Afr. J. Agric. Res.*, 9(16): 1265–1277.

Sneh, B., Burpee, L., and Ogoshi, A. (1991). Identification of Rhizoctonia species, (pp. vi+-133).

Spencer, D. M., and Atkey, P. T. (1981). Parasitic effects of *Verticillium lecanii* on two rust fungi. *Transactions of the British Mycological Society*, 77(3): 535–542.

Sriram, S., Roopa, K. P., and Savitha, M. J. (2011). Extended shelf-life of liquid fermentation derived talc formulations of *Trichoderma harzianum* with the addition of glycerol in the production medium. *J. Crop Prot.*, 30(10): 1334–1339.

Sriwati, R., Chamzhurni, T., Alfizar, A., Soekarno, B. P., Maulidia, V., and Sahputra, I. (2019). Pellet Formulation of *Trichoderma* Isolated from Aceh with Potential for Biocontrol of *Phytophthora* Sp. On Cacao Sedlings. *Journal of Applied Agricultural Science and Technology*, 3(2): 202–212.

Suresh, P., Shanmugaiah, V., Rajagopal, R., Muthusamy, K., and Ramamoorthy, V. (2022). *Pseudomonas fluorescens* VSMKU3054 mediated induced systemic resistance in tomato against *Ralstonia solanacearum*. *Physiol. Mol. Plant Pathol.*, 119: 101836.

Tapio, E., and Pohto-Lahdenperä, A. (1991). Scanning electron microscopy of hyphal interaction between *Streptomyces griseoviridis* and some plant pathogenic fungi. *Agricultural and Food Science*, 63(5): 435–441.

Teixidó, N., Torres, R., Viñas, I., Abadias, M., and Usall, J. (2011). Biological control of postharvest diseases in fruit and vegetables. pp. 364–402. *In:* Protective Cultures, Antimicrobial Metabolites and Bacteriophages for Food and Beverage Biopreservation. Elsevier.

Teixidó, N., Usall, J., and Torres, R. (2022). Insight into a successful development of biocontrol agents: Production, formulation, packaging, and shelf life as key aspects. *Hortic.*, 8(4): 305.

Thambugala, K. M., Daranagama, D. A., Phillips, A. J., Kannangara, S. D., and Promputtha, I. (2020). Fungi vs. fungi in biocontrol: An overview of fungal antagonists applied against fungal plant pathogens. *Front. Cell. Infect. Microbiol.*, 10: 604923.

Trunk, K., Peltier, J., Liu, Y.-C., Dill, B. D., Walker, L., Gow, N. A. et al. (2018). The type VI secretion system deploys antifungal effectors against microbial competitors. *Nat. Microbiol.*, 3(8): 920–931.

Van Den Berg, F., Paveley, N. D., and Van Den Bosch, F. (2016). Dose and number of applications that maximize fungicide effective life exemplified by *Zymoseptoria tritici* on wheat – a model analysis. *J. Plant Pathol.*, 65(8): 1380–1389.

Van Loon, L. C., Bakker, P. A. H. M., and Pieterse, C. M. J. (1998). Systemic resistance induced by rhizosphere bacteria. *Annu. Rev. Phytopathol.*, 36(1): 453–483.

Van Peer, R., Niemann, G. J., and Schippers, B. (1991). Induced resistance and phytoalexin accumulation in biological control of Fusarium wilt of carnation by *Pseudomonas* sp. Strain WCS 417 r. *Phytopathol.*, 81(7): 728–734.

Verma, M., Brar, S. K., Tyagi, R. D., Surampalli, R. nY, and Valero, J. R. (2007). Antagonistic fungi, *Trichoderma* spp: Panoply of biological control. *Biochemical Engineering Journal*, 37(1): 1–20.

Viterbo, A., and Horwitz, B. A. (2014). Mycoparasitism. pp. 676–693. *In:* Borkovich, K. A., and Ebbole, D. J. (Eds.), Cellular and Molecular Biology of Filamentous Fungi. ASM Press.

Wang, J., Xiong, Z., Fan, Y., Wang, H., An, C., Wang, B. et al. (2024). Lignin/Surfactin Coacervate as an eco-friendly pesticide carrier and antifungal agent against phytopathogen. *ACS nano*, 18(33): 22415–22430.

Wang, S., Zheng, Y., Gu, C., He, C., Yang, M., Zhang, X. et al. (2018). Bacillus cereus AR156 activates defense responses to *Pseudomonas syringae* pv. tomato in *Arabidopsis thaliana* similarly to flg22. *Molecular Plant-Microbe Interactions*, 31(3): 311–322.

Wei, G., Kloepper, J. W., and Tuzun, S. (1991). Induction of systemic resistance of cucumber to *Colletotrichum orbiculare* by select strains of plant growth-promoting rhizobacteria. *Phytopathol.*, 81(11): 1508–1512.

Yu, C., Chen, H., Zhu, L., Song, Y., Jiang, Q., Zhang, Y. et al. (2023). Profiling of Antimicrobial Metabolites Synthesized by the Endophytic and Genetically Amenable Biocontrol Strain *Bacillus velezensis* DMW1. *Microbiol. Spectr.*, 11(2).

Zamora, P., Martín, A. B., San Martín, R., Martínez-Álvarez, P., and Diez, J. J. (2014). Control of chestnut blight by the use of hypovirulent strains of the fungus *Cryphonectria parasitica* in northwestern Spain. *Biocontrol.*, 79: 58–66.

Zehra, A., Dubey, M. K., Meena, M., and Upadhyay, R. S. (2017). Effect of different environmental conditions on growth and sporulation of some *Trichoderma* species. *J. Environ. Biol.*, 38(2): 197.

Zhang, X., Zhang, G., Li, P., Yang, Q., Chen, K., Zhao, L. et al. (2017). Mechanisms of glycine betaine enhancing oxidative stress tolerance and biocontrol efficacy of *Pichia caribbica* against blue mold on apples. *Biocontrol.*, 108: 55–63.

Zhang, X.-F., and Qu, F. (2016). Cross protection of plant viruses: recent developments and mechanistic implications. pp. 241–250. *In:* Wang, A., and Zhou, X. (Eds.). *Current Research Topics in Plant Virology*. Springer International Publishing.

Zhang, Y., Li, Y., Liang, S., Zheng, W., Chen, X., Liu, J. et al. (2022). Study on the preparation and effect of tomato seedling disease biocontrol compound seed-coating agent. *Life*, 12(6): 849.

Zhao, G., Sun, T., Zhang, Z., Zhang, J., Bian, Y., Hou, C. et al. (2023). Management of take-all disease caused by *Gaeumannomyces graminis* var. *Tritici* in wheat through *Bacillus subtilis* strains. *Front. Microbiol.*, 14: 1118176.

Zhao, X., Zhai, Y., Wei, L., Xia, F., Yang, Y., Yi, Y. et al. (2024). Isolation and identification of a novel *Bacillus velezensis* strain JIN4 and its potential for biocontrol of kiwifruit bacterial canker caused by *Pseudomonas syringae pv. actinidiae*. *Front. Plant Sci.*, 15: 1513438.

Zhou, C., and Zhou, Y. (2012). Strategies for viral cross protection in plants. Antiviral Resistance in Plants: Methods and Protocols, 69–81.

Zin, N. A., and Badaluddin, N. A. (2020). Biological functions of *Trichoderma* spp. For agriculture applications. *Ann. Agric. Sci.*, 65(2): 168–178.

4

Biofungicides

Innovative Formulations and Development

Mohammadhassan Gholami-Shabani,[1,2] *Katayoon Nematpour,*[2] *Masoomeh Shams-Ghahfarokhi,*[3] *Farnoush Asghari-Paskiabi,*[1] *Shahrzad Saeidpour,*[1] *Mehrad Eslamifar,*[4] *Zeynab Gholami-Shabani,*[5] *Ali Eslamifar,*[6] *Fatemehsadat Jamzivar*[1] and *Mehdi Razzaghi-Abyaneh*[1,*]

1. Introduction

Fungal diseases have become major production obstacles for agricultural plants that endanger global food supply systems. Farmers have used chemical fungicides in agricultural practices at an advanced rate throughout several decades to ensure food production alongside fungal disease protection for their crops. Analysis of global dataset collections shows that microbial diseases reduce crop yields by 16%, with mold diseases causing up to 70–80% of this total (Rodrigo et al. 2022). The market offers a wide variety of budget-friendly chemical fungicides named Bavistin, which agricultural operators utilize throughout all production periods and during harvest handling. Chemical fungicides are synthesized in industries with many chemicals,

[1] Department of Mycology, Pasteur Institute of Iran, Tehran 1316943551, Iran.
[2] Department of Labor Environment, Health, Safety, Environment and Energy, Ministry of Industry, Mine and Trade, Tehran, Iran.
[3] Department of Mycology, Faculty of Medical Sciences, Tarbiat Modares University, Tehran, 14115-331, Iran.
[4] I.M. Sechenov First Moscow State Medical.
[5] Department of Aerospace, Faculty of Mechanical Engineering, Tarbiat Modares University, Tehran, Iran.
[6] Department of Clinical Research, Pasteur Institute of Iran, Tehran 1316943551, Iran.
* Corresponding author: mrab442@yahoo.com, mrab442@pasteur.ac.ir

DOI: 10.1201/9781003603023-5

including heavy metals (copper, zinc, iron), that harm the natural ecosystem. Chemical fungicides are sprayed on crops, seeds, and fruits to kill fungal pathogens; however, residues are left behind in the plant material. Frequent eating of food materials contaminated with toxins or treated with chemicals can instigate severe health complications in humans, including food poisoning, cancer, and formation of benign or malignant tumors, genetic disorders, blood disorders, and nerve disorders leading to paralysis, birth defects, and fetotoxicity. Similar toxic impacts have been observed on a variety of organisms standing at different trophic levels in ecosystems, such as other mammals, sea creatures, and birds. Nowadays, chemical fungicides have been acknowledged for causing immense environmental hazards, thereby leading the international scientific community to search for different, more environmentally friendly ways of combating fungal pathogens. Biofungicides relate to biological formulations or products produced from a consortium of organisms that suppress fungal pathogens. Biofungicides are biological agents responsible for crop protection, also known as biocontrol, biologicals, biological pesticides, or biopesticides. These are pure products from living organisms and natural compounds that protect crops from the action of many damaging pests and diseases (Lahlali et al. 2022). In effect, biofungicides are biological agents that eliminate or suppress fungi and their spores (Fenta and Mekonnen 2024). These biological products are mainly derived from plants, microorganisms, and naturally occurring compounds (Fenta and Mekonnen 2024). Biofungicides are products based on living organisms and their by-products for the management of severe fungal pathogens. They fall into two main categories: (1) microbial biofungicides, which contain active ingredients that function as biological control agents, effectively targeting or competing with pathogens or pests, and (2) plant biofungicides, also known as plant-incorporated protectants, which are pesticidal substances that plants generate through the integration of additional genetic material (Subba and Mathur 2022). They are isolated and then enhanced to offer growers new options in their planning of disease management that differ from conventional products (Soković et al. 2013). Biofungicides are environmentally friendly, nontoxic, and free of residues. Plant products treated with biofungicides are safe for human consumption and possess better quality with increased shelf life. Research has shown that growth-promoting biofungicides improve the stress-tolerating ability of plants. As with every sector of human life, so with agriculture, and as farmers become more enlightened at the ground level, greater use has been attained in organic farming practices as well as incorporation into the IPM approach of integrated pest and disease management. Several studies in the literature show these sources to have antifungal properties, and a variety of biofungicides are commercially available in the market today. Pathogenic fungi can lead to multiple problems economically and environmentally all over the world, and they need powerful fungicides for proper management (Lahlali et al. 2022). Currently, scientists from different countries are researching alternative environmentally friendly methods of managing fungal diseases since chemical fungicides prove to be serious environmental hazards (El-Baky and Amara 2021). Natural products are the most significant approach in controlling fungal pathogenic infections among economically important crops (El-Baky and Amara 2021). In the context of crop cultivation, they bring about lower toxicity levels and higher stability with fewer detrimental side effects (Subba and Mathur 2022). Most preparations in this regard are perceived as ecologically benign, primarily because of their biological sources and the relatively minor amounts of active chemical agents encompassed therein (Castillo et al. 2022). Apart from low toxicity, these agents also show a plethora of effects on various diseases, improve resistance to adverse conditions in plants/microorganisms, and appear to be more cost-effective standards (Subba and Mathur 2022). The entire concept of biofungicides includes a wide variety of microorganisms, known as microbial pesticides, along with naturally existing substances that participate in the control of plant diseases, classified as biochemical pesticides (Vero et al. 2023). The two main classes of biofungicides are plant-based and microbial. Microbial biofungicides include both bacterial and fungal varieties. Plant-based biofungicides involve various formulations for a range of plants (Subba and Mathur 2022). Biofungicides can be classified according to several criteria, such as their formulation, the resources employed, the type of crops they target, and the application methods used (Ezeorba et al. 2023). The preeminence given to organic farming practices and a greater promotion of residue-free products is expected to enhance the options available to farmers in the usage of biofungicides. Fungicides aim at loss mitigation of crop productivity. Conversely, even when synthesized quickly, chemically synthesized pesticides often have adverse effects on the environment and human health (Pathak et al. 2022). The emergence of numerous new pest species, resistance developed in fungal populations, degradation of soil fertility, and persistence of harmful substances that can affect both humans and animals upon consumption, among many others, are some of the consequences (Sánchez-Bayo 2011). Expenditure for the biofungicides market is forecast around USD 2.5 billion in 2024, with the figures expected to rise to reach USD 3 billion in 2025. The market is estimated to grow at a compound annual growth rate (CAGR) of 13.5%, reaching an estimated USD 5.5 billion by 2030. The number is forecasted by the UN (United Nations) to reach USD 9.8 billion by 2050 (Paola et al. 2023). Various scientists have researched different biological control methods used in disease management in plants (Narayanasamy and Narayanasamy 2013). The potential biological management approaches for pathogens use various microbial agents including fungi, bacteria, and endophytic organisms (Pandit et al. 2022). Plant growth-promoting rhizobacteria enhance plant development while helping plants obtain fundamental nutrients, thus promoting their growth. The antifungal properties of rhizobacteria complement their capability to enhance plant growth. Several research studies have proven that plant growth-promoting rhizobacteria demonstrate biofungicidal mechanisms that destroy fungal pathogens affecting plants (Patwardhan et al. 2022). Scientific studies have established that Bacillus along with Streptomyces and Pseudomonas microorganism strains serve as critical controllers of phytopathogenic fungi (Carmona-Hernandez et al. 2019). Scientific molecular research has discovered important metabolites and antibiotics from genera including Bacillus and Streptomyces that halt fungal growth. The development of metagenomics enables scientists to discover numerous newfound species showing strong antimicrobial properties that represent promising options for managing dangerous fungal pathogens (Carmona-Hernandez et al. 2019). Fungi function as major control agents following their proven ability to limit fungal species development, thus offering their potential in biofungicide development (Tyśkiewicz et al. 2022). Scientists demonstrate that fungal antagonists represent an essential biocontrol technique for managing powdery mildew pathogens. Research results show that fungal endophytes make a vital yet commonly overlooked impact on phytopathogenic fungal growth suppression throughout laboratory tests and field conditions (Sornakili et al. 2020). Studies demonstrate that antimicrobial qualities in natural products from fungal endophytes make them an effective resource for disease control in plants (Sornakili et al. 2020). The combination of gas chromatography–mass spectrometry (GC–MS) and liquid chromatography–mass spectrometry (LC–MS) allowed scientists to both separate and identify compounds which represent these fungal agents. Scientists identify *Trichoderma* sp. as a fundamental fungus, along with being the leading biological control agent, which boosts pest and pathogen management principles. Modern genomic and proteomic research conducted by Tyśkiewicz et al. (2022) identified the fundamental methods used by Trichoderma sp. to perform biocontrol actions. Biofungicide manufacturers formulate plant extract-based products that use different solvents such as ethanol, methanol, and ether for the preparation of fungal pathogen control solutions. The inhibitory effects of fungal growth have been noted through extracts obtained from multiple medicinal plants. These plant extracts demonstrate antifungal capabilities because they contain various essential oils that exist in specific regions such as glandular trichomes and hairs throughout the plant structure. Several essential oils extracted from different plant species have shown effective behavior in controlling fungal pathogens while treating diseases according to previous research findings. The antifungal substances, along with pest protection qualities in plant extracts, derive from their wide range of secondary metabolites, particularly phenols and glucosinolates (Mrid et al. 2021). Research shows that multiple biofungicides from plant sources have proven biostimulatory effects for developing compounds that increase plant growth and enhance nutrient availability and fruit and seed wall thickness for improved shelf life (Mrid et al. 2021). Biocontrol review literature mostly deals with plant–microbial relations and microorganisms, yet available research material is limited. The number of studies that explain biocontrol operational functions remains limited (Heydari and Pessarakli 2010). An extensive study on pathogen-targeting secondary metabolites as well as their operational principal functions exists within this chapter. This research investigates biofungicide extraction procedures and explains formulation approaches while establishing biological performance against damaging fungi before studying regulatory views about organic disease control limitations and advantages. The chapter offers a detailed investigation of contemporary biofungicide use throughout various fields.

2. Biofungicides Classification

Biofungicides exist in two primary origin-based groups, namely those obtained from microorganisms and those extracted from plants, according to Ezeorba et al. 2023. Biofungicides obtain their synthesis from numerous microbial and plant-derived substances (Figure 1). Biofungicides acquired from these natural sources enter the market through extract and suspension formulas, in addition to other formulations,

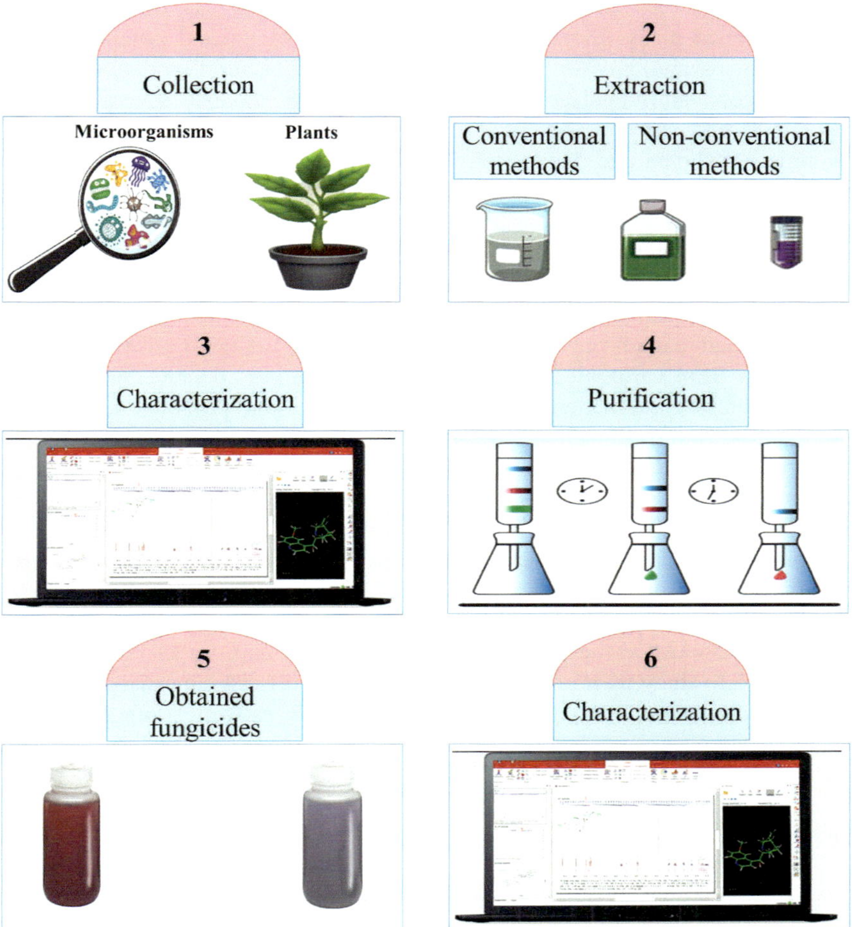

Figure 1: Schematic features of biofungicide synthesis steps through the utilization of diverse microbial and plant-derived bioactive components.

after following diverse extraction methods (Bharti et al. 2020). Scientific investigations proved that water, together with ethanol, methanol, acetone, and ethyl acetate, function as superior extraction solvents for biological materials that support biofungicide development (O'Keeffe et al. 2019).

2.1 Microorganism-based Biofungicides

Well-organized use of fungi is quite detrimental to crops through the periods of harvest to post-harvesting stages (Fenta and Mekonnen 2024). Environmental pollution, along with health-related problems, is caused by the historical use of chemical fungicides for several decades (Ons et al. 2020). According to Fenta and Mekonnen (2024), biofungicide is an effective ecological alternative that aids in dealing with fungal phytopathogens. It tops the chart of all the microbial biofungicides in the current market of biocontrol products for sustainable agriculture. Thus, writers have researched many microorganisms and their biofungicidal potentials against phytopathogens and ended up identifying numerous bacterial and fungal species that have good antifungal activities. Microorganisms produce several antifungal compounds that serve as essential ingredients of the industry for formulating biofungicides (Fenta and Mekonnen 2024). Endophytic fungi were evaluated by scientists for the production of new secondary metabolites, such as terpenoids and the related groups alkaloids, phenolics, isocoumarins, steroids, chromones, and volatile compounds, for the purpose of biofungicide development (Sornakili et al. 2020). The research study conducted by Liu et al. (2023) evaluated the antifungal action of natural/synthetic antimicrobial peptides or proteins with varying sources of origin, intended to eradicate mycotoxin-producing fungi. A lot of growth-controlling mechanisms from various ethno-bacterial microorganisms operate by competitive relations, parasitic interactions, volatile organic compound production, and biofilm formation. Many antifungal microbes were isolated from the surrounding soil of the roots, as well as from the tissues of the plants devoted to their cultivation. Roberts illustrated competition in liquid cultures between microorganisms, particularly *Penicillium glaucum* against bacteria, and coined the term antagonism in microbiological studies in the year 1874 (Jiang and Wang 2023). Hartley pioneered study during 1921 on the direct application of biological control of plant pathogens (Jiang and Wang 2023). He inoculated soils with antagonistic microorganisms. For this, he had inoculated thirteen different antagonistic fungi under testing in forest nursery soils to study their efficacy against damping-off disease caused by *Pythium debaryanum* (Jiang and Wang 2023). *Bacillus subtilis*, as well as other antagonists such as *Pseudomonas syringae* and *Bacillus amyloliquefaciens* strains, were shown to have antagonistic effects against *Fusarium* species by Ayala-Torres et al. (2023).

2.2 Bacteria-based Biofungicides

It has been reported over the years that beneficial bacteria improve the growth performance of plants and enhance disease resistance in crops (Fenta and Mekonnen 2024). Bacteria have long been used as antifungal agents against plant pathogenic fungi. Hence, these bacteria could prove to be promising natural fungicides. Up until now, a wide variety of bacteria from numerous genera, such as *Bacillus, Agrobacterium, Paenibacillus, Pseudomonas, Streptomyces, Bradyrhizobium, Azospirillum, Acinetobacter, Azotobacter,* and *Rhizobium,* have been recorded as biocontrol agents to combat different diseases in plants. The biocontrol study of *Bacillus* species has received huge attention, especially against plant pathogenic fungi. This genus is recognized for the variety of antifungal secondary metabolites produced by its members, which have gained industrial interest as potential biofungicides. Lipopeptides produced by *Bacillus subtilis* (e.g., surfactin) perforate the membranes of fungal cells. The authors of Jia et al. (2023) reported that *Bacillus amyloliquefaciens* is a good biocontrol agent against black spot disease in tomato and mycotoxin production by *Alternaria alternata. B. amyloliquefaciens* suppresses *A. alternata* by manufacturing antifungal metabolites. In another study conducted by Azish et al. (2020), a total of 110 bacterial strains had been isolated from forest soils, four of which—*Streptomyces libani, Streptomyces angustmyceticus, Bacillus subtilis,* and *Sphingopyxis* spp.—were identified to have good antifungal activity based on 16S ribosomal RNA sequencing. The dichloromethane extract obtained from the starch casein broth culture filtrate of *S. libani* was subjected to incubation for five days at 30°C and showed potent antifungal activities against *Aspergillus fumigatus, Aspergillus niger,* and *Aspergillus flavus*. In 2021, Shahid et al. studied twenty different actinomycetes for their antifungal activity in a dual culture plate assay against six phytopathogens: *Alternaria alternata, Aspergillus flavus, Fusarium oxysporum* f. sp. *lycopersici, Sarocladium oryzae, Sclerotinia sclerotiorum,* and *Rhizoctonia solani.* Among them, two promising isolates, *Streptomyces amritsarensis* and *Kribella karoonensis,* exhibited significant antifungal potential against all six fungal pathogens. Ahmadi et al. (2021) carried out research to formulate a biofungicide from sugarcane bagasse with the aid of the yeast species *Pichea membranifaciens,* and to evaluate its activity against post-harvest pathogenic fungi like *Aspergillus niger, Phytophthora capsica,* and *Penicillium digitatum.* Sujarit et al. (2020) prepared a spore suspension and alginate beads with *Streptomyces palmae,* which were mixed with sterile soil for methodical application against *Ganoderma boninense.* Further research with *Pseudomonas aeruginosa* and *Streptomyces palmae* as the two test organisms revealed antifungal activity against the pathogenic fungus *Ganoderma boninense. Pseudomonas fluorescens* fights with pathogens to occupy space and nutrients and develops antifungal substances. Research by Lin et al. (2021) analyzed antifungal effects and plant growth improvements of *Lysobacter* strains found in soils known to inhibit *Rhizoctonia* growth. The *Lysobacter* strains extracted from *Rhizoctonia*-suppressive soils were classified into four distinct species: *Lysobacter antibioticus, Lysobacter capsici, Lysobacter enzymogenes,* and *Lysobacter gummosus,* through 16S rRNA sequencing methods. Most strains in the study showed powerful antifungal action, which effectively controlled *Rhizoctonia solani* together with *Pythium ultimum, Aspergillus niger, Fusarium oxysporum,* and *Xanthomonas campestris* pathogens. Evidence from conducted studies confirms bacteria-derived biofungicides can achieve effective utilization.

2.3 Fungal-based Biofungicides

The agricultural industry obtains effective agrochemicals through the utilization of bioactive compounds that beneficial fungi generate in significant quantities (Gholami-Shabani et al. 2019). Several research investigations into the antifungal activities of diverse fungal isolates targeting various pathogenic fungi have emphasized their potential utility in the production of biofungicides (Table 1). Among the most commonly employed fungal genera for the biological control of plant diseases caused by both bacteria and fungi are *Trichoderma, Aspergillus,* and *Penicillium. Trichoderma harzianum* parasitizes pathogenic fungi and induces systemic resistance in plants. *Gliocladium virens* produces gliotoxin, a potent antifungal compound. Studies have revealed that fungi such as *Gliocladium* and *Saccharomyces* are capable of exerting antagonistic actions against various pathogenic species (Ayaz et al. 2023). Many fungal endophytes, which are diverse microorganisms, live within

Table 1: Biological sources and detailed features of biofungicides.

Source		***Pathogenic fungi***	**References**
Bacteria	*Bacillus subtilis*	*Magnaporthe grisea*	Meng et al. (2015)
		Botrytis cinerea	Arroyave-Toro et al. (2017)
		Colletotrichum acutatum	
		Fusarium oxysporum var. orthoceras	Sidorova et al. (2020)
	Bacillus velezensis	*Botrytis cinerea*	Jiang et al. (2018)
		Fusarium graminearum	Myo et al. (2019)
		Alternaria alternate	
		Botrytis cinerea	
		Fulvia fulva	
		Fusarium oxysporum	
	Bacillus siamensis	*Alternaria alternata*	Xie et al. (2021)
	Pseudomonas fluorescens	*Rhizoctonia solani*	Pal et al. (2015)
	Pseudomonas aeruginosa	*Ganoderma boninense*	Irma et al. (2018)
	Rhodopseudomonas palustris	*Magnaporthe oryzae*	Nookongbut et al. (2020)
		Curvularia lunata	
		Bipolaris oryzae	
	Streptomyces palmae	*Ganoderma boninense*	Sujarit et al. (2020)
	Lysobacter enzymogenes	*Valsa pyri*	Yang et al. (2020)
		Macrophomina phaseolina	Nian et al. (2021)
		Sclerotinia sclerotiorum	
		Fusarium virguliforme	
		Septoria glycines	
		Cercospora sojina	
		Rhizoctonia solani	
		Phytophthora sojae	
		Pythium spp.	
Fungi	*Aureobasidium pullulans*	*Trichoderma pleuroticola*	Roberti et al. (2019)
		Trichoderma pleuroti	
	Anthracocystis flocculosa	*Fusarium graminearum*	Rojas et al. (2020)
	Beauveria bassiana	*Fusarium oxysporum*	Jaber and Alananbeh (2018)
		Fusarium moniliforme	
		Fusarium culmorum	
	Fusarium oxysporum	*Sclerospora graminicola*	Nandhini et al. (2018)
	Fusarium commune	*Fusarium fujikuroi*	Saito et al. (2021)
	Fomes fomentarius	*Gaeumannomyces graminis*	Gholami et al. (2019)

Table 1 contd. ...

...Table 1 contd.

Source		***Pathogenic fungi***	**References**
Fungi	*Metarhizium brunneum*	*Fusarium oxysporum*	Jaber and Alananbeh (2018)
		Fusarium moniliforme	
		Fusarium culmorum	
	Rhizoctonia endophytica	*Gaeumannomyces graminis*	Gholami et al. (2019)
	Trichoderma spp.	*Gremmeniella abietina*	Romeralo et al. (2015)
	Trichoderma asperellum	*Sclerospora graminicola*	Nandhini et al. (2018)
		Colletotrichum truncatum	Yadav et al. (2023)
	Trichoderma harzianum	*Fusarium graminearum*	Ferrigo et al. (2020)
		Fusarium verticillioides	
		Fusarium oxysporum	Rusita and Sasongko (2020)
			Zhang et al. (2022)
		Colletotrichum gloeosporioides	
		Colletotrichum truncatum	Yadav et al. (2023)
	Talaromyces tratensis	*Alternaria padwickii*	Eakjamnong et al. (2021)
		Bipolaris oryzae	
		Curvularia lunata	
		Fusarium moniliforme	
	Penicillium olsonii	*Fusarium graminearum*	Rojas et al. (2020)
Plants	*Cystoseira myriophylloides*	*Verticillium dahlia*	Esserti et al. (2017)
	Cuminum cyminum	*Fusarium oxysporum*	Sun et al. (2017)
	Cymbopogon citratus	*Curvularia lunata*	Naveenkumar et al. (2017)
		Sarocladium oryzae	
		Bipolaris oryzae	
		Fusarium moniliforme	
	Datura stramonium	*Fusarium* sp.	Djahra et al. (2019)
		Aspergillus niger	
		Alternaria solani	
		Septoria nodurum	
	Eucalyptus globulus	*Fusarium solani*	Iqbal et al. (2020)
	Fucus spiralis	*Verticillium dahlia*	Esserti et al. (2017)
	Foeniculum vulgare	*Fusarium oxysporum*	Soylu and Incekara (2017)
	Ipomoea batatas	*Ceratocystis fimbriata*	Liu et al. (2020)
		Fusarium sp.	Saputri and Utami (2020)

Table 1 contd. ...

...*Table 1 contd.*

Source		***Pathogenic fungi***	**References**
Plants	*Lantana camara*	*Alternaria alternata*	Singh and Srivastava (2012)
	Laminaria digitata	*Verticillium dahlia*	Esserti et al. (2017)
	Moringa peregrina	*Mycogone perniciosa*	Shokouhi and Seifi (2020)
	Myrtus communis	*Fusarium oxysporum*	Soylu and Incekara (2017)
	Origanum onites		
	Oryza staiva		Wei et al. (2020)
	Thymus vulgaris	*Mycosphaerella graminicola*	Jabeur et al. (2017)
	Schleichera oleosa	*Colletotrichum musae*	Kumar et al. (2021)

the internal tissues of plants without producing any visible symptoms (Sornakili et al. 2020). Endophytic fungi establish tight biological connections with their hosts through mutualistic interactions but rarely develop parasitic bonds. The production of diverse physiologically active secondary metabolites with different structures helps endophytic fungi provide essential defense against harmful pathogens for their host species. Endophytic fungi function as one of the most important resources for identifying new secondary metabolites with great potential for agricultural development (Segaran and Sathiavelu 2019). Researchers have identified multiple chemical-based secondary metabolites such as terpenoids, alkaloids, steroids, peptides, isocoumarins, benzopyranones, and quinones in endophytic fungi (Hashem et al. 2023). Preparatory chemical knowledge deriving from these metabolites enabled researchers to develop agrochemicals that display antibacterial, antifungal, nematicidal, insecticidal, and herbicidal activities in agricultural applications. *Trichoderma aggressivum* f. *europaeum* displayed powerful antagonistic properties that blocked the growth of *Botrytis cinerea, Sclerotinia sclerotiorum,* and *Mycosphaerella melonis* mycelia together with their sclerotial germination inhibition. Through its action against *Sclerotinia sclerotiorum,* it totally prevented sclerotia germination (Ayaz et al. 2023). The hyphal growth pattern of *Trichoderma* species develops into helical shapes to encircle the pathogens' hyphal structures. Appressorium formation occurs through this interaction process, producing structures needed for lytic enzyme production that enable penetration of pathogenic hyphae (Baron et al. 2019). The research of Zhang et al. (2024) delivered two new glycosides named fusintespyrone A and cerevisterolside A, together with ten known compounds obtained from the intestinal fungus *Fusarium* sp. These compounds demonstrated strong growth suppression capabilities against *Aspergillus fumigatus, Fusarium oxysporum,* and *Verticillium dahliae.* Gholami et al. (2019) demonstrated that certain endophytic fungi, specifically *Fomes fomentarius, Rhizoctonia endophytica, Coprinopsis urticicola*, and *Rhizoctonia zeae,* are capable of producing hydrolytic enzymes such as protease, pectinase, chitinase, and cellulase. These enzymes likely contribute to the degradation of cell walls, which in turn may inhibit the growth of *Gaeumannomyces graminis* var. *tritici*, the pathogen responsible for take-all disease in wheat.

2.4 Plant-based Biofungicides

Diseases caused by pathogenic fungi pose a serious threat to crops of commercial importance, thereby creating significant supply issues within the international food chain (Davies et al. 2021). The recent surge in research surrounding biofungicides can be attributed to their less harmful effects and lower associated costs. Long before humans actively participated in the protection of plants, these organisms had already formulated their own defenses against disease (Van Loon 2015). Various plant forms have been the subject of research aimed at assessing their potential for biofungicide production (Table 1). Neem (*Azadirachta indica*), garlic extract, and essential oils (e.g., thyme, clove) exhibit strong antifungal properties. The contribution of plants to the management of plant diseases is substantial, and they are considered a viable source of biofungicides (Shokouhi and Seifi 2020). Different extracts of biofungicides are formulated with solvents including water, acetone, buffers, and acids, and these extracts are applied using a variety of methods (O'Keeffe et al. 2019). Plants demonstrate a special capability to create distinctive secondary metabolites which include terpenes, alkaloids, and phenolic compounds (Sharma et al. 2022). Scientists extract compounds from plants by both standard and new extraction methods for biofungicide production. Plant extracts show dual antifungal properties through diverse defense systems which fight against organisms. These agents demonstrate the status of being harmless to plants and show promise in disease protection applications. Biofungicides serve actively in controlling diseases because their specifically designed active ingredients successfully fight particular pathogens without harming living biological systems. These fungicides provide a sustainable, economical agricultural solution, as the growing recognition of biofungicides demonstrates their potential for solving problems associated with synthetic fungicide use. The fungicides are applied through crop spraying procedures similar to those used for chemical fungicides (Fenta and Mekonnen 2024). These biofungicides suppress pathogens by restricting their growth process, which is essential for both disease development and specialized toxin production.

3. The Biosynthetic Pathways of Bioactive Compounds and Defense Against Pathogenic Fungi

Living organisms inherently synthesize a diverse array of compounds with varying chemical properties that are essential for their growth and development (Chiocchio et al. 2021). Primary metabolites serve as fundamental components that support essential processes such as photosynthesis, translocation, and respiration. Compounds generated through primary metabolic processes that do not play a direct role in growth and development are referred to as secondary metabolites (Chiocchio et al. 2021). According to their biosynthetic origins, secondary metabolites are divided into three principal types: (1) terpenes and terpenoids, (2) alkaloids, and (3) phenolic compounds (Bhatla and Lal 2023). The pathways responsible for the synthesis of these metabolites include four major routes: (a) the shikimic acid pathway, (b) the malonic acid pathway, (c) the mevalonic acid pathway, and (d) the non-mevalonate pathway (Suzuki et al. 2014).

3.1 Terpenoids Biosynthesis Pathway

Terpenoids are an extensive and diverse group of natural products. They are biosynthesized from isoprenyl diphosphate, which is created by the consecutive condensation of the five-carbon monomer isopentenyl diphosphate and its isomer, dimethylallyl diphosphate (Kuzuyama 2017). There are two principal pathways for the synthesis of terpenoids: the mevalonate pathway and the non-mevalonate pathway, the latter of which is identified as the 2-C-methyl-D-erythritol-4-phosphate pathway, or alternatively, the 1-deoxy-D-xylulose-5-phosphate pathway (Kuzuyama 2017). Microorganisms and plants could be potential hosts for terpenoid production. Wang et al. (2018) reported recent progress in engineering industrial model strains, *E. coli* and yeasts, for terpenoid production.

3.2 Alkaloids Biosynthesis Pathway

The compounds known as alkaloids exist with nitrogen content and low molecular weight, which make them crucial for many applications. Alkaloids form through the metabolic process that uses aromatic amino acids obtained from the shikimic acid pathway as well as aliphatic amino acids produced by the tricarboxylic acid cycle (Talapatra et al. 2015).

3.3 Phenolics Biosynthesis Pathway

Phenolics are classified as secondary aromatic metabolites that arise from the shikimate/phenylpropanoid pathway or the polyketide acetate/malonate pathway, resulting in the generation of both monomeric and polymeric phenolic compounds (Naikoo et al. 2019). These compounds are vital in plants for not only combating stress but also for the regulation of physiological activities. They significantly influence defense mechanisms that function both above and below ground. Throughout their evolutionary journey, plants have synthesized a multitude of phenolic compounds to adapt to varying environmental conditions. The production of phenolics increases in plants exposed to environmental stress factors such as high light intensity, cold temperatures and drought conditions as well as heavy metallic elements (Naikoo et al. 2019).

4. Extraction of Bioactive Compounds and the Formulation of Biofungicides

Plants and microorganisms serve as a significant reservoir of chemical compounds that contribute to the formulation of innovative products (Radulovic et al. 2013, Danquah et al. 2022). Due to the significant variability among chemical compounds and the extensive diversity of microbial and plant species, it is essential to establish a standardized and cohesive methodology for the extraction of these compounds (Angst et al. 2021). The fundamental basis for multiple extraction methods depends on both solvent extraction efficiency and the usage of heat and mechanical mixing elements. Two types of methods, conventional (Figure 2) and unconventional (Figure 3),

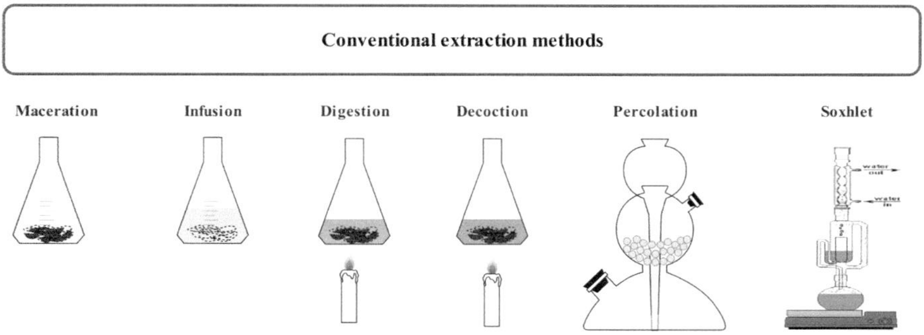

Figure 2: Multitude of traditional extraction techniques utilized for the synthesis of biofungicides.

serve to synthesize these compounds. Solvent extraction methods successfully depend on solvent selection according to Salam et al. (2019). Selection of solvents starts with evaluating the compound's polarity since it represents the most important consideration. The selection of an extraction solvent needs thorough examination of multiple crucial elements, which include solute–solvent molecular affinities together with mass transfer performance, environmental protection, costs, and human health effects (Bhadange et al. 2024). Multiple bioactive compounds illustrate extraction processes that use distinct solvents (Figure 4). Two basic types of solvents exist for chemical extraction: green solvents containing water, ethanol, glycerol, fatty oils, ionic liquids, acetic acid, isopropanol, supercritical CO_2, deep eutectic solvents, and natural deep eutectic solvents, as well as other organic solvents which include acetone, chloroform, butanol, methanol, ethyl acetate, methyl acetate, benzene, hexane, and cyclohexane (Bubalo et al. 2015). Green solvents receive their name due to their function as harmless compounds that break down quickly in nature while returning materials to nature and deriving from sustainable resources. These solvents show their signature feature as high flash points. The availability of water as a green solvent stands as the most universal and prominent among all green solvents (Clarke et al. 2018). The hydrophilic substances saponins, phenolics, and polysaccharides are efficiently separated through the non-selective nature of this solvent. Enhancements in the extraction efficiency of water can be achieved through superheating, which lowers the dielectric constant and results in better penetration capabilities. The extraction of lipophilic substances, such as essential oils, is more effectively achieved using superheated water, as the polarity of water significantly decreases within the temperature spectrum of 100 to 374°C (Cheng et al 2021). The action of ethanol is characterized by selectivity, making it suitable for the extraction of polyphenols and triterpenes. The extraction efficiency of ethanol can be adjusted by the addition of water and acid. The thermal stability of glycerol is notable, as indicated by its boiling point of 290°C (Gu and Jérôme 2010). At lower temperatures, however, its excessive viscosity contributes to lower solubility. Glycerol is applicable as an extraction solvent when utilized at temperatures greater than 60°C or alongside other co-solvents. Although it lacks efficacy in dissolving hydrophobic materials like fatty acids and oils, it is particularly selective for the extraction of polyphenolic compounds. The utilization of fatty oils as solvents is advantageous for extracting hydrophobic materials, including carotenoids, coumarins, tocopherols, and flavonoids, among

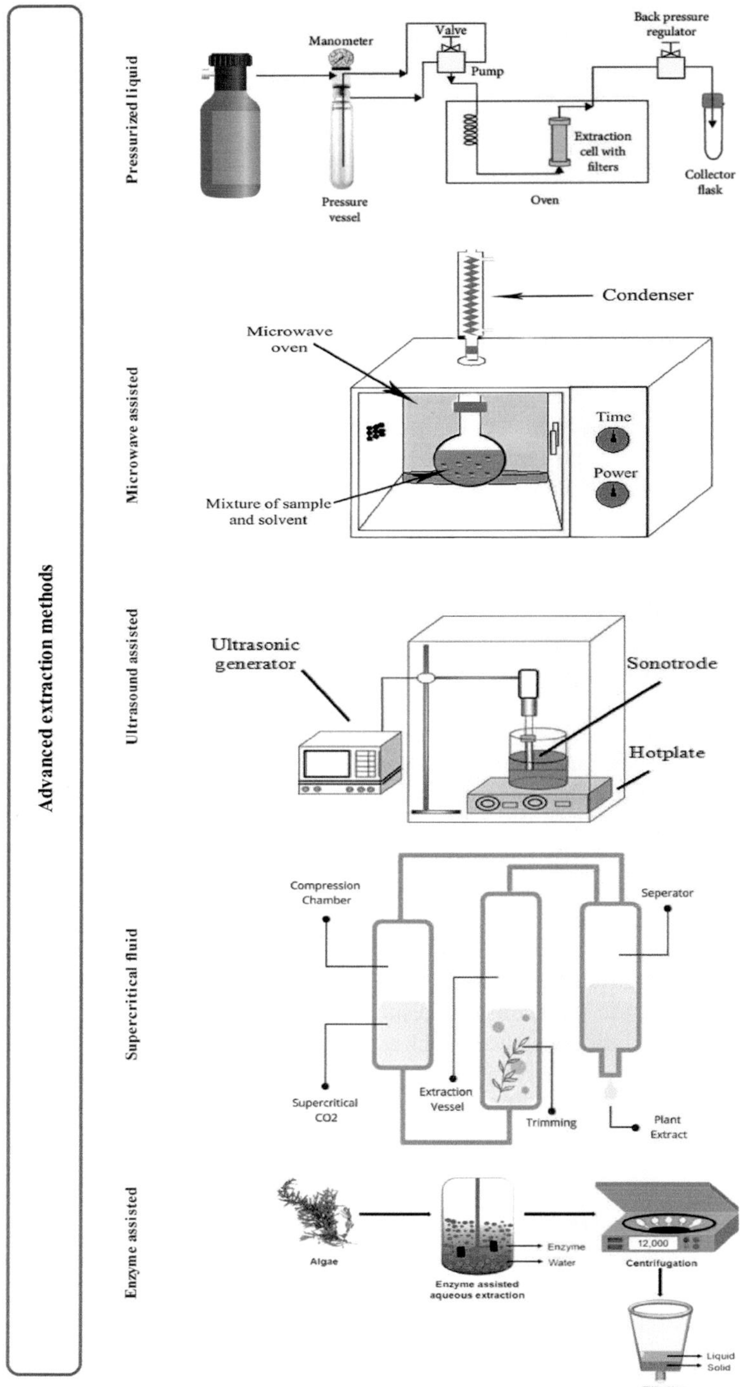

Figure 3: Advanced novel extraction techniques to produce biofungicides.

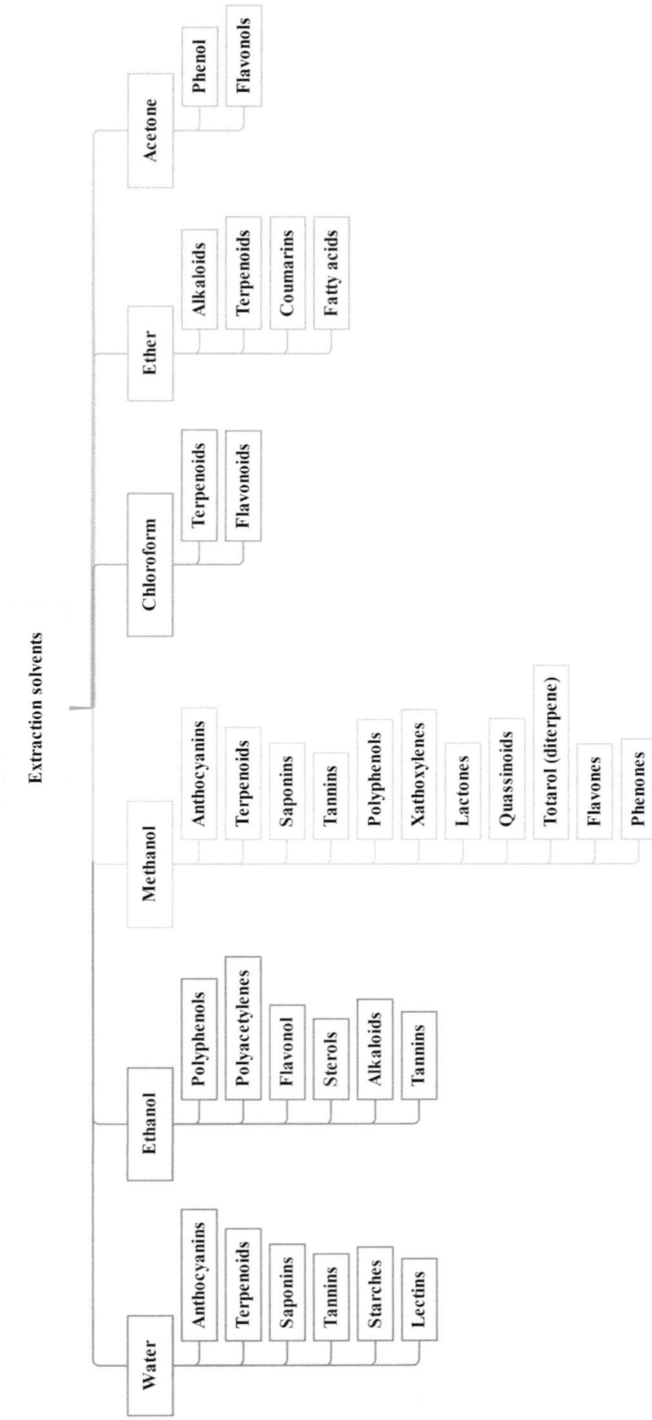

Figure 4: Detailed features of biofungicides extraction using different solvents.

others. Numerous extraction procedures use oils such as almond, soybean, olive, and sunflower as core ingredients. The extraction of phenolic compounds and anthocyanins works efficiently when using acetic acid buffers (Azman et al. 2020). Scientists view the extraction medium of isopropanol as a better choice than *n*-hexane because it successfully extracts oils, alkaloids, gums, and natural resins along with other substances (Stéphane et al. 2021). The extraction of lignans, anthocyanins, and essential oils takes place through a supercritical carbon dioxide medium (de Aguiar et al. 2023). The non-flammable and non-volatile properties of ionic liquids below 100°C define these salts, which show low melting points (Nasirpour et al. 2020). Salts convert into liquid state through melting while creating ionic solutions with excellent thermal stability, exceptional conductivity, high heat capacity, low flammability, and nearly non-existent vapor pressure. These scientific methods enable researchers to obtain diverse organic and inorganic substances, among them flavonoids, alkaloids, saponins, and lignans. A deep eutectic solvent consists of two or more pure compounds that reach a temperature point A deep eutectic solvent consists of two or more pure compounds that reach a temperature point known as the eutectic point, which exists below what ideal liquid mixtures normally exhibit. The extraction of phytochemicals containing alkaloids, flavonoids, saponins, and phenolic compounds benefits from deep eutectic solvent methods (Balaraman et al. 2021). Phytochemical extraction operations typically use type III solvents composed of choline chloride and urea at a 1:2 ratio, as well as type IV solvents made of choline chloride with zinc chloride in the same proportion (Marcus 2019). Natural deep eutectic solvents derive from different natural metabolites that include three subclasses of acids (organic, amino, and polyol) and two subclasses of choline compounds. They comprise a wide spectrum of polarities along with low volatility and minimal melting properties. Compounds can be extracted through these solvents because they dissolve substances that exist poorly in aqueous solutions. The extraction of alkaloids together with oils shows high effectiveness with acetone due to its low toxicity level when used within organic solvent categories. As a solvent, ethyl acetate exhibits no toxic effects during extraction of flavonoids and total phenolics along with multiple other substances (Riris et al. 2020). Methanol together with chloroform show toxic properties, which distinguish them from other solvents. The extraction of flavonoids together with saponins, tannins, and comparable compounds most often uses methanol as a solvent. The extraction of alkaloids and anthocyanins requires chloroform together with ethanol and fatty acids as solvent combinations according to Hasnat et al. (2023). The main function of butanol relates to the purification of fractions that contain single compounds, while it also helps extract saponins together with total phenolics and flavonoids. The solvent *n*-hexane exhibits low acute toxicity characteristics, which enable its multiple applications. Extraction through this method succeeds in retrieving flavonoids along with carbohydrates, anthraglycosides, and saponins according to Hasnat et al. (2023). Methyl acetate qualifies as a volatile solvent that results from the esterification reaction between acetic acid and methanol, and it occasionally forms as a byproduct when carboxylating methanol (de Mello et al. 2020). The substance serves as the primary agent in phytosterol and tocopherol extraction procedures. Benzene functions as a chemical substance that poses significant safety threats to users. The extraction method becomes effective for retrieving flavonoids and phytosterols and also yields alkaloids and volatile oils. The extraction of fats, waxes, and oils makes regular use of cyclohexane during these industrial processes. Increase in extraction efficiency depends heavily on optimizing the extraction conditions, including selection of solvent type, setting appropriate extraction temperature, and defining extraction duration. Research begins by breaking cell membranes, followed by extract collection. The process then follows bioactive compound purification through chromatography, ending in spectrophotometric bioactive chemical identification (Sasidharan et al. 2011).

4.1 Criteria for Choosing Solvents in Extraction Processes

A list of appropriate solvents exists in Table 2 to extract biofungicides together with their bioactive metabolites. The selection of extraction solvent needs evaluation based on key factors including selectivity, safety, cost, reactivity, recovery, viscosity, and boiling temperature (Płotka-Wasylka et al. 2017). A solvent exhibits selectivity when it extracts active components without allowing non-active substances into the mixture. Safety takes priority during solvent selection for extraction because only non-toxic and non-flammable substances qualify. Organizations should devote their resources to finding economical extraction methods. To avoid alterations in the extracted substance during the process, the solvent must stay inert. A quick and proper recovery method plus separation process of the extraction solvent from extracted materials should be performed. Smooth penetration requires solvent fluids to stay at a low viscosity level. The lowest boiling points among available solvents should be used because they help prevent thermal damage (Płotka-Wasylka et al. 2017).

4.2 Variables to Consider in the Decision-Making Process for Extraction Method Selection

Bioactive compound thermal stability determines through which extraction method a researcher should proceed (Ivanović et al. 2020). Two standard procedures for extracting heat-stable bioactive molecules include Soxhlet extraction along with microwave-assisted extraction. Bioactive compounds that break down with heat require maceration or percolation extraction methods according to Jha and Sit (2022). The characteristics of the solvent play a crucial role in the extraction process. Water selection as the solvent enables maceration to become an efficient extraction approach (Lezoul et al. 2020). Any extraction method targeted toward volatile solvents requires either the percolation extraction technique or the Soxhlet extraction method (Jha and Sit 2022). The expenses related to biofungicide production are affected by which extraction method the manufacturer uses. The main extraction method determines how costly an affordable biofungicide will be. Maceration extraction produces affordable products, whereas percolation extraction leads to more expensive biofungicide products. The extraction period depends on the method used. Microwave and ultrasound-assisted extraction methods work better

Table 2: Bioactive compounds and major extraction methods.

Bioactive compound	Solvent	Extraction method	References
Anthocyanin	Methanol (50%)	Conventional extraction backed by chromatographic techniques	Diep et al. (2020)
	Methanol	Maceration extraction	Purwaningsih et al. (2023)
Alkaloids	Aqueous ethanol	Maceration extraction	Salamah and Ningsih (2017)
	Ethanol (98%)		Justino et al. (2021)
	Methanol	Microwave-assisted extraction	Belwal et al. (2020)
	Natural deep eutectic solvents, water		Sitthisak et al. (2024)
	Ethyl acetate, chloroform with acidification	Conventional extraction techniques	Djilani et al. (2006)
	Methanoldichloromethane		Mbaoji et al. (2020)
	Petroleum ether, ethyl acetate	Maceration extraction	Harun et al. (2020)
	Methanol, ethyl acetate, 1% tartaric acid solution in methanol, dichloromethane, and cyclohexane	Pressurized liquid extraction	Dymek et al. (2021)
	Hexane, chloroform, ethanol, methanol	The combination of ordinary extraction methods with sophisticated methods of chromatographic separation	Mgbeahuruike et al. (2018)
Flavonoids	Aqueous ethanol	Conventional and microwave-assisted extraction	Alara et al. (2020)
			Salerno et al. (2014)
		Extraction by sonication	Feresin et al. (2017)
		Cold maceration after soxhlet extraction.	Arora and Itankar (2018)
	Methanol	Maceration extraction	Purwaningsih et al. (2023)
	Methanoldichloromethane	Conventional extraction	Mbaoji et al. (2020)
Tannins	Water	Pressurized hot water extraction	Mohd Jusoh et al. (2019)
		Hot water extraction (70–100°C)	Dababi et al. (2020)
	Ethanol, water	Conventional extraction	Sung et al. (2012)
Glycosides	Methanoldichloromethane	Conventional extraction	Mbaoji et al. (2020)

Table 2 contd. ...

...Table 2 contd.

Bioactive compound	Solvent	Extraction method	References
Saponins	water	Pressurized low polarity water extraction	Güçlü-Üstündağ et al. (2007)
		Hydrothermal microwave	Mohaddes-Kamranshahi et al. (2018)
Phenolics	Methanol	Microwave-assisted extraction	Belwal et al. (2020)
		Maceration extraction	Purwaningsih et al. (2023)
	Aqueous methanol (70%)	The method relies on conventional extraction together with chromatographic methods alongside mass spectrometry	Materska et al. (2018)
	Methanol (50%)	Laboratories employ combination extraction processes using chromatographic methods	Diep et al. (2020)
	Methanol (80%)	Laboratories employ combination extraction processes using chromatographic methods	Huang et al. (2007)
	Water/methanol (18:82 v:v)		Gil-Izquierdo et al. (2000)
	Methanol (100%)		Aparicio-Fernandez et al. (2018)
	Aqueous ethanol	Conventional extraction	Geng et al. (2015)
		Conventional and microwave-assisted extraction	Salerno et al. (2014)
		Microwave-assisted extraction	Alara et al. (2020)
		Extraction by sonicati	Feresin et al. (2017)
	Water	Subcritical-water extraction	Ko et al. (2020)

for short-term extraction, while maceration proves best for long-term material–solvent interaction. The required final volume performs an essential function in extraction procedures. Large-volume products that lead to the creation of tinctures use maceration technology, but concentrated products require percolation or Soxhlet extraction equipment. The preparation methods for human-completed extracts rely predominantly on maceration, but experimental preparations benefit from multiple techniques, which include maceration as one of their methods (Jha and Sit 2022).

4.3 Basic/Conventional and Emerging/Unconventional Extraction Methods

4.3.1 Maceration Extraction

The process of tissue softening uses the term maceration to describe it. The extraction method finds use for creating various dosage forms such as tinctures, extracts, and concentrated infusions. The sample immersion technique employs either water, ethanol, methanol, or ether (Morata et al. 2019, Kumar et al. 2023a) as the solvent known as the menstruum. Biochemical extraction requires leaving the mixture to rest from three to seven days while performing periodic agitation. The solid residue resulting from maceration receives pressing afterward to extract a maximum amount of liquid (Stéphane et al. 2021, Kumar et al. 2023b). The liquids proceed to clarification through filtration or decantation procedures. Petals, leaves, and flowers function as starting materials for preparing infusions. The established protocol for infusions demands water bath maceration for 15 minutes followed by 45 minutes of room temperature cooling before the solution can be utilized (Stéphane et al. 2021). The extraction technique enables researchers to obtain multiple compounds including quinones, polyphenols, tannins, flavonoids, terpenoids, saponins, polypeptides, steroids, coumarins, and glycosides through its process (Sunkar et al. 2023). Pharmaceutical organizations recognize the described extraction procedure as an efficient and secure method for crude drug extraction. The method received official approval from these pharmaceutical organizations because of its contribution to this field. The extraction efficiency depends on multiple factors that include selection of solvent, time duration, extraction temperature, as well as the composition of the sample (Sunkar et al. 2023). This method yields efficient extractions from crude materials using small amounts of solvent solution. Different bioactive compound yields increase when choosing suitable solvents for extraction. Water serves as the optimal solvent when extracting glycosides in addition to alkaloids, terpenoids, saponins, and carbohydrates. The combination of ethanol succeeds at obtaining both glycosides and carbohydrates alongside alkaloids and quinones. Methanol is particularly beneficial for flavonoids, tannins, glycosides, phenolic compounds, and amino acids (Kumar et al. 2023a). Acetone is best utilized for alkaloids, glycosides, sterols, and saponins. Ethyl acetate is effective for terpenoids, glycosides, quinones, and carbohydrates, while ether provides enhanced extraction for amino acids (Kumar et al. 2023a). Among solvents, low-polar solvents like acetone achieve a remarkable recovery rate of 100%, whereas intermediate-polar solvents, such as ethyl acetate, exhibit recovery rates ranging from 85 to 94%, and highly polar solvents, including ethanol, show recovery rates between 80–88% (Clarke et al. 2018). Employing extraction techniques that do not involve elevated temperatures is beneficial for isolating thermolabile compounds. For optimal results, a maceration time of 2 to 3 days at a temperature of 20 to 25°C is typically advised (Hidayat and Wulandari 2021).

4.3.2 Infusion Extraction

This process of extraction can be likened to maceration. The initial step involves grinding the material into a fine powder, which is then placed in a sterile container. The extraction solvent of either hot or cold temperature is added to the powdered material before the soak period begins. Botanical ingredients dissolve well when extraction methods are applied to produce this effective technique. The technique proves to be effective for making fresh extracts that need immediate use. The proportions between solvent and sample size range from 4:1 to 16:1 according to Alam and Alfath for different extraction purposes.

4.3.3 Digestion Extraction

This extraction method employs light heat during the entire extraction procedure. In the start, the extraction solvent is poured into a clean container and the powdered material is added (Hidayat and Wulandari 2021). The mixture is then placed into a water bath or oven at about 50°C. The heat decreases the viscosity of the solvent, in turn allowing for the extraction of secondary metabolites. This method has proven to be effective for materials that are highly soluble (Karaś et al. 2017).

4.3.4 Decoction Extraction

The decoction method is used to extract heat-stable, oil-soluble components like saponins, flavonoids, and tannins from resilient plant materials, bark, roots, or stems (Lezoul et al. 2020). In this method, the procedure involves boiling durable plant materials in a sufficient amount of water for 30 minutes to 2 hours and straining through 8 layers of muslin cloth. The process can be enhanced with the use of a centrifuge, which may aid in separation and extraction, after which the solvent is evaporated following the disbursement of the supernatant. Depending on the desired end quantity, the temperature, which is endothermally stable, can vary, but most commonly extraction is conducted at 65–70°C for approximately two hours. The decoction process is highly influenced by the pH, the temperature at which it is performed, and the number of ingredients. Since decoction utilizes hot water, it might favorably increase the solubility of various compounds such as alkaloids, soluble polysaccharides, flavonoids, and tannins compared to maceration (Hidayat and Wulandari 2021). The high temperature decreases glucuronidase activity, that is, the enzyme responsible for the conversion of glycosides into aglycone, so the extraction process can allow for a higher level of flavone, more specifically baicalin and wogonin, in decoction (Kumar et al. 2023a). One disadvantage of decoction, though, is finding large impurities in the crude extract, and it is not possible to extract heat-sensitive or volatile components (Lezoul et al. 2020).

4.3.5 Percolation Extraction

According to Hidayat and Wulandari (2021), percolation is commonly used in the essential oil-making process, especially in extracting oils from different seeds, barks, and leaves of plant species. The sample is immersed in excess menstruum, and then a complete seal is made on the decanter for about four hours. After that, the material is gradually introduced into the percolator for even and complete contact with the fluid. Then, the percolator is filled with liquid and sealed. This is all according to Jha and Sit (2022). Thus, the opening of the bottom outlet would depend on the establishment of a standard drop rate or percolation rate. This indicates the speed at which the sample moves through the different layers of a percolator. After this process, menstruum is added further to soak the entire content of the percolator for 24 hours. The wet mass obtained is pressed to obtain the maximum possible residual fluid; the required solvent amount is then added to achieve the desired ratio. Under percolation, the extraction rate depends on the type of solvent used and the drop rate, which then has an influence on the flow rate of the solvent. The temperature for extraction must be chosen with respect to the particular phytochemical that is to be extracted. The yield of the extract depends not only on the concentration of the solvent used but also on the part of the plant chosen for the extraction. A major limitation is the high consumption of solvent and the long time needed for extraction; this is because a high ratio of substrate to solvent is needed, and also usually 4–5 repetitions of the extraction are required, which can take from 1 to 2 days to accomplish. A significant development with the percolation process is produced by the term repercolation, which refers to using percolation multiple times. This design can have multiple (from 3 to 12) percolators connected together in a single pipeline, allowing the solvent and sample to be operated in countercurrent fashion so that fresh solvent contacts the spent plant material (Hidayat and Wulandari 2021).

4.3.6 Reflux Extraction

Reflux extraction is a solid–liquid extraction method that is performed at a constant temperature (Ji et al. 2020). The extraction procedure, or action of reflux extraction, involves both the evaporation and condensation of the solvent, which takes place over a period of 1 to 3 hours at a temperature between 60 and 80°C (Barbero et al. 2022). Reflux extraction methods are widely used for the extraction, quantification, and elucidation of various phytochemicals and essential oils from plant material. The average temperature and time required with mixed solvents such as water and ethanol is between 60 to 80°C and 80 to 120 minutes. The extraction yield from reflux methods is found to be between 70 to 90%. The boiling point of the solvent and the time of extraction are two of the most important regulated parameters in this method (Barbero et al. 2022). Depending on the set criteria, it is important to complete the extraction until either asymptotic levels are achieved or all exhaustive considerations are completely explored. This technique allows for better control of the temperature to help prevent the degradation of organic compounds. The one main drawback of this technique is that it is not capable of extracting thermolabile materials (Shi et al. 2022).

4.3.7 Soxhlet Extraction

In the Soxhlet extraction process, a small amount of dry material is placed into a thimble, which is then placed in a distillation flask containing solvents such as hexane, toluene, and petroleum ether (López-Bascón 2020). This technology is an effective process for automated continuous extraction and allows for extraction to be completed in less time and with less solvent than the alternative maceration or percolation process, and also with less chance of thermal degradation as the effect of high temperature and long extraction length is reduced. The Soxhlet extraction method uses the best of reflux and percolation, as it allows reflux and siphoning to sustain a continuous extraction of material with unspent solvent. Extraction time is approximately 24 h under extraction temperatures of 65 to 100°C. The extraction process has many variables that are pertinent to the extraction design, including the plant material type, drying rate, reduction of sample size, temperature, and solvent. The drying process is critical in how the extraction is conducted. Freshly grown plant material will contain the full complement of active enzymes, in considerable amounts of moisture, which can affect phytochemical quality. Soxhlet extraction relies on grinding (specifically, reduced size) to give the maximum surface area and therefore the best chance of effective extraction. The lower the sample size, the greater the surface area (Salam et al. 2019). Having a successful extraction means making sure the solvent you are using is inert and can be Having a successful extraction means making sure the solvent used is inert and can be removed easily afterwards. The benchmark progression is acetone, petroleum ether or ether, ethyl acetate, chloroform, methanol, ethanol, and water, which gives an indication that it is an increasingly polar number of solvents. Petroleum ether is important in extracting steroids/essential oils, chlorophyll from leaf powder, and in defatting plant materials. Methanol is used in extracting steroids and terpenoids, while acetone and ethyl acetate can extract tannins, saponins, and carbohydrates. If the sample being extracted has a high water content, in most cases it cannot be completely extracted, because water will dilute the flavonoids and expose them to degradation in the extraction method mentioned previously. The duration of the extraction process and the large volume of solvents required can make this method expensive, and the future environmental consequences of what happens to the solvent after it has been used is a point of concern. The temperature at which the extraction is performed could theoretically lead the molecule of interest to degrade per heat (Salam et al. 2019, López-Bascón 2020).

4.3.8 Pressurized Liquid Extraction

Pressurized liquid extraction makes use of high temperature and pressure to quickly and efficiently extract analytes from solid matrices using liquid solvents (Alvarez-Rivera et al. 2020). It shows great levels of solubility and improved mass transfer phenomena when employed correctly. Pressurized liquid extraction has a number of different names, which include, but are not limited to, enhanced solvent extraction, high-pressure solvent extraction, and accelerated fluid extraction (Alvarez-Rivera et al. 2020). Pressurized liquid extraction uses high pressure to keep solvents in a liquid phase despite temperatures above the boiling point of the solvent, which aids extraction (Alvarez-Rivera et al. 2020). The high temperature and pressure used in pressurized liquid extraction significantly reduce the use of solvents and the time taken to extract the analyte. The time it takes to perform pressurized liquid extraction generally ranges between 5 to 20 minutes, with extraction efficiencies ranging from 95 to 100% (Mwaurah et al. 2020). The extraction parameters for pressurized liquid extraction (PLE) are from 50 to 200°C and pressures ranging from 50 to 300 psi, leading to a great reduction in solvent use and extraction time when compared to Soxhlet extraction. PLE has been proposed as a way of using an alternative extraction method to supercritical fluid extraction, especially as CO_2 is not a good solvent for polar materials. It has been viewed as an organic version of supercritical fluid extraction, using organic solvents instead of CO_2 (Alvarez-Rivera et al. 2020). PLE uses drastically less solvent and hence is described as a green extraction method. The extraction efficiency of PLE depends on a number of variables including the pressure, temperature, and surface tension of the solvent. The use of high pressure and temperature in conjunction with low surface tension solvents results in better penetration and yield extraction efficiencies. Non-ionic surfactant solutions have gained recognition as an attractive substitute for ionic surfactant solutions when performing pressurized liquid extraction. The extraction approach of pressurized liquids efficiently extracts paraben, bronidox, and triclosan because these substances function as allergens when used as preservatives in cosmetics (Alvarez-Rivera et al. 2020). The combination of gas chromatography operated alongside pressurized liquid extraction shows accurate results when used to remove pyrethroid and organophosphorus residues from herbal plants including tea. The process of using pressurized liquid extraction effectively extracts antibiotics together with other bioactive compounds linked to suspended solid matter. The analysis of sulfonamide antibiotic residues within soil materials can be performed simultaneously through complete automatic pressurized liquid extraction systems (Alvarez-Rivera et al. 2020).

4.3.9 Liquid Gas Extraction

The batch and semi-continuous extraction process utilizes liquid gas as solvent through two main agent options, including liquefied petroleum gas made from *n*-propane and *n*-butane and dimethyl ether (Kumar et al. 2023a). The extraction procedure happens at temperatures compatible with the surroundings or pressures ranging from 200 to 1000 kPa. The extractor vessel that contains the solid sample is where the extraction takes place. The solvent introduction point can be either the upper or lower section of the extraction tanks, and it continues circulating until extraction ends (Lefebvre et al. 2021). The flow rate control uses a needle valve for the solvent while temperature maintenance occurs through heat exchangers. A storage tank devoted to extract collections receives the material before linking to other tanks in a sequence. The extraction procedure functions under two separate operational modes: isobaric or non-isobaric. Under isobaric operations, the solvent performs autonomous circulation without a compressor or pump because the pressure remains constant. The circulation pump enables the flow of liquefied gas through the material during non-isobaric operations. During the extraction process, the extracted materials undergo transit to a separating vessel. Additional volume at this moment causes a decline in pressure that enables the gas stream to efficiently vaporize due to its low boiling point. Operation at pressures of 5 bar plus allows this technique to outperform liquid gas extraction because it needs substantial pressures between 150 to 500 bar. A major challenge in the application of this technology is its financial burden, as the solvent must be stored and transported under conditions of high pressure and low temperature to remain in liquid form. This situation demands the utilization of specialized transport vehicles and specific storage facilities (Lefebvre et al. 2021).

4.3.10 Microwave Assisted Extraction

According to Destandau and Michel (2022), microwaves are part of the electromagnetic spectrum existing between radiofrequency and far-infrared waves. These electromagnetic waves operate between 0.3 to 300 GHz, sustaining wavelengths from 1 mm to 1 cm. Strong microwaves are used with solvents as an extraction method for bio-metabolite extraction. Microwave extraction is the method of food analysis in which small samples of food are extracted and dried with a single instrument while conducting the analysis of the food ingredient. This technique is used for flavor compounds and plant peel pectin extraction, along with vegetable amino acid recognition and separation of fatty acids from vegetable oils and animal fats (Yadav et al. 2020). The principle of microwave-assisted extraction is basically inside-out heating, which is a mechanism of how molecular properties are affected by ionic conduction and dipole rotation. Microwave devices operate at two distinct frequencies: 915 MHz and 2450 MHz. The 915 MHz frequency technology produces tremendous ablation zones that permit deep penetration into the electromagnetic energy. At 2450 MHz, the dipoles go through an alignment-randomization process of 4.9×10^9 times/second. The movement of molecules during the heating of water creates friction and is responsible for heating up the solution. The most distinctive feature of microwave-assisted extraction is the rapid heating of the mixture of the sample and the solvent, which makes it pertinent and quite useful to facilitate the accelerated extraction of analytes. Nor-pheynphyen stricthermic understanding is applied in this recommendation as well (Nour et al. 2021). The temperatures that can, during microwave-assisted extraction, be attained during the entire extraction process for solvents such as acetone, acetone-hexane, and dichloromethane-acetone can be 2 to 3 times the respective boiling points of these solvents, thus showing better efficiencies of extraction from the sample matrix (Kumar et al. 2023a). Solvent properties, sample-material type, target extraction components, and their respective dielectric constants have been mentioned as important contributors to the efficiency of microwave-assisted extraction (Destandau and Michel 2022). Of these, methanol, ethanol, and water are good examples of those solvents that show high capability of microwave absorption, hence heating very quickly and so requiring less time using microwave power. In spite of the above, one major limitation of this extraction method is its tendency to damage many polyphenols characterized by numerous hydroxyl substituents and those that are heat-sensitive, such as polyphenols like anthocyanins. This suggests some caution in applying this method for polymeric polyphenols such as anthocyanins and tannins (Mojzer et al. 2016).

4.3.11 Ultrasound Assisted Extraction

It is understood that ultrasonic extraction is the technique based on ultrasonic or acoustic cavitation (Wen et al. 2018). The phenomenon of acoustic coupling occurs when high-power ultrasonics are introduced in the slurry with a probe-type ultrasonic processor;

it travels through the liquid and creates high and low atmospheric pressure cycles. In the context of ultrasound-assisted extraction, the mechanical energy produced by these ultrasonic waves results in the formation of cavitation. This process facilitates thorough extraction, producing high-quality extracts in a remarkably brief period while maintaining cost efficiency. Consequently, these extracts are utilized in the food industry, pharmaceuticals, and dietary supplements. Ultrasound can be subdivided mainly into low-intensity high-frequency ultrasound (where the frequency is above 100 kHz) and high-intensity low-frequency waves (defining frequency ranging from 20 kHz to 100 kHz) (Wen et al. 2018). The low-intensity ultrasound does not change the material's physical or chemical properties, while the high-intensity shockwaves induce significant pressure and temperature shifts in the cavitation of the bubbles and disrupt the matrix structure. The phenomena of acoustic and ultrasonic cavitation give rise to extreme thermal environments characterized by abrupt temperature shifts, very rapid heating and cooling rates, high pressure differences, and extreme shear forces in the medium. These distinct phenomena include the generation of micro-jets and inter-particle collisions resulting from the collapse of cavitation bubbles against solid surfaces, causing peeling and erosion of the surface, disintegration of the particles, disruption of the cells, and sonoporation, which is the perforation of cellular walls and membranes (Wen et al. 2018). The cavitation bubble collapse in the liquid medium induces micromixing and macroturbulence. The extraction efficiency of ultrasound-assisted extraction ranges from 85 to 97% for different samples as it runs between 20 to 80°C for 10 to 60 minutes (Wen et al. 2018). The solvent-specific attributes and extraction time duration with temperature levels and ultrasonic energy output, together with gas existence levels, determine how well the extraction process works in ultrasound-assisted extraction. The yield of ultrasound-assisted extraction rises with elevated temperature but decreases when the temperature exceeds 45°C. Multiple biochemical compounds can be extracted through ultrasound-assisted extraction by using a combination of extraction solvents that include acidified water, alcohols, acetone, and water. The extraction of pectin requires acidic water solution as the preferred extraction medium. Safety measures, along with effective prevention of phenolic thermo-degradation, require the temperature to remain below 40°C. Power regulation limitations restrict ultrasound-assisted extraction because they cause poor energy distribution in extraction vessels according to Carreira-Casais et al. (2021).

4.3.12 Pulsed Electric Field Assisted Extraction

Pulsed electric field extraction utilizes non-thermal extraction that connects plant materials to direct high electric currents according to Yan et al. (2017). The extraction process takes place as either a batch setup operating between 100 to 300 V/cm or as a continuous system operating between 20 to 80 kV/cm. The short-time application of this current happens between two electrodes when placed between electrodes to maintain a specific duration of nanoseconds up to microseconds. The electrical treatment enables better conductive properties together with higher porosity and permeability of cell membranes to enable passage of encapsulated substances out of the cellular barriers. The extraction process requires a temperature interval between 20 to 50°C, while its execution spans from 5 minutes to 48 hours. A level of extraction result between 76% and 85% can be expected when performing the method. The level of electroporation achieved in pulsed electric field extraction depends heavily on both pulse parameters and cell treatment medium characteristics, as well as cellular physicochemical properties. The extent of electroporation depends on different pulse parameters, which include electrical field intensity combined with pulse length and direction as well as the number of pulses released. Tissue electroporation depends significantly on three properties of tissues: size, thickness, structural characteristics, and ionic mobility and water content (Martínez et al. 2020).

4.3.13 Enzyme Assisted Extraction

The proficient catalyst capabilities of enzymes support the extraction, modification, and synthesis of complex bioactive natural substances (Marathe et al. 2019). Enzymes demonstrate the key capability to create precise reactions with regioselective properties, which enhances the efficiency of enzyme-based extraction methods. Scientists believe that enzyme-based extraction represents an environmentally friendly solution that substitutes conventional solvent-based systems while addressing the resolution of environmental challenges (Aparicio and Marina 2023). Plants mainly use cellulases together with hemicellulases and pectinases for extracting bioactive compounds (Gligor et al. 2019). The enzymatic breakdown of the plant cell wall by enzymes leads to tissue deterioration, which weakens plant tissue structure. The extraction process using enzyme technology takes place at 33 to 67°C, while durations span from 20 to 220 minutes (Gligor et al. 2019). The total extract contains between 38 to 52% of extraction products. The hydrolytic characteristics are strongly influenced by enzymes, and the impact depends on composition and enzymatic activity together with the types and concentrations of constituents. The activity levels of enzymes increase proportionally to rising temperatures, thus improving their operational efficiency. Higher temperature levels decrease the viscosity of extraction solutions, resulting in better bioactive compound solubility. The optimal conditions for enzymatic activity require a pH level of 4.5 because it alters protein shapes and enhances substrate capture. Enzyme expenses constitute a technological challenge for extracting various samples from different sources. Current enzyme mixtures do not succeed in totally breaking down cell walls, therefore limiting the quantity of extracted biochemical compounds (Marathe et al. 2017). The extraction rates can change widely because of environmental conditions that include temperature along with nutrient availability and dissolved oxygen concentration.

4.3.14 Supercritical Fluid Extraction

Supercritical fluid extraction requires a sequential approach for extracting substances from matrices by using supercritical fluids followed by separating those biochemicals from the supercritical solvent after pressure reduction and temperature elevation (Da Silva et al. 2016). Supercritical fluid extraction functions through the use of fluids at elevated temperatures superior to their critical threshold (between 40°C and 80°C) and pressure ranges from 10 MPa to 35 MPa. Supercritical fluid extraction presents three distinguishing advantages: speedy operations, purity, and efficient natural product extraction capabilities across multiple matrix sources (Wrona et al. 2017). All essential equipment in supercritical fluid extraction needs engineering TLC to tolerate high pressure levels of 50 MPa. The needed instrumentation depends on the liquid or solid nature of the samples. During countercurrent supercritical fluid extraction, the column receives sample materials through its upper section, and the pressurized solvent arrives from the lower segment. The distribution of components occurs as the separation column runs a counter-current operation between solvent and liquid sample components. For obtaining suitable contact durations between sample and solvent, it is crucial to regulate both the sample position in extraction columns and consider extraction component separation factors (Uwineza and Waśkiewicz 2020). The most frequently used green solvent in supercritical fluid extraction is carbon dioxide (CO_2) at 31°C under 7.38 MPa pressure. Industry considers carbon dioxide a harmless substance that maintains a low cost and provides environmental protection. Supercritical CO_2 features high diffusivity along with solvent strength, making it a favorable choice across multiple fields (Wrona et al. 2017). Carbon dioxide exists as a gas at room temperature under atmospheric pressure conditions, thus enabling solvent-free extractions of materials. The extraction rate depends heavily on several vital parameters that include extraction pressure, temperature, time, solvent-to-material flow ratio, and solvent flow rate (Uwineza and Waśkiewicz 2020). Supercritical fluid density decreases through temperature elevation, which affects both its extraction ability and potentially results in heat-sensitive biochemical destruction (Uwineza and Waśkiewicz 2020). Pressurization of supercritical fluids at constant temperature will increase their density, which then leads to enhanced extractive capabilities. Raising the mass transfer rate of supercritical fluid results in proportional increases of both the sample-to-solvent ratio and flow rate. The increased flow rate of the solvent brings about shorter time periods during which the solvent remains inside the system. Through supercritical fluid extraction, many different industrial applications find successful applications, particularly in chemical and fuel production, food processing, and pharmaceutical manufacturing. A considerable majority of scientists view supercritical fluid extraction as the preferred method to extract bioactive compounds because it produces end products free of toxic residues. An essential limitation of this method occurs because it works under elevated pressure requirements and simultaneously generates lower extract yields than liquid solvent-based methods (Da Silva et al. 2016, Wrona et al. 2017).

5. Screening Approaches

The crude extract or active fraction obtained from plant and microbial materials can thereafter be subjected to biochemical screening using appropriate tests. Preliminary screenings are geared toward revealing the presence of bioactive compounds in extracts (Khade et al. 2023). A series of qualitative analyses have been developed in efforts to detect the presence of different compounds such as alkaloids, flavonoids, tannins, saponins, flavones, sterols, terpenes, glycosides, proteins, carbohydrates, and fats (Alamgir 2018).

5.1 Analysis for Alkaloids Compounds

5.1.1 Dragendorff's Reagent Screening Test

An amount of 1 mL of the extract was taken and placed into a test tube for further analysis. Next, 1 mL of potassium bismuth iodide solution, known as Dragendorff's reagent, was added into the tube while mixing the contents until they reached full homogeneity. The development of an orange-red precipitate serves as evidence for the presence of alkaloids (Chiru et al. 2020).

5.1.2 Wagner's Reagent Screening Test

A 1 mL extract quantity was taken for analysis into a test tube. The tube received 1 mL of potassium iodide (Wagner's reagent) treatment and was mixed thoroughly immediately afterward. The emergence of a reddish-brown precipitate serves as an indicator of alkaloid presence (Shaw et al. 2024).

5.1.3 Mayer's Reagent Screening Test

The extracted 1 mL volume went into a test tube to proceed with laboratory examinations. The test tube received 1 mL of Mayer's reagent containing potassium mercuric iodide solution that was properly mixed with the contents. A whitish or cream precipitate formed during the test procedure indicates that alkaloids are present in the sample (Shaw et al. 2024).

5.1.4 Hager's Reagent Screening Test

An amount of 1 mL of the extract was taken and placed into a test tube for further analysis. Following this, 1 mL of saturated ferric solution, referred to as Hager's reagent, was added into the tube and the contents were mixed. The production of a yellow precipitate denotes the existence of alkaloids (Kumar et al. 2020).

5.2 Analysis for Glycosides Compounds

5.2.1 Modified Borntrager's Screening Test

One gram of crude extract was accurately measured before placing it in a test tube where it was dissolved with 5 mL of dilute hydrochloric acid. The mixture received 5 mL of 5% ferric chloride solution. The technician mixed the test tube before placing it inside a heated water chamber. A heating process at boiling temperature lasted for 10 minutes before the mixture cooled down and filtration occurred. A subsequent benzene extraction took place as part of the method. The addition of ammonia solution in equal volume to the benzene layer revealed anthraquinone glycoside presence when the solution turned pink according to Ramya et al. (2019).

5.2.2 Legal's Screening Test

The analysis required extraction of a 1 mL solution followed by addition of 1 mL sodium nitroprusside and small volumes of sodium hydroxide solution. The detection of glycosides becomes visible by observing a pink-to-blood-red precipitate after the mixture has been mixed (Maheshwaran et al. 2024).

5.2.3 Keller's Reagent Screening Test

The researchers prepared a 2 mL extract diluted with equal parts of water. The mixture received 0.5 mL lead acetate prior to shaking, followed by filtration of the solution. After extraction of the filtrate with chloroform, the operator evaporated the liquid obtained and produced a residue that dissolved in glacial acetic acid. The solution received a small amount of ferric chloride addition. Two milliliters of sulfuric acid received the entire mixed solution. When the reaction solution shows a reddish-brown layer that transforms into a bluish-green formation, it confirms the presence of digitoxose compounds derived from digitoxin (Bartnik and Facey 2024).

5.3 Analysis of Steroids Compounds

Libermann burchard's reagent is a screening test for analysis of steroids. The procedure applies to extract substances into alcohol. The drying process using evaporation starts the method before chloroform completes the second extraction. The test tube receives a few drops of acetic anhydride before slow addition of sulfuric acid begins from the tube's wall. A violet to blue ring at the interface between the two liquids acts as a diagnostic signal of steroid occurrence (Wutsqa et al. 2021).

5.4 Analysis for Terpenoids Compounds

Salkowski's reagent screening is a test for analysis of terpenoids. The researcher added 1 mL extract solution to which 2 mL chloroform was then introduced. The mixture received agitation before filtering out the contents. The mixture received several drops of concentrated sulfuric acid while being mixed before it was allowed to rest. The formation of a golden-yellow precipitate serves as an indicator of terpenoid presence (Kale 2020).

5.5 Analysis for Tannins Compounds

5.5.1 Gold Beater's Skin Screening Test

Ox leather provides doctors with the necessary material to obtain Gold Beater's Skin. Preparations start when skin pieces absorb 2% hydrochloric acid solution followed by washing with distilled water. An extract solution bathes the skin for five minutes before another rinse with distilled water follows the procedure. After distilling the solution, the material receives 1% ferrous sulfate treatment. When tannins are present, the solution develops brown or black coloring (Singh 2019).

5.5.2 Gelatin's Screening Test

The evaluation began with the selection of 1 mL of extract material which was placed into a test tube. Subsequently, the mixture received a 1% gelatin solution with sodium chloride, which was followed by agitation. White precipitate formation in the solution indicates the existence of tannins (Singh 2019).

5.6 Analysis for Flavonoids Compounds

5.6.1 Shinoda's Screening Test

The experiment required transfer of 1 mL extract into a test tube. Several concentrated hydrochloric acid drops were added first before introducing 0.5 mg of magnesium turnings to the mixture, which was then mixed. A pink color appearance indicates that flavonoids are present in the solution (Win et al. 2019).

5.6.2 Lead Acetate Screening Test

The determination of flavonoids in test samples began with measuring 1 mL of extracted solution, which was added to a test tube. After placing drops of lead acetate into the solution, the substances were mixed through shaking. The formation of a yellow precipitate serves as an indicator of flavonoids (Win et al. 2019).

5.6.3 Ferric Chloride Screening Test

The addition of several drops of neutral ferric chloride solution to 1 mL extract solution was performed. The appearance of a blackish-red color indicates that flavonoids exist in the sample (Win et al. 2019).

5.6.4 Alkaline Reagent Screening Test

The tester obtained one milliliter of extract for measurement in a test tube. After measurement, a few sodium hydroxide solution drops were added, which required shaking. A deep yellow color in the solution at first indicates flavonoids according to Ramya et al. (2019), but the addition of dilute acid leads to the color turning colorless.

5.7 Analysis for Phenolic Compounds

5.7.1 Ferric Chloride Screening Test

The analysis started with the addition of 1 mL extract solution to a test tube. A 1% gelatin solution mixed with sodium chloride was added to the mixture, after which vigorous shaking took place. When phenolic compounds exist in the sample solution, it produces a bluish-black color change (Singh 2019, Bashmil et al. 2021).

5.7.2 Lead Acetate Screening Test

The experimental solution required 1 mL in a test tube. 1 mL of alcohol and 1 mL of extract solution were added, then the mixture received 20% sulfuric acid as a dilution. The researcher added the sodium hydroxide solution. The red-to-blue color change reaction functions as a phenol detection indicator (Singh 2019).

5.7.3 Gelatin's Screening Test

The test tube received a solution of sample extract mixed with 2 mL of 1% gelatin solution before being followed by agitation. The formation of a white precipitate shows the existence of phenolic compounds in the sample (Banu et al. 2015).

5.7.4 Mayer's Reagent Screening Test

A solution of potassium mercuric iodide at 1 mL volume was used to conduct the analyses in the acidic conditions of the extract. A white precipitate formation during the test shows that phenolic compounds exist in the sample (Nimsarkar et al. 2021).

5.8 Analysis for Proteins

5.8.1 Biuret Reagent Screening Test

A particular quantity of extract served to generate a sodium hydroxide solution containing 4% of the compound. The addition of 1% copper sulfate came after 4% sodium hydroxide solution was applied. The formation of violet color indicates the existence of peptide linkages in the extract (Singh 2019).

5.8.2 Ninhydrin Reagent Screening Test

The experiment used a 1 mL solution placed in a test tube. The solution received 0.25% ninhydrin reagent (tricyclic 1,2,3-trione), followed by thorough shaking. The mixture boiled for a few minutes under heat application. Protein measurement depends on the blue color reaction (Singh 2019).

5.8.3 Xanthoproteic Reagent Screening Test

One milliliter of extract was taken into a test tube. Then, a few drops of Xanthoproteic reagent (hot concentrated nitric acid) were added and shaken. It reacts with aromatic amino acids such as phenylalanine, tyrosine, and tryptophan. The resulting yellow color indicates the presence of proteins (Singh 2019).

6. Techniques for Fractionation and Purification

The extract resulting from the previously mentioned methods contains complex components that represent a variety of natural products. Further separation and purification are essential to obtain the active fraction or pure natural products. Separation is guided by the physical or chemical properties that distinguish one natural product from another. Fractionation refers to the method of separating extracts into distinct fractions. This process further divides these fractions into segments that contain multiple compounds. The procedure persists until a pure compound is successfully isolated. In instances where multiple solvents are necessary for fractionation, they should be introduced in a sequence that reflects increasing polarity. Fractionation techniques can be fundamentally categorized into physical and chemical methods (Jha and Sit 2022). This extraction method relies largely upon the functional groups possessed by the compound being separated in a given mixture. Chemical reactions are then normally applied with suitable reagents to effect separation or purification of the compounds (Singh 2019). Various physical techniques employed for the separation of compounds from mixtures encompass the use of separation funnels, fractional distillation, fractional liberation, fractional crystallization, sublimation, and chromatographic methods (Lefebvre et al. 2021, Jha and Sit 2022).

6.1 Separation Funnel Assay

A separatory funnel is a fundamental device frequently utilized for the purpose of liquid-liquid extractions (Srivastava et al. 2021). It comes in a variety of shapes and sizes, although the operational method remains uniform. These funnels are designed with a tap at the bottom, which allows for the drainage of their contents. Glass taps are lubricated with a small amount of grease, and a plastic screw is used to secure the tap. Teflon taps, however, do not require grease and are held in position by a nut and washer. In order to carry out the extraction process, both liquid phases must be introduced into the separatory funnel, ensuring that the tap remains closed. It is essential that the funnel is not filled beyond three-quarters of its total height to facilitate the effective mixing of the two layers. After adding the liquids, a stopper should be placed in the funnel. This technique facilitates the adjustment of the tap while simultaneously securing both the stopper and the tap firmly in position. Subsequently, the solvents are subjected to vigorous shaking for a duration of 20 to 30 seconds. This shaking action significantly enhances the surface area interaction between the two liquids, thereby promoting a more rapid establishment of equilibrium. Frequent venting of the separatory funnel is necessary during the shaking process to relieve excess vapor pressure. This is performed by inverting the funnel and opening the tap. It is crucial to point the stem of the funnel into a fume hood during venting to avoid any risk of spraying the contents onto individuals nearby. The process of venting holds significant relevance when sodium bicarbonate is employed to extract acidic solutions, primarily due to the substantial release of CO_2. Upon completion of the shaking period, the funnel is vented one last time and secured with a ring clamp to facilitate the separation of the layers. The separation of the liquids is achieved by carefully drawing off the lower layer, ensuring a slow approach as the interface nears the tap. Liquid-liquid extraction typically involves an aqueous phase as a fundamental component. It is important to note that the presence of a solute can occasionally alter the anticipated relative densities of the two liquids. To identify the respective layers, one can add water dropwise to the mixture in the separatory funnel and observe in which layer the drops become miscible. A crucial piece of advice is to retain both fractions until there is absolute certainty regarding the location of the desired product (Srivastava et al. 2021).

6.2 Fractional Distillation

Fractional distillation is employed for the separation of liquid mixtures with closely related boiling points, specifically those differing by less than 70°C. In this process, the liquid mixture is heated, and the vapors ascend through a glass apparatus known as a "fractionating column," where they undergo separation. This column is strategically positioned between the flask containing the mixture and a "Y" adaptor, enhancing the efficiency of the separation process. Because of the glass beads in such a glass fractionating column, it is more effective than simple distillation. These glass beads act like "theoretical plates." They favor the condensation and re-evaporation of vapor, thus effecting many distillations. Each theoretical plate corresponds to a single vaporization-condensation cycle, akin to one instance of simple distillation. As a result, more volatile components ascend towards the top of the column, while those with higher boiling points remain at the bottom, thereby achieving improved separation. Ultimately, the vapor reaches the condenser, where it is cooled and subsequently collected in a designated vessel. Typically, a Vigreux-type column, such as the Chemglass CG-1231 series, is utilized, with its optimal length determined by the boiling point proximity of the target substance to the impurities. It is important to note that an increase in the length of the Vigreux column may lead to a decrease in recovery efficiency. Traditionally, fractional distillation has been utilized for the purpose of separating the elements within volatile mixtures, especially those associated with volatile oils (Sunkar et al. 2022).

6.3 Fractional Crystallization

Crystallization stands as a highly effective purification technique within the realm of process engineering. The method of fractional crystallization operates as a sequential separation process that utilizes the transition between liquid and solid phases. This technique allows for the division of multicomponent mixtures into finely defined fractions, ultimately achieving high purity levels for specific components, thanks to the selectivity inherent in solid-liquid equilibria. Fractional crystallization has additional applications in the separation of organic materials like isomer separations, tar chemical mixtures, organic acids, monomers, and so on. The technique utilizes the varying solubility of the constituents within a mixture when placed in a specific solvent. Occasionally, derivatives of these components are utilized, such as picrates for alkaloids and osazones for sugars (Sunkar et al. 2022).

6.4 Sublimation Assay

Sublimation is the scientific term for this phenomenon wherein a solid changes directly into a gas without undergoing a liquid state as a transitional phase. Camphor and certain volatile oils can be heated and separated in this manner and immediately converted into a gas. Such a principle may be used during caffeine extraction from tea, and purification processes of crude extracts may yield materials (Krakowska-Sieprawska et al. 2022).

6.5 Liberation Assay

Certain groups of compounds can be isolated through fractional liberation from a mixture. For instance, when a mixture containing alkaloid bases is agitated with a sodium hydroxide solution, the phenolic alkaloids will be extracted in the form of salts (Krakowska-Sieprawska et al. 2022).

6.6 Chromatographic Techniques

This methodology involves various separation techniques (Figure 5). A common aspect of these techniques is that the components of the sample mixture are allocated between two phases, with one phase remaining stationary and the other phase moving through or over the surface of the stationary phase (Czaplicki 2013, Srivastava et al. 2021). The flow of the mobile phase causes the sample components to migrate differentially. In other terms, chromatography is characterized by the distribution of a compound between two phases: a mobile phase that flows over a stationary phase (Janson 2011). The separation mechanism is fundamentally rooted in the unique distribution characteristics of the compounds involved (Srivastava et al. 2021). The stationary phase performs either as a solid or liquid structure, whereas the mobile phase exists in liquid or gas categories (Janson 2011). Two specific combinations cannot be achieved by chromatographic methods: gas-gas and solid-solid (Sjödahl et al. 2020). The extraction method known as chromatography depends upon the mobile phase or solvent system and the stationary phase that may be silica gel or

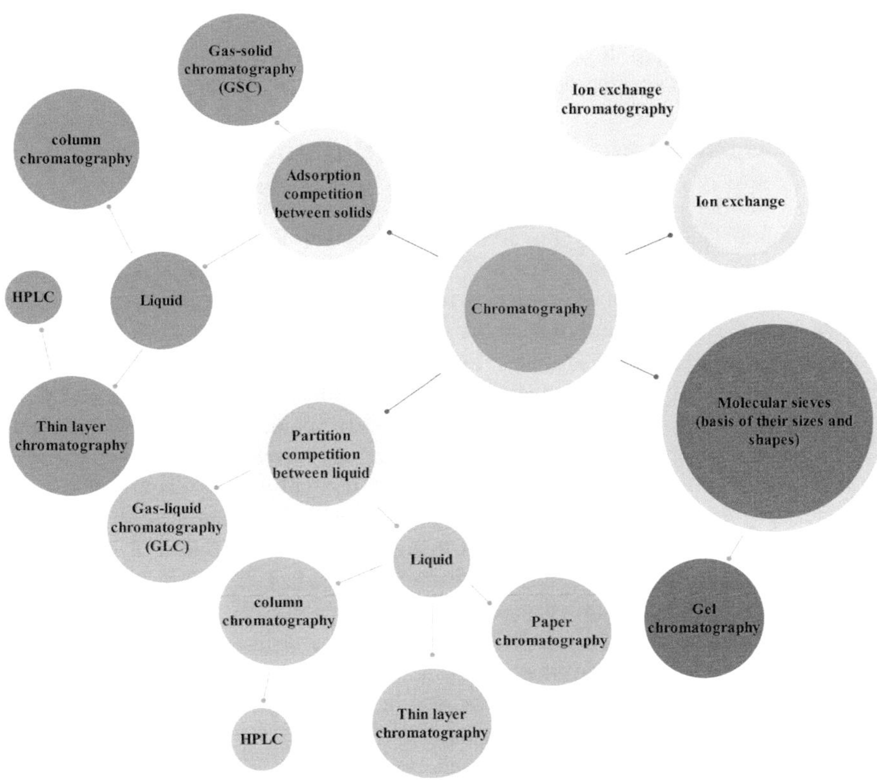

Figure 5: The main features of different chromatographic techniques utilized for biofungicide analysis.

Sephadex, alongside calcium sulfate as the binding agent (Janson 2011). Silica gel represents an optimal choice for separating amino acids, sugars, fatty acids, lipids, and alkaloids (Ingle et al. 2017). The primary purpose of the Sephadex material is to separate proteins and amino acids (Martínez-Maqueda et al. 2013). Aluminum strengthens the process that separates vitamins from carotenes, phenols, steroids, and alkaloids simultaneously. The ability of cellulose powder to separate amino acids, food dyes, and alkaloids from each other matches its performance with Celite, as this substance effectively separates organic cations and steroids (Tripathi et al. 2022). Chromatographic separation methods use adsorption as well as partition mechanisms, and affinity or ion-exchange and size-exclusion processes (Janson 2011). Common chromatographic techniques encompass paper chromatography (PC), thin-layer chromatography (TLC), column chromatography (CC), liquid chromatography (LC), gas chromatography (GC), and high-performance liquid chromatography (HPLC).

7. Identification and Analysis of Bioactive Compounds

Multiple procedures serve for determining compounds extracted from biological materials during identification processes. The identification process includes functional group analysis along with the assessment of multiple bond connections, cyclic patterns, and hydrogen and carbon atomic configuration until reaching full structural definition (Rehman et al. 2020). The analysis uses mass spectrometry (MS) along with ultraviolet spectroscopy (UV), nuclear magnetic resonance spectroscopy (NMR), and infrared spectroscopy (IR).

8. Development and Marketing of Biofungicide Products

The use of biological control methods for plant disease management in commercial sectors remains restricted by factors that include variable effectiveness between different environmental conditions and agent specificity to target host organisms (Collinge et al. 2022). Engineers need to develop new biological agents through modern biotechnological approaches in order to enhance stability, efficiency, and resilience (Youssef 2021). Multiple fundamental requirements need attention to turn biological control agents into commercial products successfully (Kagot et al. 2019). Biological control agent commercialization requires a multi-step process, which becomes the main cost driver in this industry. A systematic procedure includes microorganism isolation in pure cultures, together with agent identification and characterization, followed by formulation development, mass production, efficacy testing, storage stability inspection, identification of industrial partners for business operations, handling human and environmental safety concerns, and market registration procedures (Köhl et al. 2019). The global market has introduced many biologically derived products that aid in the management of fungal plant pathogens (Lahlali et al. 2022). Various mineral and organic carriers serve as base ingredients to create wettable powders, granules, dusts, and liquid formulations that are either oil-based or aqueous (Bharti et al. 2020). The market has accepted several products that have received patents, while experts evaluate their potential for commercial sales and manufacturers approve these agents primarily for use in horticulture (Lahlali et al. 2022). Table 3 presents specific commercial fungal biological control agents that fight plant fungal diseases with their corresponding technical characteristics.

Table 3: List of certain commercial fungal biological control agents employed in the management of plant fungal diseases.

Sources	Trademark	Pathogen(s)/Plant disease	Producer/Country
Ampelomyces quisqualis	AQ10® Bio Fungicide	Powdery mildew	Ecogen Inc., USA
Pseudozyma flocculosa (*Anthracocystis flocculosa*)	Sporodex L		Plant Products Co., Canada
Candida oleophila	Aspire	Post-harvest diseases (green and blue molds)	Ecogen Inc., USA
Geotrichum candidum			
Paraphaeosphaeria minitans (*Coniothyrium minitans*)	Contans WG	Sclerotinia diseases: *Sclerotinia sclerotiorum and Sclerotinia minor*	Prophyta Biologischer Pflanzenschutz GmbH, Germany
	KONI		
Clonostachys rosea (*Gliocladium catenulatum*)	Primastop	Damping-off disease, seed rot, root and stem rot, wilt and certain fungal foliar diseases (Powdery mildew, black mildew, sooty mold, and phyllode spot) caused by *Rhizoctonia, Phytophthora, Pythium, Verticillium, Fusarium, Didymella, Botrytis,* Alternaria, *Penicillium, Cladosporium, Helminthosporium,* and Plicaria	RegWest Co., USA
			Kemira Agro OY, Finland
			Kemira Agro OY, Chaina
	Prestop		AgBio, Inc., USA
			Danstar Ferment Ag., Switzerland
Fusarium oxysporum	Fusaclean	Wilt diseases	SIAPA, Italy
	Biofox C		Natural Plant Protection, France
Phlebiopsis gigantea	Rotstop®	Root rot fungal pathogens	Kemira Agro Oy, Finland
Trichoderma virens (*Gliocladium virens*)	Soilgard®	Soil-borne fungal pathogens; *Pythium* and *Rhizoctonia* species	Certis, USA
Trichoderma harzianum	RootShield®	Root rot fungal pathogens; *Pythium, Fusarium, Rhizoctonia solani, Thielaviopsis* and *Cylindrocladium* species	BioWorks Inc., USA
Trichoderma harzianum	Trichodex	Gray mold caused by *Botrytis cinerea* (Ascomycota); *Sclerotinia* and *Colletotrichum, Rhizoctonia* species	Makhteshim Agan,, USA
Trichoderma harzianum	Binab T	Root rot diseases and pruning wounds affecting timber trees, ornamentals, and shade trees	BINAB Bio-Innovation AB, Sweden
Trichoderma polysporum			
Trichoderma viride	*Trichoderma viride* Trieco	Soil-borne fungal pathogens: *Fusarium, Pythium, Rhizoctonia, Phytophthora, Verticillium, Rhizopus*	Ecosense Lab (I) Pvt. Ltd., India

9. Formulation Technologies for Biofungicides

The development of biofungicide formulations depends on both microorganism stabilization and field performance improvement processes. Biofungicides exist as two major formulation types pertaining to physical state classification: they can be either dry or liquid. Dry biotechnology products consist of dusts (DP), wettable powder (WP), powders for seed dressing (DS), microgranules (MG), granules (GR), and water-dispersible granules (WG). At the same time, the liquid category contains suspension concentrates (SC), capsule suspensions (CS), oil dispersions (OD), suspo-emulsions (SE), emulsions, and ultra-low volume formulations (Singh et al. 2016). Common formulation types include:

9.1 Liquid Formulations

A large number of farmers prefer liquid formulations because these formulations allow easy application, uniform distribution, and compatibility with today's agricultural equipment, including foliar sprays and drip irrigation systems. The sustainable fungicide solution known as liquid biofungicides shows efficient results at both the laboratory level and across real farming conditions. Research on formulation technology and field adaptability will improve their commercial potential (Singh et al. 2016). Table 4 shows laboratory-scale studies, and Table 5 shows large-scale field trials on liquid biofungicides.

9.1.1 Suspension Concentrates (SC) Composition

SCs are stable suspensions of fine particles containing active ingredients in water. They are designed for dilution before application. Formulations of *Trichoderma harzianum* SC are commonly employed for the suppression of soil-borne pathogens such as *Rhizoctonia solani* and *Pythium* spp.

9.1.2 Emulsifiable Concentrates (EC) Composition

ECs consist of active ingredients dissolved in organic solvents with emulsifiers, forming an emulsion when mixed with water. *Bacillus subtilis* EC formulations have been developed to manage foliar diseases such as powdery mildew.

9.1.3 Soluble Liquids (SL) Composition

SLs are formulations where active ingredients are completely dissolved in water, allowing for easy mixing and application. *Pseudomonas fluorescens* SL formulations are employed to combat damping-off diseases in seedlings.

9.1.4 Nanoemulsions Composition

Nanoemulsions are extremely fine oil-in-water dispersions differing by more than 88% from the conventional macroemulsion. The droplet size of nanoemulsions is usually in the range of about 20–200 nm, which improves bioavailability and stability of the active components. Nanoemulsion formulations of essential oils, such as thyme oil, have demonstrated efficacy against *Botrytis cinerea* in strawberries.

Table 4: Laboratory-scale studies on liquid biofungicides.

Biofungicide	Target pathogen	Key findings	References
Bacillus amyloliquefaciens (liquid SC)	*Fusarium oxysporum*	Reduced tomato wilt by 78% *in vitro*.	Mousa et al. (2021)
Trichoderma asperellum	*Sclerotinia sclerotiorum*	92% inhibition in Petri dish assays.	Tomah et al. (2023)
Pseudomonas chlororaphis (oil-based EC)	*Botrytis cinerea*	Suppressed gray mold in strawberries by 85%.	Nicot et al. (2016)
Streptomyces griseoviridis (SL)	*Pythium ultimum*	Inhibited damping-off in cucumbers by 90% in lab tests.	Carisse et al. (2003)
Beauveria bassiana (oil-based EC)	*Colletotrichum gloeosporioides* (Anthracnose)	80% reduction in mango fruit infections.	González-Fernández and Hormaza (2020)
Pantoea agglomerans (SC)	*Erwinia amylovora* (Fire blight)	Controlled bacterial infections in apple shoots by 75%.	Pusey et al. (2011)
Ampelomyces quisqualis (SL)	*Podosphaera xanthii (Powdery mildew)*	70% suppression in melon leaves.	Sundin et al. (2009)
Burkholderia spp.	*Ralstonia solanacearum (Bacterial wilt)*	88% inhibition in tomato roots.	Nion and Toyota (2008)

SC=Suspension Concentrates; EC= Emulsifiable Concentrates; SL=Soluble Liquids

Table 5: Large-scale field trials on liquid biofungicides.

Biofungicide	Crop	Results	References
Bacillus subtilis QST 713 (Serenade®)	Wheat, barley, oats and triticale	Reduced powdery mildew by 70% (equivalent to chemical fungicides).	Matzen et al. (2019)
Pseudomonas fluorescens (liquid SL)	Rice	Controlled *Rhizoctonia solani* by 65% in field trials.	Nur Mawaddah et al. (2023)
Streptomyces spp. (Mycostop®)	Cucumber	Reduced *Fusarium* wilt by 60% in greenhouse trials.	Bubici, (2018)
Trichoderma harzianum (SC)	Tomato	Decreased *Fusarium* crown rot by 55% in open-field conditions.	El-Mohamedy et al. (2014)
Coniothyrium minitans (EC)	Soybean	Reduced *Sclerotinia* stem rot by 50% in large-scale applications.	Li et al. (2006)
Bacillus velezensis (SC)	Wheat	Suppressed *Fusarium graminearum* (head blight) by 60%.	Ma et al. (2024)
Gliocladium virens (SL)	Cotton	Controlled *Rhizoctonia solani* (damping-off) by 45% in field tests.	Lewis et al. (1991)

SC=Suspension Concentrates; EC= Emulsifiable Concentrates; SL=Soluble Liquids

9.2 Dry Formulations

Dry formulations of biofungicides offer advantages in storage stability, ease of transport, and extended shelf life compared to liquid formulations. They are widely used in agriculture for seed treatment, soil application, and foliar protection. Below is a detailed classification with global case studies. Solid biofungicides are critical for sustainable agriculture, with proven efficacy in global trials. Innovations in formulation technology are expanding their use against soil-borne and foliar pathogens (Singh et al. 2016). Table 6 shows laboratory-scale studies, and Table 7 shows large-scale field trials on solid biofungicides.

9.2.1 Wettable powders (WP)

With a coating of powdery particles so fine that they will easily pass through a housefly mesh, WPs are a class of agrochemical formulations in which water is meant for further activation and application. This powder is made of active chemical constituents mixed with an inert carrier substance to ensure stability during storage and controlled dispersion upon hydration. *Trichoderma viride* WP is commonly used to control root rot diseases in various crops.

9.2.2 Granules (GR)

Granular formulations are dry preparations where the active ingredient is coated onto or absorbed into granule carriers. *Bacillus thuringiensis* GR formulations are applied to soil to manage fungal pathogens affecting root systems.

Table 6: Laboratory-scale studies on solid biofungicides.

Biofungicide	Target Pathogen	Key Findings	References
Trichoderma spp. WP	*Sclerotium rolfsii*	85% inhibition in peanut crops under controlled conditions.	Ram et al. (2020)
Bacillus amyloliquefaciens GR	*Fusarium graminearum*	Reduced wheat head blight by 70% *in vitro*.	Jamal et al. (2017)
Streptomyces spp. microcapsules	*Pythium* spp.	Enhanced UV stability and prolonged antifungal activity.	Pacios-Michelena et al. (2021)
Pseudomonas fluorescens WP	*Rhizoctonia solani*	Inhibited hyphal growth by 78% in lab assays.	Nagendran et al. (2019)
Beauveria bassiana DS	*Colletotrichum* spp.	Reduced anthracnose in chili by 80%	Yanty et al. (2024)
Coniothyrium minitans GR	*Sclerotinia sclerotiorum*	Degraded sclerotia by 90% in soil microcosms.	Zeng et al. (2012)
Gliocladium virens WP	*Pythium aphanidermatum*	Suppressed cucumber damping-off by 65%.	Wolffhechel et al. (1992)
Metarhizium anisopliae DS	*Bemisia tabaci*	Caused 75% mortality in whitefly nymphs.	Gebremariam et al. (2022)

WP = Wettable Powders; GR = Granules; DS = Dusts

Table 7: Large-scale field trials on solid biofungicides.

Biofungicide	**Crop**	**Results**	**References**
Trichoderma asperellum WP	Tomato	Reduced *Fusarium* wilt by 60% in farmer fields.	Sehim et al. (2023), Selva Amala et al. (2024)
Pseudomonas spp. GR	Rice	Controlled sheath blight (*Rhizoctonia*) by 55% in paddies.	Nagarajkumar et al. (2004)
Beauveria bassiana DS	Coffee	Reduced berry borer infestation by 65% in plantations.	Matsumoto et al. (2021)
Bacillus subtilis WP	Grapes	Suppressed powdery mildew (*Erysiphe necator*) by 50%.	Ferreira dos Santos et al. (2024)
Streptomyces griseoviridis GR	Tomato	Reduced *Fusarium* wilt by 45% in greenhouse trials.	Minuto et al. (2006)
Pantoea agglomerans WP	Pear	Controlled fire blight (*Erwinia amylovora*) by 40% in orchards.	Albastawisi et al. (2025)
Ampelomyces Quisqualis WP	Melon	Reduced powdery mildew (*Podosphaera xanthii*) by 70%.	Kimura et al. (2023)
Burkholderia spp. GR	Banana	Suppressed Panama disease (*Fusarium oxysporum*) by 60%.	Ho et al. (2015)

WP= Wettable Powders; GR= Granules; DS= Dusts

9.2.3 Dusts (DS)

Dust formulations are dry, ready-to-use powders containing the active ingredient and a carrier. *Beauveria bassiana* DS formulations are utilized in greenhouse settings to control fungal pathogens on ornamental plants.

9.3 Microencapsulated Formulations Composition

Microencapsulation allows controlled release since it includes enclosing active factors within a protective coating. Microencapsulated formulations of *Trichoderma harzianum* have been developed to enhance shelf life and efficacy against soil-borne diseases.

10. Concluding Remarks and Future Perspectives

Over the past few years, farmers globally have increasingly relied on chemical fungicides to manage infections in their agricultural ecosystems (Beckerman et al. 2023). Nonetheless, the negative effects of these fungicides on human health and the environment require the development of efficient pathogen management strategies, including the use of biofungicides (Lahlali et al. 2022). Public concern regarding the chemical nature of fungicides has contributed to the growing interest in biological control methods (Heydari and Pessarakli 2010). In light of the growing food demand and environmental concerns, the large-scale production of biofungicides has become essential in today's agricultural landscape (Bharti and Ibrahim 2020). Several studies conducted thus far have provided evidence for the potential of biofungicides to serve as replacements for traditional chemical fungicides (Fenta and Mekonnen 2024). The availability of a vast number of sources makes it more straightforward to screen for low-cost biofungicides that demonstrate superior effectiveness (Stéphane et al. 2021). The disease management capabilities of biofungicides, together with their ability to boost plant growth, indicate potential opportunities to create biofungicide-based biofertilizers (Kumar et al. 2024). There is a notable scarcity of research concerning the toxicity of biofungicides on non-target organisms; thus, it is imperative to conduct thorough investigations on biofungicides before their application in the near future (Ezeorba et al. 2023). The extension of biofungicide shelf life represents a critical factor in sustaining their viability and effectiveness across prolonged periods (Ezeorba et al. 2023). While numerous investigations into biofungicides have yielded promising results *in vitro*, there remains a limited number of assessments regarding their effectiveness in real-world agricultural conditions (Fenta and Mekonnen 2024). It is recommended that the use of biofungicides be expanded significantly within agricultural sectors. Furthermore, most biofungicides are particularly tailored to combat specific fungal pathogens. In light of the insights provided by some studies, the combined use of biocontrol agents has been advocated as a potential solution. From this standpoint, the creation of universal biofungicides holds considerable importance. Furthermore, the issue of toxicity related to biofungicides is a matter of serious concern. It is imperative that thorough studies and trials are conducted on biofungicides prior to their commercialization. The findings indicate that the concentration of biofungicide significantly influences plant growth and development, with higher concentrations appearing to have a detrimental effect (Ezeorba et al. 2023). Prior to the introduction of biofungicides, it is essential to resolve these issues to mitigate potential negative effects. Moreover, the molecular pathways triggered by the use of biofungicides are not well understood. Consequently, conducting molecular studies, including transcriptomics and proteomics, may provide insights into the mechanisms underlying the efficacy of biofungicides in combating fungal diseases in the near term. In pursuit of improved production and application techniques, scientists and researchers are anticipated to develop safe and effective products for managing plant diseases. Furthermore, future investigations will concentrate on advancing methodologies and fostering multidisciplinary research to generate high-quality, secure, and economically viable plant protection strategies.

References

Ahmadi, F., Najafpour, G., and Mohammadi, M. (2021). Production of bio-fungicide from sugarcane bagasse using *Pichia membranifaciens* yeast and its activity against post-harvest pathogenic fungi. *Biointerface Research in Applied Chemistry*, 11(3): 10435–10445. https://doi.org/10.33263/briac113.1043510445.

Alam, G., Sartini, N., and Alfath, A. (2019). Comparison of microwave assisted extraction (MAE) with variations of power and infusion extraction method on antibacterial activity of rosella calyx extract (*Hibiscus sabdariffa*). *Journal of Physics Conference Series*, 1341(7): 072002. https://doi.org/10.1088/1742-6596/1341/7/072002.

Alamgir, A. N. M. (2018). Methods of qualitative and quantitative analysis of plant constituents. pp. 721–804. *In: Progress in Drug Research.* https://doi.org/10.1007/978-3-319-92387-1_9.

Alara, O., Abdurahman, N., and Olalere, O. (2020). Ethanolic extraction of flavonoids, phenolics and antioxidants from *Vernonia amygdalina* leaf using two-level factorial design. *Journal of King Saud University - Science*, 32(1): 7–16. https://doi.org/10.1016/j.jksus.2017.08.001.

Albastawisi, E. M., Kaya, M., Fırat, Ş., and Kotan, R. (2025). Efficacy of *Pantoea agglomerans* strain RK-84 in controlling fire blight caused by *Erwinia amylovora* in pear orchards. *Journal of Plant Pathology,* 107(1): 597–603. https://doi.org/10.1007/s42161-024-01809-7.

Alvarez-Rivera, G., Bueno, M., Ballesteros-Vivas, D., Mendiola, J. A., and Ibañez, E. (2020). Pressurized liquid extraction. pp. 375–398. *In: Elsevier eBooks*. https://doi.org/10.1016/b978-0-12-816911-7.00013-x.

Ángel, D. L. Z. M., Paola, M. R. N., Lizeth, B. M. C., Athenas, S. J. T., Roberto, A. V., Cecilia, A. C. J. et al. (2023). Biofungicides for management of postharvest diseases. pp. 283–311. *In: CRC Press eBooks*. https://doi.org/10.1201/9781003287575-11.

Angst, G., Mueller, K. E., Nierop, K. G., and Simpson, M. J. (2021). Plant- or microbial-derived? A review on the molecular composition of stabilized soil organic matter. *Soil Biology and Biochemistry*, 156: 108189. https://doi.org/10.1016/j.soilbio.2021.108189.

Aparicio, M. D. P. G., and Marina, M. L. (2023). Application of enzyme-assisted extraction for food analysis and release of natural products. pp. 431–496. *In: Green extraction techniques in food analysis*. Bentham Science Publishers. https://doi.org/10.2174/9789815049459123030l.

Aparicio-Fernández, X., Vega-Ahuatzin, A., Ochoa-Velasco, C. E., Cid-Pérez, S., Hernández-Carranza, P., and Ávila-Sosa, R. (2017). Physical and antioxidant characterization of edible films added with red prickly pear (*Opuntia ficus-indica* L.) cv. san martín peel and/or its aqueous extracts. *Food and Bioprocess Technology*, 11(2): 368–379. https://doi.org/10.1007/s11947-017-2017-x.

Arora, S., and Itankar, P. (2018). Extraction, isolation and identification of flavonoid from *Chenopodium album* aerial parts. *Journal of Traditional and Complementary Medicine*, 8(4): 476–482. https://doi.org/10.1016/j.jtcme.2017.10.002.

Arroyave-Toro, J. J., Mosquera, S., and Villegas-Escobar, V. (2017). Biocontrol activity of Bacillus subtilis EA-CB0015 cells and lipopeptides against postharvest fungal pathogens. *Biological Control*, 114: 195–200. https://doi.org/10.1016/j.biocontrol.2017.08.014.

Ayala-Torres, A. M., Aranda-Ocampo, S., De León-García De Alba, C., Nava-Díaz, C., and Sánchez-Pale, J. R. (2023). Antagonistic bacteria against *Fusarium* spp. isolated from sclerotia of *Claviceps gigantea* in maize (Zea mays). *Revista Mexicana De Fitopatología(En LíNea)/Revista Mexicana De FitopatologíA*, 41(2). https://doi.org/10.18781/r.mex.fit.2208-2.

Ayaz, M., Li, C., Ali, Q., Zhao, W., Chi, Y., Shafiq, M. et al. (2023). Bacterial and fungal biocontrol agents for plant disease protection: Journey from lab to field, current status, challenges, and global perspectives. *Molecules*, 28(18): 6735. https://doi.org/10.3390/molecules28186735.

Azish, M., Shams-Ghahfarokhi, M., and Razzaghi-Abyaneh, M. (2021). Optimization of the antifungal metabolite production in *Streptomyces libani* isolated from northern forests soils in Iran. *Current Medical Mycology*. https://doi.org/10.18502/cmm.6.4.5333.

Azman, E. M., Charalampopoulos, D., and Chatzifragkou, A. (2020). Acetic acid buffer as extraction medium for free and bound phenolics from dried blackcurrant (*Ribes nigrum* L.) skins. *Journal of Food Science*, 85(11): 3745–3755. https://doi.org/10.1111/1750-3841.15466.

Balaraman, H., Selvasembian, R., Rangarajan, V., and Rathnasamy, S. (2021). Sustainable and Green Engineering Insights on Deep Eutectic Solvents toward the Extraction of Nutraceuticals. *ACS Sustainable Chemistry & Engineering*, 9(34): 11290–11313. https://doi.org/10.1021/acssuschemeng.1c03034.

Banu, K. S., and Cathrine, L. (2015). General techniques involved in phytochemical analysis. *International Journal of Advanced Research in Chemical Science*, 2(4): 25–32.

Barbero, G. F., Ferreiro-González, M., Freitas, V. C. M., Carrera, C., Espada-Bellido, E., Ruiz-Rodríguez, A. et al. (2022). Chapter 2. Extraction of Natural products: Principles and Fundamental aspects. pp. 66–116. *In: RSC green chemistry series*. https://doi.org/10.1039/9781839165894-00066.

Baron, N. C., Rigobelo, E. C., and Zied, D. C. (2019). Filamentous fungi in biological control: current status and future perspectives. *Chilean Journal of Agricultural Research*, 79(2): 307–315. https://doi.org/10.4067/s0718-58392019000200307.

Bartnik, M., and Facey, P. (2023). Glycosides. pp. 103–165. *In: Elsevier eBooks*. https://doi.org/10.1016/b978-0-443-18657-8.00001-3.

Bashmil, Y. M., Ali, A., Bk, A., Dunshea, F. R., and Suleria, H. a. R. (2021). Screening and characterization of phenolic compounds from australian grown bananas and their antioxidant capacity. *Antioxidants*, 10(10): 1521. https://doi.org/10.3390/antiox10101521.

Beckerman, J., Palmer, C., Tedford, E., and Ypema, H. (2023). Fifty years of fungicide development, deployment, and future use. *Phytopathology*, 113(4): 694–706. https://doi.org/10.1094/phyto-10-22-0399-ia.

Belwal, T., Pandey, A., Bhatt, I. D., and Rawal, R. S. (2020). Optimized microwave assisted extraction (MAE) of alkaloids and polyphenols from Berberis roots using multiple-component analysis. *Scientific Reports*, 10(1). https://doi.org/10.1038/s41598-020-57585-8.

Bhadange, Y. A., Carpenter, J., and Saharan, V. K. (2024). A comprehensive review on advanced extraction techniques for retrieving bioactive components from natural sources. *ACS Omega*, 9(29): 31274–31297. https://doi.org/10.1021/acsomega.4c02718.

Bharti, V., and Ibrahim, S. (2020). Biopesticides: production, formulation and application systems. *International Journal of Current Microbiology and Applied Sciences*, 9(10): 3931–3946. https://doi.org/10.20546/ijcmas.2020.910.453.

Bhatla, S. C., and Lal, M. A. (2023). Secondary metabolites. pp. 765–808. *In:* Plant physiology, development and metabolism. Springer, Singapore. https://doi.org/10.1007/978-981-99-5736-1_33.

Bubalo, M. C., Vidović, S., Redovniković, I. R., and Jokić, S. (2015). Green solvents for green technologies. *Journal of Chemical Technology & Biotechnology*, 90(9): 1631–1639. https://doi.org/10.1002/jctb.4668.

Bubici, G. (2018). *Streptomyces* spp. as biocontrol agents against *Fusarium* species. *CABI Reviews*, 2018, 1–5. https://doi.org/10.1079/PAVSNNR201813050.

Carisse, O., Bernier, J., and Benhamou, N. (2003). Selection of biological agents from composts for control of damping-off of cucumber caused by *Pythium ultimum*. *Canadian Journal of Plant Pathology*, 25(3): 258–267. https://doi.org/10.1080/07060660309507078.

Carmona-Hernandez, S., Reyes-Pérez, J., Chiquito-Contreras, R., Rincon-Enriquez, G., Cerdan-Cabrera, C., and Hernandez-Montiel, L. (2019). Biocontrol of postharvest fruit fungal diseases by bacterial antagonists: a review. *Agronomy*, 9(3): 121. https://doi.org/10.3390/agronomy9030121.

Carreira-Casais, A., Otero, P., Garcia-Perez, P., Garcia-Oliveira, P., Pereira, A. G., Carpena, M. et al. (2021). Benefits and Drawbacks of Ultrasound-Assisted Extraction for the Recovery of Bioactive Compounds from Marine Algae. *International Journal of Environmental Research and Public Health*, 18(17): 9153. https://doi.org/10.3390/ijerph18179153.

Castillo, N. E. T., Acosta, Y. A., Parra-Arroyo, L., Martínez-Prado, M. A., Rivas-Galindo, V. M., Iqbal, H. M. N. et al. (2022). Towards an eco-friendly coffee rust control: compilation of natural alternatives from a nutritional and antifungal perspective. *Plants*, 11(20): 2745. https://doi.org/10.3390/plants11202745.

Cheng, Y., Xue, F., Yu, S., Du, S., and Yang, Y. (2021). Subcritical water extraction of natural products. *Molecules*, 26(13): 4004. https://doi.org/10.3390/molecules26134004.

Chiocchio, I., Mandrone, M., Tomasi, P., Marincich, L., and Poli, F. (2021). Plant secondary metabolites: an opportunity for circular economy. *Molecules*, 26(2): 495. https://doi.org/10.3390/molecules26020495.

Clarke, C. J., Tu, W., Levers, O., Bröhl, A., and Hallett, J. P. (2018). Green and sustainable solvents in chemical processes. *Chemical Reviews*, 118(2): 747–800. https://doi.org/10.1021/acs.chemrev.7b00571.

Collinge, D. B., Jensen, D. F., Rabiey, M., Sarrocco, S., Shaw, M. W., and Shaw, R. H. (2022). Biological control of plant diseases – What has been achieved and what is the direction? *Plant Pathology*, 71(5): 1024–1047. https://doi.org/10.1111/ppa.13555.

Conte, R., Gullich, L. M., Bilibio, D., Zanella, O., Bender, J. P., Carniel, N. et al. (2016). Pressurized liquid extraction and chemical characterization of safflower oil: A comparison between methods. *Food Chemistry*, 213: 425–430. https://doi.org/10.1016/j.foodchem.2016.06.111.

Czaplicki, S. (2013). Chromatography in bioactivity analysis of compounds. In *IntechOpen eBooks*. https://doi.org/10.5772/55620.

Da Silva, R. P., Rocha-Santos, T. A., and Duarte, A. C. (2015). Supercritical fluid extraction of bioactive compounds. *TrAC Trends in Analytical Chemistry*, 76: 40–51. https://doi.org/10.1016/j.trac.2015.11.013.

Dababi, I., Gimello, O., Elaloui, E., and Brosse, N. (2020). Water extraction of tannins from aleppo pine bark and sumac root for the production of green wood adhesives. *Molecules*, 25(21): 5041. https://doi.org/10.3390/molecules25215041.

Danquah, C. A., Minkah, P. a. B., Duah, I. O., Junior, Amankwah, K. B., and Somuah, S. O. (2022). Antimicrobial Compounds from Microorganisms. *Antibiotics*, 11(3): 285. https://doi.org/10.3390/antibiotics11030285.

Davies, C. R., Wohlgemuth, F., Young, T., Violet, J., Dickinson, M., Sanders, J. et al. (2021). Evolving challenges and strategies for fungal control in the food supply chain. *Fungal Biology Reviews*, 36: 15–26. https://doi.org/10.1016/j.fbr.2021.01.003.

De Aguiar, A. C., Vardanega, R., Viganó, J., and Silva, E. K. (2023). Supercritical carbon dioxide technology for recovering valuable phytochemicals from *Cannabis sativa* L. and valorization of its biomass for food applications. *Molecules*, 28(9): 3849. https://doi.org/10.3390/molecules28093849.

Destandau, E., and Michel, T. (2022). Chapter 4. Microwave-assisted extraction. In *RSC green chemistry series* (pp. 144–201). https://doi.org/10.1039/9781839165894-00144.

Diep, T., Pook, C., and Yoo, M. (2020). Phenolic and Anthocyanin Compounds and Antioxidant Activity of Tamarillo (*Solanum betaceum* Cav.). *Antioxidants*, 9(2): 169. https://doi.org/10.3390/antiox9020169.

Djahra, A. B., Benkaddour, M., Zeghib, K., Benkherara, S., Shaieb, I., Ghania, A., and Bdida, S. (2019). Biofungicide activity of *Datura stramonium* leaf extract against phytopathogenic fungi. *International Journal of Biological and Agricultural Research*, 2(1): 1–5.

Djilani, A., Legseir, B., Soulimani, R., Dicko, A., and Younos, C. (2006). New extraction technique for alkaloids. *Journal of the Brazilian Chemical Society*, 17(3): 518–520. https://doi.org/10.1590/s0103-50532006000300013.

Dymek, A., Widelski, J., Wojtanowski, K. K., Vivcharenko, V., Przekora, A., and Mroczek, T. (2021). Fractionation of Lycopodiaceae alkaloids and evaluation of their anticholinesterase and cytotoxic activities. *Molecules*, 26(21): 6379. https://doi.org/10.3390/molecules26216379.

Eakjamnong, W., Keawsalong, N., and Dethoup, T. (2020). Novel ready-to-use dry powder formulation of *Talaromyces tratensis* KUFA0091 to control dirty panicle disease in rice. *Biological Control*, 152: 104454. https://doi.org/10.1016/j.biocontrol.2020.104454.

El-Baky, N. A., and Amara, A. a. a. F. (2021). Recent approaches towards control of fungal diseases in plants: An updated review. *Journal of Fungi*, 7(11): 900. https://doi.org/10.3390/jof7110900.

El-Mohamedy, R. S., Abdel-Kareem, F., and Daami-Remadi, M. (2014). Chitosan and *Trichoderma harzianum* as fungicide alternatives for controlling *Fusarium* crown and root rot of tomato. *Management*, 3(4): 5–6.

Esserti, S., Smaili, A., Rifai, L. A., Koussa, T., Makroum, K., Belfaiza, M. et al. (2016). Protective effect of three brown seaweed extracts against fungal and bacterial diseases of tomato. *Journal of Applied Phycology*, 29(2): 1081–1093. https://doi.org/10.1007/s10811-016-0996-z.

Ezeorba, T. P. C., Chukwudozie, K. I., Okoye, C. O., Okeke, E. S., Ezugwu, A. L., and Anaduaka, E. G. (2023). Biofungicides. pp. 12–39. *In: CRC Press eBooks*. https://doi.org/10.1201/9781003287575-2.

Fenta, L., and Mekonnen, H. (2024). Microbial biofungicides as a substitute for chemical fungicides in the control of phytopathogens: Current perspectives and research directions. *Scientifica*, 2024: 1–12. https://doi.org/10.1155/2024/5322696.

Feresin, R. G., Pourafshar, S., Huang, J., Zhao, Y., Arjmandi, B. H., and Salazar, G. (2017). Extraction and purification of polyphenols from freeze-dried berry powder for the treatment of vascular smooth muscle *in vitro*. *Journal of Visualized Experiments*, 125. https://doi.org/10.3791/55605.

Ferreira dos Santos, J. L., Laranjeira, D., and Gava, C. A. (2024). Selecting *Bacillus* strains antagonist to *Erysiphe necator* (Schw.) Burr. the causal agent of grape powdery mildew. *Biocontrol Science and Technology,* 34(11): 1055–1067. https://doi.org/10.1080/09583157.2024.2402394.

Ferrigo, D., Mondin, M., Ladurner, E., Fiorentini, F., Causin, R., and Raiola, A. (2020). Effect of seed biopriming with *Trichoderma harzianum* strain INAT11 on *Fusarium* ear rot and *Gibberella* ear rot diseases. *Biological Control*, 147: 104286. https://doi.org/10.1016/j.biocontrol.2020.104286.

Gebremariam, A., Chekol, Y., and Assefa, F. (2022). Extracellular enzyme activity of entomopathogenic fungi, *Beauveria bassiana* and *Metarhizium anisopliae* and their pathogenicity potential as a bio-control agent against whitefly pests, *Bemisia tabaci* and *Trialeurodes vaporariorum* (Hemiptera: Aleyrodidae). *BMC Research Notes*, 15(1): 117. https://doi.org/10.1186/s13104-022-06004-4.

Geng, S., Liu, Y., Ma, H., and Chen, C. (2015). Extraction and antioxidant activity of phenolic compounds from *Okra* flowers. *Tropical Journal of Pharmaceutical Research*, 14(5): 807. https://doi.org/10.4314/tjpr.v14i5.10.

Gholami, M., Amini, J., Abdollahzadeh, J., and Ashengroph, M. (2018). Basidiomycetes fungi as biocontrol agents against take-all disease of wheat. *Biological Control*, 130: 34–43. https://doi.org/10.1016/j.biocontrol.2018.12.012.

Gholami-Shabani, M., Shams-Ghahfarokhi, M., and Razzaghi-Abyaneh, M. (2019). Natural product synthesis by fungi: recent trends and future prospects. pp. 195–228. *In: Fungal biology*. https://doi.org/10.1007/978-3-030-14846-1_7.

Gil-Izquierdo, A., Gil, M., Conesa, M., and Ferreres, F. (2001). The effect of storage temperatures on vitamin C and phenolics content of artichoke (*Cynara scolymus* L.) heads. *Innovative Food Science & Emerging Technologies*, 2(3): 199–202. https://doi.org/10.1016/s1466-8564(01)00018-2.

Gligor, O., Mocan, A., Moldovan, C., Locatelli, M., Crişan, G., and Ferreira, I. C. (2019). Enzyme-assisted extractions of polyphenols – A comprehensive review. *Trends in Food Science & Technology*, 88: 302–315. https://doi.org/10.1016/j.tifs.2019.03.029.

González-Fernández, J., and Hormaza, J. I. (2020). *Pest and diseases in mango* (*Mangifera indica* L.). Mango Board. Retrieved from https://www.mango.org/wp-content/uploads/2020/08/Mango_Pests_and_Diseases_ENG.pdf.

Gu, Y., and Jérôme, F. (2010). Glycerol as a sustainable solvent for green chemistry. *Green Chemistry*, 12(7): 1127. https://doi.org/10.1039/c001628d.

Güçlü-Üstündağ, Ö., Balsevich, J., and Mazza, G. (2006). Pressurized low polarity water extraction of saponins from cow cockle seed. *Journal of Food Engineering*, 80(2): 619–630. https://doi.org/10.1016/j.jfoodeng.2006.06.024.

Harun, A., Aziz, N. A., Azenan, N. S. M., Kamarazzaman, N. F. M., and So'ad, S. Z. M. (2020). Antimicrobial efficacy, antioxidant profile and nine alternative active constituents from petroleum ether and ethyl acetate extract of *Entada spiralis*. *Malaysian Journal of Analytical Sciences*, 24(5): 707–718.

Hashem, A. H., Attia, M. S., Kandil, E. K., Fawzi, M. M., Abdelrahman, A. S., Khader, M. S. et al. (2023). Bioactive compounds and biomedical applications of endophytic fungi: a recent review. *Microbial Cell Factories*, 22(1). https://doi.org/10.1186/s12934-023-02118-x.

Heydari, A., and Pessarakli, M. (2010). A review on biological control of fungal plant pathogens using microbial antagonists. *Journal of Biological Sciences*, 10(4): 273–290. https://doi.org/10.3923/jbs.2010.273.290.

Hidayat, R., and Wulandari, N. P. (2021). Methods of extraction: maceration, percolation and decoction. *Eureka Herba Indonesia*, 2(1): 73–79. https://doi.org/10.37275/ehi.v2i1.15.

Ho, Y. N., Chiang, H. M., Chao, C. P., Su, C. C., Hsu, H. F., Guo, C. T. et al. (2015). *In planta* biocontrol of soilborne *Fusarium* wilt of banana through a plant endophytic bacterium, *Burkholderia cenocepacia* 869T2. *Plant and Soil,* 387(1–2): 295–306. https://doi.org/10.1007/s11104-014-2297-0.

Huang, W. Y., Cai, Y. Z., Xing, J., Corke, H., and Sun, M. (2007). A potential antioxidant resource: endophytic fungi from medicinal plants. *Economic Botany*, 61(1): 14–30. https://doi.org/10.1663/0013-0001(2007)61[14:APAREF]2.0.CO;2.

Hussain, M. K., Saquib, M., and Khan, M. F. (2019). Techniques for extraction, isolation, and standardization of bio-active compounds from medicinal plants. pp. 179–200. *In: Springer eBooks*. https://doi.org/10.1007/978-981-13-7205-6_8.

Ingle, K. P., Deshmukh, A. G., Padole, D. A., Dudhare, M. S., Moharil, M. P., and Khelurkar, V. C. (2017). Phytochemicals: Extraction methods, identification and detection of bioactive compounds from plant extracts. *Journal of Pharmacognosy and Phytochemistry*, 6(1): 32–36.

Iqbal, A., Irshad, G., Naz, F., Ghuffar, S., Khursheed, R., Bashir, A. et al. (2020). Characterization of *Fusarium Solani*, causal agent of pea wilt and its bio-management. *Pakistan Journal of Phytopathology*, 32(2): 265–271. https://doi.org/10.33866/10.33866/phytopathol.030.02.0534.

Irma, A., Meryandini, A., and Rupaedah, B. (2018). Biofungicide producing bacteria: an *in vitro* inhibitor of *Ganoderma boninense*. *HAYATI Journal of Biosciences*, 25(4): 151. https://doi.org/10.4308/hjb.25.4.151.

Ivanović, M., Razboršek, M. I., and Kolar, M. (2020). Innovative extraction techniques using deep eutectic solvents and analytical methods for the isolation and characterization of natural bioactive compounds from plant material. *Plants*, 9(11): 1428. https://doi.org/10.3390/plants9111428.

Jaber, L. R., and Alananbeh, K. M. (2018). Fungal entomopathogens as endophytes reduce several species of *Fusarium* causing crown and root rot in sweet pepper (*Capsicum annuum* L.). *Biological Control*, 126: 117–126. https://doi.org/10.1016/j.biocontrol.2018.08.007.

Jabeur, M. B., Somai-Jemmali, L., and Hamada, W. (2017). Thyme essential oil as an alternative mechanism: biofungicide-causing sensitivity of *Mycosphaerella graminicola*. *Journal of Applied Microbiology*, 122(4): 932–939. https://doi.org/10.1111/jam.13408.

Jamal, Q., Cho, J. Y., Moon, J. H., and Kim, K. Y. (2017). Purification and antifungal characterization of Cyclo (D-Pro-L-Val) from *Bacillus amyloliquefaciens* Y1 against *Fusarium graminearum* to control head blight in wheat. *Biocatalysis and Agricultural Biotechnology*, 10: 141–147. https://doi.org/10.1016/j.bcab.2017.01.003.

Janson, J. C. Protein Purification: Principles, High Resolution Methods, and Applications, 3rd ed.; *John Wiley & Sons: Hoboken*, NJ, USA, 2011; *Volume 54*, ISBN 9780471746614.

Jha, A. K., and Sit, N. (2022). Extraction of bioactive compounds from plant materials using combination of various novel methods: A review. *Trends in Food Science & Technology*, 119: 579–591. https://doi.org/10.1016/j.tifs.2021.11.019.

Ji, Y., Li, X., Wang, Z., Xiao, W., He, Z., Xiong, Z. et al. (2020). Extraction optimization of accelerated solvent extraction for eight active compounds from *Yaobitong capsule* using response surface methodology: Comparison with ultrasonic and reflux extraction. *Journal of Chromatography A*, 1620: 460984. https://doi.org/10.1016/j.chroma.2020.460984.

Jia, Q., Fan, Y., Duan, S., Qin, Q., Ding, Y., Yang, M. et al. (2023). Effects of *Bacillus amyloliquefaciens* XJ-BV2007 on growth of *Alternaria alternata* and production of tenuazonic acid. *Toxins*, 15(1): 53. https://doi.org/10.3390/toxins15010053.

Jiang, C., Liao, M., Wang, H., Zheng, M., Xu, J., and Guo, J. (2018). *Bacillus velezensis*, a potential and efficient biocontrol agent in control of pepper gray mold caused by *Botrytis cinerea*. *Biological Control*, 126: 147–157. https://doi.org/10.1016/j.biocontrol.2018.07.017.

Jiang, Y., and Wang, J. (2023). The registration situation and use of mycopesticides in the world. *Journal of Fungi*, 9(9): 940. https://doi.org/10.3390/jof9090940.

Jusoh, N. H. M., Subki, A., Yeap, S. K., Yap, K. C., and Jaganath, I. B. (2019). Pressurized hot water extraction of hydrosable tannins from *Phyllanthus tenellus* Roxb. *BMC Chemistry*, 13(1). https://doi.org/10.1186/s13065-019-0653-0.

Justino, A. B., Florentino, R. M., França, A., Filho, A. C. M. L., Franco, R. R. et al. (2021). Alkaloid and acetogenin-rich fraction from *Annona crassiflora* fruit peel inhibits proliferation and migration of human liver cancer HepG2 cells. *PLoS ONE*, 16(7): e0250394. https://doi.org/10.1371/journal.pone.0250394.

Kagot, V., Okoth, S., De Boevre, M., and De Saeger, S. (2019). Biocontrol of Aspergillus and fusarium mycotoxins in Africa: benefits and limitations. *Toxins*, 11(2): 109. https://doi.org/10.3390/toxins11020109.

Kale, S. (2020). Screening of phytochemical components of leaf extracts. *International Journal of Current Microbiology and Applied Sciences*, 9(9): 2394–2401. https://doi.org/10.20546/ijcmas.2020.909.299.

Karaś, M., Jakubczyk, A., Szymanowska, U., Złotek, U., and Zielińska, E. (2016). Digestion and bioavailability of bioactive phytochemicals. *International Journal of Food Science & Technology*, 52(2): 291–305. https://doi.org/10.1111/ijfs.13323.

Keeffe, E. O., Hughes, H., McLoughlin, P., Tan, S. P., and McCarthy, N. (2019). Methods of analysis for the in vitro and in vivo determination of the fungicidal activity of seaweeds: a mini review. *Journal of Applied Phycology*, 31(6): 3759–3776. https://doi.org/10.1007/s10811-019-01832-7.

Khade, O. S., K, S., Sonkar, R. M., Gade, P. S., and Bhatt, P. (2023). Plant secondary metabolites: Extraction, screening, analysis and their bioactivity. *International Journal of Herbal Medicine*, 11(2): 01–17. https://doi.org/10.22271/flora.2023.v11.i2a.855.

Kimura, Y., Németh, M. Z., Numano, K., Mitao, A., Shirakawa, T., Seress, D. et al. (2023). Hyperparasitic fungi against melon powdery mildew pathogens: Quantitative analysis of conidia released from single colonies of *Podosphaera xanthii* parasitised by *Ampelomyces*. *Agronomy*, 13(5): 1204. https://doi.org/10.3390/agronomy13051204.

Ko, M., Nam, H., and Chung, M. (2020). Subcritical water extraction of bioactive compounds from *Orostachys japonicus* A. berger (crassulaceae). *Scientific Reports*, 10(1). https://doi.org/10.1038/s41598-020-67508-2.

Köhl, J., Kolnaar, R., and Ravensberg, W. J. (2019). Mode of action of microbial biological control agents against plant diseases: Relevance beyond efficacy. *Frontiers in Plant Science*, 10: 854. https://doi.org/10.3389/fpls.2019.00845.

Krakowska-Sieprawska, A., Kiełbasa, A., Rafińska, K., Ligor, M., and Buszewski, B. (2022). Modern methods of Pre-Treatment of plant material for the extraction of bioactive compounds. *Molecules*, 27(3): 730. https://doi.org/10.3390/molecules27030730.

Kumar, A., P, N., Kumar, M., Jose, A., Tomer, V., Oz, E. et al. (2023a). Major phytochemicals: recent advances in health benefits and extraction method. *Molecules*, 28(2): 887. https://doi.org/10.3390/molecules28020887.

Kumar, D., Arya, S. K., Shamim, M., Srivastava, D., Tyagi, M., and Singh, K. N. (2021). *In vitro* and *in vivo* activity of essential oils against postharvest pathogen *Colletotrichum musae* of banana. *Journal of Postharvest Technology*, 9(1): 53–63. https://acspublisher.com/journals/index.php/jpht/article/view/15162.

Kumar, D., Singh, P. N., Willer, H., Sharma, S. K., Singh, U. B., Lagashetti, A. C. et al. (2023b). Fungal biofertilizers and biopesticides and their roles in sustainable agriculture. pp. 165–216. *In: Apple Academic Press eBooks*. https://doi.org/10.1201/9781003369868-9.

Kumar, J., Kaur, A., and Narang, P. (2019). Phytochemical screening and metal binding studies on floral extract of *Solanum nigrum*. *Materials Today Proceedings*, 26: 3332–3336. https://doi.org/10.1016/j.matpr.2019.09.170.

Kuzuyama, T. (2017). Biosynthetic studies on terpenoids produced by Streptomyces. *The Journal of Antibiotics*, 70(7): 811–818. https://doi.org/10.1038/ja.2017.12.

Lahlali, R., Ezrari, S., Radouane, N., Kenfaoui, J., Esmaeel, Q., Hamss, H. E. et al. (2022). Biological control of plant pathogens: a Global perspective. *Microorganisms*, 10(3): 596. https://doi.org/10.3390/microorganisms10030596.

Lefebvre, T., Destandau, E., and Lesellier, E. (2020). Selective extraction of bioactive compounds from plants using recent extraction techniques: A review. *Journal of Chromatography A*, 1635: 461770. https://doi.org/10.1016/j.chroma.2020.461770.

Lewis, J. A., and Papavizas, G. C. (1991). Biocontrol of cotton damping-off caused by *Rhizoctonia solani* in the field with formulations of *Trichoderma* spp. and *Gliocladium virens*. *Crop Protection*, 10(5): 396–402. https://doi.org/10.1016/S0261-2194(06)80031-1.

Lezoul, N. E. H., Belkadi, M., Habibi, F., and Guillén, F. (2020). Extraction processes with several solvents on total bioactive compounds in different organs of three medicinal plants. *Molecules*, 25(20): 4672. https://doi.org/10.3390/molecules25204672.

Li, G. Q., Huang, H. C., Miao, H. J., Erickson, R. S., Jiang, D. H., and Xiao, Y. N. (2006). Biological control of *Sclerotinia* diseases of rapeseed by aerial applications of the mycoparasite *Coniothyrium minitans*. *European Journal of Plant Pathology*, 114(3): 345–355. https://doi.org/10.1007/s10658-005-2232-6.

Lin, L., Xu, K., Shen, D., Chou, S., Gomelsky, M., and Qian, G. (2021). Antifungal weapons of *Lysobacter ,Lysobacter*, a mighty biocontrol agent. *Environmental Microbiology*, 23(10): 5704–5715. https://doi.org/10.1111/1462-2920.15674.

Liu, M., Gong, Y., Sun, H., Zhang, J., Zhang, L., Sun, J. et al. (2020). Characterization of a novel chitinase from sweet potato and its fungicidal effect against *Ceratocystis fimbriata*. *Journal of Agricultural and Food Chemistry*, 68(29): 7591–7600. https://doi.org/10.1021/acs.jafc.0c01813.

López-Bascón, M., and De Castro, M. L. (2020). Soxhlet extraction. pp. 327–354. *In: Elsevier eBooks*. https://doi.org/10.1016/b978-0-12-816911-7.00011-6.

Ma, J., Gao, C., Lin, M., Sun, Z., Zhao, Y., Li, X. et al. (2024). Control of *Fusarium* head blight of wheat with *Bacillus velezensis* E2 and potential mechanisms of action. *Journal of Fungi*, 10(6): 390. https://doi.org/10.3390/jof10060390.

Maheshwaran, L., Nadarajah, L., Senadeera, S. P. N. N., Ranaweera, C. B., Chandana, A. K., and Pathirana, R. N. (2024). Phytochemical testing methodologies and principles for preliminary screening/ qualitative testing. *Asian Plant Research Journal*, 12(5): 11–38. https://doi.org/10.9734/aprj/2024/v12i5267.

Marathe, S. J., Jadhav, S. B., Bankar, S. B., and Singhal, R. S. (2017). Enzyme-Assisted extraction of bioactives. pp. 171–201. *In: Springer eBooks*. https://doi.org/10.1007/978-3-319-51639-4_8.

Marathe, S. J., Jadhav, S. B., Bankar, S. B., Dubey, K. K., and Singhal, R. S. (2019). Improvements in the extraction of bioactive compounds by enzymes. *Current Opinion in Food Science*, 25: 62–72. https://doi.org/10.1016/j.cofs.2019.02.009.

Marcus, Y. (2018). The variety of deep eutectic solvents. pp. 13–44. *In: Springer eBooks*. https://doi.org/10.1007/978-3-030-00608-2_2.

Martínez-Maqueda, D., Hernández-Ledesma, B., Amigo, L., Miralles, B., and Gómez-Ruiz, J. Á. (2012). Extraction/Fractionation techniques for proteins and peptides and protein digestion. pp. 21–50. *In: Springer eBooks*. https://doi.org/10.1007/978-1-4614-5626-1_2.

Materska, M., Olszówka, K., Chilczuk, B., Stochmal, A., Pecio, Ł., Pacholczyk-Sienicka, B. et al. (2018). Polyphenolic profiles in lettuce (*Lactuca sativa* L.) after CaCl2 treatment and cold storage. *European Food Research and Technology*, 245(3): 733–744. https://doi.org/10.1007/s00217-018-3195-0.

Matzen, N., Heick, T. M., and Jørgensen, L. N. (2019). Control of powdery mildew (*Blumeria graminis* spp.) in cereals by Serenade® ASO (*Bacillus amyloliquefaciens* strain QST 713). *Biological Control*, 139: 104067. https://doi.org/10.1016/j.biocontrol.2019.104067.

Mbaoji, F. N., Onwuka, A. M., Onu, S., Peter, I. E., Nweze, J. A., and Okonta, L. E. (2020). Evaluation of Methanol-Dichloromethane Extract of *Stemonocoleus micranthus* Harms (Fabaceae) Stem Bark for Anti-Inflammatory and Immunomodulatory Activities. *Evidence-based Complementary and Alternative Medicine*, 2020(1). https://doi.org/10.1155/2020/1738163.

Meng, X., Yu, J., Yu, M., Yin, X., and Liu, Y. (2015). Dry flowable formulations of antagonistic *Bacillus subtilis* strain T429 by spray drying to control rice blast disease. *Biological Control*, 85: 46–51. https://doi.org/10.1016/j.biocontrol.2015.03.004.

Mgbeahuruike, E. E., Fyhrquist, P., Vuorela, H., Julkunen-Tiitto, R., and Holm, Y. (2018). Alkaloid-rich crude extracts, fractions and piperamide alkaloids of *Piper guineense* possess promising antibacterial effects. *Antibiotics*, 7(4): 98. https://doi.org/10.3390/antibiotics7040098.

Minuto, A., Spadaro, D., Garibaldi, A., and Gullino, M. L. (2006). Control of soilborne pathogens of tomato using a commercial formulation of *Streptomyces griseoviridis* and solarization. *Crop Protection*, 25(5): 468–475. https://doi.org/10.1016/j.cropro.2005.08.001.

Mohaddes-Kamranshahi, M., Jafarizadeh-Malmiri, H., Simjoo, M., and Jafarizad, A. (2018). Evaluation of the saponin green extraction from *Ziziphus spina-christi* leaves using hydrothermal, microwave and Bain-Marie water bath heating methods. *Green Processing and Synthesis*, 8(1): 62–67. https://doi.org/10.1515/gps-2017-0185.

Mojzer, E. B., Hrnčič, M. K., Škerget, M., Knez, Ž., and Bren, U. (2016b). Polyphenols: extraction methods, antioxidative action, bioavailability and anticarcinogenic effects. *Molecules*, 21(7): 901. https://doi.org/10.3390/molecules21070901.

Morata, A., González, C., Tesfaye, W., Loira, I., and Suárez-Lepe, J. A. (2019). Maceration and fermentation. pp. 35–49. *In: Elsevier eBooks*. https://doi.org/10.1016/b978-0-12-814399-5.00003-7.

Mousa, M. A., Abo-Elyousr, K. A., Abdel Alal, A. M., and Alshareef, N. O. (2021). Management of fusarium wilt disease in tomato by combinations of *Bacillus amyloliquefaciens* and peppermint oil. *Agronomy*, 11(12): 2536. https://doi.org/10.3390/agronomy11122536.

Mrid, R. B., Benmrid, B., Hafsa, J., Boukcim, H., Sobeh, M., and Yasri, A. (2021). Secondary metabolites as biostimulant and bioprotectant agents: A review. *The Science of the Total Environment*, 777: 146204. https://doi.org/10.1016/j.scitotenv.2021.146204.

Mwaurah, P. W., Kumar, S., Kumar, N., Attkan, A. K., Panghal, A., Singh, V. K. et al. (2019). Novel oil extraction technologies: Process conditions, quality parameters, and optimization. *Comprehensive Reviews in Food Science and Food Safety*, 19(1): 3–20. https://doi.org/10.1111/1541-4337.12507.

Myo, E. M., Liu, B., Ma, J., Shi, L., Jiang, M., Zhang, K. et al. (2019). Evaluation of *Bacillus velezensis* NKG-2 for bio-control activities against fungal diseases and potential plant growth promotion. *Biological Control*, 134: 23–31. https://doi.org/10.1016/j.biocontrol.2019.03.017.

Nagarajkumar, M., Bhaskaran, R., and Velazhahan, R. (2004). Involvement of secondary metabolites and extracellular lytic enzymes produced by *Pseudomonas fluorescens* in inhibition of *Rhizoctonia solani*, the rice sheath blight pathogen. *Microbiological Research*, 159(1): 73–81. https://doi.org/10.1016/j.micres.2004.01.005.

Nagendran, S., Kulanthaivelu, S., and Sundararajan, T. (2019). Assessment on antagonistic potential of bacterial bioagents *Pseudomonas fluorescens* and *Bacillus subtilis* against *Rhizoctonia solani* Kühn. An incitant of Sheath blight of rice. *Journal of Entomology and Zoology Studies*, 7(1): 128–142.

Naikoo, M. I., Dar, M. I., Raghib, F., Jaleel, H., Ahmad, B., Raina, A. et al. (2019). Role and regulation of plants phenolics in abiotic stress tolerance. pp. 157–168. *In: Elsevier eBooks*. https://doi.org/10.1016/b978-0-12-816451-8.00009-5.

Nandhini, M., Rajini, S., Udayashankar, A., Niranjana, S., Lund, O. S., Shetty, H. et al. (2018). Diversity, plant growth promoting and downy mildew disease suppression potential of cultivable endophytic fungal communities associated with pearl millet. *Biological Control*, 127: 127–138. https://doi.org/10.1016/j.biocontrol.2018.08.019.

Narayanasamy, P. (2013). Mechanisms of action of fungal biological control agents. pp. 99–200. *In: Springer eBooks*. https://doi.org/10.1007/978-94-007-6380-7_3.

Nasirpour, N., Mohammadpourfard, M., and Heris, S. Z. (2020). Ionic liquids: Promising compounds for sustainable chemical processes and applications. *Process Safety and Environmental Protection*, 160: 264–300. https://doi.org/10.1016/j.cherd.2020.06.006.

Naveenkumar, R., Muthukumar, A., Sangeetha, G., and Mohanapriya, R. (2017). Developing eco-friendly biofungicide for the management of major seed borne diseases of rice and assessing their physical stability and storage life. *Comptes Rendus Biologies*, 340(4): 214–225. https://doi.org/10.1016/j.crvi.2017.03.001.

Nian, J., Yu, M., Bradley, C. A., and Zhao, Y. (2020). *Lysobacter enzymogenes* strain C3 suppresses mycelium growth and spore germination of eight soybean fungal and oomycete pathogens and decreases disease incidences. *Biological Control*, 152: 104424. https://doi.org/10.1016/j.biocontrol.2020.104424.

Nicot, P. C., Stewart, A., Bardin, M., and Elad, Y. (2016). Biological control and biopesticide suppression of *Botrytis*-incited diseases. pp. 225–248. *In:* Fillinger, S., and Elad, Y. (Eds.), *Botrytis – the fungus, the pathogen and its management in agricultural systems*. Springer. https://doi.org/10.1007/978-3-319-23371-0_9.

Nimsarkar, N., Kulkarni, M., and Shaikh, F. (2021). Biochemical analysis of phenolics compounds accumulated during developing stages of wheatgrass. *International Journal for Research in Applied Sciences and Biotechnology*, 8(2): 167–172.

Nion, Y. A., and Toyota, K. (2008). Suppression of bacterial wilt and *Fusarium* wilt by a *Burkholderia nodosa* strain isolated from Kalimantan soils, Indonesia. *Microbes and Environments*, 23(2): 134–141. https://doi.org/10.1264/jsme2.23.134.

Nookongbut, P., Kantachote, D., Khuong, N. Q., and Tantirungkij, M. (2020). The biocontrol potential of acid-resistant *Rhodopseudomonas palustris* KTSSR54 and its exopolymeric substances against rice fungal pathogens to enhance rice growth and yield. *Biological Control*, 150: 104354. https://doi.org/10.1016/j.biocontrol.2020.104354.

Nour, A. H., Oluwaseun, A. R., Nour, A. H., Omer, M. S., and Ahmed, N. (2021). Microwave-assisted extraction of bioactive compounds (Review). In *IntechOpen eBooks*. https://doi.org/10.5772/intechopen.96092.

Nur Mawaddah, S., AW, M. Z., and Sapak, Z. (2023). The potential of *Pseudomonas fluorescens* as biological control agent against sheath blight disease in rice: A systematic review. *Food Research*, 7(2): 46–56. https://doi.org/10.26656/fr.2017.7(S2).11.

Ons, L., Bylemans, D., Thevissen, K., and Cammue, B. P. (2020). Combining Biocontrol Agents with Chemical Fungicides for Integrated Plant Fungal Disease Control. *Microorganisms*, 8(12): 1930. https://doi.org/10.3390/microorganisms8121930.

Pacios-Michelena, S., Aguilar González, C. N., Alvarez-Perez, O. B., Rodriguez-Herrera, R., Chávez-González, M., Arredondo Valdés, R. et al. (2021). Application of *Streptomyces* antimicrobial compounds for the control of phytopathogens. *Frontiers in Sustainable Food Systems*, 5: 696518. https://doi.org/10.3389/fsufs.2021.696518.

Pal, R., Biswas, M. K., Mandal, D., and Naik, B. S. (2015). Management of sheath blight disease of rice through bio control agents in west central table land zone of Odisha. *International Journal of Advanced Research*, 3(11): 747–753.

Pandit, M. A., Kumar, J., Gulati, S., Bhandari, N., Mehta, P., Katyal, R. et al. (2022). Major biological control strategies for plant pathogens. *Pathogens*, 11(2): 273. https://doi.org/10.3390/pathogens11020273.

Pathak, V. M., Verma, V. K., Rawat, B. S., Kaur, B., Babu, N., Sharma, A. et al. (2022). Current status of pesticide effects on environment, human health and it's eco-friendly management as bioremediation: A comprehensive review. *Frontiers in Microbiology*, 13. https://doi.org/10.3389/fmicb.2022.962619.

Patwardhan, R. B., Abhyankar, P. S., Gore, S. S., Kalekar, S. V., and Umrani, S. P. (2022). Biofungicidal properties of rhizobacteria for plant growth promotion and plant disease resistance. pp. 103–133. *In: Fungal biology*. https://doi.org/10.1007/978-3-031-04805-0_6.

Płotka-Wasylka, J., Rutkowska, M., Owczarek, K., Tobiszewski, M., and Namieśnik, J. (2017). Extraction with environmentally friendly solvents. *TrAC Trends in Analytical Chemistry*, 91: 12–25. https://doi.org/10.1016/j.trac.2017.03.006.

Purwaningsih, I., Fathiah, F., Amaliyah, N., and Kuswiyanto, K. (2023). The phenolic, flavonoid, and anthocyanin content from methanol extract of senggani fruit and its antioxidant activity. *Indonesian Journal of Chemical Research*, 10(3): 195–202. https://doi.org/10.30598//ijcr.2023.10-pur.

Pusey, P. L., Stockwell, V. O., Reardon, C. L., Smits, T. H., and Duffy, B. (2011). Antibiosis by *Pantoea agglomerans* biocontrol strain E325 against *Erwinia amylovora* on apple blossom stigmas. *Phytopathology*, 101(6): S146–S147.

Raal, A., Meos, A., Hinrikus, T., Heinämäki, J., Romāne, E., Gudienė, V. et al. (2020). Dragendorff's reagent: Historical perspectives and current status of a versatile reagent introduced over 150 years ago at the University of Dorpat, Tartu, Estonia. *Die Pharmazie-An International Journal of Pharmaceutical Sciences*, 75(7): 299–306. https://doi.org/10.1691/ph.2020.0438.

Radulovic, N., Blagojevic, P., Stojanovic-Radic, Z., and Stojanovic, N. (2013). Antimicrobial plant metabolites: structural diversity and mechanism of action. *Current Medicinal Chemistry*, 20(7): 932–952. https://doi.org/10.2174/0929867313805219136.

Ram, R. M., Rajput, R. S., and Vaishnav, A. (2020). Management of *Sclerotium rolfsii* induced diseases in crops by *Trichoderma* species. pp. 543–562. *In:* Sharma, S. K., Singh,U. B., Sahu, P. K., Singh, H. V., and Sharma, P. K. (Eds.). *Rhizosphere microbes*. Springer. https://doi.org/10.1007/978-981-15-9154-9_25.

Ramya, P., Vasanth, P. M., Prasad, P. V., and Babu, S. V. (2019). Qualitative phytochemical screening tests of *Alpinia galanga* L. *World Journal of Pharmaceutical Research*, 8(5): 1064–1077. https://doi.org/10.20959/wjpr20195-14624.

Ramya, P., Vasanth, P. M., Prasad, P. V., and Babu, S. V. (2019). Qualitative phytochemical screening tests of *Alpinia galanga* L. *World Journal of Pharmaceutical Research*, 8(5): 1064–1077. https://doi.org/10.20959/wjpr20195-14624.

Rehman, M. U., Abdullah, N., Khan, F., and Niaz, K. (2020). Introduction to natural products analysis. pp. 3–15. *In: Elsevier eBooks*. https://doi.org/10.1016/b978-0-12-816455-6.00001-9.

Riris, I. D., Juwitaningsih, T., Roza, D., Damanik, M., and Silalahi, A. (2020). Study of phytochemicals, toxicity, antibacterial activity of ethyl acetate leaf extract extract (*Paperomia pellucida* L). *Indonesian Journal of Chemical Science and Technology (IJCST)*, 3(2): 74. https://doi.org/10.24114/ijcst.v3i2.19532.

Roberti, R., Di Francesco, A., Innocenti, G., and Mari, M. (2019). Potential for biocontrol of *Pleurotus ostreatus* green mould disease by *Aureobasidium pullulans* De Bary (Arnaud). *Biological Control*, 135: 9–15. https://doi.org/10.1016/j.biocontrol.2019.04.016.

Rodrigo, S., García-Latorre, C., and Santamaria, O. (2021). Metabolites produced by fungi against fungal phytopathogens: Review, Implementation and Perspectives. *Plants*, 11(1): 81. https://doi.org/10.3390/plants11010081.

Rojas, E. C., Jensen, B., Jørgensen, H. J., Latz, M. A., Esteban, P., Ding, Y. et al. (2020). Selection of fungal endophytes with biocontrol potential against *Fusarium* head blight in wheat. *Biological Control*, 144: 104222. https://doi.org/10.1016/j.biocontrol.2020.104222.

Romeralo, C., Santamaría, O., Pando, V., and Diez, J. (2014). Fungal endophytes reduce necrosis length produced by *Gremmeniella abietina* in *Pinus halepensis* seedlings. *Biological Control*, 80: 30–39. https://doi.org/10.1016/j.biocontrol.2014.09.010.

Rusita, I., and Sasongko, H. (2020). Effectivity of *Trichoderma harzianum* as bio-fungicide against moler disease and bio-stimulator of shallot growth. *Journal of Agri-Food Science and Technology*, 1(1): 12. https://doi.org/10.12928/jafost.v1i1.1941.

Saito, H., Sasaki, M., Nonaka, Y., Tanaka, J., Tokunaga, T., Kato, A. et al. (2021). Spray application of nonpathogenic fusaria onto rice flowers controls bakanae disease (Caused by *Fusarium fujikuroi*) in the next plant generation. *Applied and Environmental Microbiology*, 87(2). https://doi.org/10.1128/aem.01959-20.

Salam, A. M., Lyles, J. T., and Quave, C. L. (2018). Methods in the extraction and chemical analysis of medicinal plants. pp. 257–283. *In: Springer protocols handbooks/Springer protocols*. https://doi.org/10.1007/978-1-4939-8919-5_17.

Salamah, N., and Ningsih, D. S. (2017). Total alkaloid content in various fractions of *Tabernaemonata sphaerocarpa* B_l. (Jembirit) leaves. *IOP Conference Series Materials Science and Engineering*, 259: 012017. https://doi.org/10.1088/1757-899x/259/1/012017.

Salerno, L., Modica, M. N., Pittalà, V., Romeo, G., Siracusa, M. A., Di Giacomo, C. et al. (2014). Antioxidant activity and phenolic content of Microwave-*AssistedSolanum melongena* extracts. *The Scientific World JOURNAL, 2014*: 1–6. https://doi.org/10.1155/2014/719486.

Sanchez-Bayo, N. F. (2011). Impacts of agricultural pesticides on terrestrial ecosystems. pp. 63–87. *In: Bentham Science Publishers eBooks*. https://doi.org/10.2174/978160805121211101010063.

Saputri, D. D., and Utami, A. W. A. (2020). The potency purple sweet potato (*Ipomoea batatas*) leaf extract as biofungicide for controlling *Fusarium* rot on chili. *Journal of Agriculture and Applied Biology*, 1(1): 1–8. https://doi.org/10.11594/jaab.01.01.01.

Segaran, G., and Sathiavelu, M. (2019). Fungal endophytes: A potent biocontrol agent and a bioactive metabolites reservoir. *Biocatalysis and Agricultural Biotechnology*, 21: 101284. https://doi.org/10.1016/j.bcab.2019.101284.

Sehim, A. E., Hewedy, O. A., Altammar, K. A., Alhumaidi, M. S., and Abd Elghaffar, R. Y. (2023). *Trichoderma asperellum* empowers tomato plants and suppresses *Fusarium oxysporum* through priming responses. *Frontiers in Microbiology*, 14: 1140378. https://doi.org/10.3389/fmicb.2023.1140378.

Selva Amala, A., Parthiban, V. K., Sudha, A., Gopalakrishnan, C., Swarnakumari, N., and Anandham, R. (2024). Antifungal and plant-growth promoting potency of *Trichoderma asperellum* against *Fusarium* wilt on tomato. *Journal of Plant Pathology, 1*–3. https://doi.org/10.1007/s42161-024-01736-7.

Shahid, M., Singh, B. N., Verma, S., Choudhary, P., Das, S., Chakdar, H. et al. (2021). Bioactive antifungal metabolites produced by Streptomyces amritsarensis V31 help to control diverse phytopathogenic fungi. *Brazilian Journal of Microbiology*, 52(4): 1687–1699. https://doi.org/10.1007/s42770-021-00625-w.

Sharma, A., Sharma, S., Kumar, A., Kumar, V., and Sharma, A.K. (2022). Plant secondary metabolites: an introduction of their chemistry and biological significance with physicochemical aspect. pp. 1–45. *In:* Sharma, A. K., and Sharma, A. (Eds.). *Plant secondary metabolites*. Springer, Singapore. https://doi.org/10.1007/978-981-16-4779-6_1.

Shaw, N., Singh, S., Paul, S., V, K., and M, R. (2023). Phytochemical Screening and Antimicrobial activity of homoeopathic dilutions of *Myristica sebifera* against *Staphylococcus aureus*. *Homœopathic Links*, 37(01): 005–010. https://doi.org/10.1055/s-0043-1775848.

Shi, L., Zhao, W., Yang, Z., Subbiah, V., and Suleria, H. a. R. (2022). Extraction and characterization of phenolic compounds and their potential antioxidant activities. *Environmental Science and Pollution Research*, 29(54): 81112–81129. https://doi.org/10.1007/s11356-022-23337-6.

Shokouhi, D., and Seifi, A. (2020). Organic extracts of seeds of Iranian *Moringa peregrina* as promising selective biofungicide to control *Mycogone perniciosa*. *Biocatalysis* and *Agricultural Biotechnology*, 30: 101848. https://doi.org/10.1016/j.bcab.2020.101848.

Sidorova, T. M., Asaturova, A. M., Homyak, A. I., Zhevnova, N. A., Shternshis, M. V., and Tomashevich, N. S. (2020). Optimization of laboratory cultivation conditions for the synthesis of antifungal metabolites by *bacillus subtilis* strains. *Saudi Journal of Biological Sciences*, 27(7): 1879–1885. https://doi.org/10.1016/j.sjbs.2020.05.002.

Singh, P. (2019) Hand Book of Phytochemical, pharmacognostical water and edible oil analysis. *New Delhi, India: Educreation Publishing*; (pp. 1–92).

Singh, P., and Srivastava, D. (2012). Biofungicidal or biocontrol activity of *Lantana camara* against phytopathogenic *Alternaria alternata*. *International Journal of Pharmaceutical Sciences and Research*, 3(12): 4818–4821.

Singh, V., Upadhyay, R. S., Sarma, B. K., and Singh, H. B. (2016). *Trichoderma asperellum* spore dose depended modulation of plant growth in vegetable crops. *Microbiological Research*, 193: 74–86. https://doi.org/10.1016/j.micres.2016.09.002.

Sitthisak, C., Nisoa, M., Chunglok, W., Prasopthum, A., Phaisan, S., Putalun, W. et al. (2023). Efficient extraction of quassinoids and alkaloids from *Eurycoma longifolia* Jack roots using natural deep eutectic solvents and microwave-assisted extraction. *Microchemical Journal*, 196: 109676. https://doi.org/10.1016/j.microc.2023.109676.

Sjödahl, J., Dilley, K. J., Eriksson, R., Pellerin, J., Parvez, S. H., and Arlinger, L. (2020). Technical aspects of biochemical high-performance column liquid chromatography. *In: Gel Permeation and Ion-Exchange Chromatography of Proteins and Peptides*, 179–217.

Soković, M. D., Glamočlija, J. M., and Ćirić, A. D. (2013). Natural products from plants and fungi as fungicides. In IntechOpen eBooks. https://doi.org/10.5772/50277.

Sornakili, A., Thankappan, S., Sridharan, A., Nithya, P., and Uthandi, S. (2020). Antagonistic fungal endophytes and their metabolite-mediated interactions against phytopathogens in rice. *Physiological and Molecular Plant Pathology*, 112: 101525. https://doi.org/10.1016/j.pmpp.2020.101525.

Soylu, E. M., and Incekara, R. (2017). Biofungicidal activities of plant essential oils against cucumber root and stem rot disease caused by *Fusarium oxysporum* f. sp. radicis-cucumerinum. *Journal of Plant Pathology*, 437–444. https://www.jstor.org/stable/44686789.

Srivastava, N., Singh, A., Kumari, P., Nishad, J. H., Gautam, V. S., Yadav, M. et al. (2020). Advances in extraction technologies: isolation and purification of bioactive compounds from biological materials. pp. 409–433. *In: Elsevier eBooks*. https://doi.org/10.1016/b978-0-12-820655-3.00021-5.

Stéphane, F. F. Y., Jules, B. K. J., Batiha, G. E., Ali, I., and Bruno, L. N. (2021). Extraction of bioactive compounds from medicinal plants and herbs. In *IntechOpen eBooks*. https://doi.org/10.5772/intechopen.98602.

Subba, R., and Mathur, P. (2022). Functional attributes of microbial and plant based biofungicides for the defense priming of crop plants. *Theoretical and Experimental Plant Physiology*, 34(3): 301–333. https://doi.org/10.1007/s40626-022-00249-x.

Sujarit, K., Pathom-Aree, W., Mori, M., Dobashi, K., Shiomi, K., and Lumyong, S. (2020). Streptomyces palmae CMU-AB204T, an antifungal producing-actinomycete, as a potential biocontrol agent to protect palm oil producing trees from basal stem rot disease fungus, *Ganoderma boninense*. *Biological Control*, 148: 104307. https://doi.org/10.1016/j.biocontrol.2020.104307.

Sun, Y., Wang, Y., Han, L., Zhang, X., and Feng, J. (2017). Antifungal activity and action mode of cuminic acid from the seeds of *Cuminum cyminum* L. against *Fusarium oxysporum* f. sp. Niveum (FON) causing *Fusarium* wilt on watermelon. *Molecules*, 22(12): 2053. https://doi.org/10.3390/molecules22122053.

Sundin, G. W., Werner, N. A., Yoder, K. S., and Aldwinckle, H. S. (2009). Field evaluation of biological control of fire blight in the eastern United States. *Plant Disease*, 93(4): 386–394. https://doi.org/10.1094/PDIS-93-4-0386.

Sung, N. S. H. (2012). Antibacterial and antioxidant activities of tannins extracted from agricultural by-products. *Journal of Medicinal Plants Research*, 6(15). https://doi.org/10.5897/jmpr11.1575.

Sunkar, S., Ranimol, G., Nellore, J., Nachiyar, C. V., Namasivayam, S. K. R., and Renugadevi, K. (2022). Extraction and purification of bioactive ingredients from natural products. *Handbook of Nutraceuticals and Natural Products: Biological, Medicinal, and Nutritional Properties and Applications*, 2: 221–257. https://doi.org/10.1002/9781119746843.ch23.

Suzuki, S., Koeduka, T., Sugiyama, A., Yazaki, K., and Umezawa, T. (2014). Microbial production of plant specialized metabolites. *Plant Biotechnology*, 31(5): 465–482. https://doi.org/10.5511/plantbiotechnology.14.1003a.

Talapatra, S. K., Talapatra, B., Talapatra, S. K., and Talapatra, B. (2015). Shikimic acid pathway. Chemistry of plant natural products: Stereochemistry, conformation, synthesis, biology, and medicine. In *Springer eBooks*. 625–78. https://doi.org/10.1007/978-3-642-45410-3.

Tomah, A. A., Zhang, Z., Alamer, I. S., Khattak, A. A., Ahmed, T., Hu, M. et al. (2023). The potential of *Trichoderma*-mediated nanotechnology application in sustainable development scopes. *Nanomaterials*, 13(17): 2475.

Tripathi, S., Prakash, V., and Morya, V. (2022). An overview of natural medicinal plant extraction, isolation and characterization methods in improvement of medicinal research. *International Journal of Scientific Development and Research*, 7: 842–856.

Tyśkiewicz, R., Nowak, A., Ozimek, E., and Jaroszuk-Ściseł, J. (2022). Trichoderma: the current status of its application in agriculture for the biocontrol of fungal phytopathogens and stimulation of plant growth. *International Journal of Molecular Sciences*, 23(4): 2329. https://doi.org/10.3390/ijms23042329.

Uwineza, P. A., and Waśkiewicz, A. (2020). Recent Advances in Supercritical Fluid Extraction of Natural Bioactive Compounds from Natural Plant Materials. *Molecules*, 25(17): 3847. https://doi.org/10.3390/molecules25173847.

Van Loon, L. C. (2015). The intelligent behavior of plants. *Trends in Plant Science*, 21(4): 286–294. https://doi.org/10.1016/j.tplants.2015.11.009.

Vero, S., Garmendia, G., Allori, E., Sanz, J. M., Gonda, M., Alconada, T. et al. (2023). Microbial biopesticides: Diversity, scope, and mechanisms involved in plant disease control. *Diversity*, 15(3): 457. https://doi.org/10.3390/d15030457.

Wei, H., Wang, Y., Jin, Z., Yang, F., Hu, J., and Gao, M. (2020). Utilization of straw-based phenolic acids as a biofugicide for a green agricultural production. *Journal of Bioscience and Bioengineering*, 131(1): 53–60. https://doi.org/10.1016/j.jbiosc.2020.09.007.

Wen, C., Zhang, J., Zhang, H., Dzah, C. S., Zandile, M., Duan, Y. et al. (2018). Advances in ultrasound assisted extraction of bioactive compounds from cash crops – A review. *Ultrasonics Sonochemistry*, 48: 538–549. https://doi.org/10.1016/j.ultsonch.2018.07.018.

Win, A., Nyo, A. M. T., Mon, S. S., and Phyu, S. (2019). Phytochemical screening, assessment of mineral content and total flavonoid content of stem bark of *Dalbergia lanceolaria* L. *International Journal of Trend in Scientific Research and Development*, 3(5): 1618–1622. https://doi.org/10.31142/ijtsrd26776.

Wolffhechel, H., and Funck Jensen, D. (1992). Use of *Trichoderma harzianum* and *Gliocladium vitens* for the biological control of post-emergence damping-off and root rot of cucumbers caused by *Pythium ultimum*. *Journal of Phytopathology*, 136(3): 221–230. https://doi.org/10.1111/j.1439-0434.1992.tb01301.x.

Wraight, S. P., Galaini-Wraight, S., Howes, R. L., Castrillo, L. A., Griggs, M. H., Carruthers, R. I. et al. (2021). Efficacy of *Beauveria bassiana* strain GHA spray applications against coffee berry borer *Hypothenemus hampei* on Hawai'i Island. *Biological Control*, 161: 104587. https://doi.org/10.1016/j.biocontrol.2021.104587.

Wrona, O., Rafińska, K., Możeński, C., and Buszewski, B. (2017). Supercritical fluid extraction of bioactive compounds from plant materials. *Journal of AOAC International*, 100(6): 1624–1635. https://doi.org/10.5740/jaoacint.17-0232.

Wutsqa, Y. U., Suratman, S., and Sari, S. L. A. (2021). Detection of terpenoids and steroids in *Lindsaea obtusa* with thin layer chromatography. *Asian Journal of Natural Product Biochemistry*, 19(2). https://doi.org/10.13057/biofar/f190204.

Xie, Z., Li, M., Wang, D., Wang, F., Shen, H., Sun, G. et al. (2020). Biocontrol efficacy of *Bacillus siamensis* LZ88 against brown spot disease of tobacco caused by *Alternaria alternata*. *Biological Control*, 154: 104508. https://doi.org/10.1016/j.biocontrol.2020.104508.

Yadav, M., Divyanshu, K., Dubey, M. K., Rai, A., Kumar, S., Tripathi, Y. N. et al. (2023). Plant growth promotion and differential expression of defense genes in chilli pepper against *Colletotrichum truncatum* induced by *Trichoderma asperellum* and *T. harzianum*. *BMC Microbiology*, 23(1). https://doi.org/10.1186/s12866-023-02789-x.

Yadav, N., Sharma, S., Joys, J. S., and Kumar, S. (2020). Microwave assisted extraction of bioactive compounds: A brief review. *Journal of the Indian Chemical Society*, 97(10a), 1751–1756.

Yan, L., He, L., and Xi, J. (2015). High intensity pulsed electric field as an innovative technique for extraction of bioactive compounds—A review. *Critical Reviews in Food Science and Nutrition*, 57(13): 2877–2888. https://doi.org/10.1080/10408398.2015.1077193.

Yang, M., Ren, S., Shen, D., Chou, S., and Qian, G. (2019). ClpP mediates antagonistic interaction of *Lysobacter enzymogenes* with a crop fungal pathogen. *Biological Control*, 140, 104125. https://doi.org/10.1016/j.biocontrol.2019.104125.

Yanty, D. P., Trizelia, T., Darnetty, D., and Trisno, J. (2024). Ability of *Beauveria bassiana* to suppress *Colletotrichum truncatum* and increase the growth of chili plants (*Capsicum annuum*). *Jurnal Proteksi Tanaman*, 8(1): 42–52. https://doi.org/10.25077/jpt.8.1.42-52.2024.

Youssef, A. S. B. (2023). Modern trends for the application of biological control and modern technologies in agricultural projects. *International Journal of Modern Agriculture and Environment*, 1(2): 26–53. https://doi.org/10.21608/ijmae.2023.215953.1013.

Zeng, W., Wang, D., Kirk, W., and Hao, J. (2012). Use of *Coniothyrium minitans* and other microorganisms for reducing *Sclerotinia sclerotiorum*. *Biological Control,* 60(2): 225–232. https://doi.org/10.1016/j.biocontrol.2011.10.009.

Zhang, H., Kong, N., Liu, B., Yang, Y., Li, C., Qi, J. et al. (2022). Biocontrol potential of *Trichoderma harzianum* CGMCC20739 (Tha739) against postharvest bitter rot of apples. *Microbiological Research*, 265: 127182. https://doi.org/10.1016/j.micres.2022.127182.

Zhang, M., Chen, B., Dai, H., Sun, J., Liu, H., and Han, J. (2024). Discovery of antifungal secondary metabolites from an intestinal fungus *Fusarium* sp. *The Journal of Antibiotics*, 77(3): 193–198. https://doi.org/10.1038/s41429-023-00692-1.

5

Carriers, Adjuvants, and Surfactants in the Formulation and Application of Biofungicides

Kamel A. Abd-Elsalam,[1,*] *Ayat F. Hashim,*[2] *Nada Essam Abo-Rashed,*[3] *Hassan Almoammar,*[4] *Rawan K. Hassan*[5] and *Toka E. Abdelkhalek*[5]

1. Introduction

Biofungicides are getting increasingly popular among growers as green options against chemical fungicides. Any progress in this field is based on a detailed knowledge of both biological and chemical concepts. This chapter provides a detailed overview of biofungicide formulation, helping readers to learn the main concepts (Galindo et al. 2013). Making biofungicides requires choosing active ingredients, typically living organisms, and pairing them with other materials for strong results, durability, and minimal effort to use them. For a formulation to work well, it must use the maximum power of its ingredients but also make sure they mix well with other substances for agriculture and meet all legal requirements (Parlakidis et al. 2023).

The chapter starts by discussing how to choose suitable biofungicidal products, looking at their actions, the ranges of fungi they treat, and their safety

[1] Plant Pathology Research Institute, Agricultural Research Center, 12619, Giza, Egypt.
[2] Fats and Oils Department, National Research Centre, Cairo, Egypt.
[3] Food Processing Technology English Program, Faculty of Agriculture, Cairo University, Giza 12613, Egypt.
[4] Wellness and Preventive Medicine Institute, Health Sector, King Abdulaziz City for Science and Technology (KACST), Riyadh, Saudi Arabia.
[5] Biotechnology English Program, Faculty of Agriculture, Cairo University, Giza 12613, Egypt.
* Corresponding author: kamelabdelsalam@gmail.com

DOI: 10.1201/9781003603023-6

for the environment. Then, it explores the various formulations found, for example, wettable powders, granules, and liquid concentrates, which are each developed to match particular needs and methods in growing crops (Adusei-Fosu et al. 2021). In formulation science, carriers, adjuvants, and surfactants are important tools. They help to stabilize, spread, and increase the power of biofungicides. This chapter discusses both properties and usages of these additives, helping readers pick the best ones for their product (Montesinos 2003). The chapter also explains the laws and regulations for making and using biofungicides. Making sure you understand these laws protects product safety, lets you enter the market, and ensures your products meet worldwide standards.

2. Biofungicide Formulation Basics

2.1 Selection of Active Ingredients

The right active ingredients must be chosen for the formulation of biofungicides. Without the active ingredients, a fungicide cannot fight disease-causing fungi (Boyetchko et al. 1999). In this part, we will explore the points that should be considered when selecting active ingredients for making biofungicides.

2.1.1 Mode of Action

The mode of action describes the specific way in which an active ingredient interacts with a fungal pathogen to prevent or kill it. Depending on the active ingredient, some substances work by targeting one stage in the life cycle of fungi, and others by targeting different ones (Köhl et al. 2019).

There are several common modes of action for biofungicides (Figure 1), including:

Antibiosis: Some active ingredients, such as those in a broad-spectrum fungicide, work by inhibiting the development of harmful fungi.

Competition: Various active ingredients may occupy important nutrients in the soil or take up space, so fungal diseases have reduced chances of starting.

Induced resistance: Active substances can enhance the plant's self-defense and hence prevent attacks by many fungi.

Interfering with the cell membranes: A few active substances can alter the structure of fungal cell membranes which can cause the fungi to die.

Enzyme inhibition: Some active compounds are able to prevent the enzymes needed for fungi to multiply and grow (Ons et al. 2020).

2.1.2 Efficacy and Spectrum of Activity

When making biofungicides, you must consider the effectiveness and range of action of the active ingredients. The efficacy of an active ingredient is how well it manages to control the fungal pathogens at suggested application amounts. Active ingredients that are known to work against the particular fungal diseases you wish

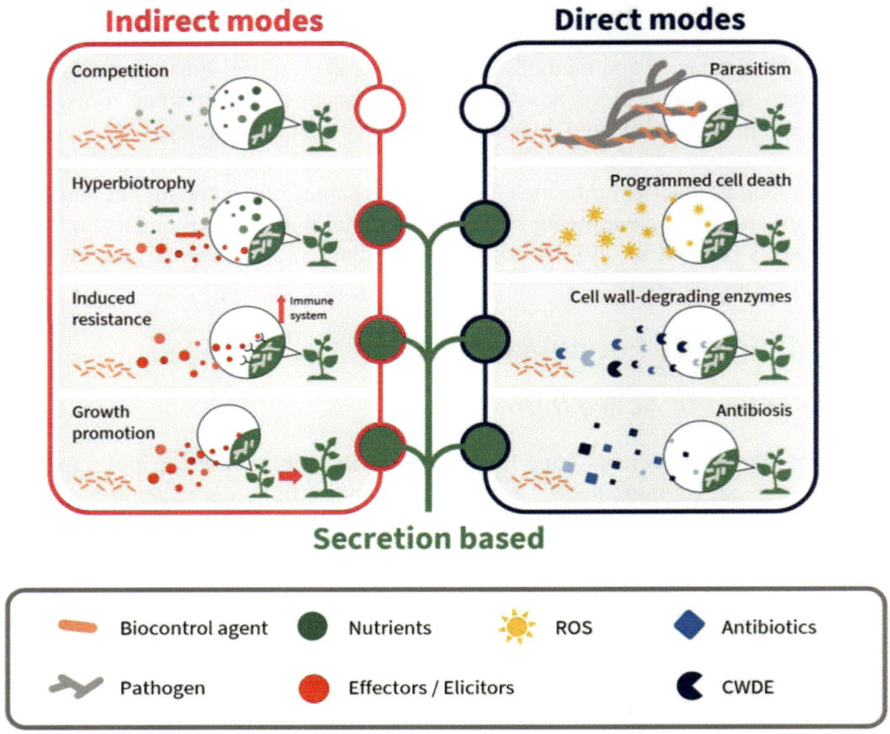

Figure 1: The main modes of action of biocontrol agents (BCAs) against plant pathogens (Srivastava et al. 2021).

to control should be used. The spectrum of activity is the array of fungal pathogens that an active substance is effective against. There are active ingredients that can work against different species of fungi, but some can only attack specific ones. It is important to use active ingredients that work against the common fungal diseases in the target area or setting.

2.1.3 Regulatory Considerations

The choice of active ingredients for biofungicide formulation is guided by regulations. Biofungicides can only be registered and used in accordance with the guidelines set by each region. Choose ingredients that are approved in your country and obey the correct regulations (Jiang and Wang 2023, Galindo et al. 2013). A close look at the law is needed to understand the rules for active ingredients before selecting a product. Reviewing MRLs and PHIs is important to protect humans and the environment from exposure to active ingredients.

2.1.4 Fitting Well with Ingredients

A good formulation depends on the formulation components, carriers, adjuvants, and surfactants all being compatible with the active ingredient. The contact between active and other components may increase the risk of instability, reduce how well the formulation works, or cause it to be in conflict with other components (Khun et al. 2021).

It's vital to make sure the main ingredients won't react negatively with the other ingredients before you begin the chemical reaction. Under compatibility testing, scientists evaluate how the active components interact with the various components in the formulation.

2.2 Formulation Types and Their Advantages

They play a key role in controlling fungal infections that affect agricultural crops (Abbey et al. 2019). They contain active materials that are sourced from bacteria, fungi, or plant extracts. But the success of biofungicides depends on the vehicle chosen, as well as the adjuvants and surfactants used in their mixture (Zamoum et al. 2022; Chen et al. 2019). This section examines the benefits of using various biofungicidal formulations in the treatment of diseases.

2.2.1 Liquid Formulations

Most biofungicide products are applied as liquids on crops (Kolombet et al. 2008). Most of the time, they are produced as suspensions, emulsions, or solutions. Using liquid formulations is more convenient, simple to spread, and works well with other products in farming.

2.2.1.1 Suspension Formulations

These formulations have small amounts of the active substance evenly spread within a liquid. Since the active substances are evenly spread, their effects tend to be dependable. Because the ingredients are not easily dissolvable, these formulations last longer and do not settle quickly. They are most beneficial when a biofungicide has problems with water stability or solubility.

2.2.1.2 Emulsion Formulations

An emulsifier creates a bond that enables oil and water to be mixed together in an emulsion. This formulation holds the molecules, maintains them evenly spread, and ensures they treat the plant effectively. Once the solution is diluted, it sprays easily from standard equipment.

2.2.1.3 Solution Formulations

Solutions are made by mixing the active ingredient with a solvent such as water. For application, they are very handy because the plants absorb them rapidly. You can use these products as sprays on plants, coatings for seeds, or apply them to soil, which makes them convenient for different uses.

2.2.2 Dry Formulations

Many biofungicides are applied in agriculture as dry materials. The most common forms of these medicines are powders, granules or dusts (Berninger et al. 2017).

For powders, the main ingredient is finely milled with other materials such as talc or clay (Allali et al. 2022). They are helpful as they remain effective for a long time and are simple to use. To apply by spray, products in powder form are mixed with water to prepare suspensions or solutions.

2.2.3 Pharmaceuticals in Granule Form

Active ingredients are mixed with inert materials such as clay or vermiculite to create granules of a desired size. They help to evenly distribute the product and apply it where it is required. The granules can be placed directly in the soil, where they release the biofungicide bit by bit and protect the crop for a long period. They are especially helpful for soil-borne diseases and can be spread either by seed drills or traditional field spreaders.

2.2.4 Advantages of Different Formulations

Making a biofungicide depends on factors such as the pathogen, the crop being treated, the method of application, and the target environment (2016). Biofungicides function better when they are formulated in various ways. Using suspensions, emulsions, and solutions in a liquid formulation covers plants with the active ingredient and prevents diseases from developing longer. These are flexible, as they can be applied in several methods and absorbed rapidly by plants. In contrast, powders, granules, or dusts don't need much storage space and can sit for months without settling.

2.3 Compatibility Testing and Formulation Stability

A biofungicide's performance depends on running compatibility and stability tests. We need to test the way various pieces of the formulation interact and determine if they remain effective over time. This section explains the need for stability testing, the factors that affect it, and the approaches to keep biofungicide products stable.

2.3.1 Compatibility Testing Plays a Key Role

Scientists test the effectiveness of mixtures by exploring the way their components—ingredients, carriers, adjuvants, surfactants, and additives—relate (Fravel et al. 2005). The goal is to ensure safety and effectiveness as the product's components are used.

Paying attention to the mixing procedure ensures important ingredients are protected, undesirable interactions are avoided, and the biofungicide is more effective.

2.3.2 Formulation's Stability

We also look at the factors that may affect the stability of a formula in this section. This means that the features of a biofungicide formulation won't change as time goes on (Gikas et al. 2022). Several factors like temperature, pH, and light conditions can affect the stability of biofungicides.

2.3.2.1 Chemical Interactions

Mixing various ingredients in a formula can cause it to become unstable. For example, certain ingredients or additives may not be compatible, causing the biofungicide to break down or form small particles. Compatibility testing helps prevent such issues.

2.3.2.2 pH and Temperature

Biofungicide formulations can be affected by changes in temperature and pH. If the pH or temperature of the solution changes, it can weaken some active ingredients. When creating and storing biofungicides, the pH and temperature should be carefully adjusted to ensure their stability.

2.3.2.3 Physical Stability

If a formulation is physically stable, its structure, particle size, and separation remain the same. Metastable conditions have been investigated to understand why different materials in formulations, like sedimentation, creaming, and phase separation, can cause instability in biofungicide products.

2.3.3 Methods for Evaluating Formulation Stability

Several strategies are needed to verify that biofungicide formulations are safe and effective for use in the short term as well as the long term (Mokhtarnejad et al. 2011). These approaches can be considered the most commonly used ways.

Visual checking: Changes in formulation color, look, or overall features are observed. When there is phase separation, sedimentation, or precipitation, the system is considered unstable.

Performing tests related to shear thinning: An analysis of particle size shows if aggregation or flocculation is occurring in the product. Researchers use laser diffraction or microscopy to determine the size of dust particles.

Testing using chemicals: Chemical testing focuses on how the formula changes its shape and flow. Viscosity, shear stress, and elasticity are monitored with the aid of rheometers and rotational viscometers. To see if the ingredients in the product will remain stable over time, this step measures the amounts of different substances in the formulation using HPLC or GC.

2.4 Regulatory Considerations for Biofungicide Formulation

2.4.1 Registration and Approval Process

A biofungicide cannot be sold until it has undergone a full registration and approval process (Montesinos 2003). In most cases, agencies such as the EPA in the United States and the EFSA in the European Union oversee the process. Its main goal is to verify that the biofungicide is an effective and safe product.

Formulators must supply information on the active ingredients present in the product, as well as facts on its ability to control fungi when registering a biofungicide. Laboratory and field tests are usually conducted to obtain the collected data.

2.4.2 Maximum Residue Limits (MRLs)

Setting Maximum Residue Limits (MRLs) for biofungicides is important for companies developing them (Ons et al. 2020). MRLs show the greatest amount of pesticide that is allowed in food. Their purpose is to keep crops that have been treated from harming people when consumed.

Any biofungicide for use on various crops needs to be below the allowed maximum residue limits for those crops. Watch the rates and deadlines for the applications you use and make sure to comply with all laws about pre-harvest results.

2.4.3 Rules for Labels

Labeling is key for biofungicide products, since it guarantees their safety and proper use. Thankfully, regulatory agencies also require that biofungicide labeling provides information on the product's active ingredients, the right way to use it, important safety rules, and any limits.

All labels on biofungicides must comply with the required regulations and guidelines. They should clearly put the ingredients on the label, list their amounts, and clearly state how to properly apply, keep, and dispose of the product. If a product does not comply with the labeling guidelines, it may be called misbranded and could be ordered off the market.

2.4.4 Environmental Impact Assessment

Chemists should have environmental considerations in mind when they formulate biofungicides. When submitting a biofungicide for registration, information on its environmental fate is expected from the formulator.

In some cases, formulators have to assess the environmental impact of their biofungicide products (Mir et al. 2023). One must consider whether beneficial insects or creatures in water may be affected and if there is a risk of polluting groundwater.

2.4.5 Good Manufacturing Practices (GMP)

Formulators of biofungicides must not only consider the points given above, but also follow Good Manufacturing Practices (GMP). GMP ensures that the quality and consistency of biofungicide production are maintained during the whole process.

Among the topics included in GMP guidelines are the environment for biofungicide formulation, safety of equipment, ingredients, batch testing, record-keeping, and other necessary requirements. Implementing GMP reduces the chance that the manufacturing of biofungicide products will become contaminated, vary in quality, or experience other issues.

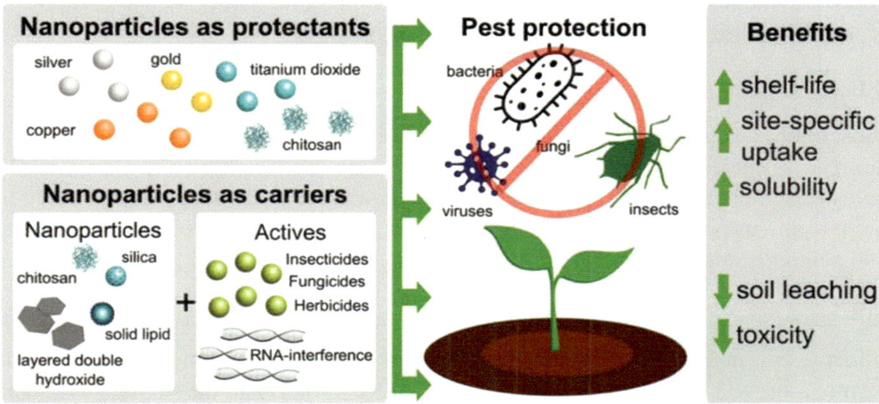

Figure 2: Nanomaterials as protectants or carriers to provide crop protection (Worrall et al. 2018).

3. Carriers for Biofungicide Formulation

3.1 Role of Carriers in Biofungicide Formulation

Carriers make it possible for biofungicides to travel to where fungal diseases are found on the plants, making their action more effective (Faruk et al. 2014). Using carriers enables you to spray biofungicide more often, which often delivers better results (Figure 2).

Carriers are important for a range of reasons. By doing this, they guard the product against harmful changes from UV rays, temperature variations, and humidity so it stays active for the whole storage period. Besides, applying the biofungicide with a carrier lets it distribute evenly in the plant, so the pathogens can be dealt with more effectively. Carriers provide better stability to the formulation since they keep all the ingredients well mixed, which improves its storage, use, and application (Miastkowska et al. 2020).

3.2 Types of Carriers and Their Properties

Carriers play a crucial role in the formulation and application of biofungicides. They serve as a medium to deliver the active ingredients to the target site effectively. The selection of an appropriate carrier is essential to ensure the stability, efficacy, and safety of the biofungicide product. In this section, we will explore the different types of carriers commonly used in biofungicide formulations and discuss their properties.

3.2.1 Liquid Carriers

Many biofungicides are formulated as liquids because they make handling simple and ensure the active chemicals dissolve easily. After blending the biofungicide, they apply it to the plant leaves in an even manner (Kolombet et al. 2008). These three are typical liquid carriers: water, oils, and organic solvents.

3.2.1.1 Water

Most often, water is used as the liquid chosen for making biofungicide solutions. The amount of nuclear power available to us is substantial; it is typically cheap, and it helps lessen damage to the environment. These water-based solutions are simple to prepare and use on plants, helping them become popular choices for seed treatments, soil solutions, and leaf sprays (Guo et al. 2023).

3.2.1.2 Oils

Mineral oil and vegetable oil are common carriers included in biofungicide formulations. They help in covering the plant evenly because they both spread and wet well on the leaves (Wurms and Chee 2018). Oils help biofungicides go deeper and be taken up better by plant cells.

3.2.1.3 Organic Solvents

Occasionally, products such as ethanol or acetone are added to biofungicides as carriers. They can dissolve lipophilic active substances that will not dissolve in water or oil. Absorbing into the plant more effectively, organic solvents can help make the chemical biofungicides work better (Purkait et al. 2022).

3.2.2 Solid Carriers

In most cases, using a solid carrier in a biofungicide formulation helps keep the ingredients stable, safe, and easy to apply. Powders, granules, and pellets are a few ways that minerals can come. Thanks to controlled release, better adhesion, and a reduced impact on nature, solid carriers are seen as beneficial (Mattedi et al. 2023, Martínez-Medina et al. 2009).

3.2.3 Inorganic Carriers

Many biofungicide preparations rely on clays and minerals as their inorganic carriers. They actively protect the ingredients inside the capsule against breakdown and improve their stability. Using inorganic carriers improves the attachment of biofungicides to plants and lengthens the products' effectiveness. Examples of inorganic carriers are kaolin clay, bentonite, and silica (Zamoum et al. 2022, Abd-El-Kareem et al. 2019).

3.2.4 Organic Carriers

Many biofungicides include organic carriers, including starch, cellulose and lignin (Thiago et al. 2020, Saberi-Riseh et al. 2023). They are gentle on active substances, and they offer controlled ways to deliver materials. Because of their water-retaining properties, organic carriers gradually release the supply of biofungicides.

3.2.5 Biodegradable Polymers

A lot of biofungicide producers now choose to use PLA and PHBV, as opposed to other biodegradable polymers (Ali et al. 2023). Polymers in this group are environmentally friendly and can be created to release biofungicides as required. The slow release of biofungicides because of biodegradable polymers means they need to be used less and do less damage to the environment (Pirzada et al. 2020).

3.2.6 Multipurpose Ships

Incorporating more carrier forms into one protection system helps make and use biofungicides more effectively (Xu et al. 2023). The purpose behind the products is to enhance stability, help the formulation stick, and ensure controlled release. A sensible approach is often to combine liquids with solids, and sometimes various solid carriers are used together.

3.2.6.1 Liquid-Solid Composites

In liquid-solid composites, water and the ingredients in oils and clays are combined with polymers to take advantage of both liquid and solid carriers (Ramakutoane et al. 2023). With these products, we get a more stable structure, equal release of ingredients, and good stickiness.

3.2.6.2 Solid-Solid Composites

To produce specific effects, solid-solid composites mix different solid materials (Wang et al. 2017, Abdelwahab et al. 2023). Blending inorganic and organic carriers can guarantee safety as well as permit precise dosing of biofungicides. The flexible structure of solid-solid composites allows them to be altered to fulfill certain uses.

3.3 Selection Considerations

There are several points to keep in mind when choosing carriers for each type of biofungicide (Faruk et al. 2014).

3.3.1 Selection

Choose a product that preserves the main ingredients and keeps them at their highest effectiveness. Researchers make sure nothing unwanted happens when different things are used together.

3.3.2 Solubility

Measuring liquid carriers' solubility and whether active substances dissolve in water is crucial. It works well if you are able to choose a carrier that allows the ingredients to spread or break down in the dish.

3.3.3 Stability

The carrier needs to ensure the biofungicide remains efficient for a long time, not allowing it to become ineffective. Performing stability tests helps decide the length of time a formulation can be stored and what conditions are best for it.

3.3.4 Application Method

Choosing the right carrier means it will work well no matter if a treatment is applied to seeds, to the soil, as a foliar spray, etc. It ought to be possible to apply the biofungicide evenly so that the coverage is uniform on all areas.

3.4 Optimizing Carrier Concentration and Compatibility

Carriers are necessary in the process of using biofungicides as they help bring the active ingredients to the important sites. The correct choice and tuning of carrier concentration influence the performance of all biofungicides. A successful combination between the carrier and other ingredients is necessary for the product to remain reliable and work as intended.

3.4.1 Optimizing Carrier Concentration and Compatibility

Delivering active ingredients in biofungicides depends heavily on the carriers used. Optimal performance of a carrier is achieved when there is enough concentration to help it stick and be moved effectively to where it should go (Purkait et al. 2022). Applying the right amount is important so you cover the area and use none in excess.

3.4.2 Importance of Carrier Concentration

The effectiveness and safety of a biofungicide depend a lot on how compatible the carrier is with other substances (Uchida et al. 2021). A product that does not dissolve correctly or works less effectively may be due to incompatible ingredients. It is important to run compatibility tests to catch any issues related to precipitation or instability.

3.4.3 Optimization Strategies

Applying different methods can improve the amount and mix of carriers used in making and using biofungicides (Palmieri et al. 2022).

3.4.3.1 Formulation Optimization

Scientists check a potential carrier's viscosity, surface tension, and size of the particles it contains to see if it is more suitable than competitors. You have to test and experiment with the carrier's amount to find out how effectively the biofungicide fights pathogens on plants.

3.4.3.2 Application Optimization

The results of a biofungicide depend on how it is applied. Equipment needs to match the type of chemical being used along with the type of crop you're caring for.

The level of deposition and distribution is influenced by the drop size, amount of spray, and type of nozzle.

3.4.3.3 Compatibility Testing

It is necessary to check how the biofungicide and soil nutrients work together regularly to ensure the system is stable and effective. Here, we look at how the carrier interacts with other essential ingredients such as active components, adjuvants, and surfactants. Tests should check appearance, pH, and how the item reacts to changes in temperature and light to match real situations.

4. Adjuvants in Biofungicide Formulation

4.1 Importance of Adjuvants in Biofungicide Performance

Using adjuvants makes biofungicides more effective because they help the products attach better to, penetrate through, and stick to plants (Adusei-Fosu et al. 2021). They help people come into contact with pathogens, making it easier to keep diseases under control. In addition, adjuvants help transport biofungicides throughout the plant (Parlakidis et al. 2023).

4.1.1 Extending Crop Life and Rain Resistance

With the help of adjuvants, biofungicides maintain their potency in the field by cutting down on evaporation and giving longer protection (Palma-Bautista et al. 2020). They form a surface that helps the product stay intact in sunlight and in rain, allowing it to keep working against fungal diseases, according to Gent et al. (2003). If biofungicides have improved rainfastness, they will still work after becoming wet (Hunsche et al. 2007).

4.1.2 Increasing the Compatibility of the Input

Many farmers combine biofungicides with fertilizers, insecticides, and herbicides for best results. Compatibility with other sprays and better performance of the spray solution are achieved thanks to the use of adjuvants (Holka and Kowalska 2023). Integrated pest management is easy to use with the biofungicide since it does not affect its usefulness.

4.2.3 Working to Protect the Environment

Applying biofungicides with adjuvants allows for a better ecological outcome by making applications more effective and using less of the main ingredient (Ryckaert et al. 2005). When performance is boosted, adjuvants lessen the dangers to both non-targeted organisms and nature while still controlling diseases properly.

4.2.4 Control Agents and Antifoaming Substances

The use of drift control agents reduces how the wind directs droplets away from the intended location (Oliveira et al. 2013). Thanks to antifoaming agents, there is less foam when mixing and applying the product, which helps it coat evenly and stay effective by lowering surface tension (Maria et al. 2023).

4.2 Types of Adjuvants and Their Functions

We will now look at different adjuvants and the roles they play.

Of note, biofungicide performance is greatly enhanced with the help of adjuvants. They are used to support the work of active ingredients in helping manage diseases and defend crops (Apostólico et al. 2016). The task of adjuvants varies with the requirements of the biofungicide and the kind of crop being grown. This section (Figure 3). looks at the different adjuvants in biofungicide formulation and what they do.

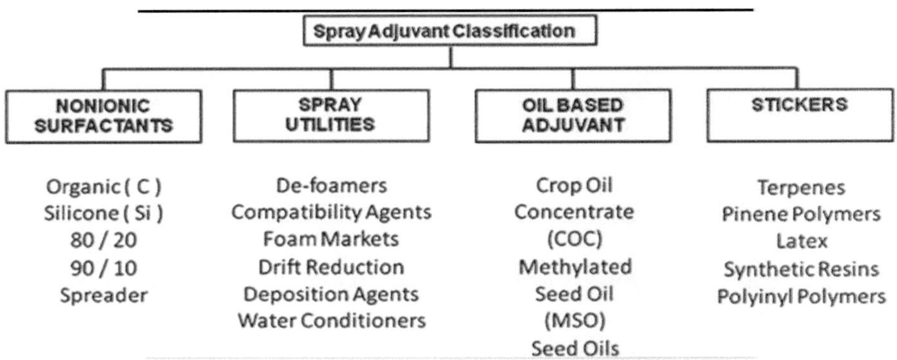

Figure 3: Classification of adjuvants by tank mix function (Adapted from Roberts 2010).

4.2.1 Wetting Agents

Spreader-stickers make biofungicides adhere better to plant leaves, since they help the droplets rest smoothly on the plant surface. As a result, these methods lower a person's risk of infection and make it easier to stop fungal diseases (Gao et al. 2022). Stickers make sure that biofungicides adhere properly to plants, so they keep working well even during rain or when the plants are watered (Gaskin and Steele 2009).

4.2.2 Penetrants and Spreader-Activators

Spray droplets often push away from the surface, but penetrants help biofungicides get into plants by lowering this repelling effect. That way, active components are concentrated in parts that need healing, leading to better treatment of fungal diseases (Orbović et al. 2007). Spreader-activators improve the application of the biofungicide on plants, as well as interact with the plant to boost its defense against diseases, thus leading to better plant health (Landim et al. 2019).

4.2.3 Buffering and Compatibility Agents

Buffer agents in biofungicide solutions help control pH, protect the reactive ingredients and improve their ability to fight fungi (Sobiech et al. 2020). Since compatibility agents keep ingredients from clashing in a product, the quality and reliability of the biofungicide remain stable (Smith 2019).

4.3 Adjuvant Selection for Enhanced Biofungicide Efficacy

Adjuvants help to boost the ability of biofungicides to be effective. These additional components help to enhance the way the main active ingredients in biofungicides work. By adding adjuvants, the biofungicide can be taken up better, stick more effectively, and penetrate well, resulting in fewer infections and healthier crops. We will now discuss why selecting the right adjuvant is crucial for better biofungicide performance.

4.3.1 Factors to Consider in Adjuvant Selection

Efficiency and compatibility in formulations can be achieved when choosing the right adjuvants for biofungicides, according to Habtom et al. (2013). For this purpose, the following issues are included:

4.3.1.1 Compatibility

Compatibility is necessary between the adjuvants and the active ingredients in any biofungicide product. They shouldn't create any problems or lower the performance of the biofungicide. Compatibility testing needs to be done to check if the biofungicide's performance will be influenced in a bad way by the adjuvant.

4.3.1.2 Crop and Target Pathogen

Because each pathogen is unique, select the right adjuvant by focusing on the crop and pathogen combination. Controlling different diseases often means using different adjuvants for each group of pathogens.

4.3.1.3 Application Method

The vaccine adjuvant has to be adaptable for the way it will be applied. Whether a biofungicide is sprayed on leaves, mixed with seeds, or drenched in soil, the adjuvant must boost the effectiveness of that method.

4.3.1.4 Environmental Conditions

The performance of adjuvants can be shaped by climate factors such as temperature, humidity, and rainfall. Choosing the right adjuvants, those effective in current weather and soil conditions, is important for the best performance of biofungicides.

4.3.2 Adjuvant Formulation and Dosage

Different types of adjuvants are developed to match the expected functions of the biofungicide and the crop type. They may be made into wet powder, concentrated liquid, or usable solutions. Adjuvant dosage must be chosen according to what the manufacturer suggests and what the chosen biofungicide requires (Holka and Kowalska 2023). Always use and apply the adjuvant as directed by the manufacturer to ensure it works best and does not harm your crops.

4.4 Adjuvant Application Methods and Dosage

Together, adjuvants and biofungicides allow the main substances to be more effective. This part looks at different application methods and the right doses for the best outcomes.

4.4.1 Application Methods

Success in adjuvant treatment depends a lot on the product and the crops being treated, which decide how the adjuvant should be applied (Bharti and Ibrahim 2020). These techniques are used quite often:

4.4.1.1 Tank Mixing

You use this method by combining the biofungicide with the adjuvant in the spray tank before spraying. If the tank mixture of the pesticide is even, it reaches all areas of the treated field and gives the same performance (Zhao et al. 2022).

4.4.1.2 Sequential Application

At times, turning the adjuvant on after you apply the biofungicide, known as sequential application, results in better protection. It gives the opportunity to carefully manage when and how much adjuvant is used, mainly when this is important during a certain growing phase or when things such as compatibility are involved (Webster et al. 2016).

4.4.1.3 Seed Treatment

Products are also applied during seed treatment by mixing them with the seeds before putting them in the soil. The coating process, which combines both a biofungicide and an adjuvant, stops soil-borne fungal diseases and helps young seedlings stay protected (Rocha et al. 2019).

4.4.2 Dosage Considerations

Properly deciding on the right quantity of adjuvants is key to having the best results. Depending on the adjuvant, target crop, biofungicide type, and fungal disease, the amount of the product may not be the same. You need to keep these factors in mind when settling on the dosage for:

4.4.2.1 Crop and Disease Factors

How much to use the adjuvant depends both on the type of crop and how severe the infection is. Some crops may respond best to higher use of adjuvants, but for other crops, a lower dose can work just as well. Depending on how badly the disease is spreading, the amount may be adjusted.

4.4.2.2 Compatibility with Biofungicide Formulation

Before setting the dosage, the fit of the adjuvant with the biofungicide must be taken into account. Taking certain adjuvants might change the stability or effectiveness of the active ingredients in a biofungicide. The best practice is to do compatibility testing before making the final decision on the dose of the adjuvant.

4.4.2.3 Environmental Factors

The levels of temperature, humidity, and sunlight can change the amount of adjuvant needed. You may have to add more biofungicide to the sprayer if the air is hot and dry, to obtain sufficient wetting and spreading. It is wise to take these factors into account when figuring out the proper dose.

5. Surfactants in Biofungicide Formulation

5.1 Role of Surfactants in Biofungicide Formulation

Surfactants are critical in the formulation and application of biofungicides. These surface-active compounds enhance the spreading and wetting properties of formulations by reducing the surface tension of liquids, allowing for more uniform distribution on target surfaces. This improvement in spreading helps increase the coverage and efficacy of active ingredients, leading to better disease control (Karamchandani et al. 2022).

5.1.1 Surfactant Selection for Improved Biofungicide Spreading and Wetting

Selecting the appropriate surfactant is vital for achieving optimal spreading and wetting properties. The choice depends on several factors, including the type of active ingredient, the target crop, and environmental conditions. Compatibility with the active ingredient and other formulation components must be considered to avoid negative impacts on efficacy. A surfactant with good spreading properties ensures broader coverage, enhancing contact with fungal pathogens (Zhang et al. 2023).

5.1.2 Optimizing Surfactant Concentration and Compatibility

The optimal concentration of surfactants is critical to ensure effective formulation. The selection should consider the same factors as above, focusing on the surfactant's spreading and wetting capabilities. Ensuring compatibility with active ingredients and other formulation components is essential to maximize biofungicide effectiveness.

5.1.3 Environmental Considerations and Surfactant Biodegradability

Surfactants also have environmental implications. Biodegradable surfactants are preferred because they decompose into non-toxic compounds over time, minimizing environmental impact (Scott and Jones 2000). Additionally, the use of surfactants in formulations must adhere to regulatory guidelines that dictate acceptable types and concentrations (Ben Khedher et al. 2020). Staying informed about these regulations is crucial to avoid legal issues and protect the environment. The development of economical surfactants (Figure 4), particularly near-natural surfactants (CTN-Surfs), allows for adjustments that enhance their performance while remaining environmentally responsible.

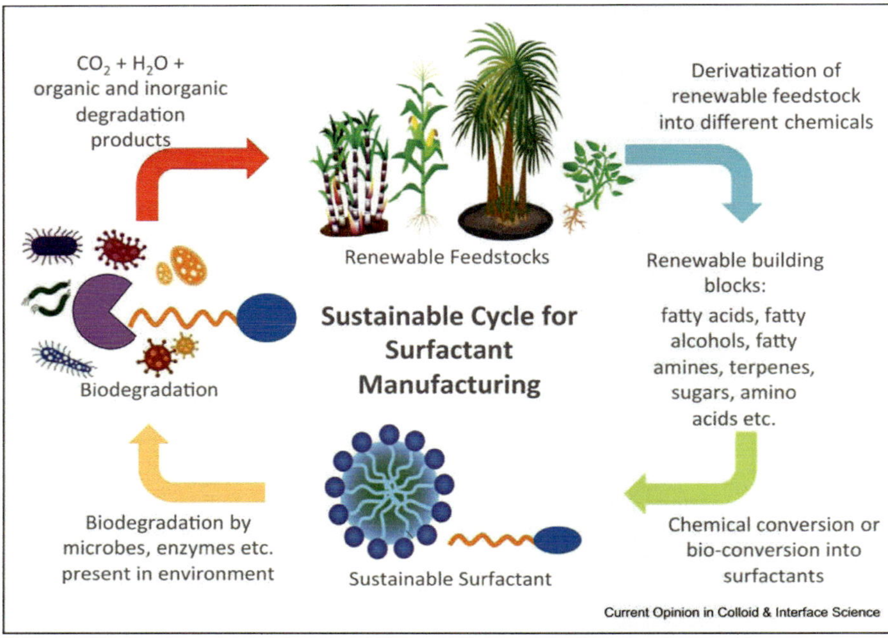

Figure 4: Ideal sustainable cycle for developing a new generation of sustainable surfactants (Bhadani et al. 2020).

5.2 Different Surfactants and the Specifics of Their Behavior

Surfactants play a key role in both developing and applying biofungicides, since they act to decrease the tension at the surface of liquids. Decreasing the volume of the formulation promotes better contact with plant leaves. Effectiveness is increased since more product stays on where it is applied (Mustafa et al. 2022).

5.2.1 Anionic Surfactants

Many biofungicides use anionic surfactants, as they carry a negative charge. They make it easier for the biofungicide to spread and absorb into plants. Almost all surfactants come from fats or petroleum, and they dissolve well in water, allowing them to mix well with oily compounds in products. Thanks to their good spreading and wetting features, biofungicides can be applied effortlessly on leaves using oils (Díaz-Blancas et al. 2016).

5.2.2 Cationic Surfactants

Anionic surfactants are typically used more often in making biofungicides than their positively charged relatives called cationic surfactants. Also, these chemicals relax the surface tension of water and can help make biofungicides stick better to plants. Most cationic surfactants are based on quaternary ammonium compounds and offer antimicrobial properties that help control diseases (Fait et al. 2019).

5.2.3 Nonionic Surfactants

Many formulators add nonionic surfactants to biofungicides since they are easy to mix with several active ingredients (Mustafa and Hussein 2020). They help spread and wet biofungicide solutions by reducing water's surface tension. Most nonionic surfactants are produced from fatty alcohols or ethoxylated compounds and are safe for plants. They can be used in formulas with many levels of pH, so they fit both acidic and alkaline products (Alamri 2014).

5.2.4 Amphoteric Surfactants

The pH of a solution affects whether amphoteric surfactants are anionic or cationic (Castro et al. 2014). While used less in most biofungicide formulations, these compounds may have important benefits for select uses. The use of amphoteric surfactants enables biofungicide solutions to be spread more evenly at the surface and to adhere well. They are commonly chosen for products where they interact well with other ingredients and help provide stability and prolong the shelf life of biofungicide preparations (Ons et al. 2020).

5.2.5 Silicone Surfactants

Biofungicides frequently contain silicone surfactants in their making. They help to reduce water's surface tension and make it easy for water to spread and form a strong bond with many surfaces. Using silicone surfactants in a formula can ensure the protective effects last, because they coat plant surfaces, increasing the life of the application (Lizamore and Winefield 2015). Also, they are able to survive rain and remain active for a long time.

5.3 Surfactant Selection for Improved Biofungicide Spreading and Wetting

Surfactants are needed in biofungicides to help the product spread and wet surfaces properly (Guijarro et al. 2018). They make liquids easier to spread and penetrate the target, so the active ingredients are more effective and the disease is controlled better.

5.3.1 Properties of Surfactants

Certain features are very important in surfactants for use with biofungicides (Karamchandani et al. 2022):

Surface Tension Reduction: Superior surfactants reduce the pulling effect of the water's surface which allows them to easily spread and stick to plant leaves.

Wetting Ability: Surfactants are needed to easily wet plants to allow penetration past the waxy covering to reach where they are used.

Stability: Formulating surfactants in a stable state ensures they do not lose their performance which matters for products used on living organisms.

Compatibility: Surfactants must be compatible with active ingredients and transporters in order for the formula to be stable and effective.

Environmental Impact: To help the environment, we choose surfactants that breakdown naturally and are safe for living creatures that aren't intended as targets.

5.3.2 Surfactant Selection Considerations

Many important criteria must be reviewed when selecting a surfactant for biofungicide mixtures.

Compatibility with Active Ingredients: The active ingredients in a product must get along with the surfactants chosen. Certain active ingredients may require certain surfactants to work well, making sure that all ingredients are mixed for best performance.

Target Surface: The special features of the plant's surface play a key role when choosing suitable herbicides. If the surface has waxy cuticles or is covered with hair, the surfactant may not wet it as easily.

Formulation Type: The kind of biofungicide (as a liquid or in powder form) affects which surfactant should be chosen. Depending on the formulation, a surfactant's properties must be carefully considered to meet all requirements.

6. Mixing and Homogenization Techniques

Mixing and homogenizing biofungicides guarantee that all active ingredients, carriers, adjuvants, and surfactants are distributed equally. As a result, the products remain stable and effective, do not settle or clump, and keep application equipment free from blockage for reliable disease control (Tanovic et al. 2015).

6.1 Mixing Techniques

Different blending strategies can be utilized amid the definition of biofungicides. A few commonly utilized blending strategies incorporate:

6.1.1 Mechanical Mixing

Mechanical blending involves the use of mechanical equipment such as blenders or stirrers to mix the ingredients. This method is effective in achieving a uniform distribution of the components and breaking down any clumps or agglomerates. Mechanical blending can be performed using different types of equipment, including paddle blenders, ribbon blenders, or high-speed dispersers, depending on the thickness and characteristics of the formulation.

6.1.2 In-Line Mixing

In-line blending could be a persistent blending strategy that's frequently utilized in large-scale generation offices. It includes the use of pumps and inactive blenders to mix the fixings as they pass through an arrangement of blending components. In-line blending offers the advantage of effective and fast blending, guaranteeing uniform quality and less handling time.

6.1.3 Recirculation Mixing

Recirculation blending involves circulating the formulation through a closed-loop system, allowing the ingredients to be continuously mixed and homogenized. This method is particularly useful for formulations with high viscosity or those that require extended mixing times.

6.2 Homogenization Techniques

Homogenization is a fundamental step in the formulation process, particularly when dealing with ingredients that have different particle sizes or densities.

6.2.1 High-Pressure Homogenization

High-pressure homogenization is a technique used to homogenize substances of different densities by applying high pressure (Hashim et al. 2018). This method effectively breaks down particles and agglomerates, resulting in a fine and uniform dispersion of the ingredients. High-pressure homogenization is particularly useful for formulations that require micronization or reduction of particle size.

6.2.2 Ultrasonication

Ultrasonication uses high-frequency sound waves to create intense cavitation and shear forces, leading to the disruption of particles and agglomerates. This method is effective in achieving uniform dispersion and homogenization of the formulation (Long et al. 2020). Ultrasonication is commonly used for formulations with heat-sensitive ingredients or those that require gentle handling.

6.2.3 Colloid Milling

Colloid milling involves the use of a colloid mill, which consists of a rotor and a stator. The formulation is passed through the narrow gap between the rotor and stator, resulting in intense shear forces that break down particles and agglomerates. Colloid milling is particularly suitable for formulations with high viscosity or those that require fine particle size reduction (Wang et al. 2007).

6.3 Emulsification and Suspension Methods

Emulsification and suspension are two important methods used in the formulation and application of biofungicides. These methods play a crucial role in ensuring the stability and effectiveness of the biofungicide product. Emulsification involves the dispersion of one liquid phase into another immiscible liquid phase, whereas suspension refers to the dispersion of solid particles in a liquid medium.

6.3.1 Emulsification Methods

Emulsification process is a method of blending two immiscible liquids, typically oil and water to form an emulsion (Miastkowska et al. 2020).

6.3.2 Mechanical Emulsification

Mechanical emulsification may be a broadly utilized strategy that includes the use of mechanical vitality to break down the dynamic ingredient and scatter it into the carrier. This could be accomplished through the use of high-speed blenders, homogenizers, or colloid plants (Khokhlova et al. 2011).

6.3.3 Surfactant-Assisted Emulsification

Surfactants play a vital part in emulsification by decreasing the interfacial pressure between the oil and water stages (Stop et al. 2016). They help stabilize the emulsion by forming a protective layer around the dispersed droplets, preventing coalescence and ensuring long-term stability. Surfactants can also improve the wetting and spreading properties of the emulsion, enhancing its effectiveness upon application.

6.3.4 Microemulsion Formation

Microemulsions are thermodynamically stable systems that consist of oil, water, surfactants, and co-surfactants. They are transparent or translucent and have a small droplet size, typically less than 100 nm. Microemulsions offer several advantages in biofungicide formulation, including improved solubilization of hydrophobic active ingredients, enhanced bioavailability, and increased penetration into plant tissues (Schulman et al. 1959).

6.4 Suspension Methods

Suspension is a key method in biofungicide formulation, involving the dispersion of solid particles, such as active ingredients or carriers, in a liquid medium. This technique is essential for insoluble or poorly soluble components, ensuring effective application on crops.

6.4.1 Wet Grinding

A common preparation method, wet grinding uses processes like ball milling to reduce particle size and achieve uniform distribution, particularly for micronizing insoluble ingredients (Sakthivel et al. 2008).

6.4.2 Stabilizers and Dispersants

These additives prevent particle settling and maintain uniform distribution. Stabilizers increase liquid viscosity to reduce sedimentation rates, while dispersants decrease interparticle forces to prevent agglomeration, crucial for the stability and effectiveness of the suspension (Li et al. 2021).

6.4.3 Spray Drying

This technique converts liquid suspensions into dry powder by atomizing the liquid into fine droplets and drying them with hot air. Spray drying enhances stability, ease of handling, and shelf life, allowing the resulting powder to be reconstituted into a suspension before application (Azad et al. 2015).

6.5 Spray Application Techniques for Biofungicides

Spray application is one of the most common and effective methods for applying biofungicides to crops. It allows for uniform coverage and delivery of biofungicide on the plant surface, ensuring maximum effectiveness in disease control (Bateman and Chapple 2001).

6.5.1 Selection of Spray Equipment

You need the proper spray equipment to ensure both the spray and the product are applied effectively. Factors that influence sprayer choice are the intended area, the kind of crops, and the amount of liquid needed. The most prevalent types of equipment are handheld sprayers, backpack sprayers, airblast sprayers, and boom sprayers (Raetano et al. 2022).

Lightweight and easy to move, handheld and backpack sprayers are best for small-area tasks or handling spot problems in gardens. For orchards and vineyards, areas are treated by airblast sprayers, which send high-speed air to spread biofungicides properly. Because of their efficiency, large-area farms often depend on boom sprayers (Kawabata and Nakamoto 2021).

6.5.2 Spray Application Techniques

The way you apply the sprayer helps ensure biofungicides are properly spread and attached to plants (Gan-Mor and Matthews 2003). There are several important techniques to use such as:

6.5.2.1 Calibration

Before using the sprayer, set it up to find the correct amount of spray and the most suitable nozzle settings. With these settings, you can get accurate coverage and avoid both under- and over-application.

6.5.2.2 Proper Timing

The most effective results with biofungicides are achieved when they are applied as the disease is starting to appear. If you apply your solution early, as soon as you notice the first symptoms, you can minimize the damage the disease causes. Checking the crops and their diseases on a regular basis helps you choose the right time for spraying (Russouw et al. 2021).

6.5.2.3 Spray Coverage

Efficient control of diseases depends on getting the spray all over the crops. It is important to distribute biofungicides across all plant surfaces—both top and bottom leaves, stems, and fruits. You can do this by altering spray pressure, the nozzle direction, and the gun's distance from the surface. Droplets should be of a size that covers your plants without being so fine that they spread too much (Magwebu et al. 2023).

6.5.2.4 Spray Volume

You should use spray in a way that matches your target crop, how large the plant is, and the type of biofungicide you're using. Following suggested spray amounts helps you treat the area correctly; too little can be ineffective and too much could negatively impact the environment. You can find the recommended amounts of products stated on bottle or box labels or learn about them from agricultural extension services (Pertot et al. 2016).

6.6 Formulation Quality Control and Testing

Formulation quality control and testing are significant steps in the development and production of biofungicides. These processes ensure that the formulated product meets the desired standards of effectiveness, stability, and safety (Mulatu et al. 2021). By conducting thorough quality control and testing, manufacturers can identify any potential issues or irregularities within the formulation, allowing them to make necessary adjustments and improvements (Kumar et al. 2014).

6.6.1 Quality Control Measures and Parameters

Throughout producing biofungicides, several crucial factors must be checked to maintain high standards and uniformity. The active ingredient's concentration, carrier compatibility, the usefulness of adjuvants, the capability of surfactants, and how well the formulation performs over time are part of this. By paying close attention to these points, manufacturers make certain that the biofungicide keeps its effectiveness and remains stable during its use (Eid et al. 2018, Teixidó et al. 2022).

6.6.2 Compatibility of the Active Ingredient and Carrier

How well biofungicides control fungal diseases is mainly affected by the amount of active substance they contain. Precise judgment and action in measuring and mixing ingredients are essential to make food safe (Mustafa et al. 2022). Testing must be carried out on a regular basis to ensure the amount of active ingredient is at an acceptable level (He et al. 2019). Besides, the carrier delivering the active ingredient should be suitable and not change its function. To be sure there aren't any problems between the active ingredient and the carrier, it is important to perform compatibility testing (Djaenuddin et al. 2021).

6.6.3 Adjuvants and Surfactants

Biofungicide products contain adjuvants to help them adhere well, spread easily, and spread more evenly on the surface they are sprayed on. Testing is needed to find out how these adjuvants change the efficacy of the biofungicide, especially by affecting its spread, ability to get inside plant cells, and staying power on the leaves. Lowering spray solution surface tension with surfactants contributes to better coverage and deeper penetration of the main ingredient (Arjjumend and Koutouki 2020).

6.6.4 Stability and Regulatory Compliance

Ensuring that biofungicides are safe and effective for their shelf life is very important and depends on their formulation. Stability testing checks if the formulation is stable in terms of its physical, chemical, and biological properties, to see if there are issues like phase separation or a loss in effectiveness (Djaenuddin et al. 2021). In addition, abiding by regulatory standards is essential so that the products are ensured as safe tools in agriculture. You should test toxin levels, the environmental impact of the pesticide, how much stays in the soil after spraying, and how it combines with other farm supplies (Khakimov et al. 2020).

7. Future Trends in Biofungicide Formulation

7.1 Enhanced Formulation Stability

Since organisms respond quickly to changes in their environment, keeping biofungicide active ingredients stable is very difficult (Naveenkumar et al. 2017). A lot of techniques and tools have appeared to help secure and improve their growth prospects.

7.1.1 Microencapsulation

UV rays and different temperatures are prevented from harming the active materials, thanks to the coating which allows for more applications and use (Riseh et al. 2021). Because of slow-release fertilizers, farmers have to use them less frequently.

7.1.2 Nanotechnology

Using nanotechnology can support the development of improved biofungicides. Use of silver nanoparticles to coat antifungal drugs can improve their ability to target infections in parts of the plant (Luksiene et al. 2020, Zaki et al. 2022).

7.1.3 Products Containing Useful Microorganisms

When biofungicides and good microorganisms are combined, they produce a biofilm that improves how the active ingredients work.

7.1.4 Combination Products

Multi-ingredient products can control a range of plant diseases and may help reduce resistance in plants, according to Axel et al. (2012). Using a mix of methods makes it easier to manage these hard-to-control fungal diseases.

7.2 Emerging Trends in Carrier and Adjuvant Development

There are exciting new developments in the fields of carrier and adjuvant development. Advances in the area of biofungicide formulation involve the rise of new carrier and adjuvant types intended to improve disease control and decrease product instability (Thind 2021).

7.2.1 Nanotechnology in Carrier Development

Nanotechnology is recognized for its role in transforming agriculture, especially through improved biofungicide formulations. Nanoparticles and nanocapsules increase the efficiency of uptake and utilization of active ingredients. They shield biofungicides from the effects of ultraviolet light and contribute to their better performance, which helps solve stability and effectiveness problems (Kutawa et al. 2021). Efforts are being made to produce carriers that can be easily degraded by the environment. Researchers are now investigating natural biopolymers from starch and chitosan, as it has been discovered that traditional synthetic polymers may persist in the environment (Korbecka-Glinka et al. 2022, Roy et al. 2014).

7.2.2 Precision Delivery Systems

Biofungicides are starting to be directed toward specific points on plants or specific areas using new precision delivery technology. Thanks to microencapsulation, nanotechnology, and smart polymers, these systems determine both the timing and location for releasing the active ingredient. Using this targeted approach fits with precision agriculture, making it both efficient and environmentally protective (Machado et al. 2020).

7.3 Biofungicides as Part of Other Pest Management Processes

Using biofungicides helps keep horticulture free from fungal disease and usually improves disease control when used along with existing methods of pest management (Chandler et al. 2011).

7.3.1 Integrated Pest Management

When using the IPM approach, people reduce their use of chemicals by combining different approaches such as cultural, biological, and chemical methods (Pottorff and Panter 2009). Biofungicides are important in IPM because they enhance the success of other strategies, lower the use of synthetic fungicides, avoid the growth of fungal resistance, and encourage environmentally friendly farming methods.

7.3.2 Biological Control Agents (BCAs)

BCAs play a major role in controlling and suppressing fungal pathogens (Palmieri et al. 2022). The combination of biofungicides and BCAs increases their success and allows for a decrease in the amount of chemical fungicides needed for managing disease.

7.3.3 Chemical Fungicides

Even with the aim of using less, occasionally chemical fungicides must be applied. Mixing biofungicides with chemical fungicides is a successful way to treat plant diseases and avoid resistance (El-Kholy et al. 2021).

7.4 Challenges and Opportunities in the Future of Biofungicide Formulation

Recent developments in biofungicides bring both new risks and new benefits for those working in this area (Pertot et al. 2016). Outcomes from biofungicides can be improved through careful planning, and effective tools for battling diseases are now available. Yet, companies do not find it easy to meet regulations because their products must undergo difficult and time-consuming tests before being approved (Sachdev and Singh 2016). Not having recognized performance targets creates additional issues for both regulators and technology developers.

8. Conclusion

In conclusion, the formulation of biofungicides involves a complex process that requires careful consideration of various components, including the selection of active ingredients, carriers, adjuvants, and surfactants. The effectiveness of biofungicides is strongly influenced by their formulation, which must ensure stability, compatibility, and optimal delivery of active ingredients to target pathogens. Advances in formulation technology, such as microencapsulation, nanotechnology, and the development of biodegradable carriers, are paving the way for more effective and sustainable biofungicide solutions. Additionally, combining biofungicides with other pest management methods, like Integrated Pest Management (IPM) and biological control agents, makes them more effective and supports sustainable farming practices. As the field continues to advance, addressing regulatory challenges and optimizing formulations will be vital for the successful commercialization and application of biofungicides in agriculture. By leveraging new technologies and fostering collaboration among stakeholders, the future of biofungicide formulation holds great promise for advancing disease management and promoting environmental sustainability in agricultural systems.

References

Abd-El-Kareem, F., Elshahawy, I. E., and Abd-Elgawad, M. M. M. (2019). Effectiveness of silicon and silicate salts for controlling black root rot and induced pathogenesis-related protein of strawberry plants. *Bull. Natl. Res. Cent.*, 43: 91.

Abdelwahab, N., Rabie, W., and Mohamed, F. (2023). Fabrication and characterization of novel biocomposite based on *Sargassum* vulgare for controlling sugar beet root diseases. *Chem. Biol. Technol. Agric.*, 10: 52.

Adusei-Fosu, K., Rolando, C. A., Richardson, B., van Leeuwen, R., Gaskin, R., Bader, M. K. F. et al. (2021). Evaluating the efficacy of potential fungicide-adjuvant combinations for control of myrtle rust in New Zealand. *Journal of Plant Diseases and Protection*, 128(6): 1501–1515.

Alamri, S. A. (2014). The synergistic effect of two formulated biofungicides in the biocontrol of root and bottom rot of lettuce. *Biocontrol. Science*, 19(4): 189–197.

Apostólico Jde, S., Lunardelli, V. A., Coirada, F. C., Boscardin, S. B., and Rosa, D.S. (2016). Adjuvants: Classification, Modus Operandi, and Licensing. *J. Immunol. Res.*, 2016: 1459394.

Arjjumend, H., and Koutouki, K. (2020). Legal barriers and quality compliance in the business of biofertilizers and biopesticides in India. *Journal of Legal Studies*, 26(40): 81–101.

Axel, C., Zannini, E., Coffey, A., Guo, J., Waters, D. M., and Arendt, E. K. (2012). Ecofriendly control of potato late blight causative agent and the potential role of lactic acid bacteria: a review. *Applied Microbiology and Biotechnology*, 96: 37–48.

Azad, M., Arteaga, C., Abdelmalek, B., Davé, R., and Bilgili, E. (2015). Spray drying of drug-swellable dispersant suspensions for preparation of fast-dissolving, high drug-loaded, surfactant-free nanocomposites. *Drug Development and Industrial Pharmacy*, 41(10): 1617–1631.

Bateman, R., and Chapple, A. (2001). The spray application of mycopesticide formulations. pp. 289–309. *In:* Fungi as biocontrol agents: progress, problems and potential. Wallingford UK: CABI Publishing.

Ben Khedher, S., Boukedi, H., Laarif, A., and Tounsi, S. (2020). Biosurfactant produced by *Bacillus subtilis* V26: A potential biological control approach for sustainable agriculture development. *Organic Agriculture*, 10: 117–124.

Berninger, T., Mitter, B., and Preininger, C. (2017). Zeolite-based, dry formulations for conservation and practical application of *Paraburkholderia phytofirmans* PsJN. *Journal of Applied Microbiology*, 122(4): 974–986.

Bhadani, A., Kafle, A., Ogura, T., Akamatsu, M., Sakai, K., Sakai, H., and Abe, M. (2020). Current perspective of sustainable surfactants based on renewable building blocks. *Current Opinion in Colloid & Interface Science*, 45: 124–135.

Bharti, V., and Ibrahim, S. (2020). Biopesticides: Production, formulation and application systems. *International Journal of Current Microbiology and Applied Science*, 9(10): 3931–3946.

Boyetchko, S., Pedersen, E., Punja, Z., and Reddy, M. (1999). Formulations of biopesticides. *Biopesticides: Use and delivery*, 487–508.

Castro, M. J., Ojeda, C., and Cirelli, A. F. (2014). Advances in surfactants for agrochemicals. *Environmental Chemistry Letters*, 12: 85–95.

Chandler David, Bailey Alastair S., Tatchell G. Mark, Davidson Gill, Greaves Justin and Grant Wyn P. (2011). The development, regulation and use of biopesticides for integrated pest management Phil. *Trans. R. Soc. B.*, 366: 1987–1998.

Chen, W.-C., Chiou, T.-Y., Delgado, A.L., Liao, C.-S. (2019). The control of rice blast disease by the novel biofungicide formulations. *Sustainability*, 11: 3449.

Díaz-Blancas, V., Medina, D. I., Padilla-Ortega, E., Bortolini-Zavala, R., Olvera-Romero, M., Luna-Bárcenas, G. et al. (2016). Nanoemulsion formulations of fungicide tebuconazole for agricultural applications. *Molecules.*, 21(10): 1271.

Djaenuddin, N., Sebayang, A., Nonci, N., and Muis, A. (2021, November). Compatibility of biocontrol agent formulas and synthetic fungicides in controlling maydis leaf blight on corn caused by *Bipolaris maydis. In IOP Conference Series: Earth and Environmental Science.*, (911(1): 012062. IOP Publishing.

Eid, H. T., Abbas, M. S., Soliman, A. S., Mahmoud, N. A., and Abou-Zeid, N. M. (2018). Induction of systemic resistance against tomato root rot disease under greenhouse conditions. *Zagazig Journal of Agricultural Research*, 45(3): 931–944.

El-Kholy, R. M., El-SamadesyA, A. M., Helalia, A. A., and Hassuba, M. M. (2021). Efficacy of several chemical fungicides and biofungicides for controlling damping-off and root rot diseases in common bean under field conditions. *Al-Azhar Journal of Agricultural Research*, 46(2): 154–167.

Fait, M. E., Bakas, L., Garrote, G. L., Morcelle, S. R., and Saparrat, M. C. N. (2019 Jan). Cationic surfactants as antifungal agents. *Appl. Microbiol. Biotechnol.* 103(1): 97–112.

Faruk, M. I., Rahman, M. L., Mustafa, M. M. H., Rahman, M. M., and Rahman, M. A. (2014). Screening of carrier materials to formulate *Trichoderma harzianum* based bio-fungicide against foot and root rot disease of tomato (*Lycopersicon esculentum* L.). *Bangladesh Journal of Agricultural Research*, 39(2): 197–209.

Fravel, D. R., Deahl, K. L., and Stommel, J. R. (2005). Compatibility of the biocontrol fungus *Fusarium oxysporum* strain CS-20 with selected fungicides. *Biological Control.*, 34.2: 165–169.

Galindo, E., Serrano-Carreón, L., Gutiérrez, C. R., Allende, R., Balderas, K., Patino, M. C. et al. (2013). The challenges of introducing a new biofungicide to the market: A case study. *Electronic Journal of Biotechnology*, 16: 5–5.

Gan-Mor, S., and Matthews, G. A. (2003). Recent developments in sprayers for application of biopesticides—An overview. *Biosystems Engineering*, 84(2): 119–125.

Gao, X., Wang, D., Jiang, Z., Li, X., and Chen, G. (2022). Effect of adjuvants on the wetting behaviors of bifenthrin droplets on tea leaves. *Applied Sciences.*, 12(9): 4217.

Gaskin, R. E., and Steele, K. D. (2009). A comparison of sticker adjuvants for their effects on retention and rainfastening of fungicide sprays. *New Zealand Plant Protection*, 62: 339–342.

Gent, D. H., Schwartz, H. F., and Nissen, S. J. (2003). Effect of commercial adjuvants on vegetable crop fungicide coverage, absorption, and efficacy. *Plant Disease*, 87(5): 591–597.

Gikas, G. D., Parlakidis, P., Mavropoulos, T., and Vryzas, Z. (2022). Particularities of fungicides and factors affecting their fate and removal efficacy: A review. *Sustainability*, 14(7): 4056.

Guijarro, B., Larena, I., Melgarejo, P., and De Cal, A. (2018). Surfactant effects on wettability of *Penicillium frequentans* formulations to improve brown rot biocontrol. *Journal of the Science of Food and Agriculture*, 98(15): 5832–5840.

Guo, X., Sun, Z., Zhao, R., Shang, H., Liu, J., Xu, Y. et al. (2023). Water-based environmentally friendly pesticide formulations based on cyclodextrin/pesticide loading system. *Journal of Renewable Materials*, 11(2).

Habtom, B., and Tesfagiorgis, Harold J. (2013) Annegarn, Evaluation of adjuvants for enhanced management of *Podosphaera xanthii* using potassium silicate and biocontrol agents. *Crop Protection*, 48: 69–75.

Hashim, A. F., Youssef, K., and Abd-Elsalam, K. A. (2018). The role of nanoemulsions as antimicrobial agents in plant protection. *Nanobiotechnology Applications in Plant Protection*, 137–153.

He, L., Li, X., Gao, Y., Li, B., Mu, W., and Liu, F. (2019). Oil adjuvants enhance the efficacy of pyraclostrobin in managing cucumber powdery mildew (*Podosphaera xanthii*) by modifying the affinity of fungicide droplets on diseased leaves. *Plant Disease*, 103(7): 1657–1664.

Holka, M., and Kowalska, J. (2023). The potential of adjuvants used with microbiological control of insect pests with emphasis on organic farming. *Agriculture*, 13(9): 1659.

Jiang, Y., and Wang, J. (2023). The registration situation and use of Mycopesticides in the World. *J. Fungi*, 9: 940.

Karamchandani, B. M., Pawar, A. A., Pawar, S. S., Syed, S., Mone, N. S., Dalvi, S. G., et al. (2022) Biosurfactants' multifarious functional potential for sustainable agricultural practices. *Front. Bioeng. Biotechnol.* 10: 1047279.

Kawabata, A. M., and Nakamoto, S. T. (2021). Spraying to Suppress Coffee Leaf Rust (*Hemileia vastatrix*) in Hawai'i. Plant Disease.

Khakimov, A. A., Omonlikov, A. U., and Utaganov, S. B. U. (2020). Current status and prospects of the use of biofungicides against plant diseases. *GSC Biological and Pharmaceutical Sciences*, 13(3): 119–126.

Khokhlova, V. A., Canney, M. S., Bailey, M. R., Hwang, J. H., Khokhlova, T. D., Kreider, W. et al. (2011). A method of mechanical emulsification in a bulk tissue using shock wave heating and millisecond boiling. *The Journal of the Acoustical Society of America*, 129(4): 2476–2476.

Khun, K. K., Ash, G. J., Stevens, M. M., Huwer, R. K., and Wilson, B. A. (2021). Compatibility of *Metarhizium Anisopliae* and *Beauveria Bassiana* with Insecticides and Fungicides Used in Macadamia Production in Australia. *Pest Manag. Sci.*, 77: 709–718.

Köhl, J., Kolnaar, R., and Ravensberg, W. J. (2019). Mode of action of microbial biological control agents against plant diseases: relevance beyond efficacy. *Frontiers in plant science*, 845.

Kolombet, L.V., Zhigletsova, S. K., Kosareva, N. I., Bystrova, E.V., Derbyshev, V. V., Krasnova, S.P. (2008). Development of an extended shelf-life, liquid formulation of the biofungicide *Trichoderma asperellum*. World J. Microbiol. Biotechnol., 24: 123–131.

Korbecka-Glinka, G., Piekarska, K., and Wiśniewska-Wrona, M. (2022). The use of carbohydrate biopolymers in plant protection against pathogenic fungi. *Polymers*, 14(14): 2854.

Kumar, S., Thakur, M., and Rani, A. (2014). Trichoderma: Mass production, formulation, quality control, delivery and its scope in commercialization in India for the management of plant diseases. *African Journal of Agricultural Research*, 9(53): 3838–3852.

Kutawa, A. B., Ahmad, K., Ali, A., Hussein, M. Z., Abdul Wahab, M. A., Adamu, A. et al. (2021). Trends in nanotechnology and its potentialities to control plant pathogenic fungi: A review. *Biology*, 10(9): 881.

Landim, T. N., da Cunha, J. P., Alves, G. S., Marques, M. G., and Silva, S. M. (2019). Interactions between adjuvants and the fungicide azoxystrobin+ benzovindiflupyr in hydraulic spraying. *Engenharia Agrícola*, 39: 600–606.

Li, J., Wang, Z., Zhang, H., Gao, J., and Zheng, A. (2021). Progress in the development of stabilization strategies for nanocrystal preparations. *Drug Delivery*, 28(1): 19–36.

Lizamore, D., and Winefield, C. (2015). The addition of an organosilicone surfactant to *Agrobacterium suspensions* enables efficient transient transformation of *in vitro* grapevine leaf tissue at ambient pressure. Plant Cell, Tissue and Organ Culture (PCTOC), 120: 607–615.

Long, Y., Huang, W., Wang, Q., and Yang, G. (2020). Green synthesis of garlic oil nanoemulsion using ultrasonication technique and its mechanism of antifungal action against *Penicillium italicum*. *Ultrasonics Sonochemistry*, 64: 104970.

Luksiene, Z., Rasiukeviciute, N., Zudyte, B., and Uselis, N. (2020). Innovative approach to sunlight activated biofungicides for strawberry crop protection: ZnO nanoparticles. *Journal of Photochemistry and Photobiology B: Biology*, 203: 111656.

Machado, T. O., Beckers, S. J., Fischer, J., Müller, B., Sayer, C., de Araújo, P. H. et al. (2020). Bio-based lignin nanocarriers loaded with fungicides as a versatile platform for drug delivery in plants. *Biomacromolecules*, 21(7): 2755–2763.

Magwebu, S., Meitz-Hopkins, J. C., Pott, R.W.M., and Lennox, C. L. (2023). Efficacy of the cyclolipopeptides fengycin and iturin A against postharvest pome fruit pathogens. *Front. Hortic.*, 2: 1175251.

Martínez-Medina, A., Roldán, A., and Pascual, J. A. (2009). Performance of a *Trichoderma harzianum* bentonite–vermiculite formulation against Fusarium wilt in seedling nursery melon plants. *HortScience*, 44(7): 2025–2027.

Mary, Leggett., Jarrod Leland, Ken Kellar and Beth Epp (2011) Formulation of microbial biocontrol agents – an industrial perspective. *Canadian Journal of Plant Pathology*, 33: 2, 101–107.

Mattedi, A., Sabbi, E., Farda, B., Djebaili, R., Mitra, D., Ercole, C. et al. (2023). Solid-State Fermentation: Applications and Future Perspectives for Biostimulant and Biopesticides Production. *Microorganisms*. 11(6): 1408.

Miastkowska, M., Michalczyk, A., Figacz, K., and Sikora, E. (2020). Nanoformulations as a modern form of biofungicide. *J. Environ. Health Sci. Eng. Jan.*, 31; 18(1): 119–128.

Mokhtarnejad, L., Etebarian, H. R., Fazeli, M. R., and Jamalifar, H. (2011). Evaluation of different formulations of potential biocontrol yeast isolates efficacy on apple blue mold at storage condition. *Archives of Phytopathology and Plant Protection*, 44(10): 970–980.

Montesinos, E. (2003). Development, registration and commercialization of microbial pesticides for plant protection. *Int. Microbiol.*, 6: 245–252.

Mulatu, A., Alemu, T., Megersa, N., and Vetukuri, R. R. (2021). Optimization of culture conditions and production of bio-fungicides from *Trichoderma* species under solid-state fermentation using mathematical modeling. *Microorganisms*, 9(8): 1675.

Mustafa, I. F., and Hussein, M. Z. (2020). Synthesis and technology of nanoemulsion-based pesticide formulation. *Nanomaterials (Basel). Aug* 17; 10(8): 1608.

Mustafa, I. F., Hussein, M. Z., Idris, A. S., Hilmi, N. H. Z., Ramli, N. R., and Fakurazi, S. (2022). The effect of surfactant type on the physico-chemical properties of hexaconazole/dazomet-micelle nanodelivery system and its biofungicidal activity against *Ganoderma boninense*. *Colloids and Surfaces A: Physicochemical and Engineering Aspects*, 640: 128402.

Ncama, K., Mditshwa, A., Tesfay, S. Z., Mbili, N. C., and Magwaza, L. S. (2019). Topical procedures adopted in testing and application of plant-based extracts as bio-fungicides in controlling postharvest decay of fresh produce. *Crop Protection*, 115: 142–151.

Ons, L., Bylemans, D., Thevissen, K., and Cammue, B. P. (2020). Combining biocontrol agents with chemical fungicides for integrated plant fungal disease control. *Microorganisms*, 8(12): 1930.

Orbović, V., Achor, D., and Syvertsen, J. P. (2007). Adjuvants affect penetration of copper through isolated cuticles of citrus leaves and fruit. *HortScience*, 42(6): 1405–1408.

Palma-Bautista, C., Vazquez-Garcia, J. G., Travlos, I., Tataridas, A., Kanatas, P., Domínguez-Valenzuela, J. A. (2020). Effect of adjuvant on glyphosate effectiveness, retention, absorption and translocation in *Lolium rigidum* and *Conyza canadensis*. *Plants (Basel). Mar.*, 1;9(3): 297.

Palmieri, D., Ianiri, G., Del Grosso, C., Barone, G., De Curtis, F., Castoria, R. et al. (2022). Advances and perspectives in the use of biocontrol agents against fungal plant diseases. *Horticulturae*. 8(7): 577.

Parlakidis, P., Adamidis, G., Alexoudis, C., Pythoglou, P., Papadopoulos, S., and Vryzas, Z. (2023). Adjuvant Effects on Pyraclostrobin and Boscalid Residues, Systemic Movement, and Dietary Risk in Garlic under Field Conditions. *Agriculture*, 13(8): 1636.

Pertot, I., Puopolo, G., Giovannini, O., Angeli, D., Sicher, C., and Perazzolli, M. (2016). Advantages and limitations involved in the use of microbial biofungicides for the control of root and foliar phytopathogens of fruit crops. *Italus Hortus*, 23(3): 3–12.

Pirzada, T., de Farias, B. V., Mathew, R., Guenther, R.H., Byrd, M. V., Sit, T. L. et al. (2020 Aug) Recent advances in biodegradable matrices for active ingredient release in crop protection: Towards attaining sustainability in agriculture. *Curr. Opin. Colloid Interface. Sci.*, 48: 121–136.

Pottorff, L. P., and Panter, K. L. (2009). Integrated pest management and biological control in high tunnel production. *HortTechnology*, 19(1): 61–65.

Purkait, A., Rana, D., Banerjee, A., Hazra, D. K., Biswas, P. K., and Kole, R. K. (2022). Selection of green carrier solvent for the effective creation of bio-inspired liquid formulation and anti-fungal potency evaluation. *Archives of Microbiology*, 204(11): 669.

Raetano, C. G., Negrisoli, M. M., da Silva, F. N., Rodrigues, D. M., and Prado, E. P. (2022). Optimization of application technology for plant protection products in soybean crops in Brazil. In Soybean-recent advances in research and applications. IntechOpen.

Ramakutoane, T. L., Roux-van der Merwe, M. P., Badenhorst, J., Kesavan Pillai, S., and Ray, S. S. (2023). Antifungal properties of nanosilver clay composites against fungal pathogens of *Agaricus bisporus. ChemistrySelect.*, 8(36): e202301696.

Roberts, J. (2010). Helena Chemical Company, Presentation to EPA.

Rocha, I., Ma, Y., Souza-Alonso, P., Vosátka, M., Freitas, H., and Oliveira, R. S. (2019). Seed coating: a tool for delivering beneficial microbes to agricultural crops. *Frontiers in Plant Science*, 10: 1357.

Roy, A., Singh, S. K., Bajpai, J., and Bajpai, A. K. (2014). Controlled pesticide release from biodegradable polymers. *Central European Journal of Chemistry*, 12: 453–469.

Russouw A., Meitz-Hopkins J., Den Breeyen A., and Lennox C. (2021). Postharvest applications of fludioxonil and pyrimethanil to control *Phlyctema vagabunda* on apple in south Africa. *Crop Protect.*, 141: 105451.

Ryckaert, B., Spanoghe, P., Steurbaut, W., Heremans, B., Haesaert, G., de Coen, W. (2005). Environmental impact of adjuvants in crop protection. *Commun. Agric. Appl. Biol. Sci.*, 70(4): 1087–91. PMID: 16628960.

Saberi-Riseh, R., Hassanisaadi, M., Vatankhaha, M., and Kennedy, J. F. (2023). Encapsulating biocontrol bacteria with starch as a safe and edible biopolymer to alleviate plant diseases: A review. *Carbohydr. Polym.*, 302: 120384. doi: 10.1016/j.carbpol.2022.120384.

Sachdev, S., and Singh, R. P. (2016). Current challenges, constraints and future strategies for development of successful market for biopesticides. *Climate Change and Environmental Sustainability*, 4(2): 129–136.

Sakthivel, S., Venkatesan, V., and Krishnan, B. (2008). Pitchumani, Influence of suspension stability on wet grinding for production of mineral nanoparticles. *Particuology*, 6(2): 120–124

Schulman, J. H., Stoeckenius, W., and Prince, L. M. (1959). Mechanism of formation and structure of micro emulsions by electron microscopy. *The Journal of Physical Chemistry*, 63(10): 1677–1680.

Scott, M. J., and Jones, M. N. (2000). The biodegradation of surfactants in the environment. *Biochimica et Biophysica Acta (BBA)-Biomembranes*, 1508(1–2): 235–251.

Smith, K. (2019). Development of bio-adjuvants for conventional and organic farming (Doctoral dissertation, Lancaster University).

Sobiech, Ł., Grzanka, M., Skrzypczak, G., Idziak, R., Włodarczak, S., Ochowiak, M. et al. (2020). Effect of Adjuvants and pH Adjuster on the Efficacy of Sulcotrione Herbicide. *Agronomy*. 10(4): 530.

Srivastava, D.A., Harris, R., Breuer, G., and Levy, M. (2021). Secretion-based modes of action of biocontrol agents with a focus on *Pseudozyma aphidis. Plants*, 10(2): 210.

Tanovic, B., Gasic, S., Hrustic, J., Mihajlovic, M., Grahovac, M., Delibasic, G. et al. (2015). Development of a thyme essential oil formulation and its effect on *Monilinia fructigena. Pesticides and Phytomedicine*, vol. 28.

Teixidó, N., Usall, J., and Torres, R. (2022). Insight into a successful development of biocontrol agents: Production, formulation, packaging, and shelf life as key aspects. *Horticulturae*, 8(4): 305.

Thiago, O., Machado, Sebastian J., Beckers, Jochen Fischer, Beate Müller, Claudia Sayer, Pedro H. H. de Araújo et al. (2020). Bio-based lignin nanocarriers loaded with fungicides as a versatile platform for drug delivery in plants. *Biomacromolecules.*, 21(7): 2755-2763.

Thind, T. S. (2021). Changing trends in discovery of new fungicides: a perspective. *Indian Phytopathology*, 74(4): 875–883.

Uchida, D. T., Siqueira, G. F., Dos Reis, E. M., Hegeto, F. L., Medina Neto, A., and Reis, A. V. (2021). Design of nanostructured lipid carriers containing *Cymbopogon martinii* (Palmarosa) essential oil against *Aspergillus nomius*. *Molecules.*, Aug 10; 26(16): 4825.

Wang, D., Jia, M., Wang, L., Song, S., Feng, J., and Zhang, X. (2017). Chitosan and β-Cyclodextrin-epichlorohydrin Polymer Composite Film as a Plant Healthcare Material for Carbendazim-Controlled Release to Protect Rape against *Sclerotinia sclerotiorum* (Lib.) de Bary. *Materials* (Basel). 2017 Mar 26; 10(4): 343. doi: 10.3390/ma10040343.

Wang, L., Li, X., Zhang, G., Dong, J., and Eastoe, J. (2007). Oil-in-water nanoemulsions for pesticide formulations. *Journal of Colloid and Interface Science*, 314(1): 230–235.

Webster, G. R., Bisset, N. B., Cahill, D. M., Jones, P., Killick, A., Hawley, A. et al. (2016). Kinetic resolution of the interactions between agrochemical products and adjuvant systems upon mixing. *Journal of Agricultural and Food Chemistry*, 64(31): 6139–6147.

Worrall, EA., Hamid, A., Mody, K. T., Mitter, N., and Pappu, H. R. (2018). Nanotechnology for Plant Disease Management. *Agronomy*. 8(12): 285.

Xu, G., Yan, D., Fang, W., Xu, D., Xu, L., Wang, Q., and Cao, A. (2023). Synergistic effect of orange oil adjuvant on acetamiprid in the control of *Edentatipsylla shanghaiensis*. *Sustainability*, 15(13): 10113.

Zaki, S. A., Ouf, S. A., Abd-Elsalam, K. A., Asran, A. A., Hassan, M. M., Kalia, A. et al. (2022). Trichogenic silver-based nanoparticles for suppression of fungi involved in damping-off of cotton seedlings. *Microorganisms*, 10(2): 344.

Zamoum, M., Allali, K., Benadjila, A., Zitouni, A., and Goudjal, Y. (2022). Formulation of biofungicides based on *Streptomyces caeruleatus* strain ZL-2 spores and efficacy against *Rhizoctonia solani* damping-off of tomato seedlings. *Arch. Microbiol.*, Sep. 17; 204(10): 629.

Zhao, R., Yu, M., Sun, Z., Li, L. J., Shang, H. Y., Xi, W. J. et al. (2022). Using tank-mix adjuvant improves the physicochemical properties and dosage delivery to reduce the use of pesticides in unmanned aerial vehicles for plant protection in wheat. *Pest Management Science*, 78(6): 2512–2522.

6

Development of Biofungicide Formulations

From General Methods to Carrier Selection and Compatibility

Niranjana Rajesh,[1] *Gauriparvathy K. B.,*[1] *Dileep Francis*[1,2] and *Sabu Abdulhammed*[3,*]

1. Introduction

In the present-day world, with a huge surge in the growing population, ensuring food security by increasing crop productivity is an ongoing struggle, and fungal infections are considered a leading cause (Zubrod et al. 2019). Research indicates that fungal infections can have detrimental effects on crops, resulting in both yield decreases and the production of toxic substances, which in turn lead to health and financial issues (Schneider and Dickert 1994). Although fungicides play a key role in protecting food crops, their excessive utilisation can lead to various challenges. The major concern is the development of fungal resistance within a few years of continuous exposure, leading to a resurgence of the targeted pathogen. This has driven the ongoing need for the formulation of new fungicides and treatment strategies (Fenta and Mekonnen 2024). Chemical bioaccumulation eventually breaks down beneficial soil microbes and their dependent species, thus negatively impacting the soil environment and plant growth (Yang et al. 2011). Various studies have established that chlorothalonil is among the most toxic fungicides, which trigger cancer development in both

[1] Department of Life sciences, Kristu Jayanti (Deemed to be University), Bengaluru, Karnataka-560077, India.
[2] St. Chavara Centre for Advanced Materials Research, Kristu Jayanti (Deemed to be University), Bengaluru, Karnataka-560077, India.
[3] Dr. Janaki Ammal Campus, Kannur University, Kannur - 670661, Kerala, India.
* Corresponding author: sabua@kannuruniv.ac.in; drsabu@gmail.com

DOI: 10.1201/9781003603023-7

animal models and human cell lines (Tao et al. 2020, Van Scoy and Tjeerdema 2014). These conditions have created a compelling need to develop credible and eco-friendly biocontrol approaches to control fungal diseases. Biological fungicides, normally called biofungicides, take over the role of biocontrol agents for controlling phytopathogenic fungal and bacterial diseases. The creation of biofungicides occurs by using microbial agents consisting of bacteria, actinomycetes, and fungi, in addition to their extracts or biological components derived from plants and animals. The diverse range of their fungicidal activity, depends on what specific agent or compound is applied (Figure 1). Several factors may be involved, including the production of antibiotics, the production of lytic enzymes against fungal membranes and hyphae, the competition for nutrients and space, inducing resistance in the host plant, mycoparasitism, spore inhibition, and ATP production (Cenobio-Galindo et al. 2024, Fenta et al. 2023).

An emphasis is generally placed on the biocontrol agent when formulating a biofungicide. Properties and activities emitted by these agents play a vital role in the control of pathogens. Currently, different biocontrol agents are being evaluated for screening. However, the underlying concern is that the agent employed here involves a living organism or a biomolecule; thus, their viability, stability, antifungal activity, and delivery must be optimal for the agent to be adapted for mass production.

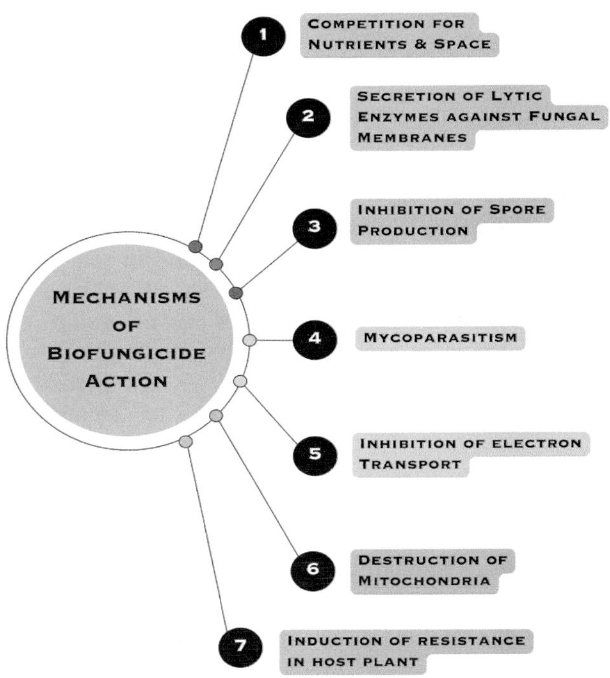

Figure 1. Illustrated presentation of various mechanisms of biofungicide on a phytopathogen. The selected biocontrol agent can inhibit the growth of phytopathogens by the competition for nutrients and space in the same environment, production of lytic enzymes against fungal membranes and hyphae, inhibition of spore production, mycoparasitism activity, inhibition of electron transport, disruption in ATP synthesis due to mitochondrial destruction, and by inducing resistance in the host plant.

Biocontrol agents cannot be directly applied to the field or stored without a decline in their efficacy (Leggett et al. 2011). The formulation of a system or a carrier that can retain the viability and activity of the biofungicide is necessary to ensure effective control of the phytopathogen in the field. Carriers are components that allow the biocontrol agent to be delivered effectively to the host plant or its surroundings. Their ingredients must ensure the stability of the formulation for its administration, the viability of the organisms or derivatives used, durability against environmental variations, a longer shelf life, and specific lethality against only the pathogen concerned (Khan et al. 2023, Vassilev et al. 2020). The use of biofungicides is often considered a preferable alternative to chemical fungicides due to their origin, but they currently pose significant challenges in formulation, development, and shelf life (Fenta and Mekonnen 2024). Special emphasis needs to be placed on the selection of carrier agents for different types of biocontrol agents and their compatibility with the agent by measuring viability, shelf life, field, and formulation studies. The chapter presents a common framework for biofungicide design and stresses the need for screening biocontrol agents, using proper carriers, and performing compatibility studies to overcome the tough challenges involved in promoting biofungicides into the market for commercial use.

2. Classification of Biofungicides

Biofungicides, which are derived from natural organisms, provide a sustainable alternative to chemical fungicides in managing plant diseases. Through the recent advancement of biological control strategies, their classification has also evolved. The concept of using natural agents to combat plant diseases dates back centuries. In the mid-17th century, practices like brining grain with salt water followed by liming were employed to control bunt in wheat, and based on these observations, wheat seeds salvaged from the sea were free of bunt infestation. These early practices laid the foundation for developing present-day biofungicides (Morton and Staub 2008). The proper application of biofungicides in sustainable agriculture depends on a full comprehension of their classification system and operational mechanisms. The use of natural biological processes enables biofungicides to provide environmentally friendly disease control methods, thus decreasing agricultural reliance on synthetic chemical controls. Three main biopesticide groups exist according to the classification system adopted by the United States Environmental Protection Agency (EPA) (Figure 2) (US EPA 2015), which are accepted all over the globe. These are listed below in detail.

2.1 *Microbial Biofungicides*

Biocontrol agents used in microbial biofungicides derive from microorganisms, which include bacteria, fungi, viruses, as well as protozoa, that target specific pest populations. We can categorise them according to their specific niches. For instance, the particular niche of endophytes includes microorganisms that can establish a symbiotic relationship with plant tissues and contribute to overall plant health by the production of bioactive compounds. (Fadiji and Babalola 2020) demonstrated

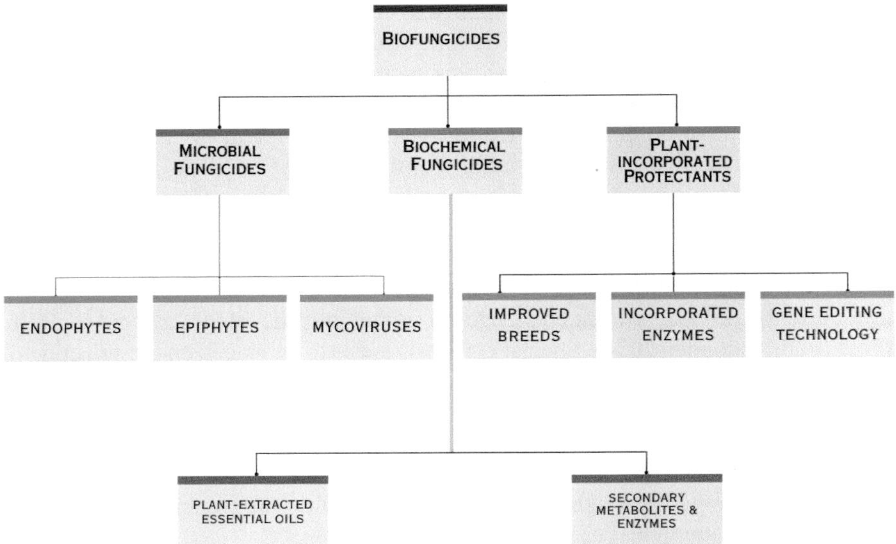

Figure 2. Schematic representation of the classification of biofungicides, which includes microbial fungicides, biochemical fungicides, and plant-incorporated biofungicides.

that such active compounds contribute to increased disease resistance and promote plant growth. Similarly, (Yu et al. 2018) reported that endophytic microbes belonging to the *Pestalotiopsis* genus have antifungal properties. Epiphytes and phylloplane microorganisms, a separate type of microorganisms, have proven capability to attach to the aerial surfaces of plants and protect them from foliar pathogens (Köhl et al. 2019). The specialised attachment to plants allows these organisms to overcome adverse environmental conditions such as UV radiation and temperature variations, thus helping them survive (Sivakumar et al. 2020, Chaudhry et al. 2021, Lahlali et al. 2022). A tertiary group of fungicides involves mycoviruses—viral agents that can prevent toxicity and spread of detrimental fungi species. New studies investigating the development of biofungicides from mycoviruses are ongoing, as this strategy displays efficient control over fungal diseases (Fenta and Mekonnen 2024). It has been proven that the mycovirus *Sclerotinia sclerotiorum* negative-stranded RNA virus 1 (SsNSRV-1) reduces the virulence of the pathogenic fungus *Sclerotinia sclerotiorum* (Kotta-Loizou 2021).

2.2 Biochemical Biofungicides

Biochemical biofungicides are formulated using natural compounds that manage pathogens through safe and non-toxic mechanisms. Antifungal activities of secondary metabolites and enzymes from plant or microbial origin regulate the behaviour and physiology of pathogens (Cenobio-Galindo et al. 2024). The available evidence shows that keeping neem and tea tree essential oils suppresses the growth of several pathogenic fungi attacking plants. (Hernández et al. 2021) reported showing fungicidal activity of isolated polyphenols against phytopathogens *M. fructicola* and *A. alternata*. Alkaloids, phenolics, and terpenoids, which are all secondary metabolites, have shown their antifungal potential, as they damage fungal cell walls and interfere with metabolic pathways. Studies with pathogens show neem and cinnamon essential oils have positive antifungal activity against fungal pathogens of chayote fruit (García-Ramírez et al. 2023, Hernández et al. 2021). Various investigations present that hydrolytic enzymes produced by microorganisms, plants, and insects show antifungal activity (Kumar et al. 2018, Nagpure et al. 2014, Veliz et al. 2017).

2.3 Plant-Incorporated Protectants (PIPs)

The approach reveals a method to genetically engineer the biocontrol constituent from plants or microbes to be transferred into host plants as a means to achieve specific biocontrol attributes. The development of better disease-resistant plants requires traditional crossbreeding and genetic modification approaches to generate novel advanced varieties. The incorporation of this method lowers the need for applying fungicides from outside sources (Dong and Ronald 2019). Plant resistance to pathogens can be increased through genetic modification that enables cells to produce antifungal enzymes naturally. In addition to degrading fungal cell walls, these enzymes can also inhibit their growth. Research into chitinases extracted from different microorganisms as fungicidal enzymes exists extensively because they specifically degrade fungal chitin cell wall structures without damaging the host plant (Kumar et al. 2018). Scientists are presently researching in utilising CRISPR/Cas9 to edit plant genomes so that they develop resistance against particular fungal pathogens. This exact method enables scientists to develop targeted improvements for plant defence systems (Wang et al. 2022).

3. Selection and Screening Method for a Biofungicide Agent

Biocontrol agents must be selected and screened before the development of biofungicides. Some common protocols and methods followed for screening and selection of biocontrol for the formulation of biofungicides are listed below:

3.1 Identification of Potential Antifungal Agent

The process to select microbial candidates requires sampling either from environmental niches that display antagonistic interactions or through random sampling procedures. To isolate distinct microbial colonies, soil samples are serially diluted and plated on a selective medium. The isolates are subcultured separately and identified using DNA sequencing of the 16S rRNA for bacteria and the 18S rRNA for fungi (Khan et al. 2023). Experts select plants for extraction using ethnobotanical surveys, metagenomic analysis, random screening, and traditional knowledge (Fabricant and Farnsworth 2001). Bioactive secondary metabolites are extracted using extraction techniques such as the Soxhlet apparatus and separated by fractionation. The bioactive compound profiles are determined using chromatography.

3.2 Screening Method for Antifungal Activity By in vitro Assays

The antifungal activity of the biocontrol agent can be confirmed through a variety of *in vitro* tests. A conventionally employed technique in identifying antagonist microorganisms (bacteria, fungi, actinomycetes) is the dual culture assay. The protocol is as follows: prepare solid media plates, commonly potato dextrose agar (PDA) plates. The plates are divided into equal halves. Inoculate the specific fungal pathogen in one half of the plate, and the biocontrol organism in the other half, or 3 cm away from the fungal inoculum (Figure 3A). Maintain a control plate containing only the pathogen for comparison. Incubate the plates at 25°C for seven days. The observation of any clear zone or bacteriostatic activity in the culture plates indicates the antagonistic activity of the inoculated organism (Maruyama et al. 2020, Ribera et al. 2017, Zeng et al. 2021). One typical method of assessing biocontrol organisms or agents utilises the well-diffusion method. The method is as follows: prepare PDA plates with 20 mL of media containing fungal spores at 10^6 spores/mL. Prepare 4 mm wells followed by the addition of varying concentrations of plant extract or microorganism suspension to each plate, with one plate of phytopathogen serving as a reference. For 48–72 hours, place the plates at the desired temperature range (see Figure 3B for more information). The observation of an inhibition zone around the well in the test indicates that the analysed agent displayed antagonistic activities on the target microorganisms (Aqueveque et al. 2017, Magaldi et al. 2004, Vassilev et al. 2020). The disc diffusion method enables the evaluation of the antagonistic effects

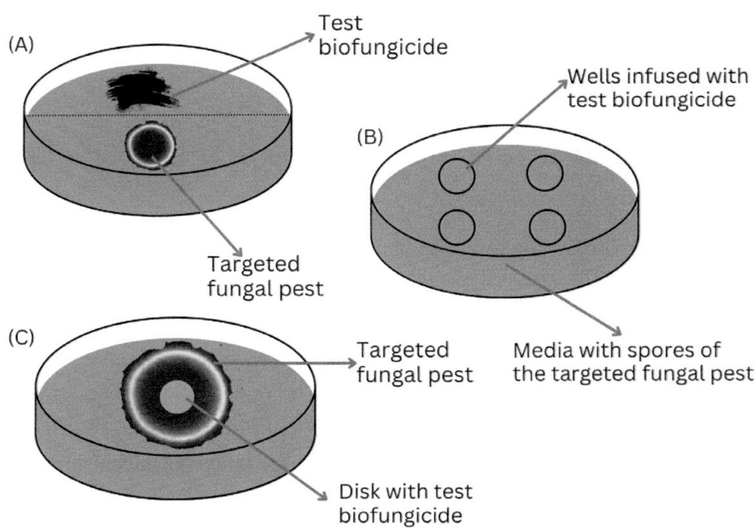

Figure 3. Screening setup for fungicidal activity using *in vitro* assays. In dual culture assay, the fungal pathogen and test organism are grown nearby in solid agar (A). Well diffusion assay is performed by generating wells of uniform size in a solid agar containing fungal inoculum and adding the test agent in each well (B). For disc diffusion assay, plant extract-soaked filter paper discs are placed on the surface of the culture plate containing the fungal pathogen (C).

of plant extracts and enzymes. The protocol is as follows: obtain all the necessary PDA plates, and inoculate them with the specified fungal pathogen, followed by incubation at 35–37°C for 2 days. Soak sterile filter paper discs (about 6 mm in diameter) in different concentrations of the plant extract. Place the moistened disc on the surface of the culture and incubate at the same temperature for seven days (Figure 3C). Maintain a positive and negative control, respectively. The presence of any fungus growth inhibition demonstrates that the plant extract possesses antifungal properties (Lee et al. 2022, Tanapichatsakul et al. 2020).

4. Substrate Medium for Growing Biocontrol Organism

Biofungicide production at a commercial scale depends on substrates that offer essential nutrients to microorganisms while also improving sporulation, metabolic processes, and activating biofungicide mechanisms responsible for effective pathogen control. Agro-based substrates from agricultural waste products create a sustainable environment in which biocontrol agents can grow through solid-state fermentation (SSF). Using agro-based substrates enables both cost reduction during production runs and efficient waste management of agricultural waste to support environmentally friendly agricultural practices. Substrates provide necessary nutrients while promoting microbe development to achieve high amounts of biocontrol organisms, thus enabling sustainable plant disease control alternatives. Wheat bran stands out among many agro-based substrates because its nutritious content helps biocontrol fungi such as *Trichoderma harzianum*, *Trichoderma viride*, and *Coniothyrium minitans* to thrive (Cavalcante et al. 2008, Chen et al. 2005). Rice husk and rice bran serve as good media for enhancing the sporulation and biomass development of entomopathogenic fungi such as *Lecanicillium lecanii* and *Trichoderma harzianum* (Feng et al. 2000, Naeimi et al. 2020). The sugarcane processing for alcohol creates sugarcane bagasse as a by-product, which supports fungal fermentation by providing a medium with a porous structure and great water retention capacity that enhances spore production of fungi such as *Metarhizium anisopliae* and *Beauveria bassiana* (Arzumanov et al. 2005, Dalla Santa et al. 2004). Similarly, coffee husks and bagasse of corn have been recognised as valuable substrates for producing *Trichoderma* spp. at a commercial level since they exhibit antagonistic activity against numerous fungal plant pathogens (Doni et al. 2014, Mousumi Das et al. 2021).

5. Carrier Selection and its Compatibility with Biocontrol Agents

The process of selecting carriers stands as a vital requirement that commonly depends on both the chosen control agent and its operating efficiency within the carrier system. Carriers that are effective for the growth and viability of one agent may not be suitable or effective for another agent. It is, therefore, essential to conduct appropriate experimental studies to decide the carrier agent and verify its compatibility with the biocontrol agent before introducing it for mass production (Wu et al. 2014). Several types of carriers are discussed (Figure 4) in this section, along with the composition

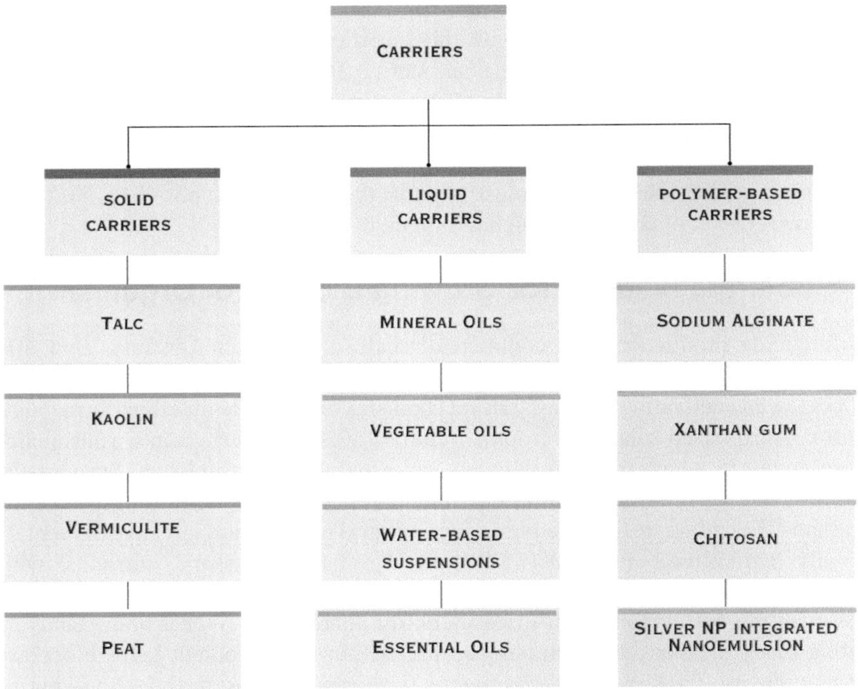

Figure 4. Graphical representation of the common types of carriers utilised in biofungicides, which include talc, kaolin, vermiculite, peat, mineral oil, vegetable oil, water-based suspensions, essential oil, sodium alginate, xanthan gum, chitosan, and silver nanoparticle integrated nanoemulsion.

of different carriers, selection criteria for carriers, compatibility studies of the carrier with the agent, and examples of compatible biofungicide formulations.

5.1 Types of Carriers in Biofungicide Formulation

5.1.1 Solid Carriers

Solid carriers are inert dry materials that stabilise, protect, and effectively deliver microbial biofungicides. A well-selected solid carrier adsorbs, protects, and gradually releases biofungicide, thus ensuring prolonged field efficacy while enhancing its shelf life. The group of common solid carriers features talc together with kaolin, vermiculite, and peat, each of which has separate advantages. The powdered talc or $Mg_3Si_4O_{10}(OH)_2$ is favoured over other materials because it demonstrates inert behaviour along with hydrophobicity and high surface area, which supports microbial stability. Besides adsorbing bacterial and fungal spores, it also protects them from desiccation and UV radiation and can be applied as either a dry formulation or a suspension. Reports on talc carrier-based *Bacillus subtilis* formulation succeeding in reducing *Fusarium oxysporum* occurrences by 44% demonstrate its high effectiveness (Jangir et al. 2021, Şenol Kotan et al. 2023). Vermiculite is a lightweight, porous silicate mineral known for its ability to retain moisture and support microbial viability. The encapsulation process allows biocontrol agents to release gradually while maintaining their effectiveness longer. Experimental studies have reported that *Trichoderma harzianum* preparations utilising vermiculite sustained their CFU counts at a high level for up to eight weeks, leading to significant reductions in Fusarium wilt (Martínez-Medina et al. 2009). The decomposed sphagnum moss substance known as peat establishes a suitable environment for maintaining microbial existence because it represents a porous material with high moisture-holding capacity. Experiments done in biocontrol formulation containing antibiotic-resistant *Pseudomonas fluorescens* mixed in peat demonstrated its capacity to inhibit the growth of the fungal pathogen *Rhizoctonia solani*, which is known to cause rice sheath blight in plants. The outcome of field trials confirmed successful disease management (Rabindran and Vidhyasekaran 1996).

5.1.2 Liquid Carriers

Liquid carriers act as a stabilising and dispersing medium, ensuring the biocontrol agent's enhanced adherence, viability, and efficacy in the field. Unlike solid carriers, liquid formulations exist in the form of suspensions or emulsions, which improve biofungicide applications (Kara and Tozlu 2024). These carriers include plant-based oils, emulsions, and aqueous suspensions, each of which offers specific advantages. Oil-based carriers derived from vegetable oils, such as neem, castor, soybean, and coconut, function as hydrophobic reservoirs that prevent microbial desiccation and extend shelf life. The carriers use water-in-oil or oil-in-water emulsion technology to safeguard biofungicide agents through protective microcapsules, which also promote the dispersal and surface adherence of microorganisms. The use of oil-based *Beauveria bassiana* formulations proved effective in suppressing the pathogen *Helicoverpa armigera* while simultaneously strengthening defence mechanisms in tomato crops (De Medeiros et al. 2024). Emulsions are a stable combination of plant-derived oils and water formulated with surfactants, such as Tween 80 and lecithin, to enhance biofungicide delivery. The use of emulsions improves all three aspects: microbial protection, solubility, and controlled release of the product. The wide application of emulsions in disease control programmes for harvested produce and field crops originates from their ability to decompose safely and operate at peak effectiveness (Syed-Ab-Rahman et al. 2020).

Biofungicides achieve improved delivery when formulated using novel techniques like nanoemulsions because these carrier systems minimise particle dimensions to nanometric levels and enable enhanced stability, together with precise release. The application of chitosan-based nanoemulsions and plant oil nanoformulations shows extended antifungal properties against *Fusarium oxysporum* and *Botrytis cinerea* according to studies done by (Naserzadeh et al. 2019, Saberi-Riseh and Moradi-Pour 2021). The utility of nanoemulsions to improve biofungicide efficiency and lower chemical use simultaneously protects plants in modern agricultural operations, establishing them as crucial agricultural tools. Recent advancements demonstrate that incorporating silver nanoparticles into nanoemulsions enhances both stability and overall efficacy of the emulsion (Chowdappa et al. 2014). Research efforts now focus on amplifying production capacity and establishing safe environmental standards for sustainable agricultural nanoemulsions (Teixidó et al. 2022).

5.1.3 Polymer-Based Carriers

Polymer-based carriers are biodegradable and biocompatible matrices that encapsulate and stabilise biofungicides, thereby ensuring their protection and controlled release. Hydrogel-based and nanoparticle-based carriers made from materials like alginate, chitosan, starch, and gelatin serve to enhance both biological viability and effectiveness of biofungicide agents (Pereira et al. 2023). Protection of microbial cells happens through microencapsulation methods, which incorporate the use of ionic gelation, spray drying, and emulsion formation technologies to create a protective polymer matrix. Biofungicides embedded inside protective barriers can withstand desiccation, UV radiation, and environmental stress because this barrier allows slow microbial release while maintaining effectiveness (Saberi Riseh et al. 2021). Studies have documented that biofungicides enclosed in polymers demonstrate better antifungal properties than free-formula biofungicides. According to (Estefania and Ligia 2018), alginate-based microcapsules containing biocontrol agent *Bacillus subtilis* exhibited significant suppression of the pathogen *Fusarium oxysporum* because they maintained bacterial viability and antifungal metabolite production. Polymer-based carriers are currently being applied extensively in the form of seed coatings, foliar sprays, and soil inoculants. They provide extended storage time and safe to nature, which makes them more attractive for sustainable agricultural applications (Pereira et al. 2023).

5.2 Carrier Selection for a Biofungicide

An ideal carrier must support the survival of microorganisms in addition to promoting effective delivery and controlled distribution of biofungicides. The selection of a carrier agent and its compatibility studies are essential steps in the preparation of biofungicides (Figure 5). The selection procedure relies on the carrier agent's physical, chemical, and biological properties, economic practicality, as well as environmental sustainability standards, which are described below.

5.2.1 Physical Properties

Carrier physical attributes, starting from particle dimension through surface exposure and porosity structure, heavily impact how well microorganisms bind to the surface as well as how well they retain moisture and control product release. Lower particle size increases microbial dispersion throughout the product. Increased porosity and higher surface area support effective absorption and stable attachment of biocontrol agents. Talc, kaolin, and vermiculite, which are valued for their exceptional adsorptive properties, are used in the formulation of biofungicides. Employing carriers extracted from these materials provides two major advantages to biocontrol microorganisms: protecting them from dehydration and ensuring favourable physiological conditions. The carrier system augments the effectiveness of biocontrol microorganisms in the fight against pathogens (Jangir et al. 2021, Zamoum et al. 2017).

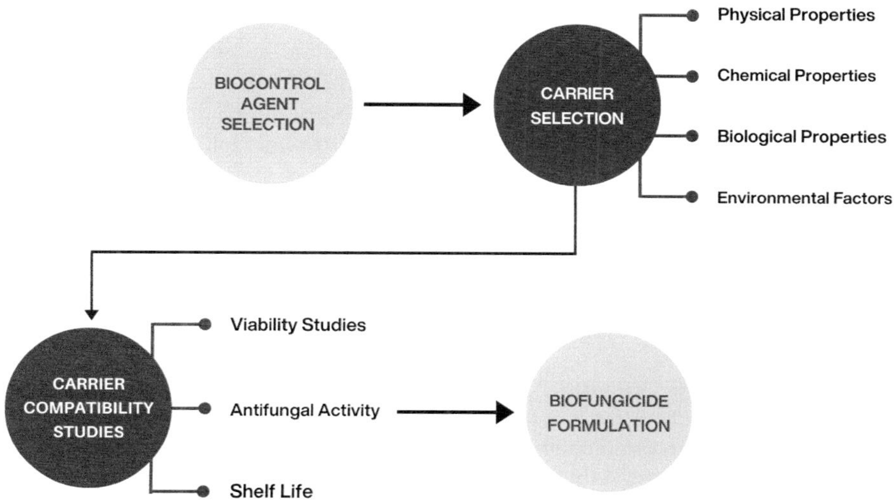

Figure 5. Illustrated step-by-step process of biofungicide formulation, including biocontrol agent selection, carrier selection, carrier compatibility studies, and formulation. The selection of carriers is based on physical, chemical, biological, and environmental factors. Studies of carrier compatibility are based on viability, antifungal activity, and shelf-life parameters.

5.2.2 Chemical Properties

It is advisable that the carrier systems maintain a pH of neutral to slightly acidic for best results to protect microbial viability and avoid enzyme-mediated breakdown. The stability of microbial activity is maintained through minimum moisture levels, which prevents early microbial growth while improving shelf stability. Some carriers can provide added protective benefits: kaolin, which generates a UV shield that protects microbes from radiation, and talc, which improves stability by reducing water loss as a consequence of its hydrophobic nature. The effects of these chemical properties on biofungicide performance and shelf life make them suitable for use in field conditions (Eigenbrode et al. 2006).

5.2.3 Biological Properties

To minimise microbial and chemical contamination, a biologically inert carrier avoids colonisation by pathogenic or competitive microbes. Sterile conditions safeguard beneficial microorganisms, keeping them viable during storage and when applied in the field. Materials obtained from chitosan and vermiculite act as biopolymer carriers that protect microorganisms against contamination. Since these carriers are inert, excellent support for protecting biofungicides in various environments is ensured. Maintaining microbial viability and preventing contamination in the process of biofungicide production and storage is key to successful production (Pereira et al. 2023).

5.2.4 Economic and Environmental Considerations

Both economic feasibility and environmental sustainability play a pivotal role in determining the preferred type of carrier selection. Their widespread use is facilitated by the integration of economic benefits and availability of talc and kaolin carriers. Oil and polymer-based carriers promote microbial protection and efficacy at a great financial cost (Eigenbrode et al. 2006, Naveenkumar et al. 2017, Saberi Riseh et al. 2021). The use of plant-based oils, chitosan-based, and starch-based polymers substantially reduces chemical residues and carbon emissions in pest control operations, making them eco-friendly carriers. Others like kaolin and vermiculite, with their natural origin and low environmental impact, make them appropriate for sustainable agricultural management (Pereira et al. 2023).

5.3 Carrier Compatibility Studies

To develop an effective and commercially viable formulation, it is necessary to have a comprehensive understanding of the biocontrol agent, target pathogen, its selected carrier agent, as well as their interactions within the environment in which they are intended to be used. The compatibility studies are designed to determine whether the selected carrier is suitable for the biocontrol agent by monitoring its viability and efficiency throughout its shelf life (Figure 5). Different aspects, mainly ensuring the viability and stability of the biocontrol agent, must be considered during compatibility testing. The biocontrol agent employed is a living organism or its by-product; thus, maintaining its stability outside the biological system for a long period is essential. In contrast, synthetic fungicide formulations offer a more stable formulation, which makes the latter more attractive to farmers (Fenta and Mekonnen 2024, Khan et al. 2023).

The viability of a biofungicide formulation containing a microorganism is normally evaluated by colony-forming unit (CFU) estimation and compared with that of a control containing only the microorganism. The general overview of CFU estimation is characterised by inoculating the formulated biofungicide in suitable agar media and incubating it at an optimised temperature for a standardised time. The number of colonies was calculated by counting the colonies present on the media plate with respect to the control plate (Kawakita et al. 2021, Maruyama et al. 2020, Wu et al. 2014, Zamoum et al. 2017). The results should show that integration of carriers into the treatment does not impair the rate of survival of the biocontrol agents. Researchers observed that encapsulation of *Raoultella planticola* Rs 2 with sodium alginate retained bacterial viability (Wu et al. 2014). It was shown in the research that viable spores of *Trichoderma harzianum* remained high throughout 10^8–10^7 CFU/mL concentration after formulation (Maruyama et al. 2020). The spores were serially diluted in a saline-surfactant solution containing Tween 80, which was then plated onto PDA and incubated at 28°C for 18 hours. Generally, the effectiveness of a biocontrol agent's metabolite is determined by its antifungal activity (Nandini et al. 2019, Naserzadeh et al. 2019). Principally, antifungal activity tests are performed concurrently with survivability assays, which ensures the carrier will not affect the biological activity of the agent. The evaluation of antagonist activity is done through methods such as disc diffusion and dual culture assay, which are described under the selection of the biocontrol agent in this chapter (section 3.2). In the disc diffusion method, sterile discs are prepared by soaking in 10–15 μL of emulsion and placing them on agar plates with 2–3 days grown target organism. Observation of the inhibition zone after incubation determines the antifungal activity of the formulation (Hou et al. 2021). Alternatively, antifungal activity can be assayed by mixing standardised aliquots of the nanoemulsion in selected media, commonly PDA. Even discs of 5–6 mm procured from the target fungi are placed in the mixed medium and incubated at 25°C for 24 hours, and the growth of the pathogen can be estimated. Dual culture assays are performed by placing the formulated organism on one side of the fungal media plate, and the other side is inoculated with the target pathogen. The plate is incubated at room temperature for 24–72 hours, and pathogen inhibition is estimated. The ability of a living organism or its derivative to maintain viability over an extended period under different environmental conditions is one of the most important functions of biofungicides. Carriers are selected to extend the shelf life of the biocontrol agent, and their ability to maintain the stability of the biocontrol agent can assist in further optimisation for improving its stability. Biofungicides are considered to have the greatest disadvantage of short shelf life (Fenta and Mekonnen 2024). Shelf-life studies of the formulation are generally analysed by subjecting the organism to various environmental conditions, including altered temperature and humidity levels, for 3 to 6 months, ideally to simulate long-term storage conditions. The stability of the long-term stored bioformulation is estimated by evaluating its viability and antagonist activity monthly, employing the techniques discussed previously. Throughout storage conditions, the CFU value should maintain 10^4 to 10^6 CFU/mL of the biocontrol organism for a formulation to be considered viable (Kumar et al. 2022, Pathirana and Yapa 2020). Shelf-life stability studies done by (Kawakita et al. 2021) reported that biocontrol agent *Collimonas arenae* encapsulated in alginate microcapsules with modified starch exhibited high cell viability of nearly 10^{10} CFU/mL and retained its antifungal activity three weeks into storage. Work done by (Tamreihao et al. 2016) reported that talc formulation containing the antagonist agent *Streptomyces corchorusii* retained its viability at 10^4 CFU/mL for 6 months at a storage temperature of 4°C. Optimisation and modification of carriers in bioformulations and their studies are required for the production of viable and potent products for commercial applications.

6. Key Techniques for Biofungicide Formulation

Formulation of a biocontrol agent can employ certain complex techniques for the production of a carrier system and characterisation of the bioformulation. Given below are a few of the most commonly used techniques in the preparation of biofungicides. Lyophilisation, or freeze-drying, is a dehydration process used to preserve biofungicides, particularly microbial-based formulations. This technique involves freezing the bioactive material, followed by sublimation of ice under low pressure, resulting in a dry, stable powder that retains its biological activity (Borges et al. 2018). This method allows biocontrol organisms to be encapsulated as dried forms, which remain viable. Emulsification allows mixing of two liquid materials that cannot dissolve in each other (e.g., oil and water) in the form of an emulsified mixture that is stable. Serving as a basic step in the evolution of biofungicides, emulsification is recognised as being of utmost significance in nanoemulsion preparation domains (Fernandez et al. 2004). Sonication is one of the commonly used techniques for the preparation of nanostructured emulsions for biofungicide application. Sonication using high-frequency sound can break the larger droplets into nanometre-sized droplets, improving their stability and bioavailability of biofungicides (Campolo et al. 2020). A UV-visible spectrophotometer is utilised to quantify and confirm the presence of nanoparticles within an emulsion. Researchers employ the UV-visible spectrophotometer to determine the ability of samples to absorb ultraviolet and visible light in the range between 190 and 900 nm in the electromagnetic spectrum. The instrument determines the UV light absorption and transmission level of the target sample at its specific wavelength without altering the sample to validate its presence (Khalid et al. 2024). Dynamic light scattering (DLS) method is used to quantify the sizes of objects in dispersions—from 1 nm to 1000 nm in size. The method is capable of determining the size distribution and proving a homogeneous structure of nanoparticles. The Doppler Effect principle is the basis for understanding the associated coupling of nanoparticles scattering light with their velocity and size (Jia et al. 2023, Ruseva et al. 2018).

7. Formulation Techniques of Biofungicide

Biofungicides are developed by combining a living organism or its by-products with an inert carrier agent to enhance their effectiveness. This is done to improve the application effectiveness, storage capacity, and handling of biofungicides. The inert ingredient functions primarily to deliver the active biocontrol ingredient, which presents fungicidal properties. The main purpose of a formulation is to provide a base for live cells and metabolites to interact with the phytopathogen and inhibit its effect on the host plant (Khan et al. 2023). As discussed in the previous sections, the active agent of a biofungicide can vary from various microorganisms or plants to their secreted enzymes and metabolites; thus, depending on the active ingredient used, the formulations can vary. An overview of the different types of biofungicides that can be produced utilising a variety of biocontrol agents and carriers is provided in Table 1. It is important to note that the methods discussed below depend on the most well-known biofungicide formulations. Hence, the steps can vary depending on the type, materials, and agent employed. Various techniques can be used to further modify and optimise a formulation method for the selected agent.

7.1 Liquid Formulation

7.1.1 Encapsulation by Liposome

One of the most common characteristic attributed to previously studied biocontrol agents like *Trichoderma* spp. is the secretion of hydrolytic enzymes, which have been reported to have antagonist properties against several phytopathogenic fungal species like *Rhizoctonia solani*, *Fusarium oxysporum*, *Sclerotinia sclerotiorum*, *Botrytis cinerea*, *Pseudocercospora* spp. and *Colletotrichum* spp. by breaking down the fungal cell wall, conidia and hyphae (Yao et al. 2023). Major hydrolysing

Table 1: Examples of carrier-based biofungicides and their successful formulations, with their effect on biocontrol.

Type of Carrier	Target Pathogen	Biocontrol Agent	Effect	References
Talc Based (Solid)	*Fusarium oxysporum*	*Bacillus subtilis*	Reported a 44% decrease in the disease incidence	(Jangir et al. 2021, Şenol Kotan et al. 2023)
Talc Based (Solid)	*Rhizoctonia solani*	*Streptomyces rochei*	Enhanced shelf life at room temperature for 1 year and enhanced biocontrol activity	(Zamoum et al. 2017)
Peat and Chitin (Solid)	*Fusarium udum* and *Aspergillus niger*	*Baciullus subtilis*	Efficient control against disease and promote seed germination in groundnut and pigeon pea	(Manjula and Podile 2001)
Vermiculite Based (Solid)	*Fusarium oxysporum*	*Trichoderma harzianum*	Efficient disease suppression	(Martinez et al. 2023)
Oil Based (Liquid)	*Phytophthora infestans*	*Trichoderma* spp.	Reported an efficiency of 83.08% in disease inhibition in the potato plant	(Abadi et al. 2024)
Oil Based (Liquid)	*Penicillium, Aspergillus, Colletotrichum* spp.	Essential oils like lemongrass oil, castor oil, and oleic acid	Effective disease suppression and increased plant defence response	(De Medeiros et al. 2024)
Liposome (Liquid)	*Fusarium oxysporum*	Chitinase enzyme	Effective inhibition of the growth of the fungal pathogen	(Anna et al. 2013)
Nanoemulsion (Liquid)	*Botrytis cinerea*	Cinnamon essential oil	Reported a reduced level of infection rate (5.43%) in strawberry fruit	(Naserzadeh et al. 2019)
Sodium Alginate (Polymer Based)	*Sclerotinia sclerotiorum*	*Trichoderma harzianum*	Increased the shelf life of the fungi and enhanced its efficiency by 20-fold	(Maruyama et al. 2020)
Sodium Alginate (Polymer Based)	*Fusarium oxysporum*	*Bacillus subtilis* NCD-2	Effective control with stability for 2 years in Chrysanthemum	(Chen et al. 2018)
Polyurethane (Polymer Based)	*Rhizoctonia solani*	*Trichoderma longibrachiatum*	Reduced infection and enhanced seed germination in tomato plants	(Mahde et al. 2019)

enzymes isolated and used as a biocontrol agent include chitinase and laminarinase (Anna et al. 2013). As a rule, it is considered more effective to isolate a hydrolysing enzyme from *Trichoderma* spp. than other enzyme-secreting bacterial species, such as *Bacillus subtilis*, *Streptomyces alfalfae*, and *Streptomyces griseus*. These species are capable of synthesising chitinase which has shown inhibitory activity against *Fusarium oxysporum*, *Lasiodiplodia theobromae*, and *Trichoderma reesei*, among others (Lv et al. 2021, Veliz et al. 2017). Enzymes and secondary metabolites are commonly formulated using the encapsulation method, primarily by liposomes. It has been demonstrated that the use of a nanocarrier system such as liposomes, improves both the bioavailability and stability of different bioactive compounds inside its core, thereby extending their shelf life and representing a good option as a carrier system (Anna et al. 2013, Borges et al. 2018). A general overview of the procedure employed includes preparing a thin lipid film layer by mixing phospholipid components, such as lecithin, with a suitable solvent in a round-bottom flask, followed by evaporation. To ensure the removal of any residual solvents, flasks need to be kept in the oven at 50°C for 24 hours. Hydrate the films by adding a suitable buffer. Different concentrations of the extracted enzymes are mixed with a hydrated film solution, followed by vortex mixing at constant temperature and pH. Different temperatures can be set to select the most stable liposomes. Continue the vortex mixing until the films detach from the flask and form liposomes (Anna et al. 2013, Cano-Salazar et al. 2011). The liposomes obtained are characterised and taken up for further antifungal assays. (Anna et al. 2013) has formulated a carrier using soya lecithin to encapsulate chitinase enzyme combined with a low concentration of chemical fungicide—thiabendazole (6 μg), which is reported to be effective against the phytopathogen *F. oxysporum*. Here, the phospholipid component used consists of 33% phosphatidylcholine, 14.1% phosphatidylethanolamine, 16.8% phosphatidylinositol, and 0.4% phosphatidylserine. The solvent and buffer used are a methanol solution and 0.1 M sodium acetate respectively. Antifungal activity of the enzyme-trapped liposomes resulted in lower CFU values of the fungal pathogen in a soil setting, and encapsulation has been reported to contribute to a higher stability of the enzyme.

7.1.2 Encapsulation by Spray Drying

Liposome encapsulation is not the only form of liquid formulation employed in biocontrol. In light of the demand for scalability of biocontrol agents, spray-dried encapsulation is considered a promising approach in enclosing microorganisms antagonistic to the fungal pathogen, all the while preserving their integrity (Schuck et al. 2013). The method is followed by inoculating and culturing the specific bacterial culture in a nutrient-rich medium of the required volume until a high biomass concentration is obtained. Procure the cells by centrifugation at 4000–5000 rpm for 15 minutes. Suspend the bacterial cell pellet in a specific solution for encapsulation. The suspended bacterial culture was mixed with an autoclaved encapsulating polymer, such as alginate, chitosan, and cellulose acetate, along with other components, such as starch, to stabilise the encapsulation and form the feed. The feed must be properly homogenised to ensure encapsulation by a homogeniser. Dispense the feed formulation into the spray drier. Optimise the various parameters in the spray drier such as inlet temperature, aspirator flow, and nozzle pressure to produce dry microcapsules containing the bacteria, which are collected at the bottom of the dryer's drum and stored at 4°C. Antifungal assays must be performed to ensure that the antagonist activity of the encapsulated bacteria is retained (Kawakita et al. 2021, Saberi-Riseh and Moradi-Pour 2021). Spray-dried form of encapsulated bacteria is demonstrated by (Kawakita et al. 2021) with the antifungal bacterium

Collimonas arenae Cal35 using *in-situ* cross-linked alginate microcapsules (CLAMs) as the carrier. Their work reported an optimised strategy for procuring high-yield and viable capsules during spray drying while maintaining the strain's antifungal properties.

7.1.3 Encapsulation by Extrusion

The most common approach of entrapment is by encapsulation with alginate. A significant reason for the success of alginate gel encapsulation is that it provides a protective environment for the entrapped material, is inexpensive, easy to perform, and biocompatible. The entrapment protocol employed involves mixing the cell suspensions or conidia (in the case of fungi) with sodium alginate solution for 1 hour with continuous agitation. The solution was dripped through a pipette or nozzle at high pressure into a solution containing cations of bivalency (typically Ca^{2+} in $CaCl_2$). The droplets transform into gel spheres instantly after cells become entrapped through ionically cross-linked alginate in the alginate and cation solution. The recovered beads were dried to obtain dry microcapsules. The factors affecting the size of the gel sphere include the distance of the nozzle or pipette from the $CaCl_2$ solution, nozzle or pipette diameter, viscosity, and alginate solution concentration (Maruyama et al. 2020, Saberi Riseh et al. 2021). According to (Maruyama et al. 2020), when *T. harzianum* was entrapped as wet microparticles in sodium alginates, its ability to control the plant pathogen *Sclerotinia sclerotiorum* was 20 times higher than that of the free fungal form or dry microcapsules, with no sclerotia (a dormant structure of fungus) formation. It was also noted that compared to the dry microparticles, wet microparticles demonstrated high cell viability at 5°C for 120 days while maintaining their hydrolysing activity, suggesting the possibility of their large-scale applications.

7.2 Nanoemulsion

Nanoemulsions are becoming popular as a biofungicide formulation due to their ability to enhance the fungicidal efficiency of bioactive compounds. Early studies investigated nanoemulsion formulation's ability to enhance biofungicidal activity of *Trichoderma harzianum* and *Pseudomonas fluorescens*. When incorporating biofungicides into nanoemulsions, there is an increased efficacy against fungal pathogens compared to the control (Chowdappa et al. 2014). Due to the smaller size of their droplets, nanoemulsions offer enhanced access and constant release of bioactive agents, hence supporting their function as active carriers of microbial biocontrol agents. The method followed entails: The oil phase is prepared mainly from plant-derived oils or essential oils such as cinnamon or clove oil, serving as a carrier for biofungicide. Later, a surfactant is added, commonly polysorbate 80 or Tween 20, since it reduces the surface tension between the oil and the aqueous layers. Furthermore, a co-surfactant like propylene glycol or ethanol is employed to improve stability and reduce the droplet size of the nanoemulsion. Steep ultrasonication operated at high energy is used in the preparation of nanoemulsions. The surfactant and co-surfactant are first mixed with the oil phase and then sonicated at a constant temperature. High-energy sonication is applied to break down large oil particles into nanoscale droplets (20–200 nm) to form stable nanoemulsions. Furthermore, the low-energy phase inversion composition (PIC) method can be used to prepare nanoemulsions. Subsequently, water is progressively added to the uniform solution of surfactant and oil while the temperature is maintained at 25 ± 1°C. The continuous stirring function is utilised to propel the nanoemulsion production procedure (Glare et al. 2012, Miastkowska et al. 2020). Optimisation of nanoemulsion formulation (i.e., oil-to-surfactant ratios, co-surfactant concentrations) is required for optimising droplet size and stability, which requires experimentation. Using Response Surface Methodology or similar optimisation techniques is found to be ideal for attaining high encapsulation efficiency and maximum antifungal activity (Llinares et al. 2018, Naserzadeh et al. 2019, Chowdappa et al. 2014). Treating strawberries with a nanoemulsion containing cinnamon essential oil has resulted in significantly lower infection of *B. cinerea* compared with chemical fungicides. This may imply that the use of cinnamon essential oil could be an alternative to traditional fungicides (Naserzadeh et al. 2019).

Organism-derived natural antifungal membrane lipids can form part of the components in producing nanoemulsions. Researchers have described their method of making nanoemulsions from membrane lipids obtained from *Trichoderma* strains (Nandini et al. 2019). The nanoemulsion was developed to enhance the plant defence against *Sclerospora graminicola*, the organism that causes downy mildew in pearl millet. Micro-droplets contained in the nanoformulation increase its bioactivity and efficacy in controlling fungi. Methodology followed: the cultivation of the biomass also required the *Trichoderma* strains to be cultured under harnessed conditions. After growing the *Trichoderma*, the biomass was collected and washed using a buffer solution. To isolate lipids from the biomass, the washed biomass was subjected to a chloroform-methanol solvent (2:1, v/v) based on the Folch method. After mixing at room temperature, the mixture was gently stirred for 2 hours. The organic phase was separated from the aqueous phase by centrifugation at 5000 rpm for 10 mins. The removal of the solvent resulted in the extraction of the lipid, which was done by evaporation using a nitrogen stream, thus giving a pure lipid fraction. The extracted lipids were dissolved in a polar solvent like chloroform to prepare a lipid solution with the desired concentration. The lipid solution was thoroughly mixed to ensure its uniformity. Tween 80 (or polysorbate 80), a non-ionic surfactant, at a concentration of 1% (w/v), was added to the lipid solution. This was performed to stabilise the emulsion and reduce the surface tension between the oil and water phases. The polar solvent was removed by evaporation under reduced pressure using a rotary evaporator at 40°C to form a thin, uniform lipid film on the inner surface of the rotary flask. This process removes all traces of the solvent, preventing any residual chemicals from affecting the nanoemulsion formulation. The lipid film was hydrated by slowly adding distilled water (pre-warmed to 60°C), followed by stirring at 500 rpm for 1 hour to allow the lipids to swell and form a lipid suspension. A nanoemulsion was formed (with a size of < 200 nm) by treating the lipid suspension with sonication to break down the lipid aggregates. A probe sonicator with set conditions at 40% amplitude and 5–second burst pulses with equal pauses was used to produce nanoemulsions for 10 minutes (Nandini et al. 2019). The developed formulation demonstrated 40 days of stability at room temperature and maintained stability for 60 days at lower temperatures during storage conditions. Additionally, it showed an 82% reduction in disease incidence during field trials.

7.3 Dried Formulation

One of the oldest and most common forms of biocontrol formulation technique is the drying of the antagonist organism and its conversion into a powder (Martinez et al. 2023). The preparation of powder formulations using antagonist organisms includes spore exploitation as powder or freezing treatment of the antagonist cell, followed by blending procedures. Certain spore-producing *Bacillus* species have been confirmed to work as biocontrol agents against well-known pathogens such as *R. solani* (Pane et al. 2012).

7.3.1 Dried Formulation by Bacterial Species

The bacterial strain capable of producing spores was inoculated and cultivated on media plates for a minimum of five days. The spores were collected by scraping the culture's surface and recovered in suspension using either Tween 20 or saline solution. The spore suspension (25 mL) was mixed with the optimal quantity of autoclaved carrier agents, such as talc powder, kaolin, and calcium carbonate, along with a dispersant, such as carboxymethyl cellulose and polyvinyl alcohol. The formulation was dried overnight in an oven at 50°C or under laminar air flow to obtain the powdered form. Typically, these formulations exist as wettable powders which need to be diluted before application to ensure efficacy (Zamoum et al. 2017, Zeng et al. 2021). As reported by (Zamoum et al. 2017), formulated wettable talcum powder containing endophytic *Streptomyces rochei* strain PTL2 spores maintained viability for more than one year when stored at room temperature. Moreover, this formulation demonstrated maximum effectiveness in the biocontrol of the pathogen *Rhizoctonia solani* damping off while enhancing the germination of tomato seeds.

Additionally, some bacterial species are more suitable to be lyophilised to be converted into a powdered form (Zeng et al. 2021). A general overview of the protocol begins by isolating the specific bacterial species and culturing them in media broth for six days. The culture broth was centrifuged at 10,000 rpm for 5 minutes to obtain the cell pellet. The pellet was washed three times with sterile distilled water and filtered through Whatman filter paper to obtain a concentrated pellet. The concentrated cell pellet was lyophilised at –55°C. The lyophilised formulation was mixed with sterilised and dried carrier agents like talc powder and kaolin of optimised values to obtain the concoction, which was further passed through a mesh/nylon screen of desired pore size to obtain a homogenous powder formulation (Tamreihao et al. 2016, Yao et al. 2023). According to (Tamreihao et al. 2016), the isolated *Streptomyces corchorusii* UCR3–16 formulated with talc exhibited significant biocontrol potential against rice fungal pathogens, showing high inhibition activity against *Rhizoctonia solani* and was also proven to impart a positive effect on crop growth during field trials. The study also reported a high shelf life of *Streptomyces* spp. after 6 months at 4°C storage.

7.3.2 Dried Formulation of Fungus

Trichoderma spp. is a leading biocontrol agent against pathogenic fungi, with its dry formulation being highly investigated and tested in different experiments. Many of the species in this group have undergone extensive research with a view to finding a suitable biocontrol agent. *Trichoderma harzianum* restricted the growth of *B. cinerea* through spatial competition, and the growth of pathogenic *Fusarium oxysporum* was restricted by the antibiotics produced by *Trichoderma pseudokoningii* (Yao et al. 2023). *Trichoderma* isolates were cultured onto PDA plates or millet seeds stocked in the dark for 5 days. In order to produce spore suspensions, surface spores of the culture plate were rinsed with 5 mL of sterile water or Tween 80 solution. Meanwhile, a suitable amount of carrier material such as talc powder or biochar was mixed with a particular quantity of spore fluid and allowed to incubate for two weeks to enable colonisation to occur. The mixture was filtered through sieves with small pore sizes to achieve uniform particle size (Lang and Chen 2023, Mahde et al. 2019, Ribera et al. 2017, Martinez et al. 2023). Studies carried out by the researchers revealed that *Trichoderma harzianum* enriched with biochar inhibited the secretion of oxalic acid in wood decay caused by basidiomycetous fungi (Ribera et al. 2017). Field trials of *T. longibrachiatum* applied with polyurethane (PU) reduced the incidence of *R. solani* while enhancing the seed germination and overall growth of tomato plants. The shelf stability of the tested material was maintained six months after the testing was done (Mahde et al. 2019).

7.4 Nanobiofungicides

A new class of fungicides combines nanoparticles with biological control agents. These formulation consist of biocontrol agents in combination with nanoparticles formed from silver, copper or zinc (Cruz-Luna et al. 2021). When bioagents are coupled with nanoparticles, biofungicide efficacy is increased, stability enhanced, and controlled release of the active compounds is provided (Miastkowska et al. 2020). Many studies have proved that silver nanoparticles (AgNPs) have prominent antifungal characteristics and can be used against a wide range of plant pathogens such as *Fusarium oxysporum*, *Alternaria solani*, and *Botrytis cinerea* (Cruz-Luna et al. 2021). It has been revealed that the synthesis of silver nanoparticles and their formulation as a nanoemulsion improved stability, bioavailability, and fungicidal activity (Chowdappa et al. 2014) Nanoemulsions offer a strong mechanism for distributing bio-active components, not only enhancing the formulation life but also making them suitable for agricultural needs. The preparation methodology is as follows: dissolve 1 mM $AgNO_3$ in distilled water to prepare the solution; this solution is used for the synthesis of silver nanoparticles. Chitosan solution is prepared by dissolving chitosan in an acidic buffer such as acetic acid, to aid the reduction of silver ions. Mix the chitosan solution with the silver nitrate solution while constantly stirring. Heat the mixture at 90°C for 1 hour, which results in the formation of silver nanoparticles. The colour change from colourless to pale yellow indicates the formation of AgNPs. The synthesised nanoparticles are characterised and confirmed using the UV-Vis spectrophotometry technique. Silver nanoparticles can be confirmed through their absorption peak at 420 nm wavelength. A TEM (Transmission Electron

Microscopy) analysis detects both the shape and dimensional characteristics of nanoparticles to verify their appropriate 10–50 nm dimensions. Dynamic light scattering (DLS) analysis provides the necessary tool to both verify size distribution and confirm the monodispersity status of the nanoparticles. (Chowdappa et al. 2014) asserted the effectiveness of silver nanoparticle incorporation in the nanoemulsion by demonstrating the improved bioavailability of antifungal polymer chitosan in them, which inhibited the growth of the pathogen *Colletotrichum gloeosporioides*.

8. Challenges in the Formulation of Biofungicide

Biofungicides are a promising solution for reducing the environmental effects of chemical pesticides, promoting sustainable agriculture practices, and protecting biodiversity. However, despite their potential, the widespread adoption and commercial success of biofungicides are hindered by several challenges. These include a limited range of action, high production costs, stability in field conditions, environmental adaptability, storage limitations, and a lack of research on novel formulations. For biofungicides to be used effectively and practically in current agricultural practices, these obstacles must be addressed.

8.1 Storage

Their storability is one of the main challenges in creating and using biofungicides. Most biofungicides, especially those designed from live microorganisms, have short shelf lives because of their sensitivity to environmental variables, including temperature, humidity, and light. Although these biofungicides are designed with a carrier system to preserve their stability and effectiveness, they sometimes need particular and costly storage conditions (e.g., refrigeration or sun protection). Improper storage can lead to a loss of microbial viability, reducing their effectiveness as fungicides. Maintaining the stability of biofungicides during storage is therefore critical for their commercial scale-up (Glare et al. 2012).

8.2 Sustainability in Different Environments

Biofungicides might not work consistently across different settings, as discussed earlier. Environmental variables, including temperature, humidity, soil pH, and the presence of other organisms, greatly influence their activity. For instance, some kinds of beneficial bacteria or fungi could only survive under particular soil conditions, thereby restricting their use in different climate situations or modified farming practices. Thus, the difficulty lies in ensuring the biofungicides are sufficiently effective to function under different environmental conditions and act consistently in agroecosystems, which is crucial for their broad use (Fravel 2005).

8.3 Stability in Real-Time Field Conditions

Although many biofungicides have demonstrated stability and efficacy in laboratory settings, converting this into real-time field conditions is usually more challenging and less successful. These products in the field are exposed to a variety of elements that can lower their effectiveness, including UV radiation, rain, temperature changes, and soil pH variations. Furthermore, field conditions may allow for the growth of non-target microorganisms that could compete with, or even outgrow, the biofungicidal organisms and further reduce their efficiency. Thus, a major difficulty is the consistent performance and repeatable effectiveness of biofungicides in the field (Chandler et al. 2011).

8.4 Preventive Mode of Action

Biofungicides are typically regarded as a preventive rather than curative measure, in contrast to chemical fungicides, which are used as a treatment option for fungal infections that already exist. They frequently prevent disease by boosting the plant's defences or by competing with harmful fungi. Because of this restriction, biofungicides are most effective when applied as part of a disease management strategy before an infection. This differs from the wide spectrum of action of present chemical fungicides, making them disadvantaged in terms of acceptance in agriculture. Therefore, when quick action is needed to contain a fungal outbreak, this preventive approach may be detrimental (Pal and McSpadden Gardener 2006).

8.5 Expensive Production

The limited range of antagonist activity that biofungicides frequently display suggests that they are only effective against particular pathogens or diseases. Because they must use multiple biofungicide products to control different types of phytopathogens, farmers may incur higher costs as a result of this limited target range. Additionally, the production and application of biofungicides are highly specialised due to their specificity, particularly when they belong to a particular strain of microorganisms, which raises costs. Moreover, strain-specific formulations are more costly to research and develop than conventional chemical fungicides because they require a significant investment in production and research (Fenta and Mekonnen 2024, Khan et al. 2023).

8.6 Lack of Research in Formulation Techniques

Although interest in biofungicides is rising, there is still a dearth of thorough research and formulation technique development in the field. Novel formulation techniques like nanoformulations, although still in their infancy, can improve the stability and efficacy of biofungicides. Upscaling production, improving formulations for greater viability and efficacy, and incorporating biofungicides into current agricultural practices all require more research. Without steady improvements in formulation methods, the adoption and commercialisation of biofungicides will continue to be restricted (Kumar et al. 2022).

8.7 Commercialisation of Biofungicides

For biofungicides to be safe, environmentally friendly, and compliant, regulation is a crucial step in their development and approval. Because biofungicides must be registered in each country, marketing them internationally requires navigating complicated regulatory frameworks. A major obstacle for businesses pursuing global commercialisation is that biofungicides are typically categorised under the same regulations as chemical fungicides, which may mean higher costs and longer product market entry times (Cenobio-Galindo et al. 2024, Galindo et al. 2013, Palmieri et al. 2022).

9. Conclusion

The pursuit of new, affordable methodologies to decrease the damaging impacts of chemical fungicides has been ongoing for the last few decades. Biofungicides represent a superior, sustainable, environmentally friendly, and secure method beyond traditional chemical fungicides, though their market implementation falls short of achieving broad adoption. The limited market influence of biofungicides stems from ongoing challenges, such as poor formulation methods, expensive manufacturing, and failures under field conditions, as well as their primary use as a means of protection instead of treatment. Due to their sensitivity to environmental conditions, biological agents and their components remain unable to fully replace synthetic precursors at present (Fenta and Mekonnen 2024). As a result of recent advancements in technology and research, more and more novel strains and formulation strategies for effective biocontrol activity in a lab setting have been identified. For the development of effective biofungicides, it is imperative to select a compatible carrier for an efficient biocontrol agent. Subsequent studies on the compatibility and optimisation strategies of formulations, taking into account external factors such as environmental conditions, organisms, and agrochemicals, will contribute to the commercial success of such formulations in the future. Several biofungicide formulations have been developed recently that combine one or more biological agents with a lower concentration of fungicide, which have shown promising results (Anna et al. 2013). As further research into formulation strategies and technological advancements is conducted, a bridge between laboratory-scale success and successful field trials can be created, resulting in an agricultural strategy that will be more ecologically responsible, safer, and more effective.

References

Abadi, A. L., Trianti, I., Choliq, F. A., Sektiono, A. W., and Yulianti, N. D. (2024). Efficacy of Biofungicide with Active Ingredients Trichoderma sp. Against Late Blight Disease (Phytophthora infestans) in Potato Plants. *Agro Bali : Agricultural Journal*, 7(1): 81–91. https://doi.org/10.37637/ab.v7i1.1591.

Anna, I., Erick, L. N. J., Cecilia, B. N. G. A., Julio, C. M. E. S. E., Moacute Nica, M. R. G. G., Elda, P. S. C. et al. (2013). Free and encapsulated chitinase and laminarinase as biological agents against *Fusarium oxysporum*. *African Journal of Microbiology Research*, 7(36): 4501–4511. https://doi.org/10.5897/AJMR12.2056.

Borges, D. F., Lopes, E. A., Fialho Moraes, A. R., Soares, M. S., Visôtto, L. E., Oliveira, C. R. et al. (2018). Formulation of botanicals for the control of plant-pathogens: A review. *Crop Protection*, 110: 135–140. https://doi.org/10.1016/j.cropro.2018.04.003.

Cavalcante, R. S., Lima, H. L. S., Pinto, G. A. S., Gava, C. A. T., and Rodrigues, S. (2008). Effect of moisture on *Trichoderma conidia* production on corn and wheat bran by solid state fermentation. *Food and Bioprocess Technology*, 1(1): 100–104. https://doi.org/10.1007/s11947-007-0034-x.

Cenobio-Galindo, A. D. J., Hernández-Fuentes, A. D., González-Lemus, U., Zaldívar-Ortega, A. K., González-Montiel, L., Madariaga-Navarrete, A. et al. (2024). Biofungicides based on plant extracts: on the road to organic farming. *International Journal of Molecular Sciences*, 25(13): 6879. https://doi.org/10.3390/ijms25136879.

Chandler, D., Bailey, A. S., Tatchell, G. M., Davidson, G., Greaves, J., and Grant, W. P. (2011). The development, regulation and use of biopesticides for integrated pest management. *Philosophical Transactions of the Royal Society B: Biological Sciences*, 366(1573): 1987–1998. https://doi.org/10.1098/rstb.2010.0390.

Chen, H., Zhao, S., Zhang, K., Zhao, J., Jiang, J., Chen, F. et al. (2018). Evaluation of soil-applied chemical fungicide and biofungicide for control of the *Fusarium wilt* of chrysanthemum and their effects on rhizosphere soil microbiota. *Agriculture*, 8(12): 184. https://doi.org/10.3390/agriculture8120184.

Chen, X., Li, Y., Du, G., and Chen, J. (2005). Application of response surface methodology in medium optimization for spore production of *Coniothyrium minitans* in solid-state fermentation. *World Journal of Microbiology and Biotechnology*, 21(4): 593–599. https://doi.org/10.1007/s11274-004-3492-6.

Chowdappa, P., Shivakumar, G., C., S. C., and S., M. (2014). Antifungal activity of chitosan-silver nanoparticle composite against *Colletotrichum gloeosporioides* associated with mango anthracnose. *African Journal of Microbiology Research*, 8(17): 1803–1812. https://doi.org/10.5897/AJMR2013.6584.

De Medeiros, A. O., Da Silva, M. D. G. C., Converti, A., De Almeida, F. C. G., and Sarubbo, L. A. (2024). Development of natural fungicidal agricultural defensives using microbial glycolipid and vegetable oil blends. *Surfaces*, 7(4): 879–897. https://doi.org/10.3390/surfaces7040058.

Doni, F., Isahak, A., Che Mohd Zain, C. R., Mohd Ariffin, S., Wan Mohamad, W. N., and Wan Yusoff, W. M. (2014). Formulation of *Trichoderma* sp. SL2 inoculants using different carriers for soil treatment in rice seedling growth. *SpringerPlus*, 3(1): 532. https://doi.org/10.1186/2193-1801-3-532.

Fenta, L., and Mekonnen, H. (2024). Microbial biofungicides as a substitute for chemical fungicides in the control of phytopathogens: current perspectives and research directions. *Scientifica*, 2024, 1–12. https://doi.org/10.1155/2024/5322696.

Fenta, L., Mekonnen, H., and Kabtimer, N. (2023). The exploitation of microbial antagonists against postharvest plant pathogens. *Microorganisms*, 11(4): 1044. https://doi.org/10.3390/microorganisms11041044.

Fravel, D. R. (2005). Commercialization and Implementation of Biocontrol. *Annual Review of Phytopathology*, 43(1): 337–359. https://doi.org/10.1146/annurev.phyto.43.032904.092924.

Galindo, E., Serrano-Carreón, L., Gutiérrez, C. R., Allende, R., Balderas, K., Patiño, M. et al. (2013). The challenges of introducing a new biofungicide to the market: A case study. *Electronic Journal of Biotechnology*, 16(3). https://doi.org/10.2225/vol16-issue3-fulltext-6.

Glare, T., Caradus, J., Gelernter, W., Jackson, T., Keyhani, N., Köhl, J. et al. (2012). Have biopesticides come of age? *Trends in Biotechnology*, 30(5): 250–258. https://doi.org/10.1016/j.tibtech.2012.01.003.

Jangir, M., Sharma, S., and Sharma, S. (2021). Development of next-generation formulation against *Fusarium oxysporum* and unraveling bioactive antifungal metabolites of biocontrol agents. *Scientific Reports*, 11(1): 22895. https://doi.org/10.1038/s41598-021-02284-1.

Khan, A., Singh, A. V., Gautam, S. S., Agarwal, A., Punetha, A., Upadhayay, V. K. et al. (2023). Microbial bioformulation: A microbial assisted biostimulating fertilization technique for sustainable agriculture. *Frontiers in Plant Science*, 14: 1270039. https://doi.org/10.3389/fpls.2023.1270039.

Kumar, S., Diksha, Sindhu, S. S., and Kumar, R. (2022). Biofertilizers: An ecofriendly technology for nutrient recycling and environmental sustainability. *Current Research in Microbial Sciences*, 3: 100094. https://doi.org/10.1016/j.crmicr.2021.100094.

Lee, S. H., Oh, Y. T., Lee, D.-Y., Cho, E., Hwang, B. S., and Jeon, J. (2022). Large-scale screening of the plant extracts for antifungal activity against the plant pathogenic fungi. *The Plant Pathology Journal*, 38(6): 685–691. https://doi.org/10.5423/PPJ.NT.07.2022.0098.

Lv, C., Gu, T., Ma, R., Yao, W., Huang, Y., Gu, J., and Zhao, G. (2021). Biochemical characterization of a GH19 chitinase from *Streptomyces alfalfae* and its applications in crystalline chitin conversion and biocontrol. *International Journal of Biological Macromolecules*, 167: 193–201. https://doi.org/10.1016/j.ijbiomac.2020.11.178.

Mahde, B. Y., Fayyadh, M. A., and Al-Luaibi, S. S. (2019). Evaluation of biofungicide formulation of *Trichoderma longibrachiatum* in controlling of tomato seedling damping-off caused by *Rhizoctonia solani*. *Basrah Journal of Agricultural Sciences*, 32(2): 135–149. https://doi.org/10.37077/25200860.2019.204.

Manjula, K., and Podile, A. R. (2001). Chitin-supplemented formulations improve biocontrol and plant growth promoting efficiency of *Bacillus subtilis* AF 1. *Canadian Journal of Microbiology*, 47(7): 618–625. https://doi.org/10.1139/cjm-47-7-618.

Martinez, Y., Ribera, J., Schwarze, F. W. M. R., and De France, K. (2023). Biotechnological development of Trichoderma-based formulations for biological control. *Applied Microbiology and Biotechnology*, 107(18): 5595–5612. https://doi.org/10.1007/s00253-023-12687-x.

Maruyama, C. R., Bilesky-José, N., De Lima, R., and Fraceto, L. F. (2020). Encapsulation of *Trichoderma harzianum* preserves enzymatic activity and enhances the potential for biological control. *Frontiers in Bioengineering and Biotechnology*, 8: 225. https://doi.org/10.3389/fbioe.2020.00225.

Mousumi Das, M., Aguilar, C. N., Haridas, M., and Sabu, A. (2021). Production of bio-fungicide, *Trichoderma harzianum* CH1 under solid-state fermentation using coffee husk. *Bioresource Technology Reports*, 15: 100708. https://doi.org/10.1016/j.biteb.2021.100708.

Naserzadeh, Y., Mahmoudi, N., and Pakina, E. (2019). Antipathogenic effects of emulsion and nanoemulsion of cinnamon essential oil against *Rhizopus rot* and grey mold on strawberry fruits. *Foods and Raw Materials*, 210–216. https://doi.org/10.21603/2308-4057-2019-1-210-216.

Pal, K. K., and McSpadden Gardener, B. (2006). Biological control of plant pathogens. *The Plant Health Instructor*. https://doi.org/10.1094/PHI-A-2006-1117-02.

Palmieri, D., Ianiri, G., Del Grosso, C., Barone, G., De Curtis, F., Castoria, R. et al. (2022). Advances and perspectives in the use of biocontrol agents against fungal plant diseases. *Horticulturae*, 8(7): 577. https://doi.org/10.3390/horticulturae8070577.

Pane, C., Villecco, D., Campanile, F., and Zaccardelli, M. (2012). Novel strains of *Bacillus,* isolated from compost and compost-amended soils, as biological control agents against soil-borne phytopathogenic fungi. *Biocontrol Science and Technology*, 22(12): 1373–1388. https://doi.org/10.1080/09583157.2012.729143.

Pereira, J. F., Oliveira, A. L. M., Sartori, D., Yamashita, F., and Mali, S. (2023). Perspectives on the use of biopolymeric matrices as carriers for plant-growth promoting bacteria in agricultural systems. *Microorganisms*, 11(2): 467. https://doi.org/10.3390/microorganisms11020467.

Rabindran, R., and Vidhyasekaran, P. (1996). Development of a formulation of *Pseudomonas fluorescens* PfALR2 for management of rice sheath blight. *Crop Protection*, 15(8): 715–721. https://doi.org/10.1016/S0261-2194(96)00045-2.

Ribera, J., Gandía, M., Marcos, J. F., Bas, M. D. C., Fink, S., and Schwarze, F. W. M. R. (2017). Effect of Trichoderma-enriched organic charcoal in the integrated wood protection strategy. *PLOS ONE*, 12(8), e0183004. https://doi.org/10.1371/journal.pone.0183004.

Şenol Kotan, M., Dikbaş, N., and Kotan, R. (2023). Solid carrier bacterial formulations against *Fusarium* root and stem rot disease in cucumber. *Journal of Plant Pathology*, 106(1): 153–163. https://doi.org/10.1007/s42161-023-01530-x.

Syed-Ab-Rahman, S. F., Carvalhais, L. C., and Omar, D. (2020). Development of plant-based emulsion formulations to control bacterial leaf blight and sheath brown rot of rice. *Heliyon*, 6(1): e03151. https://doi.org/10.1016/j.heliyon.2019.e03151.

Tanapichatsakul, C., Pansanit, A., Monggoot, S., Brooks, S., Prachya, S., Kittakoop, P. et al. (2020). Antifungal activity of 8-methoxynaphthalen-1-ol isolated from the endophytic fungus *Diatrype palmicola* MFLUCC 17-0313 against the plant pathogenic fungus *Athelia rolfsii* on tomatoes. *PeerJ*, 8: e9103. https://doi.org/10.7717/peerj.9103.

Tao, H., Bao, Z., Jin, C., Miao, W., Fu, Z., and Jin, Y. (2020). Toxic effects and mechanisms of three commonly used fungicides on the human colon adenocarcinoma cell line Caco-2. *Environmental Pollution*, 263: 114660. https://doi.org/10.1016/j.envpol.2020.114660.

Teixidó, N., Usall, J., and Torres, R. (2022). Insight into a successful development of biocontrol agents: Production, formulation, packaging, and shelf life as key aspects. *Horticulturae*, 8(4): 305. https://doi.org/10.3390/horticulturae8040305.

Van Scoy, A. R., and Tjeerdema, R. S. (2014). Environmental fate and toxicology of chlorothalonil. pp. 89–105. *In:* Whitacre, D. M. (Ed.). *Reviews of Environmental Contamination and Toxicology Volume* 232 (Vol. 232). Springer International Publishing. https://doi.org/10.1007/978-3-319-06746-9_4.

Veliz, E. A., Martínez-Hidalgo, P., and Hirsch, A. M. (2017). Chitinase-producing bacteria and their role in biocontrol. *AIMS Microbiology*, 3(3): 689–705. https://doi.org/10.3934/microbiol.2017.3.689.

Wang, Y., Zafar, N., Ali, Q., Manghwar, H., Wang, G., Yu, L. et al. (2022). CRISPR/Cas genome editing technologies for plant improvement against biotic and abiotic stresses: Advances, limitations, and future perspectives. *Cells*, 11(23): 3928. https://doi.org/10.3390/cells11233928.

Zamoum, M., Goudjal, Y., Sabaou, N., Mathieu, F., and Zitouni, A. (2017). Development of formulations based on *Streptomyces rochei* strain PTL2 spores for biocontrol of *Rhizoctonia solani* damping-off of tomato seedlings. *Biocontrol Science and Technology*, 27(6): 723–738. https://doi.org/10.1080/09583157.2017.1334257.

Zeng, Y., Liu, H., Zhu, T., Han, S., and Li, S. (2021). Preparation of nanomaterial wettable powder formulations of antagonistic bacteria from *Phellodendron chinense* and the biological control of brown leaf spot disease. *The Plant Pathology Journal*, 37(3): 215–231. https://doi.org/10.5423/PPJ.OA.02.2021.0020.

7

Biofungicides

Scale-up, Production, and Emerging Innovations in Sustainable Disease Management

Muhammad Umair Rafiq,[1] *Akhtar Hameed,*[1,*] *Muhammad Waqar Alam,*[2] *Abdul Rehman,*[3] *Rana Binyamin,*[1] *Muhammad Arslan Khan*[1] and *Subhan Ali*[1]

1. Introduction

1.1 Importance of Biofungicides

Fungal diseases present serious threats to crop production and food security. To overcome these diseases, synthetic fungicides are commonly used having extraordinary effects. Despite the rapid killing action, these synthetic chemical fungicides deteriorate the plant quality and environment. To mitigate the side-effects of synthetic chemical fungicides, biofungicides were evaluated (McLaughlin et al. 2023). Biofungicides are comprised of living organisms as active ingredients mostly used for managing phytopathogenic fungi and bacteria. The concept of biofungicides emerges from naturally observed processes in which beneficial microbes are typically extracted from the soil to control plant diseases. These beneficial microbes are known as biocontrol microorganisms, like free-living bacteria, fungi, or actinomycetes, that are found in leaves, soil, and root habitats. These microbes work antagonistically

[1] Institute of Plant Protection, MNS University of Agriculture Multan 61000, Pakistan.
[2] Department of Botany, University of Okara.
[3] Department of Plant Pathology, University of Agriculture Faisalabad 38000, Pakistan.
* Corresponding author: akhtar.hameed5490@gmail.com

DOI: 10.1201/9781003603023-8

through processes of competition and parasitism against the phytopathogenic fungi and develop a wide range of antibiotics (Fenta and Mekonnen 2024). The various active ingredients in biofungicides are target-specific, as they are not harmful to the environment and biodiversity. The major importance of their use is their contribution to the conservation of biodiversity, as the use of pesticides is reduced. The number of natural product biofungicides on the market is expected to rise for these reasons, as well as the fact that they often have minimal detrimental effects on the environment and public health (Verma et al. 2024).

1.2 Advantages Over Chemical Fungicides

The use of biofungicides regarding the conservation of biodiversity possesses some impactful advantages over synthetic chemical fungicides.

Advantages:

- The toxic effects are minimized as compared to chemical fungicides on both plants and environment.
- The management effects on crops are broad-spectrum and long-lasting in contrast to synthetic fungicides.
- The uptake of nutrients from rhizosphere to plants is improved by using biofungicides, as beneficial microbes make nutrients more accessible.
- They don't alter the pH level of soil, and as a result, the soil microbiota are not disturbed in performing their activities.
- Biofungicides can be used rotationally with synthetic fungicides in order to confer microbial resistance (Rana et al. 2024). The pictorial explanation of comparison is given in the following Figure 1.

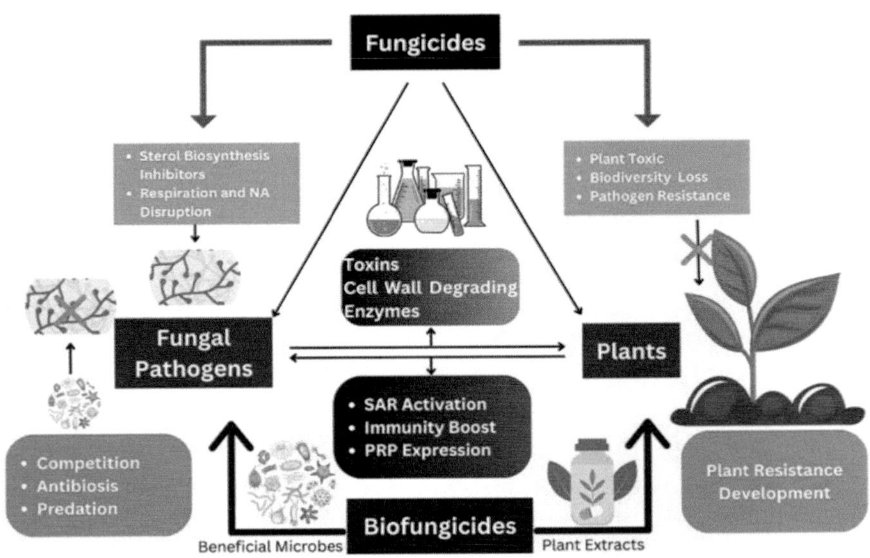

Figure 1: Comparison of biofungicides with synthetic fungicides.

1.3 Market Trends and Demand for Biofungicides

The agricultural industry needs eco-friendly pest control measures, among which biofungicides are a key component of integrated pest management (IPM). Biofungicides are unconventional organic fungicide products based on living organisms, reflecting consumer trends toward safe products. Several sources have indicated recent trends and market needs for biofungicides, as the market expanded at a fast rate in the recent past and is expected to continue doing so in the future (Zhou et al. 2024). In 2022, the market was valued at approximately USD 1.8 billion and is estimated to reach nearly USD 4.9 billion by 2030, with a projected CAGR of 15.5% during the forecast period. Another report similarly estimates that the market will grow from USD 2.3 billion in 2023 to USD 10.0 billion by 2033, with a CAGR of 15.8%. These figures indicate a rising use of biofungicides in agriculture.

1.3.1 Driver of Market Demands

Several key factors, which serve as the impetus for increasing demand for biofungicides, are the following:

1.3.1.1 Shift toward Organic Farming

The transition toward organic cultivation has increased the demand for biofungicides because of increased consumer demand for organic products, and therefore integration becomes essential in agricultural practices (Khakimov et al. 2020).

1.3.1.2 Health and environmental concerns

Due to health and environmental concerns, chemical fungicides cause adverse impacts on human health and the environment; thus, this enforces the use of safer alternatives via a set of practical regulations. Being of organic origin, biofungicides offer solutions for sustainable strategies aimed at keeping chemical residues on food to a minimum while ensuring minimal ecological impact (Mir et al. 2023).

1.3.1.3 Technological advancements

Technological advancements in the field of biotechnology and microbial research have fostered better efficacy and scalability of biofungicides to match synthetic counterparts. These advancements have enhanced their utilization across different varieties and in almost all regions (Thind et al. 2021).

1.3.1.4 Regional Market Insights

North America has dominated the biofungicide market with a very large contribution in the development of sustainable agriculture and the consumption of organic foods in this region. In 2023, North America held a significant compound share of 42.5%, driven mainly by the United States. Europe is another major part of the market, driven by tough environmental regulations along with a high consumer base for organic production. The Asia-Pacific region raised another profitable market by increasing farmers' concerns and government support initiatives toward sustainable farming (European Parliament 2024).

1.3.1.5 Application and Crop Type

Biofungicides are applied in vegetable and fruit cultivation to manage fungal diseases without toxic residues. This leads to high-quality produce for consumers. They accounted for 84.1% of the world market for biofungicides in 2022, primarily in barley, corn, sunflower, soybean, wheat, rapeseed, rye, and rice. The market will continue to grow with the move toward sustainable agriculture and organic food production. Even with production issues, continued research and technological development are likely to improve their effective application, making them increasingly prevalent in the future of agriculture. Some known biofungicides with microbial origins are listed below in the Table 1.

2. Selection and Optimization of Biofungicidal Strains

2.1 Screening of Potential Microbial Candidates

The initial step in the development of biofungicides consists of isolating useful microorganisms from highly diverse microbial habitats like soil, compost, or the rhizosphere of plants, which carry out antagonistic activities against plant pathogens. In the subsequent step, the isolated and identified microbes are studied for their antagonistic capabilities against corresponding plant pathogens (Ezeorba et al. 2023). Conventional screening techniques comprise dual-culture assays that co-cultivate the potential biofungicidal strains with the target pathogen and monitor the inhibitory effects on the latter. The widespread application of new technologies for high-throughput screening, including microfluidic droplets, in bacterial paradigms allows the effective and speedy screening of large microbial candidate collections (Akhtar and Kumar 2025).

2.1.1 Optimization of Biofungicidal Strains

After such screening, optimal conditions must be applied to enhance the efficacy and adaptability of potential biofungicidal strains. The optimization process may involve genetic, physiological, or process-level modifications.

2.1.1.1 Genetic Optimization

Genetic engineering techniques can be applied to enhance the antifungal properties of microbial strains. For example, modifications in metabolic pathways may increase the production of antifungal compounds. Furthermore, the heterologous expression of beneficial genes can add further functionalities to the strains (Mulatu et al. 2021). An example is the heterologous expression of recombinant transglutaminase in *Bacillus subtilis*, which exhibits efficient gelatin properties and demonstrates the added advantages of genetically modified microbes exerting extraordinary functionalities (Zalila-Kolsi et al. 2022).

Table 1: Biofungicides products available in the market.

Biofungicides	**Active Ingredients**	**Crop**	**Target Pathogens**	**References**	
Actinovate	*Streptomyces lydicus*	All vegetables crops	Soil-borne fungi, Powdery and Downy Mildews	(Quintana-Jones 2011)	IndiaMART (n.d.)
Serenade, Rhapsodi	*Bacillus subtilis*	Annual, Perennial, Horticultural and Edible Crops	*Botrytis cinerea, powdery mildew, Foliar spots*	(Abbasi and Weselowsk 2014)	
PRESTOP	*Gliocladium catenulatum*	Vegetables, Ornamentals, Seedlings, Plants	Seed-borne Fungi, Storage Diseases	(Lahdenperä 2020)	
Howler™	*Pseudomonas chlororaphis* strain AFS009	Grapes, Berries, Small Fruits, Vegetables	Alternaria, Botrytis, Colletotrichum, etc.	(Wesche et al. 2025)	
BotryStop	*Unocladium oudemansii* (U3 strain)	Ornamental Crops, Canola, Soybean and Pea	*Botrytis cinerea*, *Sclerotinia sclerotiorum* and *Monilinia* spp.	(Brownbridge n.d.)	
Rootshield Home, Garden Fungicides	*Trichoderma harzianum*	Herbaceous Plants	*Pythium, Rhizoctonia and Fusarium* spp.	(Hashem et al. 2023)	
BlightBan A506	*Pseudomonas fluorescens* A506	Apple, Pear, Cherry, Almonds	*Erwinia amylovora*	(Koirala et al. 2023)	
Zonix™	*Pseudomonas aeruginosa*	Citrus, Tobacco, Vegetables and Ornamental crops	*Botrytis* spp., Powdery Mildew,	(Malakar and Deka 2021)	
Tenet® WP - 16 oz.	*Trichoderma asperellum*, *Trichoderma gamsii*	Alfalfa, Maize, Chilli and Turf grasses	Soil-borne fungi, *Rhizoctonia, Phytophthora*, etc.	(Celar and Kos 2022)	
Revitalize	*Bacillus amyloliquefaciens* strain D747	Vegetables, Fruits, Nuts, Ornamental trees	Blights, Mildew, Molds, Rots	(Biondi et al. 2021)	

Note: All the products are available at https://www.indiamart.com/biofungicides. 26220448312.html. (accessed on April 15, 2025).

2.1.1.2 Physiological Optimization

The growing conditions of the biofungicidal strains are very important for enhancing their efficacy. Therefore, a proper understanding and optimization of the growth conditions, such as nutrient availability, pH, temperature, and oxygen concentration, are required for optimum growth of strains. When the biofungicidal strains are in favorable conditions, their work efficiency is ensured (Sidorova et al. 2020).

2.1.1.3 Process Optimization

The production of biofungicidal strains on an industrial level faces various challenges that require methodical optimization from the laboratory to large-scale production. Factors involved in the optimization process, such as fermentation parameters, medium composition, and bioreactor design, must be adapted to preserve the potency and viability of the strains. Innovative micro-fermenter systems enable high-throughput screening and fermentation optimization, ensuring that the biofungicidal strains retain their desired properties during industrial-scale production (Du et al. 2022).

2.2 Genetic and Phenotypic Stability

Genetic and phenotypic stability are absolutely critical to the development and optimization of biofungicidal strains. Genetic stability defines the stability of the genetic constitution of microorganisms during replication and under variable environmental conditions. Only genetically stable characteristics will allow successful expression of antifungal metabolites and mechanisms by biofungicidal agents (Silveira et al. 2024). For example, Bacillus subtilis, a widespread soil bacterium, is employed as a biofungicide as it synthesizes and secretes the antifungal lipopeptides fengycins and iturins. Since the genetic pathways involved in compound synthesis are well characterized, this increases the likelihood of the strain being developed as a biofungicide (Xiao et al. 2021). Genetic instability arises due to mutations, horizontal gene transfers, or environmental stresses, all of which lead to the eventual loss of biofungicidal activity of a supplied strain. Molecular techniques to monitor and ensure the genetic integrity of biofungicidal strains during development and application are utilized by researchers to minimize this disadvantage (Hashem et al. 2021).

Phenotypic stability refers to the consistent expression of visible qualities, including antifungal activity, under varying environmental conditions. A biofungicidal strain should be capable of strongly expressing its antifungal properties while growing under different environments (Ristić et al. 2024). *Trichoderma harzianum*, another fungus commonly used for its biofungicidal properties, primarily combats various plant pathogens. Its potency relies on the consistent synthesis of antifungal enzymes and secondary metabolites under various environmental factors such as temperature, pH, and nutrient availability. Environmental conditions affect the phenotypic expression of a strain, leading to variability in biofungicidal activity (Manzar et al. 2022). Hence, it is imperative to isolate strains that are phenotypically robust and can retain their antifungal properties under diverse environmental situations.

The genetic and phenotypic stability of biofungicidal strains are prime considerations during selection and enhancement. Various strategies, including careful strain selection, genetic engineering, control of cultivation conditions, and proper storage, contribute to stabilizing these strains. This, along with the resolution of environmental stresses and ecological interactions, will enhance the dependability and effectiveness of biofungicides in sustainable agriculture (Li et al. 2024a).

2.3 Enhancement of Bioactivity through Strain Improvement

Improvement in biofungicidal strains through enhancement of bioactivity is extremely important, with the aim of sustainable agriculture that reduces reliance on chemical fungicides and minimizes their adverse environmental impact. Overall, there are many means of strain improvement: classical mutagenesis, genetic engineering, and adaptive evolution, all of which augment the antifungal efficacy of microbial agents (Fenta and Mekonnen 2024).

2.3.1 Classical Mutagenesis

Classical mutagenesis exposes microbes to random genetic mutations through either physical or chemical mutagens, followed by the screening of strains for favorable characteristics. This method has successfully developed many industrial strains with enhanced production capabilities. For example, UV radiation and chemical agents such as ethyl methane sulfonate have developed mutants that are often better at producing bioactive compounds, thus enhancing antifungal properties. Unfortunately, as the mutations produced are random, a long process of screening is needed to isolate positive strains with higher bioactivity that would also be more stable (Chen et al. 2022).

2.3.2 Genetic Engineering

Advanced molecular biology allows for the targeted modification of microbial agents for biocontrol. Methods of manipulation include overexpression of genes, deletion, and promoter engineering for precise alterations in the metabolic pathways responsible for compound production (Bhattacharya and Seth 2023). The replacement of the native promoter of the mycosubtilin operon in *Bacillus subtilis* with a constitutive promoter increased mycosubtilin production 15–fold; importantly, this overproduction correlates with a heightened antagonistic effect against pathogenic fungal strains and improved efficacy in the biocontrol of *Pythium* on tomato seedlings (Leclère et al., 2005). Genome editing tools, such as CRISPR-Cas9, have also made a significant impact on strain improvement, allowing for precise genomic edits. Using CRISPR-Cas9, silent biosynthetic gene clusters have been activated in filamentous fungi to yield novel antifungal compounds. This method enhances both the bioactivity of the strains and greatly broadens the pool of bioactive metabolites available for use in agricultural applications (Ansori et al. 2023).

2.3.3 Adaptive Laboratory evolution (ALE)

Adaptive laboratory evolution (ALE) permits the natural selection process to advance mutant strains by protracted exposure of microbial populations to environmental pressures. This process may involve a number of features, such as improved microbial and stress tolerance to the antifungal compounds produced (Vandenberghe et al. 2021). For example, *Trichoderma* has been exposed to sub-lethal concentrations of fungicides to induce mutants with higher resistance and bioactivity. Through this process, ALE provides a strong tool to develop strains that are functional in varying environmental conditions without losing antifungal propagation (Meher et al. 2020).

2.3.4 Metabolic Engineering

Metabolic engineering involves strategic modification of metabolic pathways for maximized production of the compounds sought. Strain optimizations may occur on account of the redirection of precursor metabolites and the elimination of competing pathways for the biosynthesis of antifungal agents (Kumar et al. 2024). For example, engineering of the polyketide synthesis pathway in *Streptomyces* increased the production of antifungal polyketides and enhanced bioactivities against plant pathogens. The improvement of biofungicidal strains in terms of bioactivity stems from a series of classical and modern biotechnological inventions based on a number of approaches. State-of-the-art technologies could be leveraged to meet existing challenges by growing healthy biofungicidal agents that contribute to sustainable and eco-friendly agricultural practices (Risdian et al. 2019).

3. Fermentation and Upstream Processing

The term fermentation defines the controlled growth of fungi in a given environment for the production of the desired antifungal compounds, while upstream processing refers to all steps prior to fermentation, such as the selection of the fungal strain, preparation of the growth medium, inoculation of the fermenter, optimized biopesticides production, and others (Ramírez-Guzmán et al. 2024). Upstream processing in biotechnology describes the preparation of cell cultures intended for fermentation. However, the main goal of upstream bioprocessing is to be in a position to carry out large amounts of cell-growth from very small amounts of a variety of cell lines. The volume needed may vary, and single-use technologies serve well for different needs and volumes, whereas titers form the main benchmark for characterizing upstream manufacturing efficiency and imply a larger amount of a certain product produced from the same or less fluid or bioreactor volume (Prakash and Basu 2021).

3.1 Submerged vs. Solid-State Fermentation

Each method has a different profile of benefits and weaknesses, which in turn affect how efficiently biofungicides can be produced as well as how far this has advanced so far. Submerged fermentation refers to the growth of microorganisms in a liquid nutrient medium under controlled conditions such that pH, temperature, and oxygen can be manipulated with ease (Mejía et al. 2024). It generally finds a wide range of applications in the production of various bioactive compounds, including antibiotics and enzymes, among the main ones being the culturing of *Streptomyces* species, which has proved indispensable for antibiotic manufacture employing SmF as a specialization for improving metabolite yields. One of the most prominent advantages of SmF is the ease of scaling up to industry-wide applications; the liquid state of the medium is homogeneous, thus enabling equal distribution of nutrients, efficient mixing, and consistent product quality. On the contrary, SmF is resource-intensive and uses large amounts of water and energy for agitation and aeration. The chances of contamination in this case are also high due to the moisture content of the medium, necessitating rigorous sterilization (Portela et al. 2022).

On the other hand, solid-state fermentation (SSF) occurs as microorganisms grow on solid materials without or with only small amounts of free water. The method imitates the natural environment of many fungi and is usually employed to produce fermented foods. With respect to biofungicides, SSF has gained importance owing to its ability to utilize agro-industrial residues as substrates, thereby reducing production costs and providing waste valorization opportunities. There is an improved yield of products while at the same time reducing the risk of contamination due to very low levels of moisture. Further, it is very energy-efficient due to the nature of the process, which does not require intensive mixing or aeration (Mulatu et al. 2021).

3.1.1 Comparative Analysis in Biofungicide Production

The decision to use either SmF or SSF in the production of biofungicides will largely depend on the type of microorganism used to achieve certain products, available substrates, and economic considerations. For instance, in a study pertaining to cashew *Fusarium* wilt disease, different organic substrates for SSF showed that, in comparison to other substrates, rice husks and rice bran are highly suitable for large-scale production, proving the suitability of SSF in utilizing low-cost materials for biofungicide production. On the contrary, SmF has been optimized for specific antifungal bioactive compound production. In this context, one can cite optimization of submerged batch fermentation in *Streptomyces yanglinensis*, which increased production levels of Reveromycin A and B, active against certain plant diseases (Martău et al. 2021).

3.1.1.1 Optimization of Growth Media and Conditions

The optimization of growth media and fermentation conditions is very important in the upstream processes of biofungicide production since these strongly affect the yield, efficacy, and cost-effectiveness of the final product. Biofungicides are the green alternative to chemical fungicides, derived from natural organisms, requiring efficient production methods to satisfy agricultural demands (Yang et al. 2025).

3.1.1.2 Optimization of Growth Media

The growth medium directly affects microbial growth and the metabolic production process. The regulation of crucial components of this growth medium, mainly carbon and nitrogen sources, minerals, and growth factors, needs to be emphasized. Sugars such as glucose, sucrose, or molasses have been used as carbon sources and are the primary energy substrates in microbial culture. Factors such as growth rate and metabolite synthesis depend on medium composition (Bonnet et al. 2019). Nitrogen sources include peptone, yeast extract, or ammonium salts; they are necessary for protein synthesis and enzymatic function. The most important consideration is the C:N ratio. Optimal carbon-to-nitrogen ratios result in enhanced metabolite production; however, imbalances will likely lead to undesired by-products. Minerals such as magnesium, potassium, and trace elements act as cofactors in enzymatic reactions for numerous metabolic pathways. Depending on the specific microbial strain, growth factors including vitamins and amino acids may also be required. Another very important parameter to consider is the pH of the medium, which directly affects nutrient solubility and microbial enzyme activity. The optimal pH for growth also ensures maximal secondary metabolism (Cui et al. 2022).

3.1.1.3 Bioreactor Design and Process Parameters

Environmental parameters such as temperature, dissolved oxygen rate, and stirring speed are critical in fermentation procedures. Most of the microorganisms involved in the production of biofungicides have certain preferred temperature ranges; deviations impair growth and metabolite synthesis. For aerobic microbes, the main requirement is dissolved oxygen. Proper aeration and agitation will ensure that the optimal level is met. Poor oxygen transfer will constrain microbial activity, while high levels of agitation will impose shear stress on sensitive cells. It is thus of paramount importance to balance these parameters if optimum productivity is desired (Das et al. 2022).

4. Downstream Processing and Formulation

Biofungicides contain naturally derived actives and are, therefore, sustainable alternatives to chemical fungicides for controlling plant diseases. The effectiveness and market success of biofungicides depend on thorough downstream processing and formulation strategies. This detailed discussion describes their important aspects—from careful biomass separation and recovery to stabilization and the processes of freeze drying and spray drying—and all formulation strategies for extending shelf stability and field efficacy (Khan and Ahmad 2024).

4.1 Biomass Separation and Recovery

The first step in biofungicide production is to properly separate and recover microbial biomass from the cultivation medium. This process is key in assuring product purity and concentration; hence its potency with regard to biofungicide finality (Cenobio-Galindo et al. 2024).

4.1.1 Centrifugation

Centrifugation separates particles based on density differences by the use of centrifugal force. A high-speed centrifuge acts by separating fungal spores or mycelia from the liquid culture medium. It is the preferred acceleration technique because it can process large amounts of material rapidly, with very high recoveries, and is hence applicable to bulk production. However, the high energy consumption associated with centrifugation and shear forces in the process can damage more delicate fungal structures and potentially compromise the viability of the biofungicide (Foka et al. 2023).

4.1.2 Filtration

Filtration techniques such as microfiltration and ultrafiltration use semi-permeable membranes to separate biomass according to particle size. The methods are suitable, especially for growing organisms that cannot withstand high temperatures. With mild processing, the viability and efficacy of biofungicidal agents will be preserved. The choice between microfiltration or ultrafiltration is based on the size of fungal cells or spores and the purity target level of the filtrate (Ezeorba et al. 2023).

4.1.3 Flocculation

Flocculation consists of incorporating additives, namely flocculants, which induce the clumping of microbial cells into larger flocs so that they can be removed through sedimentation or filtration. It is a very economical and efficient way of harvesting biomass on a large scale. Choice of flocculants is an important consideration; they must be non-toxic and should not affect the ability or effectiveness of the biofungicide (Matter et al. 2019).

4.1.4 Sedimentation

Sedimentation relies on gravity to separate suspended particles from the liquid medium. Although simple and economical, sedimentation is often slow and not suitable for microorganisms showing a very low settling velocity. To improve efficiency, sedimentation is combined with flocculation, which increases the particle size in order to speed up sedimentation (Li et al. 2024b).

4.2 Stabilization and Drying Techniques

After biomass recovery, stabilization is important to sustain the viability and bioactivity of the biofungicidal agents during periods of storage and transport. Drying is a crucial phase of stabilization since this decreases moisture levels, inhibiting microbial growth and enzymatic activities that may destroy the product. Freeze drying and spray drying are the two most utilized drying methods (Fenta and Mekonnen 2024).

4.2.1 Freeze Drying (Lyophilization)

Freeze drying is a dehydration approach in which water is removed from frozen samples through sublimation under reduced pressure. The method is very useful in maintaining the structural integrity and viability of biofungicidal agents (Nowak and Jakubczyk 2020). The freeze-drying operation contains three major stages:

Freezing: The biomass is cooled below its eutectic point and allows the water to crystallize into ice but no liquid phase forms. It is important to preserve the structural integrity of sensitive biomolecules (Han et al. 2021).

Primary Drying (Sublimation): Ice sublimation under vacuum conditions by applying heat, mainly removing most of the water content. This requires exact controls to preserve quality and prevent structural collapse (Krzymińska and Kowalska 2024).

Secondary Drying (Desorption): Remaining water molecules often attached to the product matrix are removed by increasing the temperature to the target low moisture content to stabilize the product. In particular, freeze drying is praised not only for maintaining but also for being the unparalleled technology for preserving the bioactivity and structural integrity of heat-sensitive products. This makes it an undisputed technology in the pharmaceutical and food industries; however, it is also a costly, energy- and time-consuming process that must be considered with respect to cost (Nwankwo et al. 2023).

4.2.2 Spray Drying

The spray drying plant converts the liquid feed into dry powder through the atomization of the feed into a hot drying medium. The process is made up of three steps:

Atomization: The liquid feed is atomized into very tiny droplets through the atomizers.

Drying: The hot air stream quickly evaporates moisture from the droplets, forming dry particles.

Collection: The dried particles are separated from the drying air in cyclones or filters. The powder thus collected has a particle size measured in microns. Spray drying presents advantages such as fast drying, ease of scaling up, and known particle size of the powders, but the high temperatures involved can also be a limiting factor with respect to thermo-sensitive materials, with the possibility of denaturing or losing bioactivity (Saberi-Riseh and Moradi-Pour 2021).

4.3 Formulation Strategies for Enhanced Shelf Life and Field Efficacy

Biofungicides are produced from natural organisms, including bacteria, fungi, and plants, as an alternative to synthetic chemical fungicides in managing plant diseases and are generally considered eco-friendly agents. However, their real application is usually limited due to short shelf life and inconsistent field efficacy. To deal with these problems, various formulation strategies have been employed to prolong shelf life and enhance the efficacy of biofungicides (Fenta and Mekonnen 2024).

4.3.1 Microbial Biofungicides: Formulation and Stability

Microbial biofungicides employ beneficial microorganisms to combat phytopathogens. Viability of these beneficial microorganisms during storage and application is critical for their successful control. For example, the mycoparasitic fungus *Simplicillium lamellicola* BCP was formulated into a wettable powder (WP) to control gray mold diseases of tomato and ginseng. Production of conidia was scaled up in bioreactors, followed by optimized freeze-drying techniques, which included additives such as lactose, trehalose, soybean meal, and glycerin, to help maintain conidial viability (Shin et al. 2017). The WP formulation was effective in controlling gray mold under field conditions, achieving control efficacies of 65% in tomato and 78% in ginseng and attaining a shelf life of at least 2 years at room temperature (Hashem et al. 2021). In a similar study, *Streptomyces lincolnensis* strain SZ03 was formulated into a wettable powder to combat *Rhizoctonia solani*-induced damping-off in tomato seedlings. It maintained spore viability and biocontrol efficacy after one year of storage; this formulation reduced disease severity compared to untreated controls while performing comparably to chemical treatments (Zamoum et al. 2022).

4.3.2 Plant Extract-Based Biofungicides: Extraction and Formulation

Biofungicides composed of plant material refer to secondary metabolites such as terpenes, alkaloids, and phenolic compounds, which are involved in antifungal activities. The extraction methods and formulations are important to producing effective and stable biofungicides. These secondary metabolites are obtained using different extraction methods, both conventional and unconventional, and formulated into biofungicides. These are considered non-phytotoxic in practice, efficient in disease control, and as such are sustainable and economically viable alternatives in agriculture (Cenobio-Galindo et al. 2024). Some common biofungicides with botanical origin are listed in the following Table 2.

4.3.3 Nanotechnology in Biofungicide Formulation

Nanotechnology has opened up novel approaches to improving the stability and effectiveness of biofungicides. Nanomaterials act as carriers of antimicrobial agents, and they provide controlled release and targeted delivery in order to improve the persistence and performance of biofungicides. For instance, encapsulating fungicides in nanoparticles can protect active ingredients from environmental degradation while enabling a slow release and extended activity against pathogens. This would improve the efficacy of biofungicides along with the reduction of application rates, hence reducing environmental risks (Slavin and Bach 2022).

4.3.4 Application Methods and Compatibility

Biofungicides could be efficient depending on the application method and their compatibility with other agricultural inputs. The user should select the appropriate application method—for example, foliar sprays, soil drenches, or seed treatments—

Table 2: List of botanical biofungicides.

Plant Extract	Target Pathogens	Diseases on Crops	References
Thyme oil of *Thymus vulgaris* L.	*Botrytis cinerea, Erwinia amylovora*	Powdery Mildew, Grey Mold and Fire blight of Fruits	(Ouhaibi-Ben Abdeljalil et al. 2024)
Azadirachta indica	*Pythium aphanidermatum*	Damping off of Tomato	(Thangaraj et al. 2023)
Fruit extract of *Withania somnifera*	*Fusarium oxysporum* f.sp. *lycopersici*	Crown and Root rots of Tomato	(Nefzi et al. 2019)
Geraniol, a natural geranium oil plant extract	*Podosphaera fusca, Erysiphe nector, Oidium lycopersicum*	Powdery and Downy Mildew of Crops like Grapes, Cucumber, Cucurbits etc.	(Tryfon et al. 2023)
Turmeric	*Alternaria solani*	Early blight of Potato	(Kumar and Chandra 2024)
Cinnamon oil, Linseed Oil and Thyme Oil	*Podosphaera* spp.	Powdery Mildew	(Mostafa et al. 2021)
Extract of the Giant Knotwood plant (*Reynoutria sachalinensis*)	*Colletotrichum* spp., *Uromyces appendiculatus, Botrytis squamosal, Peronospora destructor, Alternaria porri*	Anthracnose, Bean Rust, Botrytis Leaf Blight, Onion Downy Mildew, Onion Purple Blotch	(Mahawer et al. 2022)

Note: All the products are available at https://www.indiamart.com/biofungicides. 26220448312.html. (accessed on April 17, 2025).

which should be evaluated concerning the target pathogen and the crop. Understanding how biofungicides may fit with other biological or chemical products is important to avoid any antagonistic interactions and enhance overall disease management techniques (Subba and Mathur 2022).

Elongated shelf life and field efficacy for biofungicides involve a multidimensional approach of formulation optimization, application methods, and detailed information on microbial and plant-derived bioactive compounds. There is an urgent need for continued research and development in this area to enable biofungicides to realize their full potential as sustainable alternatives for the management of plant diseases (Sutthisa et al. 2024).

5. Scale-up Challenges and Solutions

There are many challenges to face before transitioning biofungicides from the lab to commercial production, such as ensuring efficacy, stability, and cost-effectiveness. Bioactive compounds of concern are particularly at risk of degradation through environmental conditions like air, light, and temperature, potentially diminishing the biofungicide's efficacy during large-scale production and storage. Furthermore, developing a viable and stable formulation is also complex; it includes various stages in the course of patent application and regulatory approval that could delay commercialization. Optimization of culture conditions and protective formulation is, therefore, very critical in terms of improving biofungicide stability and efficacy during the scale-up phase (Pai et al. 2022).

5.1 Transition from Lab-Scale to Industrial Production

Ensuring the efficacy, stability, and cost-effectiveness of biofungicide production processes requires scaling up these practices from laboratory to industrial levels through quite complicated challenges.

5.1.1 Challenges in Scale-up Production of Biofungicides

5.1.1.1 Microorganism Viability and Consistency

Maintaining the viability and performance consistency of beneficial microorganisms is essential during large-scale production. Variability in fermentation conditions such as pH, temperature, and nutrient availability affects microbial viability and biofungicide effectiveness (Díaz-Rodríguez et al. 2025).

5.1.1.2 Formulation Stability

To provide effective and stable formulations, bioactive compounds need to be maintained during storage and application. Factors like moisture content, temperature fluctuation, and exposure to light can lead to degradation of bioactive compounds. This greatly affects product shelf life and field application effectiveness (Rezagholizade-Shirvan et al. 2024).

5.1.1.3 Cost-effective Production

Scaling up should be economically feasible. The high costs of substrates, fermentation, and subsequent downstream processing act as barriers to market viability compared to chemical fungicides (Mascaren et al. 2024).

5.1.1.4 Regulatory Compliance

The complicated regulatory systems require extensive documentation and maintenance of strict quality standards, making them highly resource-intensive (Mennella et al. 2024).

5.1.2 Solutions to Scale-up Challenges

5.1.2.1 Fermentation Process Optimization

Continuous and close monitoring of the fermentation process supports a high yield of active microbial populations. Bioreactors with modern monitoring equipment allow controlled conditions throughout microbial growth, ensuring that products are consistently manufactured (Palladino et al. 2024).

5.1.2.2 Novel Formulation Approaches

Incorporation of carriers like agro-industrial residues has given higher stability of formulation and increased shelf life. For example, supplementation of corn stover with date molasses as a substrate for *Trichoderma harzianum* produced a biofungicide with an effective shelf life of more than eight months and good disease suppression (Lima et al. 2024).

5.1.2.3 Cost-cutting Techniques

Using local, readily available substrates, particularly by-products from agriculture lessens, production cost dramatically. This either lessens the costs involved or adds value to wastes.

5.1.2.4 Regulatory Readiness

A comprehensive management system created to live up to the international quality, environment, and safety standards can facilitate the regulatory process. Such preparedness fast -tracks market entry and compliance issues.

The successful industrial production of biofungicides as sustainable alternatives to chemical fungicides and components of environmentally sustainable agricultural practices can only be achieved by tackling the challenges through targeted interventions (Ragasruthi et al. 2024).

5.2 Process Standardization and Quality Control

The transition from lab-scale to industrial production in biofungicides brings about major problems in terms of standardization of protocols and quality control. Addressing these problems is critical to validation, safety, and market acceptance of these biofungicidal products.

5.2.1 Challenges in Process Standardization and Quality Control

Maintaining genetic stability with desired traits of a microbial strain is often a challenge during mass-scale production. Variations and fluctuations in performance are often governed by mutation or environment-related stress, giving rise to nonconformance in biofungicidal efficacy. Scaling up of fermentation requires stringent control of parameters such as pH, temperature, and auxiliary nutrients. Irregularities in any of the above stages during this transition period can lower microbial growth and production of bioactive compounds. The formulation and stable yet effective performance during storage and application is another complicated process. Relative proportions of moisture content, particle size, and carrier materials must be maintained under strict control (Tan et al. 2022). In mass-scale production, the chance of catastrophic contamination with unwanted microorganisms may greatly increase, compromising product safety and efficacy. Uniformity across production batches is imperative because variations can arise from uncontrolled initial circumstances or other unidentifiable differences. Conformance to stringent regulations always causes a series of problems, since it demands comprehensive documentation but also requires an auditable measure of product quality (Lorenzo et al. 2018).

5.2.2 Solutions for Challenges

Implementation of advanced technologies can maintain the integrity of active organisms for extended periods, thus reducing strain variability. Real-time monitoring and control systems can greatly reduce variability of fermentation parameters, ensuring consistency throughout the process. Each stage of production must have SOPs developed that, through their detail, will help to ensure consistent output and reduce human error (Sawant et al. 2025).

Systematic quality assurance in production according to Good Manufacturing Practices (GMP) guidelines, such as those detailed by the World Health Organization (WHO), is essential. Routine training of staff enhances knowledge and respect for quality standards, helping to reduce errors and contamination risk. It is vitally important that the problems are resolved via effective standardization of processes for industrial mushroom biopesticides and through strict quality control measures, assuring them as safe, effective, and compliant with rules and regulations. Such challenges will have to be tackled through a multi-disciplined approach that brings together microbiology, engineering, and analytical chemistry. Joint research between research institutes and industry is key to the development and implementation of effective scale-up strategies and quality control measures. Furthermore, there is a need to align regulatory frameworks with regard to biofungicides so that they can be commercialized and adopted (Tang et al. 2024).

5.3 Contamination Risks and Their Management

The transition from lab-scale to industrial-scale production of biofungicides brings forth major contamination risks, with the potential to harm product effectiveness and safety. Risks associated with contamination sites are important since decoding these risks can help control them more effectively. Unwanted microorganisms introduced during large-scale fermentation might compromise the purity and effectiveness of the biofungicide. These factors might accompany the desired microbial strains, which could reduce yield and, thus, product consistency and quality. Sometimes, differences in environmental parameters like temperature and humidity might create variations in the dynamics of microbial growth. An uncontrolled environment will optimize the growth of contaminants and complicate the production process. The use of non-sterilized equipment and poor handling protocols may introduce contamination at various stages of production. This risk increases in scale-up due to the increasing complexity of industrial equipment and processes (Fenta and Mekonnen 2024).

5.3.1 Management Strategies for Contamination

Cleaning and sterilizing all equipment and facilities is a must. The stage of contaminants' introduction is monitored by periodic checking to ensure that cleaning and sterilization protocols work. If constant and suitable fermentation parameters (e.g., pH, temperature, oxygen levels) are maintained, the chances of contaminant growth are minimized. Control systems that monitor and adjust these parameters in real-time could be sophisticated and capable of immediate interventions. Sampling throughout the production process and microbial testing would allow for the early detection and mitigation of contaminants, enabling rapid responses with simple actions before problems escalate (Verma et al. 2024). A notable example is the development of a biofungicide based on the mycoparasitic fungus *Simplicillium lamellicola* BCP. The researchers scaled conidia production up from a 5–liter jar to a 5,000–liter bioreactor. They perfected the freeze-drying processes and developed a wettable powder type of product. Notably, stringent contamination control measures were important during the scale-up to maintain disease control efficacy for gray mold diseases on crops such as tomato and ginseng. The focus of combating contamination risks in industrial biofungicide production involves the application of broad sanitation protocols, fermentation environment control, and stringent quality control practices. Success stories such as the development of *Simplicillium lamellicola* BCP prove the feasibility of effective contamination management during scale-up to deliver safe and effective biofungicides (Custódio and Pereira (2024).

6. Regulatory Framework and Compliance

Regulations about biofungicides ensure their safety and effectiveness before they are put into the marketplace. The process for registration is under the Federal Insecticide, Fungicide, and Rodenticide Act (FIFRA) by the Environmental Protection Agency (EPA), which primarily regulates biopesticides along with biofungicides. While biopesticides generally require less data than traditional pesticides for the registration process, they may be exempted from tolerance levels for residues. Moreover, a comprehensive federal policy for the safety of biotechnology products, including biofungicides, is set forth by the U.S. Coordinated Framework for the Regulation of Biotechnology. Such a framework has numerous implications, carrying regulatory requirements that must be met prior to any commercial activity and use of biofungicides in agriculture (Omoyajowo 2024).

6.1 Registration Requirements for Biofungicides

The registration of biofungicides safeguards their efficacy, safety, and low impact on the environment before they are eventually registered for use. While procedures vary in different nations, those procedures basically involve stringent evaluation by regulatory agencies. In the United States, the Environmental Protection Agency (EPA) manages the review of all pesticides, including biofungicides, under the Federal Insecticide, Fungicide, and Rodenticide Act (FIFRA). The EPA has divided biopesticides mainly into three classes: microbial pesticides, biochemical pesticides, and plant-incorporated protectants. Each class has its own data requirements. It is the applicant's duty to provide extensive data showing that the biofungicide is not going to cause unreasonable adverse effects on human health or the environment. This includes studies on composition, toxicity, persistence, degradation, and efficacy. The guidelines set up by the EPA help applicants prepare these data packages (Mawcha et al. 2024).

1. **Data Submission:** The applicant needs to supply detailed information concerning the biofungicide's chemical chattel, production process, and intended use.
2. **Toxicological Studies:** Studies should be conducted to assess the acute, sub-chronic, and chronic toxicity to humans and non-target organisms to ascertain the level of danger directed toward health.

3. **Environmental Impact Assessments:** Data concerning how the biofungicide is moving through environmental media, such as persistence, degradation, and likelihood for water contamination, is required to obtain an ecological risk assessment.
4. **Efficacy Data:** Evidence of efficiency against the target fungal pathogens shows that the biofungicide is doing its job.
5. **Labeling Requirements:** The correct label must adequately inform users how to apply a treatment properly, take safety precautions and store it safely.

The courses related to biofungicides regulation must be in accordance with the requirements specifically laid out by recognized authorities. Early-stage interaction with regulatory agencies and general adherence to their guidelines can make this regulatory process smoother while ensuring that efficient and safe biofungicidal products reach the market.

6.2 Safety and Environmental Considerations

Biofungicides are usually regarded as safe for humans and animals, mostly due to their origins in nature and specificity to target pathogens. They pose virtually no risk to non-target organisms, including beneficial insects and soil microbiota. Providing caution in their handling and application is still required. For example, some biofungicides can contain proteins that may cause allergic responses in sensitive users. Therefore, it is advisable for users of biofungicides to wear designated personal protective equipment, like gloves and masks, to lower the chances of exposure to certain constituents of biofungicides. They are also generally safe; however, their living nature requires proper storage for effectiveness. Improper storage could lead to decreased viability of the microbial agents, variable efficacy, and the potential for inadvertent microorganism proliferation (Gupta et al. 2024). Another notable benefit of biofungicides is eco-friendly disposal. Because biofungicides can biodegrade, they do not persist in the environment, leaving only limited chemical residues in soil and water bodies. These characteristics support sustainable farming practices that preserve biodiversity. Yet, biofungicides' effectiveness can be modified by environmental factors, including temperature, humidity, and soil pH. These variables can influence microbial agents' survival and activity, leading to inconsistent disease control under field conditions. Given the ecological needs of biofungicides and their interactions, understanding them is vital for successful applications. Moreover, the specificity of biofungicides for target pathogens tends not to disturb beneficial microbial communities. This specificity allows for the maintenance of soil health and fertility, contributing to resilience in agro-ecosystems (Ferreyra-Suarez et al. 2024).

6.3 International Standards and Guidelines

Various international standards and guidelines have been set into existence to emphasize the safety and efficacy of biopesticides, spanning thematic concerns from registration processes to quality control measures. The registration framework for these agents can vary regionally; yet, they still adhere to certain core principles. As an example, one of the many challenges in registration is the fact that the biopesticide registration guidelines depend heavily on chemical fungicide criteria for reference purposes and thus require information that may not be immediately available for biofungicides. Because of this, the lengthy and costly nature of registration might create disincentives for the development and commercialization of biofungicides (Mehra and Avanthi 2024). In addition, toxicological and environmental risk assessments are required, which, in some countries with particularly small markets, become prohibitively expensive to conduct. In the USA, the OMRI is key to the approval of biofungicides, and OMRI standards are a synthesis of the USA National Organic Program (NOP), the Canadian Organic Regime (COR), and the Mexican Organic Products Act (LPO). The time from testing to approval is generally months. When it does fit into the program, it gets OMRI-listed® and added to the OMRI© products lists, with farmers or consumers guaranteed that it meets the organic standards. After that, it must be renewed each year for listing (Cenobio-Galindo et al. 2024). The International Organization for Biological Control (IOBC) is a global platform for promoting biological pest control and integrated pest management. In that regard, IOBC has established standards for measuring the quality control of commercially produced biological controllers, including biofungicides. Such guidelines are used worldwide to guarantee effectiveness and safety in biological control products. The IOBC addresses questions on access and benefit-sharing under the Convention on Biological Diversity, allowing for the use of genetic resources in biological control alongside fair benefit-sharing considerations (Colmenarez, and Vasquez 2024).

7. Commercialization and Market Adoption

The commercialization and market adoption of biofungicides have gained immense traction over the past few years, propelled by an ongoing global transition toward sustainable agricultural practices and a rising demand for organic crops. These biofungicides, sourced from bacteria and fungi and related to other organisms, provide an eco-friendly alternative to synthetic chemical fungicides, being more agreeable to consumers seeking safer food products. The biofungicide formulations' segments that affect the global market are described in Figure 2. The global biofungicides market is anticipated to reach around USD 8.7 billion by 2030, representing a 13.93 percent compound annual growth rate (CAGR) from 2024 to 2032, underpinned by regulatory frameworks tending to favour sustainable inputs and increased adoption of integrated pest management practices. Nevertheless, some challenges, such as difficult registration processes and higher costs than synthetic fungicides, will inhibit their speedy penetration into the market. However, new strategies being developed with microbial formulations and a greater ongoing emphasis on environmental conservation are expected to encourage further growth of biofungicides on a global scale (Ragasruthi et al. 2024).

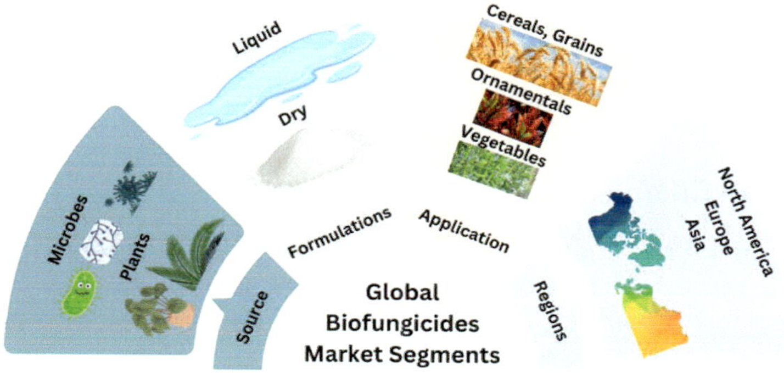

Figure 2: Global biofungicides market segments.

7.1 Cost Analysis and Economic Viability

The economic viability of biofungicides is a very complex problem dependent on production costs, market performance, and effectiveness relative to chemical fungicides. All these factors provide an indication to the varied stakeholders when deciding whether to use biofungicides in agricultural practices.

7.1.1 Production and Development Costs

Development of biofungicides involves considerable investments in research and development (R&D), primarily focusing on the selection of effective microbial strains and the formulation of stable products. Challenges of shorter shelf life and loss of efficacy in some environmental conditions necessitate much more R&D on the stabilization and performance of these products. Besides, production processes must guarantee high-quality biomass along with adequate formulation, which could be quite expensive (Rosa et al. 2025).

7.1.2 Market Growth and Adoption

There is substantial growth projected in the biofungicides market, despite the apparently high investment; this is due to increasing consumer demand for organic and sustainably produced food. The global biofungicides market size stood at USD 2.3 billion in 2023; it will reach USD 10.0 billion by 2033, expanding at a CAGR of 15.8% during the forecast period. Governments are increasingly regulating to support ecologically benign agricultural inputs along with integrated pest management (IPM) (Marrone 2024).

7.1.3 Cost-Benefit Analysis

While they may often come with a higher upfront price than chemical fungicides, biofungicides can provide an economic payoff in the long run. These include costs related to environmental cleanup, enhancing soil health, and meeting organic certification requirements, which in return can catapult agricultural products into premium-grade markets. The resistance of pathogens may also be reduced, lowering the costs associated with controlling more virulent pests (Cenobio-Galindo et al. 2024).

7.2 Supply Chain and Distribution Networks

The biofungicide supply chain and distribution networks are essential steps for the products to penetrate and be adopted into sustainable agriculture. It typically covers the journey from production to end-users, which involves several key steps (Fusar Poli and Fontefrancesco 2024).

7.2.1 Production and Formulation

First, beneficial microorganisms are cultured or plant-based compounds are extracted for biofungicide production. The formulation process is developed to enhance stability, shelf life, and application. The common challenge in this stage is that of consistent quality and the shorter shelf life relative to synthetic fungicides.

7.2.2 Regulatory Approval

Biofungicides should undergo thorough testing before market entry and must obtain approvals from the relevant regulatory bodies to ensure that the products are both safe to use and effective against the pathogens they target. However, stringent regulations can be detrimental—mostly to small and medium-sized enterprises—as they may delay market entry.

7.2.3 Distribution Network

A strong distribution network makes biofungicides available to farmers. Manufacturers usually work with regional distributors to supply products. For example, Marrone Bio Innovations partnered with local distributors in Southeast Asian countries to promote its product line of biofungicides. Yet, farmers' lack of awareness and limited distribution infrastructure in developing countries contribute to low adoption.

7.2.4 Market Dynamics

The global biofungicides market is witnessing substantial growth due to the increasing demand for organic produce and sustainable farming practices. The market is expected to reach about USD 10.0 billion by 2033, registering a compound annual growth rate (CAGR) of 15.8% between 2023 and 2033. Government regulations favoring eco-friendly agri-inputs, as well as increasing adoption of integrated pest management (IPM) practices, have supported this rapid growth (Pharmiweb.com 2024).

8. Future Perspectives and Emerging Technologies

The development of biofungicides is crucial for sustainable agriculture to provide eco-friendly alternatives to chemical fungicides. New technologies, including nanotechnology, are being integrated to enhance the efficacy and stability of biofungicides. For example, nano-biofungicides have been identified as generating high delivery rates in practice and maximizing the efficiency of active agents against fungal pathogens. In parallel developments, the search for new microbial strains and natural compounds is ongoing, exponentially diversifying the mechanisms of action of biofungicidal agents against phytopathogens. These innovations are expected to help address current limitations of biofungicides with short shelf life and variable field performance, thus facilitating their adoption along with other pest management strategies (Ghani et al. 2024).

8.1 Advances in Bioprocessing and Formulation Technologies

Technological advancement in bioprocessing and formulation of biofungicides has improved their effectiveness as well as their commercial viability. Modern bioprocessing techniques are known to optimize growth conditions for microbial agents for higher yield and potency. For example, controlling fermentation parameters significantly improved antifungal metabolite production aimed at increased biofungicide efficacy. Innovative formulations, however, have been introduced to address specific challenges, mainly short shelf life and sensitivity to environmental conditions. Various encapsulation methods, such as microencapsulation, have been developed that embody other methodologies for protecting active ingredients from environmental degradation—to improve the shelf life of products and ensure sustained release of the bioactive compounds (Ferreyra-Suarez et al. 2024). Another methodology for formulation design makes use of stable inert carriers combined with adjuvants, further improving the stability and efficiency of biofungicides and therefore their competitiveness with synthetic fungicides. Other processes of technology development have included freeze-drying and spray-drying for better storage and transport of microbial agents. These latter methods improve reliability against field conditions for biofungicides by retaining their potency through to application. Also, consortium-based formulations containing various microbial strains have been developed to deal with a broader spectrum of pathogens and reduce the possibility of developing resistance. That involves using the synergistic action of different microbes to create more powerful biofungicidal products. Technologically, these areas ensure what was previously unachievable by biofungicides: enhanced, easier manufacturing processes and use in their sprayed forms—all to increase uptake in organic farming (Kariyannae et al. 2024).

8.2 Integration with Nanotechnology and AI for Precision Application

Though the addition of nanotechnology and AI in biofungicide application is still in the early stages, it represents significant advancement in sustainable agriculture. It aims at enhanced precision, optimum efficiency, and environmental safety compared to chemical fungicides.

8.2.1 Nanotechnology in Biofungicides

Nanotechnology provides opportunities for innovative formulation and delivery systems for biofungicides. Various nanomicron carriers can serve as agents for protecting the active materials from degrading effects in the environment, assuring controlled release of bioactive molecules while improving the stability and efficiency of biofungicides. Nano-encapsulated pesticides exhibit greater lasting stability and efficacious performance in pest control with decreased environmental toxicity. Nanocarriers make targeted delivery possible so that biofungicides can be delivered directly to specific tissues of the plant or pathogens, increasing treatment efficiency and limiting off-target effects as described in Figure 3 (Romero and Costa 2025).

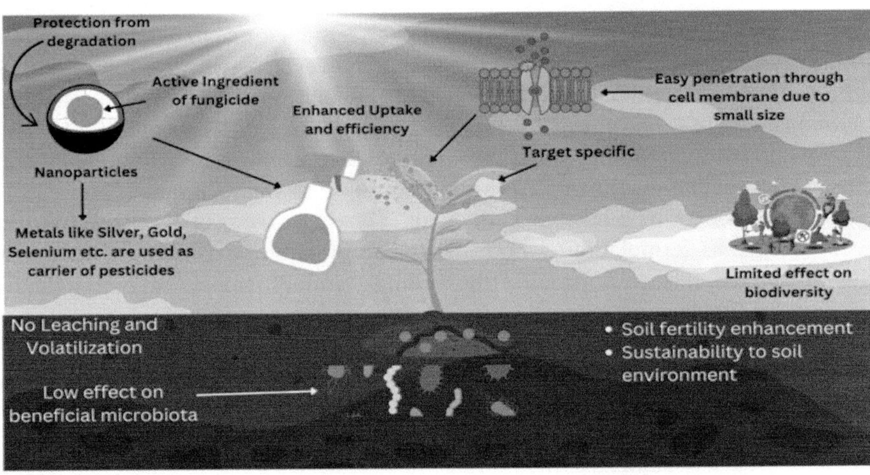

Figure 3: Application of NMs-based Biofungicides.

8.2.2 Artificial Intelligence for Precision Agriculture

AI technologies have transformed precision agriculture with the advent of real-time monitoring and decision-making capabilities. AI enables the utilization of vast amounts of data collected from sensors, drones, and satellites to assess crop health status, predict disease outbreaks, and optimize the timing and dosage of biofungicide applications. This ensures that biofungicides are applied only when and where needed with more efficiency and less waste (Agrawal and Arafat 2024).

8.2.3 Synergetic Integration of Nanotechnology and AI

The merger of nanotechnology and AI appears immensely potent for advanced precision application to biofungicides. Real-time parameters indicating the existence of fungal infections are captured and fed into AI algorithms for determining the optimal application dynamics. At the same time, delivery of biofungicides will be ensured to be effective at the right parameters. This synergism strengthens strategies of disease management by decreasing crop losses and environmental pollution. For instance, the addition of artificial intelligence and nanotechnology has helped create precision medicine by customizing treatments based on data from individual patients, an identical approach that can be adopted for the management of plant diseases (Pan et al. 2024).

While integration of nanotechnology and AI promises great advantages, the challenges must be overcome in order to achieve the full possibilities of applying biofungicides. Some of the challenges include long-term environmental impacts of nanoparticles, low-cost and easily scalable production methods, and ensuring that AI systems are friendly to access and use by farmers. Future research should consider such interdisciplinary collaborations to design safe nanomaterials and robust AI models that are agriculture-specific. As a result, the regulatory regime must modernize to accommodate the nuanced complexities surrounding nanotechnology and AI in agriculture. Thus, the incorporation of nanotechnology and AI into the precision application of biofungicides provides a radical alternative in sustainable agriculture. Enhancing the specificity and efficiency of disease management practices through the integration of these tools will, therefore, translate to great strides toward food security and environmental conservation (Khan et al. 2024).

8.3 Prospects for Next-Generation Biofungicides

Next-generation biofungicides have the potential to reduce the impact of synthetic chemical fungicides on crops and promote more sustainable agriculture. Microbial antagonists, plant-derived compounds, and RNA interference (RNAi) technologies can be used to create innovative biofungicides. Microbial biofungicides make use of beneficial microorganisms like bacteria and fungi to reduce the impact of plant pathogens. For example, the species *Trichoderma harzianum* and *Bacillus subtilis* are effective against the pathogen *Botrytis cinerea* that causes gray mold disease of numerous plants (affecting more than 1000 plant species). These microbes use competition for nutrients, antifungal compounds, and other approaches to help plants. There are still issues with their shelf life and performance in the field that need research (Vishnu et al. 2024). Fungi that come from plants use secondary metabolites such as terpenes, alkaloids, and phenolic compounds, which have antifungal activities. Use of standard and advanced techniques helps extract these compounds that furnish promising biofungicide formulations eventually controlling a range of phytopathogens. It has been observed that the essential oils of clove, cinnamon, and lemongrass exhibit strong inhibitory action against *B. cinerea*. These biofungicides are eco-friendly, decay easily, and have low toxicity for humans. However, challenges with formulation stability and standardization remain, which need continued research and development. RNAi-based biofungicides are an innovative new method of control that uses double-stranded RNA (dsRNA) to silence key genes in target pathogens. Spray-induced gene silencing (SIGS) using dsRNA molecules has been effective in combating gray mold disease by targeting virulence genes of *B. cinerea*. However, challenges remain, such as the stability of dsRNA in the field and its delivery to the target. Next-generation biofungicides will make a great addition to crop protection programs as they can replace chemical fungicides that cause environmental pollution and biodiversity loss. But to help make them more widely used, we must address issues related to efficacy, stability, and regulatory approval. Biocontrol agents need to be enhanced and optimized through concerted research and interactions between various fields in order to modernize agriculture while protecting human health and the environment (Islam et al. 2020).

9. Conclusion

The production of sustainable biofungicides is believed to give agriculture a better way to produce fungicidal effects at lower costs while not harming the environment. Biofungicide is a term that refers to a wide array of products that utilize naturally occurring organisms like bacteria or fungi. However, development and application depend on many factors like efficacy, production scalability, and regulatory procedures, among others. Integration with other agricultural practices would also depend on the above points. Biofungicides work in several different ways, including competing with pathogens for nutrients and space, producing bioactive agents, and inducing resistance in plants. *Trichoderma* species are known to parasitize plant pathogens and produce bioactive agents, which antagonize plant pathogens, causing healthier plants. In the same way, *Bacillus subtilis* has been used against powdery mildew on roses with foliar applications. This shows that microbial biofungicides can control diseases. Technological developments in fermentation have made it easier to produce biofungicides on a mass scale so that their commercial value can possibly be enhanced. Biofungicide-making organisms have recently been genetically mapped, making more useful biocontrol agents possible. These agents will be able to work in different climate circumstances. This is important for sustainability in agriculture in tough areas. The regulations that mandate the safety and efficacy of biofungicides can influence their use. Today, the majority of biofungicides available for controlling plant-pathogenic fungi are derived from organisms existing in soil or on plant surfaces. Due to their very specific nature and safe components, they cause very little harm to the environment, agricultural producers, and consumers.

Biofungicides are integral to organic farming systems, offering sustainable solutions to crop protection and soil health maintenance. They are products used to enhance the agriculture system. They are allowed for use in organic farming as they are natural products that have little side effects on the environment. These products minimize the use of synthetic chemicals in food and cut down residues in the product as well. Biofungicides can be utilized in standard pesticide management programs in regular farming to lessen the use of chemical fungicides and lower the problems of resistance in pathogens. Adding biofungicides in combination with chemical fungicides/plant extracts may help in better control of diseases in crops. Biofungicides can also be added in the process of plant breeding, which improves the cultivation of gene-modified fungi. Even though biofungicides have many benefits, they still face the problem of variability in field performance, narrow spectrum of activity, and shorter shelf life in comparison with chemical fungicides. In the future, research should work on the enhancement of stability and effectiveness of biofungicides, their interactions in the agroecosystem, and farmer-friendly formulations. As man-made organisms that can deliver numerous advantages, versatility, and benefits, they will help communities complete agricultural processes. In the coming years, solid biofungicide production will require overall collaboration between various fields related to science, technology, regulation, and farming. We can utilize biofungicides in sustainable farming and environmental safety through biotechnological contributions to sustain bio-resources.

References

Abbasi, P. A., and Weselowski, B. (2014). Influence of foliar sprays of *Bacillus subtilis* QST 713 on development of early blight disease and yield of field tomatoes in Ontario. *Canadian Journal of Plant Pathology*, 36(2): 170–178.

Agrawal, J., and Arafat, M. Y. (2024). Transforming farming: a review of AI-powered UAV technologies in precision agriculture. *Drones* (2504-446X). 8(11).

Akhtar, R., and Kumar, T. S. (2025). Phytoconstituent profiling and antifungal efficacy of *Artemisia absinthium* L.: Towards Sustainable Bio-Fungicide Development. *Applied Biochemistry and Biotechnology*. 197(1): 288–312.

Ansori, A.N., Antonius, Y., Susilo, J., Hayaza, S., Kharisma, V.D., Parikesit, A.A. et al. (2023). Application of CRISPR-Cas9 genome editing technology in various fields: A review. *Narra J.*, 3(2): 184.

Bautista Pinzón, J. F. (2024). Preparation of a biofungicide from Streptomyces for control of phytopathogens (*Fusarium oxysporum* and *Botrytis cinerea*) in basil (*Ocimum basilicum* L.) (Master's thesis, Universidad de La Sabana). Available at: http://hdl.handle.net/10818/63868. (accessed on April 05: 2025).

Bhattacharya, S., and Seth, A. (2023). 18 improvement in the activity of biofungicides through the application of advanced nanotechnology and genetic engineering. *Applications of Environmental Biotechnology for Global Sustainability*. 151.

Biondi, E., Schiavi, D., Perez Fuentealba, S., Kuzmanovic, N., Minardi, P., and Balestra, G. M. (2021). *Bacillus amyloliquefaciens* subsp. plantarum strain D747 to control the kiwifruit bacterial canker disease. In *X International Symposium on Kiwifruit* 1332. (95–102).

Bonnet, M., Lagier, J. C., Raoult, D., and Khelaifia, S. (2019). Bacterial culture through selective and non-selective conditions: the evolution of culture media in clinical microbiology. *New Microbes New Infect.*, 34: 100622.

Brownbridge, M. (n.d.). *Technical white paper: BotryStop® WP*. Certis Biologicals. Available at: https://www.certisbio.com/products/botrystop (accessed on April 10, 2025).

Celar, F. A., and Kos, K. (2022). Compatibility of the commercial biological control agents *Trichoderma asperellum* (ICC 012) and *Trichoderma gamsii* (ICC 080) with selected herbicides. *Journal of Plant Diseases and Protection*, 129(1): 85–92.

Cenobio-Galindo, A. D. J., Hernández-Fuentes, A. D., González-Lemus, U., Zaldívar-Ortega, A. K., González-Montiel, L., Madariaga-Navarrete, A. et al. (2024). Biofungicides based on plant extracts: On the road to organic farming. *International Journal of Molecular Sciences*. 25(13): 6879.

Chen, S., Daly, P., Zhou, D., Li, J., Wang, X., Deng, S. et al. (2022). The use of mutant and engineered microbial agents for biological control of plant diseases caused by Pythium: Achievements versus challenges. *Fungal Biology Reviews.*, 40: 76–90.

Colmenarez, Y. C., and Vasquez, C. (2024). Benefits associated with the implementation of biological control programmes in Latin America. *BioControl.*, 69(3): 303–320.

Cui, J., Zhu, R., Wang, X., Xu, X., Ai, C., He, P. et al. (2022). Effect of high soil C/N ratio and nitrogen limitation caused by the long-term combined organic-inorganic fertilization on the soil microbial community structure and its dominated SOC decomposition. *Journal of Environmental Management*., 303: 114155.

Custódio, F. A., and Pereira, O. L. (2024). New treasures in Cordycipitaceae: Fungicolous fungi associated with *Pseudocercospora fijiensis* and *P. musae* in Brazil, including Matutinistella gen. nov. *Fungal Systematics and Evolution*.

Das, M. M., Haridas, M., Leza, H. A. R., and Sabu, A. (2022). Customization of bioreactor technology for on-site production of biocontrol agents. In *Biocontrol Systems and Plant Physiology in Modern Agriculture* (pp. 23–43). Apple Academic Press.

Díaz-Rodríguez, A. M., Parra Cota, F. I., Cira Chávez, L. A., García Ortega, L. F., Estrada Alvarado, M. I., Santoyo, G. et al. (2025). Microbial Inoculants in Sustainable Agriculture: Advancements, Challenges, and Future Directions. *Plants*, 14(2): 191.

Du, Y.-H., Wang, M.-Y., Yang, L.-H., Tong, L.-L., Guo, D.-S., and Ji, X.-J. (2022). Optimization and Scale-Up of Fermentation Processes Driven by Models. *Bioengineering*, 9(9): 473. https://doi.org/10.3390/bioengineering9090473.

Ezeorba, T. P. C., Chukwudozie, K. I., Okoye, C. O., Okeke, E. S., Ezugwu, A. L., and Anaduaka, E. G. (2023). Biofungicides: Classification, Applications and Limitations. In *Biofungicides: Eco-Safety and Future Trends* (pp. 12–39). CRC Press.

Fenta, L., and Mekonnen, H. (2024). Microbial biofungicides as a substitute for chemical fungicides in the control of phytopathogens: Current perspectives and research directions. *Scientifica*. 2024(1): 5322696.

Ferreyra-Suarez, D., García-Depraect, O., and Castro-Muñoz, R. (2024). A review on fungal-based biopesticides and biofertilizers production. *Ecotoxicology and Environmental Safety*. 283: 116945.

Foka, E. T., Tayo, P. M. T., Pacha, T. F., Dzelamonyuy, A., Sado, S. L. K., Onomo, P. E. et al. (2023). Biofungicide based calcium, *Azadirachta indica* and *Sida acuta* Against *Phytophthora megakarya*. *Journal of Plant Sciences*., 11(6): 171–181.

Fusar Poli, E., and Fontefrancesco, M. F. (2024). Trends in the implementation of biopesticides in the Euro-Mediterranean region: A narrative literary review. *Sustainable Earth Reviews*, 7(1): 14.

Ghani, A., Bokhari, S. A. I., Ali, S. A., Faheem, M., Alomar, S. Y., Basheir, M. M. et al. (2024). Assessment of a nanobiofungicide with antimicrobial potential against mycopathogenic *Fusarium* species. *Journal of King Saud University-Science*, 36(1): 103036.

Gupta, R., Malik, P., Rani, R., Solanki, R., Ameta, R. K., Malik, V. et al. (2024). Recent progress on nanoemulsions mediated pesticides delivery: insights for agricultural sustainability. *Plant Nano Biology*. 100073.

Han, P., Ni, L., Wei, Y., Jiang, S., Xu, F., Wang, H. et al. Optimization of the freeze-drying of marine yeast *Sporidiobolus pararoseus* ZMY-1 for its application in biocontrol of fungal infections. *Biological Control*., 161: 104707.

Hashem, A. H., Abdelaziz, A. M., and Attia, M. S. (2023). Trichoderma a promising biofungicide. In *Biofungicides: Eco-Safety and Future Trends*. CRC Press. 166–189.

Hashem, M., Mostafa, Y. S., Alamri, S., Abbas, A. M., and Eid, E. M. (2021). Exploitation of agro-industrial residues for the formulation of a new active and cost effective biofungicide to control the root rot of vegetable crops. *Sustainability*., 13(16): 9254.

Islam, M.T., and Sherif, S.M. (2020). RNAi-Based biofungicides as a promising next-generation strategy for controlling devastating gray mold diseases. *Int. J. Mol. Sci*., 21(6): 2072.

Kariyanna, B., Ratnala, S., Panda, S., and Sainath, G. (2024). Formulations of Biopesticides: Techniques, Applications, Challenges and Future Prospects. *HEXAPODA*. 97-113.

Khakimov, A. A., Omonlikov, A. U., and Utaganov, S. B. U. (2020). Current status and prospects of the use of biofungicides against plant diseases. *GSC Biological and Pharmaceutical Sciences*., 13(3): 119–126.

Khan, E., and Ahmad, I. Z. (2024). Production and commercialization of biocontrol products. In *Biocontrol Agents for Improved Agriculture*. Academic Press. 165–180.

Khan, H., Jan, Z., Ullah, I., Alwabli, A., Alharbi, F., Habib, S. et al. (2024). A deep dive into AI integration and advanced nanobiosensor technologies for enhanced bacterial infection monitoring. *Nanotechnology Reviews*, 13(1): 20240056.

Koirala, S., Myers, B., Shin, G. Y., Gitaitis, R., Kvitko, B. H., and Dutta, B. (2023). Evaluating options to increase the efficacy of biocontrol agents for the management of *Pantoea* spp. under field conditions. *Plant Disease*, 107(9): 2701–2708.

Krzymińska, J., and Kowalska, J. (2024). The protective effect of trehalose and monosodium glutamate on yeast viability and antagonistic properties during freeze-drying. *Journal of Plant Protection Research*. 362–372.

Kumar, P., Singh, S., Sankhyan, S., and Ray, S. (2024). Metabolic engineering of lactic acid bacteria for antimicrobial peptides production. In *Antimicrobial Peptides from Lactic Acid Bacteria: Diversity, Biosynthesis and Applications*. Springer Nature Singapore. 67–95.

Kumar, S., and Chandra, R. (2024). Integrated field application of T. viride, botanicals, and fungicides for managing early blight (*Alternaria solani*) and enhancement of plant growth, tuber nutritional quality, and potato yield. *Journal of Natural Pesticide Research*. 8: 100070.

Lahdenperä, M. L. (2020). A case study: Use of prestop® mix biofungicide in entomovectoring on apple against storage rot diseases. *Entomovectoring for Precision Biocontrol and Enhanced Pollination of Crops*. 137–145.

Leclère, V., Béchet, M., Adam, A., Guez, J. S., Wathelet, B., Ongena, M. et al. 2005. Mycosubtilin overproduction by Bacillus subtilis BBG100 enhances the organism's antagonistic and biocontrol activities. *Applied and Environmental Microbiology*, 71(8): 4577–4584.

Li, H., Kalunke, R., Tetorya, M., Czymmek, K. J., and Shah, D. M. (2024). Modes of action and potential as a peptide-based biofungicide of a plant defensin MtDef4. *Molecular Plant Pathology*, 25(4): 13458a.

Li, Y., Xu, Z., Zhan, X., and Zhang, T. (2024). Summary of Experiments and Influencing Factors of Sediment Settling Velocity in Still Water. *Water*. 16(7): 938b.

Lima, P. C., Karimian, P., Johnston, E., and Hartley, C. J. (2024). The Use of *Trichoderma* spp. for the Bioconversion of Agro-Industrial Waste Biomass via Fermentation: A Review. *Fermentation*. 10(9): 442.

Lorenzo, J. M., Munekata, P. E., Dominguez, R., Pateiro, M., Saraiva, J. A., and Franco, D. (2018). Main groups of microorganisms of relevance for food safety and stability: General aspects and overall description. In *Innovative technologies for food preservation*. Academic Press. 53–107.

Mahawer, S. K., Arya, S., Kabdal, T., Kumar, R., Prakash, O., Chitara, M. K. et al. (2022). Plant defense systems: mechanism of self-protection by plants against pathogens. *Plant Protection: from Chemicals to Biologicals*. 115.

Malakar, C., and Deka, S. (2021). Biosurfactants against drug-resistant human and plant pathogens: Recent advances. *Biosurfactants for a Sustainable Future: Production and Applications in the Environment and Biomedicine*. 353–372.

Manzar, N., Kashyap, A. S., Goutam, R. S., Rajawat, M. V. S., Sharma, P. K., Sharma, S. K. et al. (2022). Trichoderma: advent of versatile biocontrol agent, its secrets and insights into mechanism of biocontrol potential. *Sustainability*, 14(19): 12786.

Marrone, P. G. (2024). Status of the biopesticide market and prospects for new bioherbicides. *Pest Management Science*, 80(1): 81–86.

Martău, G.-A., Unger, P., Schneider, R., Venus, J., Vodnar, D. C., and López-Gómez, J. P. (2021). Integration of solid state and submerged fermentations for the valorization of organic municipal solid waste. *Journal of Fungi*, 7(9): 766.

Mascarin, G. M., Golo, P. S., de Souza Ribeiro-Silva, C., Muniz, E. R., de Oliveira Franco, A., Kobori, N. N. et al. (2024). Advances in submerged liquid fermentation and formulation of entomopathogenic fungi. *Applied Microbiology and Biotechnology*, 108(1): 451.

Matter, I. A., Bui, V. K. H., Jung, M., Seo, J. Y., Kim, Y.-E., Lee, Y.-C. et al. (2019). Flocculation Harvesting Techniques for Microalgae: A Review. *Applied Sciences*, 9(15): 3069.

Mawcha, K. T., Malinga, L., Muir, D., Ge, J., and Ndolo, D. (2024). Recent advances in biopesticide research and development with a focus on microbials. *F1000Research*, 13: 1071.

McLaughlin, M.S., Roy, M., Abbasi, P.A., Carisse, O., Yurgel, S.N., and Ali, S. (2023). Why do we need alternative methods for fungal disease management in plants? Plants (Basel). 12(22): 3822.

Meher, J., Rajput, R. S., Bajpai, R., Teli, B., and Sarma, B. K. (2020). Trichoderma: A globally dominant commercial biofungicide. *Trichoderma: Agricultural Applications and Beyond*. 195–208.

Mehra, M. K., and Avanthi, A. (2024). Challenges and Safety Regulations for Employing Fungi in Environmental Applications. In *Bioprospecting of Multi-tasking Fungi for a Sustainable Environment: Volume I.* Springer Nature Singapore. 315–341.

Mejía, C., Bautista, E. J., García, L., Barrios Murcia, J. C., and Barrera, G. (2024). Assessment of fungal lytic enzymatic extracts produced under submerged fermentation as enhancers of Entomopathogens' biological activity. *Current Microbiology*, 81(7): 217.

Mennella, C., Maniscalco, U., De Pietro, G., and Esposito, M. (2024). Ethical and regulatory challenges of AI technologies in healthcare: A narrative review. *Heliyon*. 10(4): 26297.

Mir, S. A., Padhiary, A., Ekka, N. J., Baitharu, I., and Nayak, B. (2023). Environmental impacts of synthetic and biofungicides. In *Current Developments in Biotechnology and Bioengineering*. Elsevier. 487–504.

Mitrović, I., Grahovac, J., Dodić, J., Jokić, A., Rončević, Z., and Grahovac, M. (2020). Production of plant protection agents in medium containing waste glycerol by *Streptomyces hygroscopicus*: Bioprocess analysis. *Acta Alimentaria.*, 49(3): 270-277.

Mostafa, Y. S., Hashem, M., Alshehri, A. M., Alamri, S., Eid, E. M., Ziedan, E. S. H. et al. (2021). Effective management of cucumber powdery mildew with essential oils. *Agriculture*, 11(11): 1177.

Mulatu, A., Alemu, T., Megersa, N., and Vetukuri, R. R. (2021). Optimization of culture conditions and production of bio-fungicides from *Trichoderma* species under solid-state fermentation using mathematical modeling. *Microorganisms*, 9(8): 1675.

Nefzi, A., Abdallah, R. A. B., Jabnoun-Khiareddine, H., Ammar, N., and Daami-Remadi, M. (2019). Ability of endophytic fungi associated with *Withania somnifera* L. to control Fusarium Crown and Root Rot and to promote growth in tomato. *Brazilian Journal of Microbiology*, 50: 481–494.

Nowak, D., and Jakubczyk, E. (2020). The freeze-drying of foods—the characteristic of the process course and the effect of its parameters on the physical properties of food materials. *Foods*, 9(10): 1488.

Nwankwo, C. S., Okpomor, E. O., Dibagar, N., Wodecki, M., Zwierz, W., and Figiel, A. (2023). Recent developments in the hybridization of the freeze-drying technique in food dehydration: A review on chemical and sensory qualities. *Foods*, 12(18): 3437.

Ouhaibi-Ben Abdeljalil, N., Makhlouf, L., and Boughalleb-M'Hamdi, N. (2024). Assessment of the *in vitro, in planta* and *in orchard* efficacy of three plant extracts against *Erwinia amylovora*, responsible agent of fire blight disease in Tunisia. *Eur. J. Plant Pathol.*, 170: 929–941.

Omoyajowo, K. (2024). Sustainable Environmental Policies: An Impact Analysis of US Regulations on Pesticides and Chemical Discharges. *Available at SSRN* 4943299.

Pai, S., Hebbar, A., and Selvaraj, S. (2022). A critical look at challenges and future scopes of bioactive compounds and their incorporations in the food, energy, and pharmaceutical sector. *Environ. Sci. Pollut. Res. Int.*, 29(24): 35518–35541.

Palladino, F., Marcelino, P. R. F., Schlogl, A. E., José, Á. H. M., Rodrigues, R. d. C. L. B., Fabrino, D. L. et al. (2024). Bioreactors: Applications and innovations for a sustainable and healthy future—a critical review. *Applied Sciences*, 14(20): 9346.

Pan, Y., Xue, X., and Liang, X. J. (2024). Nanotechnology-empowered combination cancer immunotherapies: mechanisms, synergies, and perspectives. *Advanced NanoBiomed Research*, 4(4): 2300129.

Portela, V. O., Santana, N. A., Balbinot, M. L., Antoniolli, Z. I., de Oliveira Silveira, A., and Jacques, R. J. S. (2022). Phytotoxicity optimization of fungal metabolites produced by solid and submerged fermentation and its ecotoxicological effects. *Applied Biochemistry and Biotechnology*, 194(7): 2980–3000.

Prakash, V., and Basu, K. (2021). Mass Multiplication of Trichoderma. *Trichoderma: Agricultural Applications and Beyond.*, 61: 113.

Quintana-Jones, T. A. (2011). Evaluation of drip applications and foliar sprays of the biocontrol product Actinovate on powdery mildew and other fungal diseases of tomato. California Polytechnic State University.

Ragasruthi, M., Balakrishnan, N., Murugan, M., Swarnakumari, N., Harish, S., and Sharmila, D. J. S. (2024). Bacillus thuringiensis (Bt)-based biopesticide: Navigating success, challenges, and future horizons in sustainable pest control. *Science of The Total Environment*, 176594.

Ramírez-Guzmán, N., Martínez-Medina, G., Rodriguez-Gonzalez, L., Chávez-González, M. L., Hernández-Almanza, A., and Aguilar, C. N. (2024). Biofungicide production by solid-state fermentation as sustainable biotechnology for agroindustrial waste management and fresh crop production. *In*: Benkeblia, N. (eds.). Recent Advances in Postharvest Technologies, Volume 1. Springer, Cham.

Rana, A. K., Kaur, K., and Vyas, P. (2024). Biofungicides and plant growth promoters: Advantages and opportunities in entrepreneurship. In *Entrepreneurship with Microorganisms*. Academic Press. 259–277.

Rezagholizade-Shirvan, A., Soltani, M., Shokri, S., Radfar, R., Arab, M., Shamloo, E. et al. (2024). Bioactive compound encapsulation: Characteristics, applications in food systems, and implications for human health. *Food Chem X*, 24: 101953.

Risdian, C., Mozef, T., and Wink, J. (2019). Biosynthesis of Polyketides in *Streptomyces*. *Microorganisms*, 7(5): 124.

Ristić, D., Vučurović, I., Blagojević, J., Keserović, V., Starović, M., Trkulja, N. et al. (2024). Efficacy evaluation of synthetic fungicides and biofungicides against Fusarium wilt in blueberry. In *XIII International Symposium on Agricultural Sciences" AgroReS 2024" 27–30 May 2024, Trebinje, Bosnia and Herzegovina*. University of Banja Luka, Faculty of Agriculture. 61.

ROMERO, C. M., and Costa, J. S. D. (2025). Nano-Biofungicides and Bio-Nanofungicides: State of the Art of Innovative Tools for Controlling Resistant Phytopathogens.

Rosa, J. M., Toledo, M. Z., Dupas, E., and Turtt, J. C. T. (2025). Can biologicals lower input costs in soybean? *Observatório DE LA Economía Latinoamericana*. 23(1): 8773.

Saberi-Riseh, R., and Moradi-Pour, M. (2021). A novel encapsulation of *Streptomyces fulvissimus* Uts22 by spray drying and its biocontrol efficiency against *Gaeumannomyces graminis*, the causal agent of take-all disease in wheat. *Pest Management Science*, 77(10): 4357–4364.

Salazar, B., Ortiz, A., Keswani, C., Minkina, T., Mandzhieva, S., Pratap Singh, S. et al. (2023). *Bacillus* spp. as bio-factories for antifungal secondary metabolites: Innovation beyond whole organism formulations. *Microbial Ecology*, 86(1): 1–24.

Sawant, S. S., Park, H.-Y., Sim, E.-Y., Kim, H.-S., and Choi, H.-S. (2025). Microbial fermentation in food: impact on functional properties and nutritional enhancement—A review of recent developments. *Fermentation*, 11(1): 15.

Shin, T. S., Yu, N. H., Lee, J., Choi, G. J., Kim, J. C., and Shin, C. S. (2017). Development of a biofungicide using a mycoparasitic fungus *Simplicillium lamellicola* BCP and its control efficacy against gray mold diseases of tomato and ginseng. *Plant Pathol.*, 33(3): 337–344.

Sidorova, T. M., Asaturova, A. M., Homyak, A. I., Zhevnova, N. A., Shternshis, M. V., and Tomashevich, N. S. (2020). Optimization of laboratory cultivation conditions for the synthesis of antifungal metabolites by *Bacillus subtilis* strains. *Saudi Journal of Biological Sciences*, 27(7): 1879–1885.

Silveira, R. D., Veras, F. F., Hernandes, K. C., Bach, E., Passaglia, L. M. P., Zini, C. A. et al. (2024). Genomic analysis reveals genes that encode the synthesis of volatile compounds by a *Bacillus velezensis*-based biofungicide used in the treatment of grapes to control *Aspergillus carbonarius*. *International Journal of Food Microbiology*, 415: 110644.

Slavin, Y. N., and Bach, H. (2022). Mechanisms of antifungal properties of metal nanoparticles. *Nanomaterials*, 12(24): 4470.

Subba, R., and Mathur, P. (2022). Functional attributes of microbial and plant based biofungicides for the defense priming of crop plants. *Theoretical and Experimental Plant Physiology*, 34(3): 301–333.

Sutthisa, W., Hompana, W., and Yutthasin, R. (2024). Enhancing Biocontrol Potential: Development and Efficacy Assessment of a Liquid Formulation of *Trichoderma Asperellum* MSU007 against *Sclerotium Rrolfsii*. *Trends in Sciences*, 21(4): 7550–7550.

Tan, Y. S., Zhang, R. K., Liu, Z. H., Li, B. Z., and Yuan, Y. J. (2022). Microbial adaptation to enhance stress tolerance. *Frontiers in Microbiology*, 13: 888746.

Tang, Q., Yan, F., Yuan, L., Tang, Y., Chen, H., Sun, Y. et al. (2024). Enhancing laboratory biosafety management: a comprehensive strategy from theory to practice. *Frontiers in Public Health*, 12: 1439051.

Thangaraj, P., Subbiah, K. A., Sevugapperumal, N., Uthandi, S., Damodarasamy, A., and Shanmugam, H. (2023). Activity of volatiles induced by microbes and natural plants stifled the growth of Pythium aphanidermatum-the damping off in Tomato. *BMC Plant Biology*, 23(1): 384.

Thind, T. S. (2021). Changing trends in discovery of new fungicides: a perspective. *Indian Phytopathology*, 74(4): 875–883.

Tryfon, P., Kamou, N. N., Pavlou, A., Mourdikoudis, S., Menkissoglu-Spiroudi, U., and Dendrinou-Samara, C. (2023). Nanocapsules of ZnO nanorods and geraniol as a novel mean for the effective control of Botrytis cinerea in tomato and cucumber plants. *Plants*, 12(5): 1074.

Vandenberghe, L. P., Pandey, A., Carvalho, J. C., Letti, L. A., Woiciechowski, A. L., Karp, S. G. et al. (2021). Solid-state fermentation technology and innovation for the production of agricultural and animal feed bioproducts. *Systems Microbiology and Biomanufacturing*, 1: 142–165.

Verma, J., Kumar, C., Sharma, M., Shukla, A. C., and Saxena, S. (2024). Exploitation of microbial consortia for formulating biofungicides, biopesticides, and biofertilizers for plant growth promotion. In *Entrepreneurship with Microorganisms*). Academic Press. 227–257.

Vishnu, M., Malaisamy, K., Soundararajan, R. P., Suganthi, A., Subramanian, A., Senthilkumar, M. et al. (2024). Nano-bioformulations: emerging trends and potential applications in next generation crop protection. *Environmental Science: Nano*.

Wesche, J., Repp, J., Hu, M., Faust, J., and Schnabel, G. (2025). Cross-resistance between *Pseudomonas chlororaphis* strain AFS009 metabolites (Howler EVO) and fludioxonil in Botrytis cinerea. *Plant Disease*.

Xiao, J., Guo, X., Qiao, X., Zhang, X., Chen, X., and Zhang, D. (2021). Activity of Fengycin and Iturin A Isolated From *Bacillus subtilis* Z-14 on *Gaeumannomyces graminis* Var. *tritici* and Soil Microbial. Diversity. *Front. Microbiol.*, 212: 682437.

Yang, X., Yuan, L., Zeeshan, M., Yang, C., Gao, W., Zhang, G. et al. (2025). Optimization of fermentation conditions to increase the production of antifungal metabolites from *Streptomyces* sp. KN37. *Microbial Cell Factories*, 24(1): 26.

Zalila-Kolsi, I., Kessentini, S., Tounsi, S., and Jamoussi, K. (2022). Optimization of *Bacillus amyloliquefaciens* BLB369 culture medium by response surface methodology for low cost production of antifungal activity. *Microorganisms*, 10(4): 830.

Zamoum, M., Allali, K., Benadjila, A., Zitouni, A., and Goudjal, Y. (2022). Formulation of biofungicides based on Streptomyces caeruleatus strain ZL-2 spores and efficacy against *Rhizoctonia solani* damping-off of tomato seedlings. *Archives of Microbiology*, 204(10): 629.

Zhang, M. (2022). Microorganisms and Bioprocessing, General. In *Iron Ores Bioprocessing*. Cham: Springer International Publishing. 53–87.

Zhou, W., Arcot, Y., Medina, R. F., Bernal, J., Cisneros-Zevallos, L., and Akbulut, M. E. (2024). Integrated pest management: an update on the sustainability approach to crop protection. *ACS omega*, 9(40): 41130–41147.

8

Development of a Biofungicide Formulation to Control Wheat Blast Disease

Methods and Protocols

Lamya Muazzeda Medha, Dipali Rani Gupta and *Tofazzal Islam**

1. Introduction

The fungus *Magnaporthe oryzae Triticum* (MoT) causing wheat blast first appeared in Brazil in 1985. It turned up in Bangladesh decades later, in 2016, when it devastated hundreds of hectares of wheat in eight of 64 districts, leaving crops destroyed in parts of a few zones (Islam et al. 2016). The disease has now spread across more than 35 districts, becoming a serious threat to the country's food and nutritional security (Islam et al. 2020). Wheat blast has already affected about 2.8 million hectares of wheat fields in Brazil, Argentina and Paraguay, and Bolivia (Igarashi et al. 1986). More recently, the disease has been reported in Uruguay, and in 2018 it was detected in Zambia, underscoring its ongoing geographical expansion from South America into the African continent (Tembo et al. 2020, Silva et al. 2025). This disease is dangerous because it is so fast to spread and to adapt. Few resistant wheat varieties exist, and it evolves quickly, so it breaks through resistance in no time (Cruz et al. 2016). As such, wheat blast has become a major threat to food security on a worldwide scale (Islam et al. 2020).

Institute of Biotechnology and Genetic Engineering, Gazipur Agricultural University, Gazipur 1706, Bangladesh.

* Corresponding author: tofazzalislam@bsmrau.edu.bd

DOI: 10.1201/9781003603023-9

The primary method to control wheat blast through chemical fungicides has proven inefficient but remains expensive and harmful to the environment (Zubrod et al. 2019). Frequent documentation of chemical treatment resistance is a common finding in research conducted by Castroagudín et al. (2015). Wheat blast control requires urgent development of new biorational solutions that protect the environment, according to published findings from Islam et al. (2020). Biotechnology has developed beneficial plant probiotic bacteria into commercial biofungicides that provide a sustainable alternative to dangerous synthetic chemical fungicides (Chen et al. 2019a, Chakraborty et al. 2021, Bhattacharjee et al. 2025). Biocontrol studies have demonstrated that microorganisms from *Pseudomonas*, *Bacillus*, *Streptomyces,* and *Lysobacter* genetic groups can effectively fight fungal and oomycetous diseases (Islam et al. 2005, Robin and Marchand 2019, Surovy et al. 2023, 2024).

Four types of endophytic probiotic bacteria, including *Bacillus, Streptomyces, Lysobacter*, and *Pseudomonas* along with rhizobacteria, have proven to be effective biocontrol agents against plant pathogens, including *Magnaporthe oryzae*, based on findings by Islam et al. (2005) and Chakraborty et al. (2021). Bacteria in this category block pathogens through the generation of antimicrobial substances, together with aggressive competition for nutrients and disruption of pathogen virulence elements, including quorum sensing mechanisms (Kaila et al. 2019). The bacteria cause plants to activate defense mechanisms as well as improve their health (Köhl et al. 2019). The success of biocontrol agents depends upon their dosage rate while being susceptible to environmental conditions. Biocontrol strategies require optimized sustainability by evaluating key factors (Bonaterra et al. 2022).

Bacillus species are well known as biological control agents in both soil and on plant surfaces (Chakraborty et al. 2020, Dame et al. 2021, Surovy et al. 2024). Their ecological versatility and ability to form endospores make them adaptable to diverse environments. They show antagonistic properties toward several plant pathogens and possess bioactive compounds useful for agriculture. In Ongena et al. (2008) and Bonaterra et al. (2022), they are able to synthesize antimicrobial metabolites and stimulate plant defense mechanisms. Isolated from 'Rangabinni' rice seeds, *B. subtilis* BTS-3 and *B. velezensis* BTS-4, along with *B. velezensis* BTLK6A from 'Kanchan' wheat seeds in Bangladesh (Surovy et al. 2017, Chanclud et al. 2017, Dutta et al. 2018, Surovy et al. 2022a, 2023, 2024), include antioxidant enzymes, antimicrobial peptides, enzymes of cell wall degradation, compounds inducing systemic resistance (ISR) in plants, and volatile organic compounds (VOCs) (Surovy et al. 2018, Dutta et al. 2018, Surovy et al. 2023). *Bacillus* spp. secretes the antifungal VOCs, e.g., acetoin, suppress phytopathogen growth (Chen et al. 2018c, Kumar et al. 2021, Sasani and Ahmadzadeh 2021, Surovy et al. 2023). Marine *B. subtilis* also produces linear lipopeptides that inhibit MoT by blocking conidia germination or causing nonnormal germ tube and appressoria formation (Chakraborty et al. 2020). *In silico* studies revealed that potent natural compounds such as camptothecin and chaetoviridin A readily inhibit *M. oryzae* enzymes participating in melanin biosynthesis and appressorium formation even more comparably versus commercial fungicides azoxystrobin and strobilurin (Khan et al. 2022). The marine *Streptomyces* spp. secondary metabolites bonactin and feigrisolide C also inhibit MoT by disrupting the mycelial development, conidia production, germination, and conidial morphology (Rabby et al. 2022).

Fluorescent pseudomonads are one of the beneficial microorganisms prevailing in plant environments. They have unique characteristics that help them control plant diseases (Khatun et al. 2018, Chakraborty et al. 2021). Plants gain immunity through their resistance to phytopathogens and their strong adaptability to the environment. Some *Pseudomonas* spp. can colonize the plant surface. They use exudates of plants for their nutrition, and their growth rate is also high, which helps them outcompete other microorganisms present in the plant environment (Oso et al. 2019, Bonaterra et al. 2022). Strains identified as *P. wayambapalatensis* 'Amana' and *Pseudomonas* sp. nov. 'Yara' (in the *P. putida* group), together with *Trichoderma koningiopsis* Cachara, inhibited *in vitro Pyricularia oryzae triticum* (PoT) mycelial growth and reduced wheat blast disease severity through biofilm production, antagonistic processes, and mycoparasitism mechanisms (Pereira et al. 2022). The application of two BCAs (*P. stutzeri* AN10 and *T. harzianum* Th56) analyzed salt-stressed seedlings that underwent infection from two *P. oryzae* isolates at different aggressiveness levels. Plants developed lower blast infections following salt-induced priming because the treatment activated early expression of defense genes (Cabot et al. 2018).

The potential of biocontrol of wheat blast by plant probiotic *Bacillus* spp., *Pseudomonas* spp., *Streptomyces* spp., *Trichoderma* spp., and metabolites from these bacteria and other microorganisms has been demonstrated in both greenhouse and field conditions (Surovy et al. 2018, Dutta et al. 2018, Surovy et al. 2023, Cabot et al. 2018, Pereira et al. 2022, Rabby et al. 2022, Surovy et al. 2022a). However, no bacterial biocontrol agents or their metabolites have yet been commercialized as biofungicides for the biocontrol of wheat blast. Development of an effective formulation for field application is critical for the commercialization of any bacterial or microbial biofungicide. This chapter provides a comprehensive and systematic guide on the development of biofungicide formulations using biological control agents to effectively manage wheat blast disease.

2. Sources of Microorganisms for the Development of Biofungicide

Living organism–based biofungicides are being used as secure, less expensive alternatives to conventional chemical fungicides. Biofungicides are comprised of beneficial microorganisms such as *Bacillus* and *Streptomyces* as well as plant extracts of algae and flowering plants used to bolster the plant's preventive defense system against fungal infections (Kekuda et al. 2018, Soliman et al. 2018, O'Keeffe et al. 2019, Ferrigo et al. 2020, Liu et al. 2020, Nian et al. 2021). Some biofungicides available today use extract-based (Kanwal et al. 2015, Falade et al. 2018) processes as well as suspension (Sujarit et al. 2020) and formulated (Eakjamnong et al. 2021) methods with different extraction methods. The biofungicide manufacturing processes have been tested using different solvents (i.e., water; ethanol together with methanol, acetone, and ethyl acetate) and have shown that biological compounds can be extracted (Wei et al. 2020). Bacterial and fungal antagonists are the main reservoirs of the collective antagonists from soils, rhizosphere, plant roots, stems, and seeds.

2.1 Microbial Biofungicides

Microbial biofungicides have gained attention in recent years for their low toxicity and cost-effectiveness (Ayilara et al. 2023). They target specific pathogens while being safe for non-target organisms and offer prolonged disease control by utilizing living organisms. These biofungicides suppress pathogens through competition, toxin production, nutrient competition, biofilm formation, and plant resistance induction (Lahlali et al. 2022). Unlike chemical fungicides, which primarily inhibit respiration and sterol biosynthesis, microbial biofungicides are environmentally friendly and safer for consumers (Gikas et al. 2022). However, challenges in storage and transport persist, as different bacterial strains require specific conditions. Key microorganisms used include *Pseudomonas, Bacillus,* and *Trichoderma* spp., with *Trichoderma harzianum* being promoted in Cambodia to manage fungal diseases and enhance crop yields (Chou et al. 2020). Various application methods like spray drying and lyophilization are employed, though further research is needed for longer shelf life (de Oliveira Filho et al. 2021). Although not many effective biofungicides have been developed yet to control wheat blast, there are some other microbial biofungicides used to control plant diseases (Table 1).

3. Development of Biofungicide: A Flowchart

The formulation of biofungicide is prepared by isolating and screening enough antagonists followed with tests for antifungal nature. Microorganisms that show promise are further characterized in terms of their mode of action and potency against specific pathogens. Toxicological testing ensures safety. Selected candidates are then scaled up for mass production of specific strains, thereby refining the growth and cultivation. When grown in bulk, they are processed into commodities that go on to be commercialized through packaging, labeling, and regulatory approval. This process ensures a scientific approach by following the right trail: from isolation to market availability through efficacy, safety, and readiness of the marketplace (Figure 1).

4. Isolation of Target Microorganism

The first step in creating biofungicides involves isolating and screening bacteria that can inhibit specific plant pathogens and reduce disease incidence (Figure 2). Choosing the right sampling sites, culture media, and isolation techniques is crucial for identifying various probiotic bacteria (Montesinos et al. 2003). These beneficial bacteria can be found in the phyllosphere, rhizosphere, endosphere, and soil. Diverse sources serve as reservoirs for discovering potential biological control agents (BCA). Beneficial microorganisms often reside in suppressive soils, healthy plants in areas affected by outbreaks, or near pathogen infection sites (Montesinos et al. 2003, Köhl et al. 2011). For instance, soil containing *Pseudomonas* biocontrol agents helps

Table 1: Some commonly used microbial agents as biofungicides for biocontrol of fungal plant diseases.

Biocontrol agent	Fungal pathogen	Name of disease	Host plant	References
Fomes fomentarius M40	*Gaeumannomyces graminis* var. *tritici*	Take-all disease	Wheat	Gholami et al. (2019)
Coprinopsis urticicola M2				
Rhizoctonia zeae M32				
Penicillium olsonii ML37	*Fusarium graminearum*	Wheat head blight	Wheat	Rojas et al. (2020)
Anthracocystis flocculosa P1P				
Sarocladium strictum C113L				
Trichoderma harzianum BTB 022	*Pyricularia oryzae*	Rice blast	Rice	Chou et al. (2020)
Pseudomonas fluorescens	*Rhizoctonia solani*	Sheath blight	Rice	Pal et al. (2015)
Talaromyces tratensis KUFA 0091	*i. Alternaria padwickii,* *ii. Curvularia lunata*	Dirty panicle	Rice	Eakjamnong et al. (2021)
Rhodopseudomonas palustris KTSSR54	*i. Magnaporthe oryzae* PTRC63 *ii. Bipolaris oryzae* NPT0508		Rice	Nookongbut et al. (2020)
Bacillus subtilis 5	*Magnaporthe oryzae*	Rice blast	Rice	
Pseudomonas fluorescens 10S2				
Talaromyces tratensis	*i. Fusarium moniliforme* *ii. Bipolaris oryzae*	Dirty panicle	Rice	Eakjamnong et al. (2021)
B. amyloliquefaciens	*Ralstonia solanacearum*	Bacterial wilt	Tomato	Chen et al. (2019a)
Trichoderma polysporum	*Magnaporthe oryzae*	Rice blast	Rice	Khanzada et al. (2012)
T. pseudokoningii,				
Gliocladium virens				
Paecilomyces variotii				
B. subtilis strain T429	*Magnaporthe grisea*	Rice blast	Rice	Meng et al. (2015)
Lysobacter enzymogenes strain	i. *Macrophomina phaseolina* ii. *Sclerotinia sclerotiorum* iii. *Fusarium virguliforme* iv. *Rhizoctonia solani* vi. *Phytophthora sojae*	i. Charcoal rot, ii. *Sclerotinia* stem rot, iii. Sudden death syndrome iv. Root rot, vi. *Phytophthora* stem and root rot	Soybean	Nian et al. (2021)
B. siamensis LZ88	*Alternaria alternata*	Brown spot disease	Tobacco	Xie et al. (2021)

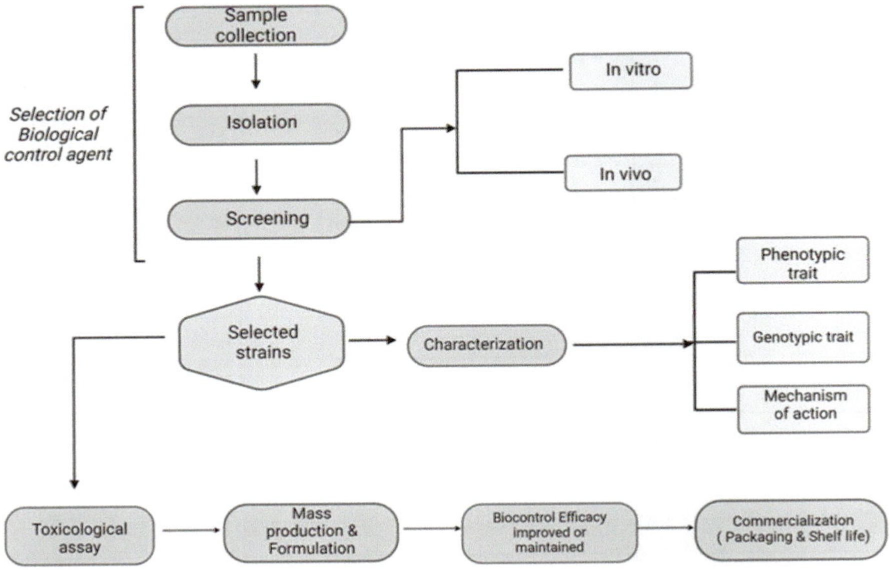

Figure 1: Major steps of biofungicide formulation.

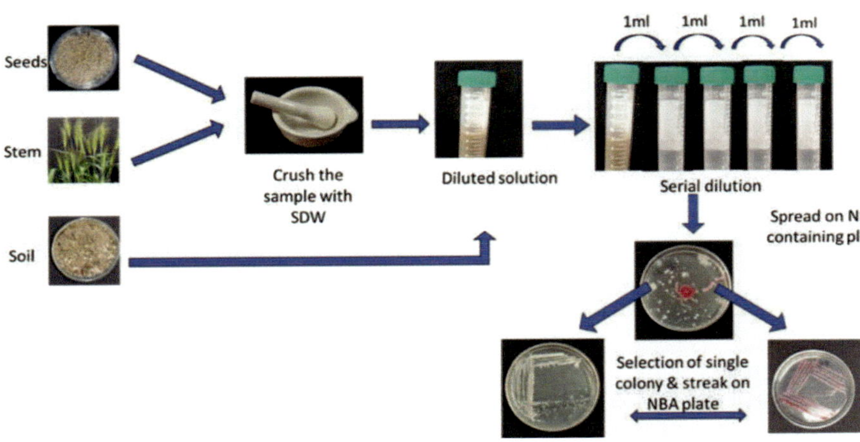

Figure 2: Major steps of isolation of plant probiotic bacteria.

suppress Take-all disease in wheat (Kwak and Weller 2013). Additionally, alternative environments beyond plant surroundings can also be sources of beneficial bacteria. Selective culture media assist in successfully isolating target microorganisms, though this method may restrict the diversity of isolated organisms (Bonaterra et al. 2022). Endophytic bacteria from seeds or stems should be isolated by washing the sample with sterile distilled water (SDW) and soaking it overnight to activate the bacteria. The sample should then be air-dried on sterile tissue paper and surface sterilized with 70% ethanol for 5 minutes, followed by soaking in NaOCl for 1 minute and rinsing with SDW. After air-drying again on sterile tissue paper, the sample is crushed in 10 ml of SDW. The solution should be vortexed to mix and filtered using tissue or a filter. Dilutions are prepared by adding 1 ml of the solution to a 15 ml falcon tube containing 9 ml of SDW, resulting in 5–6 dilutions. Each dilution should be plated on NBA plates with 100–200 μl applied. After 24 hours, single colonies should appear on the plates. A single colony is picked and streaked for culture. The identified bacteria should be tested against MoT. For isolating from soil samples, serial dilutions should be the first step (Fatema et al. 2024).

5. Screening of Potential Biocontrol Bacteria

After collecting isolates, the targeted biological control agent is selected based on specific traits, necessitating a screening process. Screening is a vital step in biofungicide formulation to isolate an effective biological control agent (Montesinos et al. 2003, Köhl et al. 2011, Raymaekers et al. 2020). Various *in vitro* assays are used to determine pathogen growth inhibition by testing the target pathogen along with biocontrol agents in liquid or solid media conditions. These assays are rapid, consistent, and reliable, facilitating the evaluation of numerous isolates. However, they primarily select for antagonistic bacteria, potentially overlooking microorganisms that act through mechanisms like competitive exclusion or inducing plant resistance (Köhl et al. 2019, Bonaterra et al. 2022, Pliego et al. 2011).

5.1 *In vitro Screening*

The dual culture methods serve as basic screening tools for BCAs due to their ability to produce rapid and consistent results. These assays measure inhibition zones and spore germination, making results easily accessible and quantifiable (Sabu et al. 2018, Anisha et al. 2018, Surovy et al. 2024). Antimicrobial compounds produced by microbes can be isolated and identified using modern techniques (Kashyap et al. 2023).

A fresh bacterial colony should be carefully streaked across PDA medium, starting at least 2 cm inside from the Petri dish edge. A 7-day-old culture of MoT produces one mycelial disc which is placed on the plate surface 9 cm away from the bacterial streak. The documented procedure by Lefèvre et al. (2020) showed that plates are incubated in temperature-controlled conditions at 25°C for 14 days. The experiment includes an additional control plate which only contains the MoT mycelial plug without bacterial addition. The researchers measure mycelial growth inhibition percentage after 14 days through the following method: MGI (%) = $[(C - T) / C] \times 100$. The calculation involves dividing the difference between control plate MoT radial growth (C in cm) and T in the bacterial presence to determine the MGI percentage. A light microscope examination of MoT hyphal structures close to the bacterial streak generates images through a digital camera system (Figure 3) (Surovy et al. 2024).

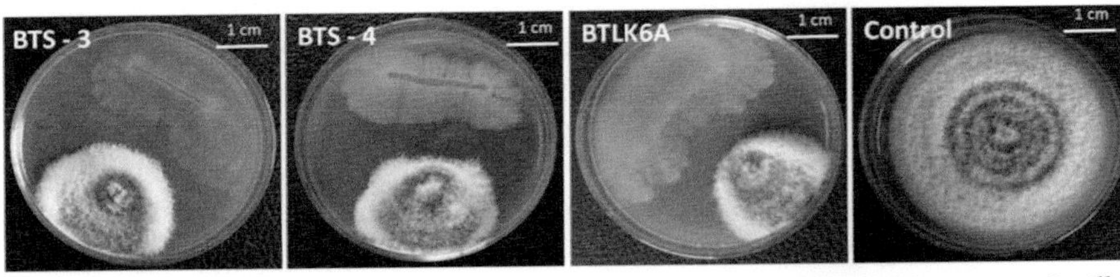

Figure 3: Growth suppression of *M. oryzae Triticum* by *Bacillus* spp. in dual culture assay. *Bacillus subtilis* BTS-3, *Bacillus velezensis* BTS-4 and *Bacillus velezensis* BTLK6A (Surovy et al. 2024).

5.2 *In vivo* Screening for Biocontrol Activity of Bacteria

The *in vivo* method stands as a dependable system for biocontrol screening. This analysis requires the inoculation of microbial antagonists alongside pathogens directly inside fruits or plant components. To determine the effectiveness of biocontrol agents, scientists examine the lesion diameters which help evaluate the severity of the disease. Researchers set specific time points to introduce pathogens and microbial antagonists during this procedure (Raymaekers et al. 2020, Raymaekers et al. 2020, Al-Rashdi et al. 2023).

Fifteen days after seedling emergence (DAE) perform two different experiments (preventive and curative) using a completely randomized design for all pots. Using sprayers, apply 100 milliliters of bacteria-containing solutions at 1×10^9 CFU/ml to wheat seedlings at day 15 after emergence. Allow the solution to dry for 24 hours. After bacterial inoculation, the plant received a conidial spray of MoT at (ca. 1×10^5 conidia/ml) which was followed by polythene bag coverage for 24 hours at 25°C under dark conditions. The curative control procedure starts by inoculating wheat seedlings with a solution containing MoT spores at 1×10^5 conidia/ml concentration followed by polythene bag coverage to maintain humidity levels above 90% for 24 hours at 25°C under dark conditions for fungal infection to occur. Spray the seedlings with a bacterial solution containing 1×10^9 CFU/ml. Researchers should move seedlings to a controlled growth chamber operating at 25°C with 12 hours of daily light exposure and maintaining humidity above 85%. Wheat seedlings maintained as both healthy controls and untreated controls received no applications of MoT or bacterial treatment. The observation time will be 4:00 pm during which seedlings will receive daily sprays of sterilized water to maintain high humidity levels throughout the day. The researchers measured disease progression at seven days post-inoculation according to Hettiarachchige et al. (2019), and Surovy et al. (2024).

6. Characterization of Selected Strain

Development of a biofungicide formulation from a biocontrol agent requires the in-depth study of some of the strains. Physical and genetic characteristics determination helps to assess if a microbial strain can be effective as a BCA. The biocontrol agent has some important aspects: production of antimicrobial extracellular compounds like enzymes, antibiotics, bacteriocins, etc., which fight harmful microorganisms or induce systemic resistance in plants (Compant et al. 2005). This allows figuring out the best way to use biocontrol in certain situations. Genome sequencing and comparison with previously sequenced genomes (Martinez-Garcia et al. 2015) is one way to identify traits. This allows putative gene candidates for controlling disease to be identified in comparison to existing genomes. This is also useful for biosciences to understand the biological control mechanism and to develop new strategies in disease management (De Vrieze et al. 2020, Xu et al. 2022). Thus, the 16S rRNA gene is used as a basis for identification of bacterial isolate procedures (Chanclud et al. 2017, Surovy et al. 2017, Dutta et al. 2018). Genomic DNA is extracted from the bacterial isolates, then whole genome sequencing is performed for genomic characterization. Then, bioinformatics tools are used to process the sequencing data and identify the isolates. Surovy et al. (2024) used *B. subtilis* as BTS-3, *B. velezensis* both BTS-4 and BTLK6A. The genomic study elucidated antimicrobial function gene clusters together with the information on infection-related signaling response clusters, as well as strain genetic and evolutionary elements of antimicrobial function gene clusters and cell wall degradation sequences. The information provided here is important for assessing the potential of these strains in the biological control of wheat blast disease (Surovy et al. 2024).

7. *In vitro* Bacterial Compatibility Test

The agar plate described by Tabacchioni et al. (2021) should be used in the *in vitro* compatibility test of the bacterial strains. For this, one hundred microliters of approximately 1×10^8 colony-forming units per milliliter (CFU/mL) of the test bacteria (biocontrol bacterium 1) is spread on the surface of fresh Nutrient Agar (NA) plates. Dip sterilized filter paper disks (5–mm diameter) into the medium suspension of the other test endophytic bacteria (bacterium 2) and then air-dry the disks for 30 min and gently press onto the bacterium 1 plated agar surfaces at four equidistant positions. Incubate at 26°C and observe at 24 h intervals over a period of 4 days (Figure 4).

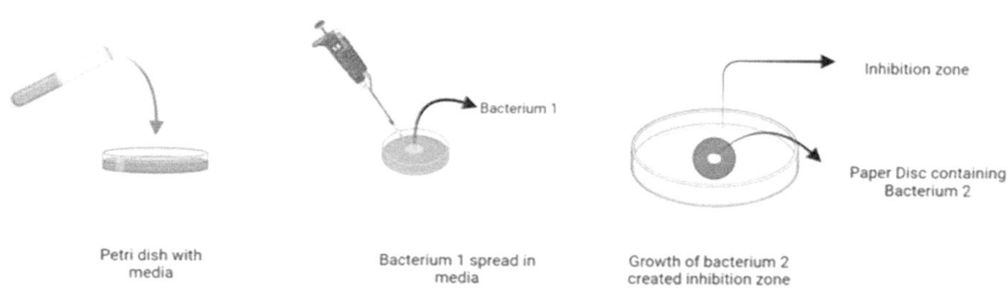

Figure 4: Diagrammatic presentation of disk diffusion method for bacterial compatibility test.

8. Greenhouse and Field Evaluation of Single and Consortium of Bacteria for Biocontrol of Blast Disease in Wheat Seedlings

Before a bacterial strain can be considered for microbial biopesticide development, pilot trials, including greenhouse and field bioassays, should be conducted across various pathosystems and environmental conditions. These trials ensure broad applicability and consistent efficacy under real-world conditions (Montesinos 2003). Optimizing the dose and frequency of applications is crucial, as the relative amounts of pathogen and biocontrol agent significantly affect efficacy and reliability. In greenhouse conditions at 25 ± 3°C, wheat seeds are surface-sterilized and treated with a bacterial suspension of approximately 1×10^9 CFU/mL. The seeds undergo a 24-hour treatment in bacterial suspensions containing either single strains or consortia followed by planting (Surovy et al. 2018, Gupta et al. 2021). At the 5–6 leaf stage, seedlings are inoculated with MoT fungus, and control plants are treated with sterilized distilled water as previously described (Surovy et al. 2018). One day before fungal inoculation, seedlings are sprayed with a single or consortium of bacteria (approximately 1×10^9 CFU/mL) and monitored for disease development and severity using a 0–5 scale (Gupta et al. 2021). Biocontrol of wheat blast by the application of *Bacillus* sp. in field conditions was also demonstrated by Islam et al. (2025, unpublished).

9. Optimizing Conditions for Large-Scale Cultivation to Develop Biopesticide Formulations

The final steps in the formulation of biofungicides include industrial-scale production, formulation, and preservation. To maximize cell yield in the shortest time, an efficient and cost-effective mass production system at an industrial scale must be meticulously designed. Additionally, the production method must preserve the biocontrol characteristics of the strains. The population density and viability of microbes during formulation are significantly influenced by culture conditions, which vary for different microbes and are highly specific. Therefore, careful screening is essential to enhance the performance of microorganisms in the field (Bonaterra et al. 2022). Then, microbial revival commences with microbial recovery under standard media, i.e., nutrient broth or LB medium, and then 1–10% of this inoculum is used to transfer onto a selective growth medium. The composition of this medium depends upon general growth, sporulation, or metabolite production (Shaikh and Sayyed 2014). Effective growth media with suitable carbon, nitrogen sources, inorganic salts, and growth factors must be used in the growth media (Carvalho et al. 2010, Abhyankar et al. 2011). *B. thuringiensis* can produce more spores if glucose at 20 g/L and ammonium sulfate at 2.5 g/L are provided, whereas *B. subtilis* requires only 2 g/L glucose (Elsayed et al. 2014, Kardziani et al.). Sporulation of *Bacillus* species normally occurs under environmental stress, especially during depletion of glucose under the stationary phase (Castillo et al. 2013, Reis et al. 2005). It was found that sporulation efficiency can be further influenced by adding vitamins, minerals, and specific ions such as calcium and ammonium (Monteiro et al. 2014, Leo Daniel et al. 2013, Monteiro et al. 2005, Brul et al. 2011). Solid substrates obtained from waste materials represent a good alternative for fermentation based on environmentally friendly production at low cost (Singhania et al. 2009). These growth substrates for *Bacillus* species have been agricultural waste, namely wheat bran, rice bran, or cotton stalks. In addition to spontaneous fermentations of agricultural waste products resulting in high-quality spores, the products produced are resistant to environmental stresses and have better adhesion to plant roots (Vassilev et al. 2015). *Bacillus* spp. should be grown in defined medium (Davis minimal medium) with appropriate carbon and nitrogen sources, incubated at 37°C for 72 hours to refine culture conditions for developing biofungicide on a large scale. The medium should also be supplemented with mineral salts, as is the case in commercial production of *Bacillus* spp. (Chung et al. 2007). Culture of *Bacillus* spp. requires optimal culture conditions that can be obtained via utilization of a fermenter. *Bacillus* spp. should be incubated in the large-scale culture medium at 37°C for three days to improve spore formation. Dipicolinic acid and $CaCl_2$ should be introduced during the spore induction phase to increase endospore production (Lim and Kim 2010). $MnSO_4$, $CaCl_2$, $ZnSO_4$, and KCl also help improve the sporulation process and increase spore yield (Omer et al. 2010). As the carbon source, glycerol should be used, combined with an equal mix of peptone and yeast extract for a nitrogen source, to enhance the growth and biocontrol efficacy of *Pseudomonas* strains (Pereira et al. 2003).

10. Development of Biofungicide Formulation

The formulation is a combination of active ingredients with functional and inert substances with the aim of improving the quality of a product (Malusá et al. 2012). Formulated products contain organisms that are suspended in a carrier solution along with additives to extend the shelf life of the organisms, facilitate ease of application, and guarantee life following application. Unlike chemical compounds, these formulated organisms are granular and often biological or protein-based, rendering them more susceptible to environmental factors and storage conditions (Bharti et al. 2020). A suitable formulation, whether in dry or liquid form, is crucial for prolonging shelf life, improving delivery, ensuring the effectiveness of biocontrol agents in agricultural fields, and maintaining their viability and biocontrol effectiveness (Segarra et al. 2015). An ideal formulation should boast an extended shelf life and avoid any harm to crop plants, ensuring it is not phytotoxic. It must demonstrate resilience against adverse environmental conditions while remaining cost-effective and consistently effective in controlling plant diseases. Good solubility in water is essential, alongside the use of inexpensive and easily accessible carriers for formulation development. Compatibility with other agrochemicals is also crucial for optimal performance (Kumar et al. 2014). Hence, protective additives such as sucrose, Arabic gum, glycerol, and compatible adjuvants like emulsifiers, surfactants, dispersants, stabilizing agents, and coupling agents are commonly used in formulations. These additives can be incorporated at different stages of the production–formulation process. Formulations could be solid or liquid and prepared in various ways. However, liquid formulations are easier to manufacture but have a shorter shelf life, and it becomes difficult to package and store them. On the other hand, solid formulations can prolong the stability of microorganisms (Teixidó et al. 2022).

10.1 Solid Formulation

Dried products may be economical for storage and transportation, although with loss of viability by thermal and dehydration stresses during the drying (Melin et al. 2007, Abadias et al. 2005). For this reason, each strain should be treated with a customized approach. The specific biological control agent (BCA) and its intended environmental application determine formulation requirements (Teixidó et al. 2022). To produce dry formulations which incorporate functional additives (e.g., binders, dispersants and wetting agents), spray drying, freeze drying or air-drying technologies are employed to improve product qualities (Knowles 2008). Examples of such species of Bacillus are well suited for solid formulations containing carriers, stabilizers, protectants and supplements. Low density polypropylene bags are packed with carrier materials, peat or talc, sealed and sterilized in solid formulations. The *Bacillus* cultures are then grown to a range of maximum cell density (E > 10^8 cells/mL) and mixed with sterilized carrier (CMC, sodium benzoate, $CaCO_3$) and nutrient, like glucose or sucrose, then incubated at room temperature (28–30°C) for one week before use (Tittabutr et al. 2007, Omer et al. 2010). Schematic of solid formulation development is presented in Figure 5.

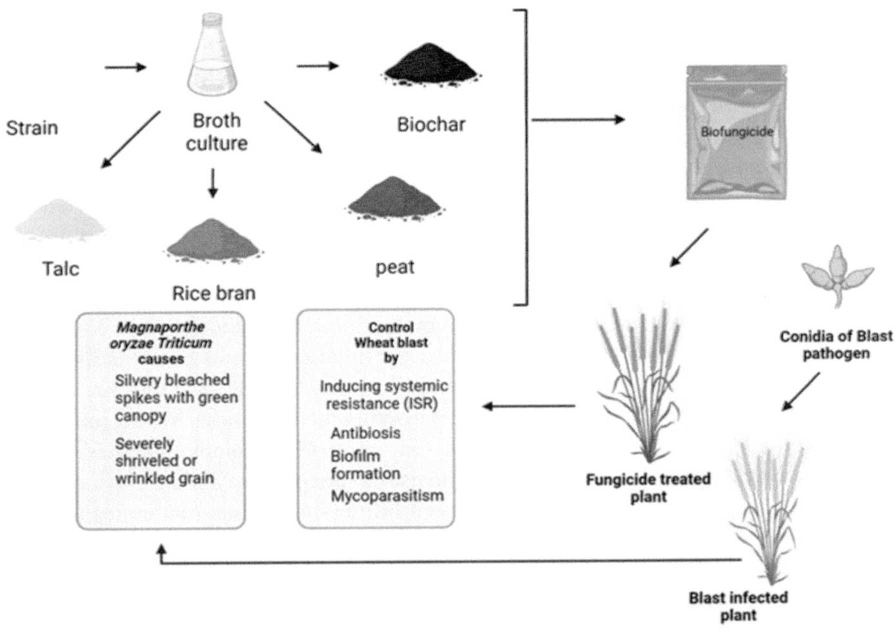

Figure 5: Flow diagram showing the developmental process of solid formulation of *Bacillus* sp. and effects on controlling wheat blast.

10.2 Liquid Formulation

Liquid culture involves whole cultures or cell suspensions combined with stabilizers, surfactants, colorants, or additional nutrients to maximize product viability (Teixidó et al. 2011) and enhance the adhesion, surfactant properties, and dispersion capabilities of biocontrol agents. These formulations are straightforward to produce and generally more cost-effective than solid alternatives. In liquid formulations, suspensions of particles are prepared in various liquids, often with suspending agents, dispersants, surfactants, and carrier liquids such as oil or water. Oil-based formulations involve mixing processed cultures with mineral or vegetable oils, along with emulsifiers and surfactants to aid dispersion in water. It is crucial to ensure that the oils used are non-toxic to microorganisms, plants, humans, and animals. Under dry conditions, oil-based foliar spray formulations are preferred because they offer better microorganism survival by providing greater protection against desiccation than water-based alternatives (Brar et al. 2006). For example, *Trichoderma harzianum* is formulated effectively against *Botrytis* through emulsification (Batta 2004, Teixidó et al. 2011). In developing liquid formulations, Davis minimal medium (DMM) serves as a base with selected concentrations of additives like Gum Arabic, PVP, chitosan, and sodium alginate (Postma et al. 2009). The viability of *Bacillus* spp. in liquid formulations should be evaluated using methods described by Lim and Kim (2010).

11. Toxicological Safety Assay of the Formulated Biofungicides

Ensuring the safety and environmental standards of biofungicides is crucial in agriculture, where regulatory preferences often favor these products over chemical fungicides due to their reduced environmental impact and perceived safety benefits. However, the high costs associated with rigorous toxicological studies for registration present a significant barrier for manufacturers and developers (Kumar et al. 2014). To assess the safety of the formulated biofungicide, a brine shrimp lethality test should be carried out using the procedures described by Solis et al. (1992). The biofungicide toxicity is evaluated by a brine shrimp mortality test using varied concentrations, saline solution and methanol as negative control, and Nativo WG75 as the positive benchmark.

12. Application System of Biocontrol Agent

The product delivery system should be efficient, cost-effective, and timely to deliver the product to a target site effectively and be integrated with existing agricultural practices and equipment (Bharti and Ibrahim 2020). Formulated microbes can be applied to various targets through techniques such as seed priming; soil application using drenches, mixes, and microirrigation; or aerial application via foliar spray (Figure 6) or through endotherapy (Jambhulkar et al. 2016, Bonaterra et al. 2022).

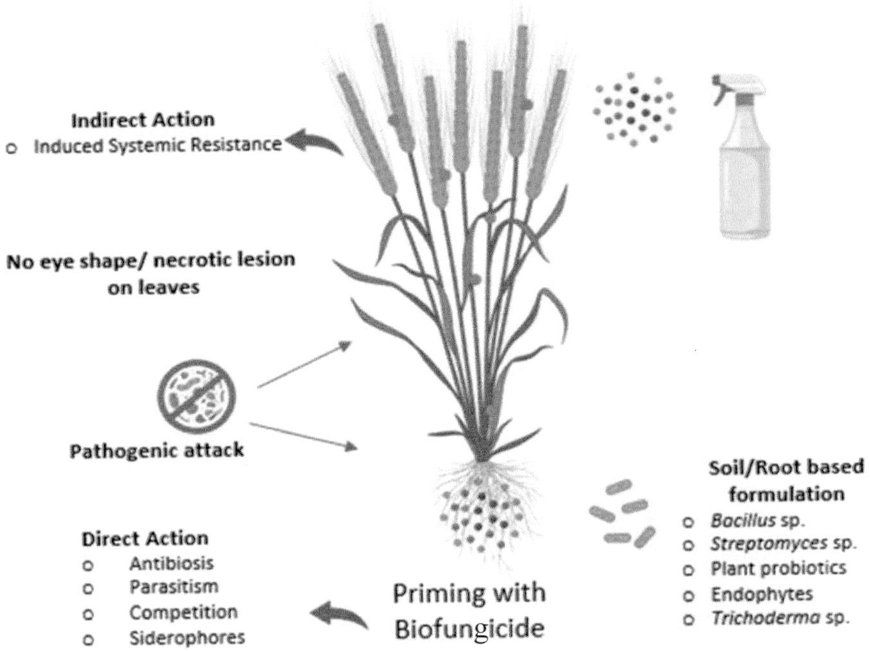

Figure 6: Different Application Methods of Biocontrol Agents.

12.1 Bio-priming

Bio-priming is the seed treatment method where seeds are inoculated with beneficial microorganisms before planting (Waqas et al. 2019). This technique is very useful in preventing early plant development as it induces plant resistance (Amruta et al. 2019, Rajaput et al. 2019). Seed bio-priming also has multiple agricultural benefits including faster germination, better plant growth attributes, improved metabolic functioning, and higher tissue nutrients in the crops (Singh et al. 2016). These bacterial and fungal bio-primers use two main mechanisms, namely activating defense pathways (Ferrigo et al. 2020, Yadav et al. 2021) or secreting hydrolytic enzymes (Durairaj et al. 2018, Myo et al. 2019), making them versatile plant protection agents. Microbial agents for seed bio-priming against fungal pathogens are commonly used with *Bacillus* sp., *Pseudomonas* sp., *Serratia* sp. (Tumpa et al. 2016, Rajaput et al. 2019, Amruta et al. 2019, Sufyan et al. 2020). The fungal biocontrol agents known to act as effective seed bio-priming organisms include species of *Trichoderma, Rhizobium meliloti,* and *Paenibacillus dendritiformis* (Rafi et al. 2016, Mona et al. 2017, Yadav et al. 2021).

Studies where various types of seed priming have been used have observed significant increases in plant defense mechanisms. Bio-priming of *Capsicum annuum* with *Bacillus* sp. BSp.3/aM enhanced the levels of defense enzymes (LOX, PPO, PAL, POX) and resulted in higher phenolic compounds, thereby conferring better pathogen resistance (Jayapala et al. 2019). The synergistic effects of a combination of *Ochrobactrum* sp. BHU PB-1, *Trichoderma asperellum* BHU P–1, and ascorbic acid on PPO, POX, and PAL enzyme activities showed marked increases (Singh et al. 2020). The application of *T. erinaceum* in *Solanum lycopersicum* resulted in increased CAT and SOD levels under *Fusarium oxysporum* infection (Aamir et al. 2019). Tomato seeds treated with *T. pseudokoningii* BHUR2 displayed increased defense against *Sclerotium rolfsii* by accumulation of phenolic compounds and increases in activities of defense-related enzymes such as PAL, POX, and SOD (Rajput et al. 2019). Antifungal plant-derived extracts are widely used as seed priming agents against different phytopathogenic fungi (Abdelkhalek et al. 2020). The methanolic extracts of *Ammi visnaga* have been found to increase resistance against *Rhizoctonia solani* in maize through induction of pathogen related gene expression, especially the presence of thaumatin like proteins and chitinases (Rashad et al. 2018). Extracts from *Balanites aegyptiaca* and *Eclipta alba* increased yield and antifungal activity against *Curvularia lunata* in sorghum (Zida et al. 2018). Treatments with *Toddalia asiatica* extract improved vigor and increased germination of maize seeds without altering the time or intensity of infection by *Aspergillus flavus* and *Fusarium verticillioides* (Aiyaz et al. 2015). Proteolytic enzymes present in papaya extract such as chymopapain and papain can be used to combat fungal pathogens including *Macrophomina phaseolina*, *Fusarium* species, and *Rhizoctonia solani* (Rafi et al. 2015). Aegle marmelos leaf extract treatment improves the growth parameters of tomato seeds when challenged with *Fusarium oxysporum* (Prabha et al. 2016). The root rot fungi control has been accomplished through extracts of *Sapindus mukorossi* and *Acacia nilotica* (Rafi et al. 2015).

12.2 Foliar Application

A traditional biofungicide delivery technique is foliar spraying in which microbial preparations or plant-based extracts are applied directly to above-ground plant structures and effectively control fungal diseases (Pertot et al. 2017, Jiang et al. 2018). Recently, many endophytic bacteria, including *B. velezensis* (BTLK6A, BTS-4) and *B. subtilis* (BTS-3), applied as foliar sprays, have been shown to control wheat blast through a variety of defense mechanisms, including cell wall degradation, antioxidant defenses, antimicrobial activity, and induced systemic resistance (Surovy et al. 2024). The fundamental reason foliar applications are effective results from their ability to boost host defensive capabilities through one or more mechanisms. These treatments induce transcriptional changes in host plants and increase the production of defense-related enzymes as well (Jiang et al. 2018). This also increases the expression of pathogenesis-related proteins (Xie et al. 2021) and the synthesis of protective compounds (Figure 6). Antifungal metabolites, including phenazine-1-carboxamide derivatives (Chen et al. 2018a, b), lipopeptides (Ji et al. 2013), volatile organic compounds and secondary metabolites (Hassan et al. 2021), and compounds such as cladosporin 8-methyl ether, 5′-hydroxyasperentin, and isocladosporin (Wang et al. 2013), are among these protective compounds.

12.3 Soil Application

Biofungicides have been generally recognized as a sustainable soil-based application against plant pathogens that not only increase crop health and productivity (Figure 6) but also prevent the emergence of disease resistance in host plants. This method not only controls soil borne disease but also increases plant vigor and productivity potential (Narasimhan and Shivakumar 2015). Soil incorporation of a variety of biological agents has been tested in diverse plants across many plant–pathogen systems. *Trichoderma* species have been widely selected for their biocontrol performance that is reliable and hence have been pivotal in commercial biofungicide development (Srivastava et al. 2016). Now, such fungal biocontrol agents are widely used in various commercial products across the world (Woo et al. 2014). This is particularly true for talc-based formulations of *Trichoderma viride* which showed exceptional ability in reducing plant mortality as well as enhancing the yields in crops like cotton, sunflower, and okra under *Rhizoctonia solani* infested condition (Mathivanan et al. 2000). Further example of using soil as a medium to deliver antagonists comes from *Lysobacter enzymogenes* strain LE16 when autolysates were used in soil and resulted in significant reduction of *Phytophthora capsici* blight in pepper plants (Chen et al. 2020), similar to what was shown above. As a result, application of biofungicides, i.e., *Trichoderma* and *Lysobacter*, to the soil is a promising and environmentally sound alternative to chemical pesticides. It offers dual functionality by promoting both disease suppression and growth of crops.

13. Challenges Associated with Commercialization of Biofungicides

Availability of microbial products or formulations allowing the passage of technology from the lab culture to agricultural fields represents a condition for effective use of biofungicides in disease control. Commercialization of biofungicides brings its own challenges in terms of compliance with regulatory standards, as well as in various stages of development. Selection of an effective strain is the most important factor that determines the commercial success of biofungicide formulation. A challenge is that there is a lack of proper screening protocols to pick the promising candidate. One of the major challenges in the commercialization of biofungicides is improvement of solid or liquid formulations using an appropriate medium and identifying suitable carrier materials with extended shelf life (Kumar et al. 2014). Nutrient composition of the growth medium affects productivity of biomass and the capacity of biocontrol agents to inhibit plant pathogens (Schisler et al. 1997). Therefore, the selected medium must encourage growth and efficacy of the biocontrol agent and be cost-effective in terms of maintaining the viability of the technology (Kumar et al. 2014). The major challenge in scaling up production while maintaining biocontrol strain characteristics includes meticulous process optimization and strictly rigorous procedures for quality control to ensure efficacy (Bejarano and Puopolo 2020). The control of plant diseases stems from an intricate interaction between the biocontrol agent, plant, pathogen, and surrounding environment. Developing a biofungicide is restricted due to lack of knowledge in microbial ecology (Kumar et al. 2014).

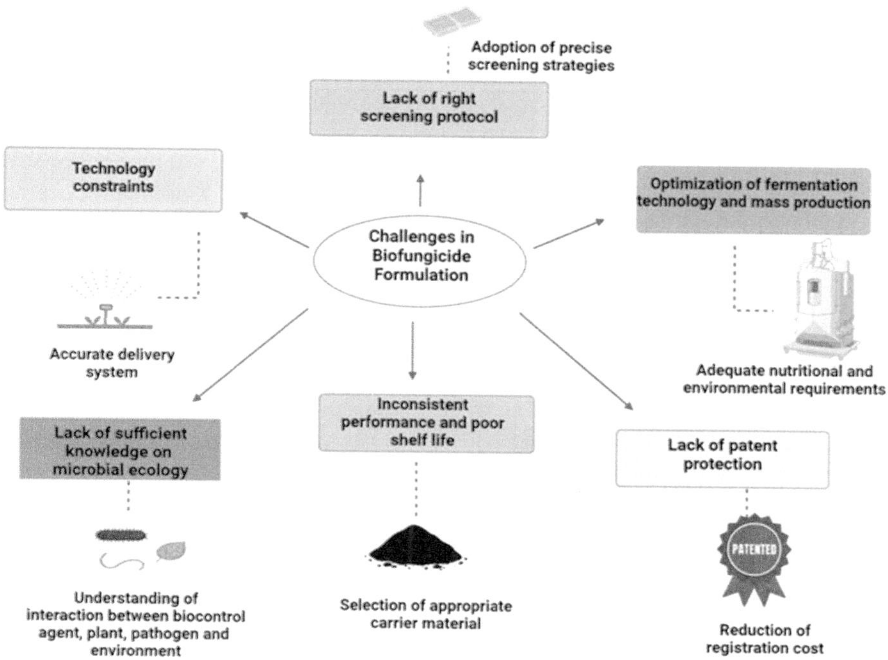

Figure 7: Challenges associated with Biofungicide Formulation.

Formulation development also includes substituting shelf life, delivery mechanisms, and efficacy with compatible biocontrol agents (Segarra et al. 2015). In addition, long-term viability and stability during storage and transport require innovative preservation techniques and packaging (Bejarano and Puopolo 2020). Potential negative effects on plants, non-target organisms, and humans, as well as risk assessments on traceability and environmental impact (Montesinos, 2003), are to be thoroughly evaluated by means of biosafety evaluations. Additionally, the rigorous registration required for market access is very time-consuming due to the extensive scientific data requested in detailed dossiers (Bonaterra et al. 2012, Montesinos 2003). Collectively, these challenges emphasize the complexity of successfully marketing biofungicides and the need to consider each challenge systematically to support commercialization (Figure 7).

14. Pros and Cons of Biofungicides

14.1 Benefits

Biofungicides have gained interest for their dual function, on the one hand in the field of plant protection and on the other in crop enhancement. Furthermore, in addition to managing plant diseases, these biological agents also make a substantial contribution towards improving seed germination and associated early planting development, yielding increased agricultural productivity in various crop species (Rajput and

Harlapur 2015). Seed priming is a reliable way to increase seedling vigor and promote healthy plant growth (Rafi et al. 2015). Improvements in germination have been reported in species including *Vigna radiata*, *Oryza sativa*, and *Cicer arietinum* after the application of biofungicides (Sarkar et al. 2018, Naveenkumar et al. 2017, Sufyan et al. 2020). Furthermore, several field experiments have validated that seeds of rice, soybean, and maize showed better results when treated with microbes such as *Bacillus* and *Pseudomonas fluorescens* (Amruta et al. 2019, Rajput and Harlapur 2015). Use of *Trichoderma* spp. has improved both yield and grain quality in maize and rice (Pal et al. 2015, Venkateswarlu and Beura 2020). Likewise, rice yield in *Talaromyces tratensis* and liquid biofungicide for tomato in brinjal-based systems increased tomato production (Upamanya et al. 2020, Eakjamnong et al. 2021). Some biofungicides have a stimulatory effect on growth via the production of plant hormones like IAA, gibberellins, and siderophores (Nandhini et al. 2018) and P and K depletion (Upamanya et al. 2020). In fact, application of bio-priming through pre-transmitted bacteria or antagonistic bacterial treatments of seeds has been found to improve seedling growth, biomass, and root–shoot development in various crops like rice, soybean, and peas (Amruta et al. 2019, Singh et al. 2016). Intercropping or addition of organic amendments like oil cakes enhances growth and resistance as they act as biofungicides, further promoting their use in sustainable agriculture (Falade et al. 2018, Rafi et al. 2016).

14.2 Limitations

The implementation of microbial-based fungal control agents as environment-friendly plant disease management solutions faces multiple constraints that limit their widespread use in agricultural systems. One major drawback of biological products originates from their short operational lifespan that extends up to three months only (Sabaratnam and Traquair 2002). The microorganisms living within biofungicides require specific nutritional conditions for survival, thus resulting in their limited storage period. Technological advancements for improving fermentation and stabilization methods attempt to prolong product shelf life (Kolombet et al. 2008). Presently, biofungicides face inferiority against synthetic chemical counterparts because synthetic fungicides retain effectiveness despite their extended shelf-life stability. Studies have revealed two main critical consequences of biofungicidal usage on target crops: seed germination may be impeded and plant seedling development can be impaired (Meepagala et al. 2019). Specific biofungicide formulations show toxic reactions to beneficial organisms according to findings from brine shrimp model studies documented by Li et al. (2014). Studies have established that essential oil derivatives from *Thymus* species, which primarily contain thymol, show inhibitory effects on unwanted plant species and lethal activities against storage pests (Ghasemi et al. 2020). The sophisticated nature of reliable biological fungicide development is illustrated by multiple factors, including formulation difficulties and environmental impact consequences, while additional stability issues for antifungal ingredients complicate this challenging task. Research must keep advancing to enable effective biological fungicide implementation in commercial farming.

15. Improvement of Biocontrol Agents

The use of bacteria to control plant diseases has been limited by variability in the effectiveness of the bacteria alone. For example, genetic traits, pathogenic traits, inoculum density, and environmental conditions greatly affect how one biocontrol agent can fight disease. BCAs can be disrupted by a variety of biotic and abiotic factors that compromise the mechanisms through which they operate or disrupt the intricate interrelationships between plants, pathogens, and bacteria, and consequently reduce their efficacy. Improvement in nutrition, adaptation of bacteria to stress, formulation refinement, and genetic manipulation are strategies that can be adopted to increase biocontrol agent performance (Table 2). Additionally, the selection of proper delivery methods suitable for the specific features of the BCA and the pathogen should be carefully employed for biocontrol programs (Bonaterra et al. 2022).

One strategy to supplement the nutritional quality of the performance profile of biological control agents (BCA) is to supply nutrients in formulations that BCAs use more efficiently compared to pathogens. For instance, the introduction of glycine and Tween 80, which also comprised *Pseudomonas fluorescens* EPS 62e, enhanced the survival and adaptability of this bacterium in plant environments (Cabrefiga et al. 2011). Another approach is to use 2-deoxy-D-glucose, a glucose analog, to increase biocontrol efficacy against blue mold on apples and pears (Janisiewicz 1994). In addition, consortia of compatible biocontrol agents are another effective approach to utilize mechanisms from different biocontrol agents and to harmonize ecological traits. It can provide an increase in efficacy and reliability across a wide range of environmental conditions and broaden the spectrum of activity through synergistic effects. An example of biocontrol improvement is achieved when *Pseudomonas*

Table 2: Different strategies to enhance the efficacy of bacterial biocontrol agents.

Biocontrol agent	Strategy for enhancement	Effect on biocontrol agent	References
Trichoderma reesei	Inter-specific protoplast fusion	Enhanced cellulose production	(Daranas et al. 2018)
Pseudomonas fluorescens EPS62e	Integration of salt-induced osmotic stress and osmolyte supplements	N-acetylglutaminylglutamine amide and enhancement of cell viability on plant	(Roselló et al. 2017)
P. fluorescens EPS817 and EPS894	Improvement of nutrition along with osmotic adaptation	Enhanced effectiveness on plant surface	(Bonaterra et al. 2022)
P. *fluorescens* EPS817 and EPS894	Generation of various bioactive metabolites by a bacterial consortium	Improved efficacy and reliability to control Phythophthora root biocontrol	(Cabrefiga et al. 2014)
Lactobacillus plantarum PM411	Integration of hyperosmotic and acidic stress adaptation	Enhanced viability on plant surfaces and upregulation of stress-related genes	(Cabrefiga et al. 2011)
L. plantarum TC92 and PM411	Integration of lactic acid in consortium of bacteria	Improved efficacy and reliability to control fire blight	(Agusti et al. 2011)

fluorescens and *Pantoea* sp. assist in their suppression of fire blight of pear (Stockwell et al. 2010). Similarly, biocontrol agents have been amended with low-toxicity antimicrobial compounds like bioregulators and essential oils, with positive effects on biological control. When lactic acid was combined with *L. plantarum* strains PM411 and TC92, the biocontrol efficacy was enhanced (Roselló et al. 2017). The same is true for essential oils together with *Bacillus amyloliquefaciens* or *L. plantarum* strains when biological control is improved (Arrebola et al. 2010, Zamani-Zadeh et al. 2014). However, genetic changes can improve biocontrol efficacy, but regulatory limitations placed on genetically modified organisms (GMOs) should be regarded. Isolating and identifying disease management property genes for enhancing biocontrol agents is very important for effective plant disease control. Alteration of native microbial strains by molecular techniques gives them disease-fighting ability. A genetic modification and recombination approach can be used to increase or modify specific biocontrol agents (Raguchander et al. 2011). Overproduction of antimicrobial polyketides in *P. fluorescens* CHA0 and increased mycosubtilin production in *Bacillus subtilis* ATCC 6633 (Girlanda et al. 2001, Fickers et al. 2009) are examples. Recent studies showed how to safely commercially control disease with beneficial microbes, but the lack of availability and efficacy of these microbes is limiting their use. Future directions should involve obtaining new BCAs, developing rapid screening approaches, and carrying out comprehensive studies to establish more efficacious strategies. Using a multi-omics approach, environmental concerns and the improvement of biocontrol efficacy can also be addressed in terms of bacterial effects on the native plant microbiome (Bonaterra et al. 2022).

16. Future Perspectives

Microbial biofungicides have gained increasing attention as a promising alternative to chemical fungicides, particularly due to growing concerns over the environmental and health risks posed by chemical pesticides. However, the production of biofungicides on an industrial scale has become crucial given the rising global food demand and environmental threats. Numerous studies support the potential of biofungicides as substitutes for chemical fungicides, and with the wide range of available microbial sources, screening for cost-effective and efficient biofungicides is becoming more feasible. Beyond disease control, microbial crop protection agents have shown capacities to enhance plant growth, making them promising agricultural inputs for sustainable farming systems of the future. Despite their promise, significant challenges remain. Thorough ecosystem impact studies are therefore necessary before the use of microbial fungicides can become widespread, because they necessitate adequate ecotoxicological assessment of microbial fungicides against non-target organisms to prevent ecological harm (Sánchez-Bayo and Tennekes 2017). At the same time, sustaining pathogen suppression efficacy of these biological formulations in commercial agricultural use (Köhl et al. 2019, Wilkes et al. 2019) requires the extension of such biological functional longevity. Although many studies have demonstrated the positive effects of biofungicides under laboratory conditions, relatively few have explored their performance in real-world agricultural settings.

It is essential that future research includes large-scale field trials to assess their true potential. Biofungicides tend to be highly specific in their action against certain fungal pathogens, which limits their broader application. To address this, some research has suggested using combinations of biocontrol agents to broaden the spectrum of efficacy. Developing universal biofungicides that are effective against a wider range of fungal pathogens could significantly enhance their utility. Another concern is the toxicity of biofungicides; studies have shown that higher concentrations can negatively affect plant growth and development. Therefore, appropriate testing and optimization of biofungicide formulations and dosages must be conducted before they are commercially released. Moreover, the molecular mechanisms behind the effectiveness of biofungicides are still poorly understood. Future research should focus on the molecular pathways involved in biofungicide-induced plant responses, using tools such as transcriptomics and proteomics to uncover the mechanisms of fungal disease control. Advances in genomics, including the availability of pathogenic fungal genome sequences, can also be leveraged to develop targeted biofungicides that exploit vulnerabilities in fungal virulence factors. By addressing these challenges, biofungicides could play a more significant role in sustainable agriculture, offering a safer and more eco-friendly solution to disease management.

17. Conclusion

Wheat blast, caused by *Magnaporthe oryzae Triticum* (MoT), poses a severe threat to global food security, necessitating urgent, sustainable alternatives to chemical fungicides. This chapter underscores the potential of biofungicides derived from plant probiotic bacteria, such as *B. subtilis*, *B. velezensis*, and *Pseudomonas* spp., which demonstrate robust antifungal activity through mechanisms like antibiosis, induced systemic resistance, and biofilm formation. *In vitro* and *in vivo* trials highlight their efficacy in suppressing MoT, with formulations developed via systematic protocols—ranging from microbial isolation and screening to optimized mass production and delivery systems (e.g., seed bio-priming, foliar sprays). Despite promising results, challenges persist in scaling production, ensuring shelf-life stability, and navigating heterogeneous regulatory frameworks. Future efforts must prioritize advanced formulation technologies (e.g., nanoparticle carriers, stress-adapted consortia), large-scale field validation, and multi-omics approaches to unravel microbial-plant-pathogen interactions. Harmonizing global regulations and fostering interdisciplinary collaboration among researchers, policymakers, and farmers will be critical to translating lab success into field-ready solutions. By bridging ecological safety with agronomic efficacy, biofungicides offer a transformative pathway to mitigate wheat blast, reduce reliance on synthetic chemicals, and safeguard food systems in an era of climate uncertainty. This chapter provides a roadmap for advancing these innovations, emphasizing their role in sustainable agriculture and resilience against evolving fungal threats.

Acknowledgement

The authors are thankful to the funding from the Bangladesh Academy of Sciences under the BAS-USDA BSMRAU MPA-15 grant to TI.

References

Aamir, M., Kashyap, S. P., Zehra, A., Dubey, M. K., Singh, V. K., Ansari, W. A. et al. (2019). *Trichoderma erinaceum* bio-priming modulates the WRKYs defense programming in tomato against the *Fusarium oxysporum* f. sp. *lycopersici* (Fol) challenged condition. *Frontiers in Plant Science*, 10: 911. https://doi.org/10.3389/fpls.2019.00911

Abadias, M., Teixidó, N., Usall, J., Solsona, C., and Viñas, I. (2005). Survival of the postharvest biocontrol yeast Candida sake CPA-1 after dehydration by spray-drying. *Biocontrol Science and Technology*, 15(8): 835–846. https://doi.org/10.1080/09583150500187041

Abbassy, M. A., Marei, G. I. K., and Rabia, S. M. (2014a). Antimicrobial activity of some plant and algal extracts. *International Journal of Plant & Soil Science*, 3(10): 1366–1373. https://doi.org/10.9734/IJPSS/2014/8440

Abbassy, M. A., Marzouk, M. A., Rabea, E. I., and Abd-Elnabi, A. D. (2014b). Insecticidal and fungicidal activity of *Ulva lactuca* Linnaeus (Chlorophyta) extracts and their fractions. *Annual Research & Review in Biology*, 4(13): 2252–2262. https://doi.org/10.9734/ARRB/2014/9511

Abdelkhalek, A., Salem, M. Z., Kordy, A. M., Salem, A. Z., and Behiry, S. I. (2020). Antiviral, antifungal, and insecticidal activities of Eucalyptus bark extract: HPLC analysis of polyphenolic compounds. *Microbial Pathogenesis*, 147: 104383. https://doi.org/10.1016/j.micpath.2020.104383

Abhyankar, W., Beek, A. T., Dekker, H., Kort, R., Brul, S., and de Koster, C. G. (2011). Gel-free proteomic identification of the *Bacillus subtilis* insoluble spore coat protein fraction. *Proteomics*, 11(23): 4541–4550. https://doi.org/10.1002/pmic.201100003

Agusti, L., Bonaterra, A., Moragrega, C., Camps, J., and Montesinos, E. (2011). Biocontrol of root rot of strawberry caused by *Phytophthora cactorum* with a combination of two *Pseudomonas fluorescens* strains. *Journal of Plant Pathology*, 363–372. https://www.jstor.org/stable/i40093868

Aiyaz, M., Divakara, S. T., Chandranayaka, S., and Niranjana, S. R. (2015). Efficacy of seed hydropriming with phytoextracts on plant growth promotion and antifungal activity in maize. *International Journal of Pest Management*, 61(2): 153–160. https://doi.org/10.1080/09670874.2015.1025116

Ali, M., Hayat, S., Ahmad, H., Ghani, M. I., Amin, B., Atif, M. J. et al. (2019). Priming of *Solanum melongena* L. seeds enhances germination, alters antioxidant enzymes, modulates ROS, and improves early seedling growth: indicating aqueous garlic extract as seed-priming bio-stimulant for eggplant production. *Applied Sciences*, 9(11): 2203. https://doi.org/10.3390/app9112203

Al-Rashdi, A., Al-Hinai, F. S., Al-Harrasi, M. M. A., Al-Sabahi, J. N., Al-Badi, R. S., Al-Mahmooli, I. H. et al. (2023). The potential of endophytic bacteria from Prosopis cineraria for the control of *Pythium aphanidermatum*-induced damping-off in cucumber under saline water irrigation. *Journal of Plant Pathology*, 105(1): 39–56. https://doi.org/10.1007/s42161-022-01237-5

Ambika, S., and Sujatha, K. (2014). Comparative studies on brown, red and green alga seaweed extracts for their antifungal activity against *Fusarium oxysporum* f. sp. udum in Pigeon pea var. CO (Rg) 7 (*Cajanus cajan* (L.) Mills.). *Journal of Biopesticides*, 7(2): 167

Amruta, N., Kumar, M. K., Narayanaswamy, S., Gowda, M., Channakeshava, B. C., Vishwanath, K. et al. (2016). Isolation and identification of rice blast disease-suppressing antagonistic bacterial strains from the rhizosphere of rice. *Journal of Pure & Applied Microbiology*, 10(2). https://doi.org/10.1016/j.cropro.2009.08.007

Amruta, N., Prasanna Kumar, M. K., Kandikattu, H. K., Sarika, G., Puneeth, M. E., Ranjitha, H. P. et al. (2019). Bio-priming of rice seeds with novel bacterial strains, for management of seedborne *Magnaporthe oryzae* L. *Plant Physiology Reports*, 24: 507–520. https://doi.org/10.1007/s40502-019-00492-6

Anisha, C., Jishma, P., Bilzamol, V. S., and Radhakrishnan, E. K. (2018). Effect of ginger endophyte Rhizopycnis vagum on rhizome bud formation and protection from phytopathogens. *Biocatalysis and Agricultural Biotechnology*, 14: 116–119. https://doi.org/10.1016/j.bcab.2018.02.015

Anitha, M. U., and Jahagirdar, S. H. A. M. A. R. A. O. (2015). Influence of seed priming agents on yield, yield parameters and purple seed stain disease in soybean. *Karnataka J. Agric. Sci.*, 28(1): 20–23.

Arrebola, E., Sivakumar, D., Bacigalupo, R., and Korsten, L. (2010). Combined application of antagonist *Bacillus amyloliquefaciens* and essential oils for the control of peach postharvest diseases. *Crop protection*, 29(4): 369–377. https://doi.org/10.1016/j.cropro.2009.08.001

Ayilara, M. S., Adeleke, B. S., Akinola, S. A., Fayose, C. A., Adeyemi, U. T., Gbadegesin, L. A. et al. (2023). Biopesticides as a promising alternative to synthetic pesticides: A case for microbial pesticides, phytopesticides, and nanobiopesticides. *Frontiers in Microbiology*, 14: 1040901. https://doi.org/10.3389/fmicb.2023.1040901

Basile, A., Rigano, D., Loppi, S., Di Santi, A., Nebbioso, A., Sorbo, S. et al. (2015). Antiproliferative, antibacterial and antifungal activity of the lichen *Xanthoria parietina* and its secondary metabolite parietin. *International journal of molecular sciences*, 16(4): 7861–7875. https://doi.org/10.3390/ijms16047861

Batta, Y. A. (2004). Postharvest biological control of apple gray mold by *Trichoderma harzianum* Rifai formulated in an invert emulsion. *Crop Protection*, 23(1): 19–26. https://doi.org/10.1016/S0261-2194(03)00163-7

Bavaresco, L. G., Osco, L. P., Araujo, A. S. F., Mendes, L. W., Bonifacio, A., and Araujo, F. F. (2020). *Bacillus subtilis* can modulate the growth and root architecture in soybean through volatile organic compounds. *Theoretical and Experimental Plant Physiology*, 32(2): 99–108. https://doi.org/10.1007/s40626-020-00173-y

Bejarano, A., and Puopolo, G. (2020). Bioformulation of microbial biocontrol agents for a sustainable agriculture. *How Research Can Stimulate the Development of Commercial Biological Control Against Plant Diseases*, 275–293. https://doi.org/10.1007/978-3-030-53238-3_16

Bharti, V., and Ibrahim, S. (2020). Biopesticides: Production, formulation and application systems. *International Journal of Current Microbiology and Applied Science*, 9(10): 3931–3946. https://doi.org/10.20546/ijcmas.2020.910.453

Bhattacharjee, Pallab, Julfikar Ali, and Tofazzal Islam. (2025). Management of Wheat Blast Disease in Bangladesh Caused by *Magnaporthe oryzae Pathotype Triticum. Plant Health Cases* 2025: phcs20250025. https://doi.org/10.1079/planthealthcases.2025.0025

Bonaterra, A., Badosa, E., Cabrefiga, J., Francés, J., and Montesinos, E. (2012). Prospects and limitations of microbial pesticides for control of bacterial and fungal pomefruit tree diseases. *Trees*, 26: 215–226. https://doi.org/10.1007/s00468-011-0626-y

Bonaterra, A., Badosa, E., Daranas, N., Francés, J., Roselló, G., and Montesinos, E. (2022). Bacteria as biological control agents of plant diseases. *Microorganisms*, 10(9): 1759. https://doi.org/10.3390/microorganisms10091759

Bonaterra, A., Cabrefiga, J., Camps, J., and Montesinos, E. (2007). Increasing survival and efficacy of a bacterial biocontrol agent of fire blight of rosaceous plants by means of osmoadaptation. *FEMS Microbiology Ecology*, 61(1): 185–195. https://doi.org/10.1111/j.1574-6941.2007.00313.x

Bonaterra, A., Camps, J., and Montesinos, E. (2005). Osmotically induced trehalose and glycine betaine accumulation improves tolerance to desiccation, survival and efficacy of the postharvest biocontrol agent *Pantoea agglomerans* EPS125. *FEMS Microbiology Letters*, 250(1): 1–8. https://doi.org/10.1016/j.femsle.2005.06.028

Borriss, R. (2020). Phytostimulation and biocontrol by the plant-associated *Bacillus amyloliquefaciens* FZB42: an update. *Phyto-microbiome in stress regulation*, 1-20. https://link.springer.com/book/10.1007/978-981-15-2576-6

Boyetchko, S. M. (1996). Impact of soil microorganisms on weed biology and ecology. In *Phytoprotection* (Vol. 77, No. 1, pp. 41–56). Érudit. https://doi.org/10.7202/706099ar

Brar, S. K., Verma, M., Tyagi, R. D., and Valéro, J. R. (2006). Recent advances in downstream processing and formulations of *Bacillus thuringiensis* based biopesticides. *Process Biochemistry*, 41(2): 323–342. https://doi.org/10.1016/j.procbio.2005.07.015

Brown, J. S., and Holden, D. W. (1998). Insertional mutagenesis of pathogenic fungi. *Current Opinion in Microbiology*, 1(4): 390–394. https://doi.org/10.1016/S1369-5274(98)80054-4

Brul, S., van Beilen, J., Caspers, M., O'Brien, A., de Koster, C., Oomes, S. et al. (2011). Challenges and advances in systems biology analysis of Bacillus spore physiology; molecular differences between an extreme heat resistant spore forming *Bacillus subtilis* food isolate and a laboratory strain. *Food Microbiology*, 28(2): 221–227. https://doi.org/10.1016/j.fm.2010.06.011

Cabot, C., Bosch, R., Martos, S., Poschenrieder, C., and Perelló, A. (2018). Salinity is a prevailing factor for amelioration of wheat blast by biocontrol agents. *Biological control*, 125: 81–89. https://doi.org/10.1016/j.biocontrol.2018.07.003

Cabrefiga, J., Francés, J., Montesinos, E., and Bonaterra, A. (2011). Improvement of fitness and efficacy of a fire blight biocontrol agent via nutritional enhancement combined with osmoadaptation. *Applied and Environmental Microbiology*, 77(10): 3174–3181. https://doi.org/10.1128/AEM.02760-10

Cabrefiga, J., Francés, J., Montesinos, E., and Bonaterra, A. (2014). Improvement of a dry formulation of *Pseudomonas fluorescens* EPS62e for fire blight disease biocontrol by combination of culture osmoadaptation with a freeze-drying lyoprotectant. *Journal of Applied Microbiology*, 117(4): 1122–1131. https://doi.org/10.1111/jam.12582

Carvalho, A. L. U. D., Oliveira, F. H. P. C. D., Mariano, R. D. L. R., Gouveia, E. R., and Souto-Maior, A. M. (2010). Growth, sporulation and production of bioactive compounds by *Bacillus subtilis* R14. *Brazilian Archives of Biology and Technology*, 53: 643–652. https://doi.org/10.1590/S1516-89132010000300020

Castillo, H. F., Reyes, C. F., Morales, G. G., Herrera, R. R., and Aguilar, C. (2013). Biological control of root pathogens by plant-growth promoting *Bacillus* spp. *Weed and Pest Control-Conventional and New Challenges*, 79–103. https://dx.doi.org/10.5772/54229

Castroagudín, V. L., Ceresini, P. C., de Oliveira, S. C., Reges, J. T., Maciel, J. L., Bonato, A. L. et al. (2015). Resistance to QoI fungicides is widespread in Brazilian populations of the wheat blast pathogen *Magnaporthe oryzae*. *Phytopathology*, 105(3): 284–294. https://doi.org/10.1094/PHYTO-06-14-0184-R

Chakraborty, M., Mahmud, N. U., Gupta, D. R., Tareq, F. S., Shin, H. J., and Islam, T. (2020). Inhibitory effects of linear lipopeptides from a marine *Bacillus subtilis* on the wheat blast fungus *Magnaporthe oryzae Triticum*. *Frontiers in microbiology*, 11: 665. https://doi.org/10.3389/fmicb.2020.00665

Chakraborty, M., Mahmud, N. U., Ullah, C., Rahman, M., and Islam, T. (2021). Biological and biorational management of blast diseases in cereals caused by *Magnaporthe oryzae*. *Critical Reviews in Biotechnology*, 41(7): 994–1022. https://doi.org/10.1080/07388551.2021.1898325

Chanclud, E., Win, J., Malone, J., Surovy, M. Z., Gupta, D. R., Islam, T. et al. (2017). Genome sequences of candidate wheat blast biocontrol bacteria. *Figshare*, 10, m9.

Chen, L. H., Cui, Y. Q., Yang, X. M., Zhao, D. K., and Shen, Q. R. (2012). An antifungal compound from *Trichoderma harzianum* SQR-T037 effectively controls *Fusarium* wilt of cucumber in continuously cropped soil. *Australasian Plant Pathology*, 41: 239–245. https://doi.org/10.1007/s13313-012-0119-5

Chen, H., Zhao, S., Zhang, K., Zhao, J., Jiang, J., Chen, F. et al. (2018a). Evaluation of soil-applied chemical fungicide and biofungicide for control of the *Fusarium* wilt of chrysanthemum and their effects on rhizosphere soil microbiota. *Agriculture*, 8(12): 184. https://doi.org/10.3390/agriculture8120184

Chen, Y., Wang, J., Yang, N., Wen, Z., Sun, X., Chai, Y. et al. (2018b). Wheat microbiome bacteria can reduce virulence of a plant pathogenic fungus by altering histone acetylation. *Nature Communications*, 9(1): 3429.

Chen, L., Heng, J., Qin, S., and Bian, K. (2018c). A comprehensive understanding of the biocontrol potential of *Bacillus velezensis* LM2303 against Fusarium head blight. *PLoS One*, 13(6): e0198560. https://doi.org/10.1371/journal.pone.0198560

Chen, M. C., Wang, J. P., Zhu, Y. J., Liu, B., Yang, W. J., and Ruan, C. Q. (2019a). Antibacterial activity against *Ralstonia solanacearum* of the lipopeptides secreted from the *Bacillus amyloliquefaciens* strain FJAT-2349. *Journal of Applied Microbiology*, 126(5): 1519–1529. https://doi.org/10.1111/jam.14213

Chen, C., Cai, N., Chen, J., and Wan, C. (2019b). Clove essential oil as an alternative approach to control postharvest blue mold caused by *Penicillium italicum* in citrus fruit. *Biomolecules*, 9(5): 197. https://doi.org/10.3390/biom9050197

Chin-A-Woeng, T. F., Bloemberg, G. V., and Lugtenberg, B. J. (2003). Phenazines and their role in biocontrol by *Pseudomonas* bacteria. *New Phytologist*, 157(3): 503–523. https://doi.org/10.1046/j.1469-8137.2003.00686.x

Choudhury, D., Dobhal, P., Srivastava, S., Saha, S., and Kundu, S. (2018). Role of botanical plant extracts to control plant pathogens-A review. *Indian Journal of Agricultural Research*, 52(4): 341–346. DOI : 10.18805/IJARe.A-5005

Commare, R. R., Nandakumar, R., Kandan, A., Suresh, S., Bharathi, M., Raguchander, T. et al. (2002). *Pseudomonas fluorescens* based bio-formulation for the management of sheath blight disease and leaffolder insect in rice. *Crop Protection*, 21(8): 671–677. https://doi.org/10.1016/S0261-2194(02)00020-0

Compant, S., Duffy, B., Nowak, J., Clément, C., and Barka, E. A. (2005). Use of plant growth-promoting bacteria for biocontrol of plant diseases: principles, mechanisms of action, and future prospects. *Applied and Environmental Microbiology*, 71(9): 4951–4959. https://doi.org/10.1128/AEM.71.9.4951-4959.2005

Cook, R. J., and Baker, K. F. (1983). *The nature and practice of biological control of plant pathogens*. American Phytopathological Society. https://www.cabidigitallibrary.org/doi/full/10.5555/19841396200

Cory, J. S., and Bishop, D. H. (1997). Use of baculoviruses as biological insecticides. *Molecular Biotechnology*, 7: 303–313. https://doi.org/10.1007/BF02740821

Cruz, C. D., Peterson, G. L., Bockus, W. W., Kankanala, P., Dubcovsky, J., Jordan, K. W. et al. (2016). The 2NS translocation from *Aegilops ventricosa* confers resistance to the *Triticum pathotype* of *Magnaporthe oryzae*. *Crop Science*, 56(3): 990–1000. https://doi.org/10.2135/cropsci2015.07.0410

Daranas, N., Badosa, E., Francés, J., Montesinos, E., and Bonaterra, A. (2018). Enhancing water stress tolerance improves fitness in biological control strains of *Lactobacillus plantarum* in plant environments. *PLoS One*, 13(1): e0190931. https://doi.org/10.1371/journal.pone.0190931

de Oliveira Filho, J. G., Silva, G. D. C., Cipriano, L., Gomes, M., and Egea, M. B. (2021). Control of postharvest fungal diseases in fruits using external application of RNAi. *Journal of Food Science*, 86(8): 3341–3348. https://doi.org/10.1111/1750-3841.15816

De Vrieze, M., Varadarajan, A. R., Schneeberger, K., Bailly, A., Rohr, R. P., Ahrens, C. H. et al. (2020). Linking comparative genomics of nine potato-associated *Pseudomonas* isolates with their differing biocontrol potential against late blight. *Frontiers in Microbiology*, 11: 857. https://doi.org/10.3389/fmicb.2020.00857

Durairaj, K., Velmurugan, P., Park, J. H., Chang, W. S., Park, Y. J., Senthilkumar, P. et al. (2018). An investigation of biocontrol activity *Pseudomonas* and *Bacillus* strains against Panax ginseng root rot fungal phytopathogens. *Biological Control*, 125: 138–146. https://doi.org/10.1016/j.biocontrol.2018.05.021

Dutta, S., Surovy, M. Z., Gupta, D. R., Mahmud, N. U., Chanclud, E., Win, J. et al. (2018). Genomic analyses reveal that biocontrol of wheat blast by *Bacillus* spp. may be linked with production of antimicrobial compounds and induced systemic resistance in host plants. *Figshare*, 17: 48. DOI: 10.6084/m9.figshare.5852661.v1

Eakjamnong, W., Keawsalong, N., and Dethoup, T. (2021). Novel ready-to-use dry powder formulation of *Talaromyces tratensis* KUFA0091 to control dirty panicle disease in rice. *Biological Control*, 152: 104454. https://doi.org/10.1016/j.biocontrol.2020.104454

Elsayed, E. A., Othman, N. Z., Malek, R., Awad, H. M., Wu, K., Aziz, R. et al. (2014). Bioprocess development for high cell mass and endospore production by *Bacillus thuringiensis* var *israelensis* in semi-industrial scale. *J Pure Appl Microbiol*, 8(4): 2773–2783. https://www.researchgate.net/publication/265385685

Falade, M., Enikuomehin, O., Borisade, O., and Aluko, M. (2018). Control of cowpea (*Vigna unguiculata* L. Walp) diseases with intercropping of maize (*Zea mays* L) and spray of plant extracts. *Journal of Advances in Microbiology*, 7(4): 1–10. https://doi.org/10.9734/JAMB/2017/38156

Fatema, K., Mahmud, N. U., Gupta, D. R., Siddiqui, M. N., Sakif, T. I., Sarker, A. et al. (2024). Enhancing rice growth and yield with weed endophytic bacteria *Alcaligenes faecalis* and *Metabacillus indicus* under reduced chemical fertilization. *Plos one*, 19(5): e0296547.https://doi.org/10.1371/journal.pone.0296547

Ferrigo, D., Mondin, M., Ladurner, E., Fiorentini, F., Causin, R., and Raiola, A. (2020). Effect of seed biopriming with *Trichoderma harzianum* strain INAT11 on Fusarium ear rot and Gibberella ear rot diseases. *Biological Control*, 147: 104286. https://doi.org/10.1016/j.biocontrol.2020.104286

Fickers, P., Guez, J. S., Damblon, C., Leclère, V., Béchet, M., Jacques, P. et al. (2009). High-level biosynthesis of the anteiso-C17 isoform of the antibiotic mycosubtilin in *Bacillus subtilis* and characterization of its candidacidal activity. *Applied and Environmental Microbiology*, 75(13): 4636–4640. https://doi.org/10.1128/AEM.00548-09

Filippi, M. C. C., Da Silva, G. B., Silva-Lobo, V. L., Côrtes, M. V. C., Moraes, A. J. G., and Prabhu, A. S. (2011). Leaf blast (*Magnaporthe oryzae*) suppression and growth promotion by rhizobacteria on aerobic rice in Brazil. *Biological Control*, 58(2): 160–166. https://doi.org/10.1016/j.biocontrol.2011.04.016

Funk, L. M., He, D. N., Pedersen, E. A., and Reddy, M. S. (1997). Optimization of product delivery for a microbial inoculant, *Burkholderia cepacia*, for commercial use in the forest industry (Abstr). *Can. J. Plant Pathol.*, 19: 108.

Ghasemi, G., Alirezalu, A., Ghosta, Y., Jarrahi, A., Safavi, S. A., Abbas-Mohammadi, M. et al. (2020). Composition, antifungal, phytotoxic, and insecticidal activities of Thymus kotschyanus essential oil. *Molecules*, 25(5): 1152. https://doi.org/10.3390/molecules25051152]

Gholami, M., Amini, J., Abdollahzadeh, J., and Ashengroph, M. (2019). Basidiomycetes fungi as biocontrol agents against take-all disease of wheat. *Biological Control*, 130: 34–43. https://doi.org/10.1016/j.biocontrol.2018.12.012

Gikas, G. D., Parlakidis, P., Mavropoulos, T., and Vryzas, Z. (2022). Particularities of fungicides and factors affecting their fate and removal efficacy: A review. *Sustainability*, 14(7): 4056. https://doi.org/10.3390/su14074056

Gindrat, D. (1979). Biocontrol of plant diseases by inoculation of fresh wounds, seeds, and soil with antagonists.

Girlanda, M., Perotto, S., Moenne-Loccoz, Y., Bergero, R., Lazzari, A., Defago, G. et al. (2001). Impact of biocontrol *Pseudomonas fluorescens* CHA0 and a genetically modified derivative on the diversity of culturable fungi in the cucumber rhizosphere. *Applied and Environmental Microbiology*, 67(4): 1851–1864. https://doi.org/10.1128/AEM.67.4.1851-1864.2001

Grewal, P. S. (2005). Formulations and quality control of entomopathogenic nematodes. *Department of Entomology, Ohio-State-University Wooster*.

Gupta, D. R., Khanom, S., Rohman, M. M., Hasanuzzaman, M., Surovy, M. Z., Mahmud, N. U. et al. (2021). Hydrogen peroxide detoxifying enzymes show different activity patterns in host and non-host plant interactions with *Magnaporthe oryzae Triticum* pathotype. *Physiology and Molecular Biology of Plants*, 27: 2127–2139. https://doi.org/10.1007/s12298-021-01057-4

Hassan, S. M., El-Bebany, A. F., Salem, M. Z., and Komeil, D. A. (2021). Productivity and post-harvest fungal resistance of hot pepper as affected by potassium silicate, clove extract foliar spray and nitrogen application. *Plants*, 10(4): 662. https://doi.org/10.3390/plants10040662

Hettiarachchige, I. K., Ludlow, E. J., Ekanayake, P. N., Brohier, N. D., Sahab, S., Sawbridge, T. I. et al. (2019). Generation of Epichloë strains expressing fluorescent proteins suitable for studying host-endophyte interactions and characterisation of a T-DNA integration event. *Microorganisms*, 8(1): 54. https://doi.org/10.3390/microorganisms8010054

Igarashi, S. (1986). Pyricularia em trigo. 1. Ocorrencia de *Pyricularia* sp noestado do Parana. *Fitopatol. Bras*, 11(3): 351–352. https://cir.nii.ac.jp/crid/1572261549262634496

Islam, M. T., Croll, D., Gladieux, P., Soanes, D. M., Persoons, A., Bhattacharjee, P. et al. (2016). Emergence of wheat blast in Bangladesh was caused by a South American lineage of *Magnaporthe oryzae*. *BMC Biology*, 14(84): 1–11. https://doi.org/10.1186/s12915-016-0309-7

Islam, M. T., Gupta, D. R., Hossain, A., Roy, K. K., He, X., Kabir, M. R. et al. (2020). Wheat blast: a new threat to food security. *Phytopathology Research*, 2(28): 1–13. https://doi.org/10.1186/s42483-020-00067-6

Islam, M. T., Hashidoko, Y., Deora, A., Ito, T., and Tahara, S. (2005). Suppression of damping-off disease in host plants by the rhizoplane bacterium *Lysobacter* sp. strain SB-K88 is linked to plant colonization and antibiosis against soilborne Peronosporomycetes. *Applied and Environmental Microbiology*, 71(7): 3786–3796. https://doi.org/10.1128/AEM.71.7.3786-3796.2005

Jaber, L. R., and Alananbeh, K. M. (2018). Fungal entomopathogens as endophytes reduce several species of Fusarium causing crown and root rot in sweet pepper (*Capsicum annuum* L.). *Biological Control*, 126: 117–126. https://doi.org/10.1016/j.biocontrol.2018.08.007

Jambhulkar, P. P., Sharma, P., and Yadav, R. (2016). Delivery systems for introduction of microbial inoculants in the field. *Microbial inoculants in sustainable agricultural productivity: Vol. 2: Functional Applications*, 199–218. https://doi.org/10.1007/978-81-322-2644-4_13

Janisiewicz, W. J. (1994). Enhancement of biocontrol of blue mold with the nutrient analog 2-deoxy-D-glucose on apples and pears. *Applied and Environmental Microbiology*, 60(8): 2671–2676. https://doi.org/10.1128/aem.60.8.2671-2676.1994

Jayamohan, N. S., Patil, S. V., and Kumudini, B. S. (2020). Seed priming with *Pseudomonas putida* isolated from rhizosphere triggers innate resistance against Fusarium wilt in tomato through pathogenesis-related protein activation and phenylpropanoid pathway. *Pedosphere*, 30(5): 651–660. https://doi.org/10.1016/S1002-0160(20)60027-3

Jayapala, N., Mallikarjunaiah, N. H., Puttaswamy, H., Gavirangappa, H., and Ramachandrappa, N. S. (2019). Rhizobacteria *Bacillus* spp. induce resistance against anthracnose disease in chili (*Capsicum annuum* L.) through activating host defense response. *Egyptian Journal of biological pest control*, 29(1): 1–9. https://doi.org/10.1186/s41938-019-0148-2

Ji, S. H., Paul, N. C., Deng, J. X., Kim, Y. S., Yun, B. S., and Yu, S. H. (2013). Biocontrol activity of *Bacillus amyloliquefaciens* CNU114001 against fungal plant diseases. *Mycobiology*, 41(4): 234–242. https://doi.org/10.1038/s41467-018-05683-7

Jiang, C. H., Liao, M. J., Wang, H. K., Zheng, M. Z., Xu, J. J., and Guo, J. H. (2018). Bacillus velezensis, a potential and efficient biocontrol agent in control of pepper gray mold caused by *Botrytis cinerea*. *Biological Control*, 126: 147–157. https://doi.org/10.1016/j.biocontrol.2018.07.017

Kalia, V. C., Patel, S. K., Kang, Y. C., and Lee, J. K. (2019). Quorum sensing inhibitors as antipathogens: biotechnological applications. *Biotechnology Advances*, 37(1): 68–90. https://doi.org/10.1016/j.biotechadv.2018.11.006

Kanwal, S., Dawar, S., and Tariq, M. (2015). Impact of seed priming with Carica papaya (l.) extract in the control of root infecting fungi and growth of mash bean and chick pea. *Int. J. Biol. Res*, 3(2): 119–123.

Kashyap, N., Singh, S. K., Yadav, N., Singh, V. K., Kumari, M., Kumar, D. et al. (2023). Biocontrol screening of endophytes: applications and limitations. *Plants*, 12(13): 2480. https://doi.org/10.3390/plants12132480

Kekuda, T. P., Vinayaka, K. S., and Sachin, M. B. (2018). Chemistry, ethnobotanical uses and biological activities of the lichen genus *Heterodermia Trevis*. (Physciaceae; Lecanorales; Ascomycota): A comprehensive review. *Journal of Applied Pharmaceutical Science*, 8(5): 148–155. DOI: 10.7324/JAPS.2018.8521

Khan, M. A., Al Mamun Khan, M. A., Mahfuz, A. M. U. B., Sanjana, J. M., Ahsan, A., Gupta, D. R. et al. (2022). Highly potent natural fungicides identified *in silico* against the cereal killer fungus *Magnaporthe oryzae*. *Scientific Reports*, 12(1): 20232. https://doi.org/10.1038/s41598-022-22217-w

Khan, M., Salman, M., Jan, S. A., and Shinwari, Z. K. (2021). Biological control of fungal phytopathogens: A comprehensive review based on *Bacillus* species. *MOJ Biol. Med*, 6: 90–92. https:// doi. org/ 10. 15406/ mojbm. 2021. 06. 00137

Khanzada, M., and Shah, G. S. (2012). *In-vitro* evaluation of fungicides, plant extracts and bio-controlagents against rice blast pathogen *Magnaporthe oryzae* couch. *Pak. J. Bot.*, 44(5): 1775–1778.

Knowles, A. (2008). Recent developments of safer formulations of agrochemicals. *The Environmentalist*, 28(1): 35–44. https://doi.org/10.1007/s10669-007-9045-4

Köhl, J., Kolnaar, R., and Ravensberg, W. J. (2019). Mode of action of microbial biological control agents against plant diseases: relevance beyond efficacy. *Frontiers in Plant Science*, 10: 454982. https://doi.org/10.3389/fpls.2019.00845

Köhl, J., Postma, J., Nicot, P., Ruocco, M., and Blum, B. (2011). Stepwise screening of microorganisms for commercial use in biological control of plant-pathogenic fungi and bacteria. *Biological control*, 57(1): 1–12. https://doi.org/10.1016/j.biocontrol.2010.12.004

Kolombet, L. V., Zhigletsova, S. K., Kosareva, N. I., Bystrova, E. V., Derbyshev, V. V., Krasnova, S. P. et al. (2008). Development of an extended shelf-life, liquid formulation of the biofungicide

Trichoderma asperellum. *World Journal of Microbiology and Biotechnology*, 24: 123–131. https://doi.org/10.1007/s11274-007-9449-9.

Kumar, M., Charishma, K., Sahu, K. P., Sheoran, N., Patel, A., Kundu, A. et al. (2021). Rice leaf associated *Chryseobacterium* species: An untapped antagonistic flavobacterium displays volatile mediated suppression of rice blast disease. *Biological Control*, 161: 104703. https://doi.org/10.1016/j.biocontrol.2021.104703

Kumar, S., Thakur, M., and Rani, A. (2014). Trichoderma: Mass production, formulation, quality control, delivery and its scope in commercialization in India for the management of plant diseases. *African Journal of Agricultural Research*, 9(53): 3838–3852. DOI: 10.5897/AJAR2014. 9061

Kwak, Y. S., and Weller, D. M. (2013). Take-all of wheat and natural disease suppression: a review. *The Plant Pathology Journal*, 29(2): 125. https://doi.org/10.5423%2FPPJ.SI.07.2012.0112

Lahlali, R., Ezrari, S., Radouane, N., Kenfaoui, J., Esmaeel, Q., El Hamss, H. et al. (2022). Biological Control of Plant Pathogens: A Global Perspective. *Microorganisms*, 10(3), 596. https://doi.org/10.3390/microorganisms10030596

Lefèvre, E., Gardner, C. M., and Gunsch, C. K. (2020). A novel PCR-clamping assay reducing plant host DNA amplification significantly improves prokaryotic endo-microbiome community characterization. *FEMS Microbiology Ecology*, 96(7): fiaa110. https://doi.org/10.1093/femsec/fiaa110

Legein, M., Smets, W., Vandenheuvel, D., Eilers, T., Muyshondt, B., Prinsen, E. et al. (2020). Modes of action of microbial biocontrol in the phyllosphere. *Frontiers in Microbiology*, 11: 544057. https://doi.org/10.3389/fmicb.2020.01619

Leo Daniel, A. E., Venkateswarlu, B., Suseelendra, D., Praveen Kumar, G., Mir Hassan Ahmed, S. K., Meenakshi, T. et al. (2013). Effect of polymeric additives, adjuvants, surfactants on survival, stability and plant growth promoting ability of liquid bioinoculants. *J. Plant Physiol. Pathol*, 1: 2. http://dx.doi.org/10.4172/jppp.1000105

Lewis, J. A. (1991). Formulation and delivery systems of biocontrol agents with emphasis on fungi. In *The Rhizosphere and Plant Growth: Papers presented at a Symposium held May* 8–11*: 1989, at the Beltsville Agricultural Research Center (BARC), Beltsville, Maryland* (pp. 279–287). Springer Netherlands. https://doi.org/10.1007/978-94-011-3336-4_55

Li, S., Xu, X., Zhao, T., Ma, J., Zhao, L., Song, Q. et al. (2022). Screening of *Bacillus velezensis* E2 and the inhibitory effect of its antifungal substances on *Aspergillus flavus*. *Foods*, 11(2): 140. https://doi.org/10.3390/foods11020140

Lim, J. H., and Kim, S. D. (2010). Biocontrol of Phytophthora blight of red pepper caused by *Phytophthora capsici* using *Bacillus subtilis* AH18 and *B. licheniformis* K11 formulations. *Journal of the Korean Society For Applied Biological Chemistry*, 53: 766–773. https://doi.org/10.3839/jksabc.2010.116

Liu, M., Gong, Y., Sun, H., Zhang, J., Zhang, L., Sun, J. et al. (2020). Characterization of a novel chitinase from sweet potato and its fungicidal effect against *Ceratocystis fimbriata*. *Journal of Agricultural and Food Chemistry*, 68(29): 7591–7600. https://doi.org/10.1021/acs.jafc.0c01813

Lugtenberg, B. J., and Dekkers, L. C. (1999). What makes *Pseudomonas* bacteria rhizosphere competent? *Environmental Microbiology*, 1(1): 9–13. https://doi.org/10.1046/j.1462-2920.1999.00005.x

Lugtenberg, B. J., Dekkers, L., and Bloemberg, G. V. (2001). Molecular determinants of rhizosphere colonization by *Pseudomonas*. *Annual Review of Phytopathology*, 39(1): 461–490. https://doi.org/10.1146/annurev.phyto.39.1.461

Luh Suriani, N., Ngurah Suprapta, D., Nazir, N., Made Susun Parwanayoni, N., Agung Ketut Darmadi, A., Andya Dewi, D. et al. (2020). A mixture of piper leaves extracts and rhizobacteria for sustainable plant growth promotion and bio-control of blast pathogen of organic bali rice. *Sustainability*, 12(20): 8490. https://doi.org/10.3390/su12208490

Lumsden, R. D., Lewis, J. A., and Fravel, D. R. (1995). Formulation and delivery of biocontrol agents for use against soilborne plant pathogens. https://doi.org/10.1021/bk-1995-0595.ch011.

Malusá, E., Sas-Paszt, L., and Ciesielska, J. J. T. S. W. J. (2012). Technologies for beneficial microorganisms inocula used as biofertilizers. *The Scientific World Journal*, 2012. https://doi.org/10.1100/2012/491206

Mannai, S., Benfradj, N., Horrigue-Raouani, N., and Boughalleb-M'Hamdi, N. (2018). Antifungal activity and growth promotion of three types of compost extracts against *Fusarium oxysporum* and *Fusarium solani* associated with peach seedling decline in nurseries. *Journal of Crop Protection*, 7(3): 349–363.

Martínez-García, P. M., Ruano-Rosa, D., Schilirò, E., Prieto, P., Ramos, C., Rodríguez Palenzuela, P. et al. (2015). Complete genome sequence of *Pseudomonas fluorescens* strain PICF7, an indigenous root endophyte from olive (*Olea europaea* L.) and effective biocontrol agent against *Verticillium dahliae*. *Standards in Genomic Sciences*, 10: 1 7. https://doi.org/10.1186/1944-3277-10-10

Mathivanan, N., Srinivasan, K., and Chelliah, S. (2000). Biological control of soil-borne diseases of cotton, eggplant, okra and sunflower by *Trichoderma viride*/Biologische Bekämpfung bodenbürtiger Krankheiten an Baumwolle, Auberginen, Okra und Sonnenblumen mit Trichoderma viride. *Zeitschrift für Pflanzenkrankheiten und Pflanzenschutz/Journal of Plant Diseases and Protection*, 235–244.

Matsumoto, H., Fan, X., Wang, Y., Kusstatscher, P., Duan, J., Wu, S. et al. (2021). Bacterial seed endophyte shapes disease resistance in rice. *Nature Plants*, 7(1): 60–72. https://doi.org/10.1038/s41477-020-00826-5

McIntyre, J. L., and Press, L. S. (1991). Formulation, delivery systems and marketing of biocontrol agents and plant growth promoting rhizobacteria (PGPR). pp. 289–295. *In: The Rhizosphere and Plant Growth: Papers presented at a Symposium held May 8–11, 1989, at the Beltsville Agricultural Research Center (BARC), Beltsville, Maryland*. Springer Netherlands. https://doi.org/10.1007/978-94-011-3336-4_56

McSpadden Gardener, B. B. (2004). Ecology of *Bacillus* and *Paenibacillus* spp. in agricultural systems. *Phytopathology*, 94(11): 1252–1258. https://doi.org/10.1094/PHYTO.2004.94.11.1252.

Meepagala, K. M., Clausen, B. M., Johnson, R. D., Wedge, D. E., and Duke, S. O. (2019). A phytotoxic and antifungal metabolite (pyrichalasin H) from a fungus infecting *Brachiaria eruciformis* (signal grass). *Journal of Agricultural Chemistry and Environment*, 8(03): 115. https:// doi. org/ 10. 4236/ jacen. 2019. 83010

Melin, P., Håkansson, S., and Schnürer, J. (2007). Optimisation and comparison of liquid and dry formulations of the biocontrol yeast *Pichia anomala* J121. *Applied Microbiology and Biotechnology*, 73: 1008–1016. https://doi.org/10.1007/s00253-006-0552-x

Meng, X., Yu, J., Yu, M., Yin, X., and Liu, Y. (2015). Dry flowable formulations of antagonistic *Bacillus subtilis* strain T429 by spray drying to control rice blast disease. *Biological Control*, 85: 46–51. https://doi.org/10.1016/j.biocontrol.2015.03.004

Mona, M. M. R., Ashour, A. M. A., El-Mohamedy, R. S. R., Morsy, A. A., and Hanafy, E. K. (2017). Seed bio priming as biological approach for controlling root rot soil born fungi on soybean (*Glycine max* L.) plant. *International Journal of Agricultural Technology*, 13(5): 771–788.

Monteiro, S. M. S., Clemente, J. J., Carrondo, M. J. T., and Cunha, A. E. (2014). Enhanced spore production of *Bacillus subtilis* grown in a chemically defined medium. *Advances in Microbiology*, 4(08): 444. https://doi.org/10.4236/aim.2014.48049

Monteiro, S. M., Clemente, J. J., Henriques, A. O., Gomes, R. J., Carrondo, M. J., and Cunha, A. E. (2005). A procedure for high-yield spore production by *Bacillus subtilis*. *Biotechnology Progress*, 21(4): 1026-1031. https://doi.org/10.1021/bp050062z

Montesinos, E. (2003). Development, registration and commercialization of microbial pesticides for plant protection. *International Microbiology*, 6: 245–252. https://doi.org/10.1007/s10123-003-0144-x

Myo, E. M., Liu, B., Ma, J., Shi, L., Jiang, M., Zhang, K. et al. (2019). Evaluation of *Bacillus velezensis* NKG-2 for bio-control activities against fungal diseases and potential plant growth promotion. *Biological Control*, 134: 23–31. https://doi.org/10.1016/j.biocontrol.2019.03.017

Nandhini, M., Rajini, S. B., Udayashankar, A. C., Niranjana, S. R., Lund, O. S., Shetty, H. S. et al. (2018). Diversity, plant growth promoting and downy mildew disease suppression potential of cultivable endophytic fungal communities associated with pearl millet. *Biological Control*, 127: 127–138. https://doi.org/10.1016/j.biocontrol.2018.08.019

Narasimhan, A., and Shivakumar, S. (2015). Evaluation of *Bacillus subtilis* (JN032305) biofungicide to control chilli anthracnose in pot controlled conditions. *Biocontrol Science and Technology*, 25(5): 543–559. https://doi.org/10.1080/09583157.2014.996737

Narusaka, M., Minami, T., Iwabuchi, C., Hamasaki, T., Takasaki, S., Kawamura, K. et al. (2015). Yeast cell wall extract induces disease resistance against bacterial and fungal pathogens in *Arabidopsis thaliana* and *Brassica* crop. *PLoS One*, 10(1): e0115864. https://doi.org/10.1371/journal.pone.0115864

Naveenkumar, R., Muthukumar, A., Sangeetha, G., and Mohanapriya, R. (2017). Developing eco-friendly biofungicide for the management of major seed borne diseases of rice and assessing their physical stability and storage life. *Comptes Rendus Biologies*, 340(4): 214–225. https://doi.org/10.1016/j.crvi.2017.03.001

Nian, J., Yu, M., Bradley, C. A., and Zhao, Y. (2021). *Lysobacter enzymogenes* strain C3 suppresses mycelium growth and spore germination of eight soybean fungal and oomycete pathogens and decreases disease incidences. *Biological Control*, 152: 104424. https://doi.org/10.1016/j.biocontrol.2020.104424

O'Keeffe, E., Hughes, H., McLoughlin, P., Tan, S. P., and McCarthy, N. (2019). Methods of analysis for the *in vitro* and *in vivo* determination of the fungicidal activity of seaweeds: a mini review. *Journal of Applied Phycology*, 31: 3759–3776. https://doi.org/10.1007/s10811-019-01832-7

Omer, A. M. (2010). Bioformulations of *Bacillus spores* for using as biofertilizer. *Life Sci. J.*, 7(4): 124–131. https://www.researchgate.net/publication/265890977

Ongena, M., and Jacques, P. (2008). *Bacillus lipopeptides*: versatile weapons for plant disease biocontrol. *Trends in Microbiology*, 16(3): 115–125. https://doi.org/10.1016/j.tim.2007.12.009

Ongena, M., Duby, F., Jourdan, E., Beaudry, T., Jadin, V., Dommes, J. et al. (2005). *Bacillus subtilis* M4 decreases plant susceptibility towards fungal pathogens by increasing host resistance associated with differential gene expression. *Applied Microbiology and Biotechnology*, 67: 692–698. https://doi.org/10.1007/s00253-004-1741-0

Oso, S., Walters, M., Schlechter, R. O., and Remus-Emsermann, M. N. (2019). Utilisation of hydrocarbons and production of surfactants by bacteria isolated from plant leaf surfaces. *FEMS Microbiology Letters*, 366(6): fnz061. https://doi.org/10.1093/femsle/fnz061

Paau, A. S. (1988). Formulations useful in applying beneficial microorganisms to seeds. *Trends in Biotechnology*, 6(11): 276–279. https://doi.org/10.1016/0167-7799(88)90124-2

Pal, R., Biswas, M. K., Mandal, D., and Naik, B. S. (2015). Management of sheath blight disease of rice through bio control agents in west central table land zone of Odisha. *Int. J. Adv. Res.*, 3(11): 747–753.

Panwar, V., Aggarwal, A., Singh, G., Verma, A., Sharma, I., and Saharan, M. S. (2014). Efficacy of foliar spray of Trichoderma isolates against *Fusarium graminearum* causing head blight of wheat. *J. Wheat Res.*, 6(1): 59–63.

Patil, S. V., and Kumudini, B. S. (2019). Seed priming induced blast disease resistance in finger millet plants through phenylpropanoid metabolic pathway. *Physiological and Molecular Plant Pathology*, 108: 101428. https://doi.org/10.1016/j.pmpp.2019.101428

Pereira, M. R. D. A., Moreira, S. I., Silva, A. G. D., Nunes, T. C., Vicentini, S. N. C., Silva, D. P. D. et al. (2022). Potential of *Pseudomonas* and *Trichoderma* from the Brazilian amazon as biocontrol agents against the wheat blast disease. *Agronomy*, 12(9): 2003. https://doi.org/10.3390/agronomy12092003

Pertot, I., Giovannini, O., Benanchi, M., Caffi, T., Rossi, V., and Mugnai, L. (2017). Combining biocontrol agents with different mechanisms of action in a strategy to control *Botrytis cinerea* on grapevine. *Crop Protection*, 97: 85–93. https://doi.org/10.1016/j.cropro.2017.01.010

Piggot, P. J., and Coote, J. G. (1976). Genetic aspects of bacterial endospore formation. *Bacteriological Reviews*, 40(4): 908–962.

Pliego, C., Ramos, C., de Vicente, A., and Cazorla, F. M. (2011). Screening for candidate bacterial biocontrol agents against soilborne fungal plant pathogens. *Plant and Soil*, 340: 505–520. https://doi.org/10.1007/s11104-010-0615-8

Postma, J., Stevens, L. H., Wiegers, G. L., Davelaar, E., and Nijhuis, E. H. (2009). Biological control of *Pythium aphanidermatum* in cucumber with a combined application of *Lysobacter enzymogenes* strain 3.1 T8 and chitosan. *Biological Control*, 48(3): 301–309. https://doi.org/10.1016/j.biocontrol.2008.11.006

Prabha, D., Negi, S., Kumari, P., Negi, Y. K., and Chauhan, J. S. (2016). Effect of seed priming with some plant leaf extract on seedling growth characteristics and root rot disease in tomato. *International Journal of Agriculture System*, 4(1): 46–51. https:// doi. org/ 10. 20956/ ijas. v4i1. 240

Rabby, S. F., Chakraborty, M., Gupta, D. R., Rahman, M., Paul, S. K., Mahmud, N. U. et al. (2022). Bonactin and feigrisolide C inhibit *Magnaporthe oryzae Triticum fungus* and control wheat blast disease. *Plants*, 11(16): 2108. https://doi.org/10.3390/plants11162108

Rafi, H., Dawar, S., and Tariq, M. (2016). Combined effect of soil amendment with oil cakes and seed priming in the control of root rot fungi of leguminous and non-leguminous crops. *Pak. J. Bot.*, 48(3): 1305–1311.

Rafi, H., Dawar, S., and Zaki, M. J. (2015). Seed priming with extracts of *Acacia nilotica* (L.) willd. ex Delile and *Sapindus mukorossi* (L.) plant parts in the control of root rot fungi and growth of plants. *Pak. J. Bot.*, 47(3): 1129–1135.

Raguchander, T., Saravanakumar, D., and Balasurramanian, P. (2011). Molecular approaches to improvement of biocontrol agents of plant diseases. *J. Biol. Control.*, 25(2): 71–84. https://www.researchgate.net/publication/302135401

Rajaput, J., Rao, M. S. L., Hegde, R. V., and Kulkarni, V. (2019). Biopriming: A novel seed treatment options to manage the seed-borne fungal infection of tomato. *Journal of Pharmacognosy and Phytochemistry*, 8(6): 659–661.

Rashad, Y. M., Aseel, D. G., and Hafez, E. E. (2018). Antifungal potential and defense gene induction in maize against *Rhizoctonia* root rot by seed extract of *Ammi visnaga* (L.) Lam. *Phytopathologia Mediterranea*, 57(1): 73–88. https:// doi. org/ 10. 14601/ Phyto pathol_ Medit err- 21366

Raymaekers, K., Ponet, L., Holtappels, D., Berckmans, B., and Cammue, B. P. (2020). Screening for novel biocontrol agents applicable in plant disease management–a review. *Biological Control*, 144: 104240. https://doi.org/10.1016/j.biocontrol.2020.104240

Reis, A., da Silva, T. L., Kent, C. A., Kosseva, M., Roseiro, J. C., and Hewitt, C. J. (2005). Monitoring population dynamics of the thermophilic *Bacillus licheniformis* CCMI 1034 in batch and continuous cultures using multi-parameter flow cytometry. *Journal of Biotechnology*, 115(2): 199–210. https://doi.org/10.1016/j.jbiotec.2004.08.005

Rishbeth, J. (1975). Stump inoculation: a biological control of *Fomes annosus*. In *Biology and Control of Soil Borne Plant Pathogens International Symposium.*

Robin, D. C., and Marchand, P. A. (2019). Evolution of the biocontrol active substances in the framework of the European Pesticide Regulation (EC) No. 1107/2009. *Pest Management Science*, 75(4): 950–958. https://doi.org/10.1002/ps.5199

Rojas, E. C., Jensen, B., Jørgensen, H. J., Latz, M. A., Esteban, P., Ding, Y. et al. (2020). Selection of fungal endophytes with biocontrol potential against Fusarium head blight in wheat. *Biological Control*, 144: 104222. https://doi.org/10.1016/j.biocontrol.2020.104222

Roselló, G., Francés, J., Daranas, N., Montesinos, E., and Bonaterra, A. (2017). Control of fire blight of pear trees with mixed inocula of two *Lactobacillus plantarum* strains and lactic acid. *Journal of Plant Pathology*, 111–120. https://www.jstor.org/stable/i40221401

Sabaratnam, S., and Traquair, J. A. (2002). Formulation of a *Streptomyces* biocontrol agent for the suppression of *Rhizoctonia* damping-off in tomato transplants. *Biological Control*, 23(3): 245–253. https://doi.org/10.1006/bcon.2001.1014

Sabu, R., Aswani, R., Prabhakaran, P., Krishnakumar, B., and Radhakrishnan, E. K. (2018). Differential modulation of endophytic microbiome of ginger in the presence of beneficial organisms, pathogens and both as identified by DGGE analysis. *Current Microbiology*, 75: 1033–1037. https://doi.org/10.1007/s00284-018-1485-1

Saito, H., Sasaki, M., Nonaka, Y., Tanaka, J., Tokunaga, T., Kato, A. et al. (2021). Spray application of nonpathogenic fusaria onto rice flowers controls bakanae disease (caused by *Fusarium fujikuroi*) in the next plant generation. *Applied and Environmental Microbiology*, 87(2): e01959–20. https://doi.org/10.1128/AEM.01959-20

Sarkar, G., Jatar, N., Goswami, P., Cyriac, R., Suthindhiran, K., and Jayasri, M. A. (2018). Combination of different marine algal extracts as biostimulant and biofungicide. *Journal of Plant Nutrition*, 41(9): 1163–1171. https://doi.org/10.1080/01904167.2018.1434201

Sasani, M., and Ahmadzadeh, M. (2021). Evaluation of antagonistic effect of several strain of Bacillus bacteria on control of crown and root rot of wheat with *Fusarium pseudograminearum*. *Biological Control of Pests and Plant Diseases*, 9(2): 115–126. https://doi.org/10.22059/jbioc.2021.326484.307.

Schaechter, M. (2009). *Encyclopedia of microbiology*. Academic Press.

Segarra, G., Puopolo, G., Giovannini, O., and Pertot, I. (2015). Stepwise flow diagram for the development of formulations of non spore-forming bacteria against foliar pathogens: the case of *Lysobacter capsici* AZ78. *Journal of Biotechnology*, *216*: 56–64. https://doi.org/10.1016/j.jbiotec.2015.10.004

Sergaki, C., Lagunas, B., Lidbury, I., Gifford, M. L., and Schäfer, P. (2018). Challenges and approaches in microbiome research: from fundamental to applied. *Frontiers in Plant Science*, 9: 387738. https://doi.org/10.3389/fpls.2018.01205

Shafi, J., Tian, H., and Ji, M. (2017). *Bacillus* species as versatile weapons for plant pathogens: a review. *Biotechnology & Biotechnological Equipment*, 31(3): 446–459. https://doi.org/10.1080/13102818.2017.1286950

Shaikh, S. S., and Sayyed, R. Z. (2014). Role of plant growth-promoting rhizobacteria and their formulation in biocontrol of plant diseases. pp. 337–351. *In: Plant microbes symbiosis: applied facets*. New Delhi: Springer India. https://doi.org/10.1007/978-81-322-2068-8_18

Shieh, T. R. (2000). Biopesticide formulations and their applications. *Proceedings of American Chemical Society*, 104–114.

Shimoi, S., Inoue, K., Kitagawa, H., Yamasaki, M., Tsushima, S., Park, P. et al. (2010). Biological control for rice blast disease by employing detachment action with gelatinolytic bacteria. *Biological Control*, 55(2): 85–91. https://doi.org/10.1016/j.biocontrol.2010.07.008

Shivanna, R., and Garampalli, R. H. (2015). Evaluation of fungistatic potential of lichen extracts against *Fusarium solani* (Mart.) Sacc. causing Rhizome rot disease in Ginger. *Journal of Applied Pharmaceutical Science*, 5(10): 067–072. DOI: 10.7324/JAPS.2015.501012

Singh, R., Ahluwalia, V., Singh, P., Kumar, N., Prakash Sati, O., and Sati, N. (2016). Antifungal and phytotoxic activity of essential oil from root of *Senecio amplexicaulis* Kunth.(Asteraceae) growing wild in high altitude-Himalayan region. *Natural Product Research*, 30(16): 1875–1879. https://doi.org/10.1080/14786419.2015.1079910

Silva, P., Cruppe, G., Pereira, F., García, R., Bentos, D., Villero, M. et al. (2025). Genomic and pathogenic insights into the geographic range expansion of wheat blast into Uruguay. *Plant Disease. Advance Online Publication.*

Singh, S., Singh, U. B., Malviya, D., Paul, S., Sahu, P. K., Trivedi, M. et al. (2020). Seed biopriming with microbial inoculant triggers local and systemic defense responses against *Rhizoctonia solani* causing banded leaf and sheath blight in maize (*Zea mays* L.). *International Journal of Environmental Research and Public Health*, 17(4): 1396. https://doi.org/10.3390/ijerph17041396

Singhania, R. R., Patel, A. K., Soccol, C. R., and Pandey, A. (2009). Recent advances in solid-state fermentation. *Biochemical Engineering Journal*, 44(1): 13–18. https://doi.org/10.1016/j.bej.2008.10.019

Soliman, A. S., Ahmed, A. Y., Abdel-Ghafour, S. E., El-Sheekh, M. M., and Sobhy, H. M. (2018). Antifungal bio-efficacy of the red algae *Gracilaria confervoides* extracts against three pathogenic fungi of cucumber plant. *Middle East J. Appl. Sci*, 8(3): 727–735.

Solis, P. N., Wright, C. W., Anderson, M. M., Gupta, M. P., and Phillipson, J. D. (1993). A microwell cytotoxicity assay using *Artemia salina* (brine shrimp). *Planta Medica*, 59(03): 250–252. DOI: 10.1055/s-2006-959661

Soylu, E. M., and Incekara, R. (2017). Biofungicidal activities of plant essential oils against cucumber root and stem rot disease caused by *Fusarium oxysporum* f. sp. *radicis-cucumerinum*. *Journal of Plant Pathology*, 437–444.

Srivastava, M., Kumar, V., Shahid, M., Sonika, P., and Singh, A. (2016). Trichoderma-a potential and effective bio fungicide and alternative source against notable phytopathogens: A review. *African Journal of Agricultural Research*, 11(5): 310–316. https:// doi. org/ 10. 5897/ AJAR2 015. 9568

Stamenković-Stojanović, S., Karabegović, I., Beškoski, V., Nikolić, N., and Lazić, M. (2019). *Bacillus* based microbial formulations: Optimization of the production process. *Hemijska industrija*, 73(3): 169–182. https://doi.org/10.2298/HEMIND190214014S

Stasz, T. E., Harman, G. E., and Gullino, M. L. (1989). Limited vegetative compatibility following intra- and interspecific protoplast fusion in Trichoderma. *Experimental Mycology*, 13(4): 364–371. https://doi.org/10.1016/0147-5975(89)90032-7

Steinke, W. E., and Akesson, N. B. (1993). Atomization and Application of Biopesticide Formulations. *ASTM Special Technical Publication*, 1146: 257–257.

Stockwell, V. O., Johnson, K. B., Sugar, D., and Loper, J. E. (2010). Control of fire blight by *Pseudomonas fluorescens* A506 and *Pantoea vagans* C9-1 applied as single strains and mixed inocula. *Phytopathology*, 100(12): 1330–1339. https://doi.org/10.1094/PHYTO-03-10-0097

Sufyan, M., Tahir, M. I., Haq, M. I., Hussain, S., and Saeed, M. (2020). Effect of seed bio priming with rhizobacteria against root associated pathogenic fungi in chickpea. *Pakistan Journal of Phytopathology*, 32(1): 89–96.

Sujarit, K., Pathom-aree, W., Mori, M., Dobashi, K., Shiomi, K., and Lumyong, S. (2020). Streptomyces palmae CMU-AB204T, an antifungal producing-actinomycete, as a potential biocontrol agent to protect palm oil producing trees from basal stem rot disease fungus, *Ganoderma boninense*. *Biological Control*, 148: 104307. https://doi.org/10.1016/j.biocontrol.2020.104307

Sun, Y., Wang, Y., Han, L. R., Zhang, X., and Feng, J. T. (2017). Antifungal activity and action mode of cuminic acid from the seeds of *Cuminum cyminum* L. against *Fusarium oxysporum* f. sp. niveum (FON) causing *Fusarium* wilt on watermelon. *Molecules*, 22(12): 2053. https://doi.org/10.3390/molecules22122053

Surovy, M. Z., Gupta, D. R., Chanclud, E., Win, J., Kamoun, S., and Islam, T. (2017). Plant probiotic bacteria suppress wheat blast fungus *Magnaporthe oryzae Triticum pathotype*. *Figshare*, 10: m9.

Surovy, M. Z., Dutta, S., Mahmud, N. U., Gupta, D. R., Farhana, T., Paul, S. K. et al. (2022a). Probiotic Bacillus species: Promising biological control agents for managing worrisome wheat blast disease. doi: 10.20944/preprints202211.0382.v1

Surovy, M. Z., Rahman, S., Dame, Z. T., and Islam, T. (2022b). Discovery of bioactive natural products from Bacillus Species: Chemistry, biosynthesis and biological activities. *Bacilli in Agrobiotechnology: Plant Stress Tolerance, Bioremediation, and Bioprospecting*, 47–87.

Surovy, M. Z., Rahman, S., Rostás, M., Islam, T., and Von Tiedemann, A. (2023). Suppressive Effects of Volatile Compounds from *Bacillus* spp. on *Magnaporthe oryzae Triticum* (MoT) Pathotype, Causal Agent of Wheat Blast. *Microorganisms*, 11(5): 1291. https://doi.org/10.1007/978-3-030-85465-2_3

Surovy, M. Z., Dutta, S., Mahmud, N. U., Gupta, D. R., Farhana, T., Paul, S. K. et al. (2024). Biological control potential of worrisome wheat blast disease by the seed endophytic bacilli. *Frontiers in Microbiology*, 15: 1336515.https://doi.org/10.3389/fmicb.2024.1336515

Tabacchioni, S., Passato, S., Ambrosino, P., Huang, L., Caldara, M., Cantale, C. et al. (2021). Identification of beneficial microbial consortia and bioactive compounds with potential as plant biostimulants for a sustainable agriculture. *Microorganisms*, 9(2): 426. https://doi.org/10.3390/microorganisms9020426

Tadros, T. F. (2006). *Applied surfactants: principles and applications*. John Wiley & Sons.

Teixidó, N., Torres, R., Viñas, I., Abadias, M., and Usall, J. (2011). Biological control of postharvest diseases in fruit and vegetables. pp. 364–402. *In: Protective cultures, antimicrobial metabolites and bacteriophages for food and beverage biopreservation*. Woodhead Publishing. https://doi.org/10.1533/9780857090522.3.364

Teixidó, N., Usall, J., and Torres, R. (2022). Insight into a successful development of biocontrol agents: Production, formulation, packaging, and shelf life as key aspects. *Horticulturae*, 8(4): 305. https://doi.org/10.3390/horticulturae8040305

Tembo, B., Mulenga, R. M., Sichilima, S., M'siska, K. K., Mwale, M., Chikoti, P. C. et al. (2020). Detection and characterization of fungus (*Magnaporthe oryzae pathotype Triticum*) causing wheat blast disease on rain-fed grown wheat (*Triticum aestivum* L.) in Zambia. *PloS one*, 15(9): e0238724. https://doi.org/10.1371/journal.pone.0238724

Thomashow, L. S., and Weller, D. M. (1990). Application of fluorescent pseudomonads to control root diseases of wheat and some mechanisms of disease suppression. *Biological Control of Soil-Borne Plant Pathogens*, 109–122.

Thomson, L. (1998). *A guide to agricultural spray adjuvants used in the United States* (No. Ed. 5). Thomson Publications.

Tittabutr, P., Payakapong, W., Teaumroong, N., Singleton, P. W., and Boonkerd, N. (2007). Growth, survival and field performance of bradyrhizobial liquid inoculant formulations with polymeric additives. *Science Asia*, 33(1): 69–77. doi: 10.2306/scienceasia1513-1874.2007.33.069

Tumpa, F. H., Sultana, A., Alam, M. Z., and Khokon, M. A. R. (2016). Bio-stimulation by seed priming with *Bacillus subtilis* for suppressing seed-borne fungal pathogens of vegetables in Bangladesh. DOI: 10.3329/jbau.v14i2.32692

Upamanya, G. K., Bhattacharyya, A., and Dutta, P. (2020). Consortia of entomo-pathogenic fungi and bio-control agents improve the agro-ecological conditions for brinjal cultivation of Assam. *3 Biotech*, 10: 1–19. https://doi.org/10.1007/s13205-020-02439-3

Valadbeigi, T., Bahrami, A. M., and Shaddel, M. (2014). Antibacterial and antifungal activities of different lichens extracts. *Journal of Medical Microbiology and Infectious Diseases*, 2(2): 71–75.

Vassilev, N., Vassileva, M., Lopez, A., Martos, V., Reyes, A., Maksimovic, I. et al. (2015). Unexploited potential of some biotechnological techniques for biofertilizer production and formulation. *Applied Microbiology and Biotechnology*, 99: 4983–4996. https://doi.org/10.1007/s00253-015-6656-4

Vehapi, M., Koçer, A. T., Yılmaz, A., and Özçimen, D. (2020). Investigation of the antifungal effects of algal extracts on apple-infecting fungi. *Archives of Microbiology*, 202: 455–471. https://doi.org/10.1007/s00203-019-01760-7

Venkateswarlu, B., and Beura, S. K. (2020). Evaluation of selective fungicides and biocontrol agents for suppression of banded leaf and sheath blight of maize (*Zea mays*). *Current Journal of Applied Science and Technology*, 39(15): 83–89. https://doi.org/10.9734/cjast/2020/v39i1530775

Vidhyasekaran, P., and Muthamilan, M. (1995). Development of formulations of *Pseudomonas fluorescens* for control of chickpea wilt. https://doi.org/10.1094/PD-79-0782

Wang, X., Radwan, M. M., Taráwneh, A. H., Gao, J., Wedge, D. E., Rosa, L. H. et al. (2013). Antifungal activity against plant pathogens of metabolites from the endophytic fungus Cladosporium cladosporioides. *Journal of Agricultural and Food Chemistry*, 61(19): 4551–4555. https:// doi. org/ 10. 1021/ jf400 212y

Waqas, M., Korres, N. E., Khan, M. D., Nizami, A. S., Deeba, F., Ali, I. et al. (2019). Advances in the concept and methods of seed priming. *Priming and Pretreatment of Seeds and Seedlings: Implication in Plant Stress Tolerance and Enhancing Productivity in Crop Plants*, 11–41. https://doi.org/10.1007/978-981-13-8625-1_2

Wei, H., Wang, Y., Jin, Z., Yang, F., Hu, J., and Gao, M. T. (2021). Utilization of straw-based phenolic acids as a biofugicide for a green agricultural production. *Journal of Bioscience and Bioengineering*, 131(1): 53–60. https://doi.org/10.1016/j.jbiosc.2020.09.007

Woo, S. L., Ruocco, M., Vinale, F., Nigro, M., Marra, R., Lombardi, N. et al. (2014). Trichoderma-based products and their widespread use in agriculture. *The Open Mycology Journal*, 8(1). http://dx.doi.org/10.2174/1874437001408010071

Xie, Z., Li, M., Wang, D., Wang, F., Shen, H., Sun, G. et al. (2021). Biocontrol efficacy of *Bacillus siamensis* LZ88 against brown spot disease of tobacco caused by *Alternaria alternata*. *Biological Control*, 154: 104508. https://doi.org/10.1016/j.biocontrol.2020.104508

Xu, S., Zhang, Z., Xie, X., Shi, Y., Chai, A., Fan, T. et al. (2022). Comparative genomics provides insights into the potential biocontrol mechanism of two *Lysobacter enzymogenes* strains with distinct antagonistic activities. *Frontiers in Microbiology*, 13: 966986. https://doi.org/10.3389/fmicb.2022.966986

Yadav, M., Dubey, M. K., and Upadhyay, R. S. (2021). Systemic resistance in chilli pepper against anthracnose (caused by *Colletotrichum truncatum*) induced by *Trichoderma harzianum, Trichoderma asperellum* and *Paenibacillus dendritiformis*. *Journal of Fungi*, 7(4): 307. https://doi.org/10.3390/jof7040307

Yu, Y. Y., Jiang, C. H., Wang, C., Chen, L. J., Li, H. Y., Xu, Q. et al. (2017). An improved strategy for stable biocontrol agents selecting to control rice sheath blight caused by *Rhizoctonia solani*. *Microbiological Research*, 203: 1–9. https://doi.org/10.1016/j.micres.2017.05.006

Zamani-Zadeh, M., Soleimanian-Zad, S., Sheikh-Zeinoddin, M., and Goli, S. A. H. (2014). Integration of *Lactobacillus plantarum* A7 with thyme and cumin essential oils as a potential biocontrol tool for gray mold rot on strawberry fruit. *Postharvest Biology and Technology*, 92: 149–156. https://doi.org/10.1016/j.postharvbio.2014.01.019

Zhang, X., Li, B., Wang, Y., Guo, Q., Lu, X., Li, S., and Ma, P. (2013). Lipopeptides, a novel protein, and volatile compounds contribute to the antifungal activity of the biocontrol agent *Bacillus atrophaeus* CAB-1. *Applied Microbiology and Biotechnology*, 97: 9525–9534. https://doi.org/10.1007/s00253-013-5198-x

Zida, P. E., Néya, B. J., Soalla, R. W., Sérémé, P., and Lund, O. S. (2018). Seed-priming of sorghum with antifungal extracts from *Balanites aegyptiaca* and *Eclipta alba* in different agro-ecological zones of *Burkina Faso*. *African Journal of Agricultural Research*, 13(44): 2516–2525. https:// doi. org/ 10. 5897/ AJAR2 018. 113431

Zubrod, J. P., Bundschuh, M., Arts, G., Brühl, C. A., Imfeld, G., Knäbel, A. et al. (2019). Fungicides: an overlooked pesticide class?. *Environmental Science & Technology*, 53(7): 3347–3365. https://doi.org/10.1021/acs.est.8b04392

9

Formulation of Biofungicides Based on Streptomyces

Khadija Goura,[1,2] *Najwa Seddiqi Kallali,*[1,2] *Abdelaaziz Farhaoui*[1] and *Rachid Lahlali*[1,*]

1. Introduction

Agriculture is consistently affected by the occurrence of pests and pathogens, which are unwanted organisms that invade crops and lead to a notable reduction in yield (Mishra et al. 2018, Farhaoui et al. 2023). A significant challenge lies in the damage inflicted by phytopathogenic fungi (Alblooshi et al. 2022). Generally, managing pests and pathogens in agriculture involves the use of chemical pesticides that interact with these undesirable organisms, thereby diminishing their populations and minimizing the damage they cause (Nicolopoulou-Stamati et al. 2016). However, when used improperly, these active substances tend to accumulate in the environment, fostering water body eutrophication, as well as the presence ofmolecules that may pose toxicity risks to human health through overexposure (Khan et al. 2014, Bonner and Alavanja 2017).

Nowadays, both consumers and producers are growing more aware of food safety and ecological issues. This trend is in line with the market's preference to privilege the use of more eco-friendly options, as well as the desire to reduce waste and lessen environmental impact by exploring alternatives that are both biodegradable and safe, while maintaining effectiveness (El-shakh et al. 2015). In this regard, products based on microorganisms represent an effective option to minimize reliance on agricultural chemicals, mixing high productivity with an approach that respects all environmental components (Umesha et al. 2018). The use of microbial inoculants

[1] Department of Plant Protection and Environment, Phytopathology Unit, Ecole Nationale d'Agriculture de Meknès, Km10, Rte Haj Kaddour, BP S/40 11201, 50001, Meknes, Morocco.
[2] Laboratory of Biotechnology and Valorization of Biological Resources, Faculty of Sciences, Moulay Ismail University, Avenue Zitoune, P.O. Box 11201, 50000, Meknes, Morocco.
* Corresponding author: rlahlai@enameknes.ac.ma

DOI: 10.1201/9781003603023-10

has grown considerably in recent decades, serving as alternatives to traditional fertilizers and pesticides. They provide an additional and complementary resource to those already in use (Shcherbakov et al. 2013). Biopesticides incorporating microbial biological control agents (MBCAs) demonstrate exceptional effectiveness against various plant pathogens (Thakur et al. 2020), particularly due to the precise targeting of their molecular mechanisms, which can reduce pest populations and restore the ecological equilibrium in the environment (Abbey et al. 2019). Among the diverse MBCAs that can be utilized in bioformulations aimed at combating plant pests and diseases, *Streptomyces* are renowned for their bioactive characteristics (Matsumoto and Takahashi 2017).

The utilization of *Streptomyces* as a biocontrol agent against pathogens has emerged as an alternative to chemical fungicides (Boukaew et al. 2011). These bacteria, found abundantly in the soil, are distinguished by their ability to produce commercially valuable molecules, including antibiotics, hormones, and enzymes. These metabolites are generally byproducts of secondary metabolism and are not essential during the key stages of the organism's development (Jakubiec-Krzesniak et al. 2018). *Streptomyces*, both endophytic and rhizospheric, demonstrate a strong capacity to release compounds of interest linked to their interactions with the host plant and the soil microbiota (Silva et al. 2022).

Therefore, this chapter aims to clarify key aspects of the exploration of *Streptomyces* and their role as agricultural inoculants and biocontrol agents. It explores the role of *Streptomyces* in stimulating plant development and offering protection to crops through their bioactive properties. It also outlines the main challenges in expanding the production of *Streptomyces* and achieving consistent formulations.

2. Streptomyces as a Source of Biofungicides

2.1 Overview of Streptomyces

The *Streptomyces* genus, a significant group of Gram-positive bacteria belonging to the phylum Actinobacteria, is universally valued for its complex life cycle and metabolic flexibility (Figure 1) (Raja and Prabakarana 2011, Law et al. 2018). Its role in soil ecology is fundamental, as it degrades organic compounds, thereby contributing substantially to the nutrient cycle and the soil's overall health (Pang et al. 2022). This highlights the importance of understanding and harnessing *Streptomyces'* potential in soil ecosystems, which, in turn, highlights the significant impact of researchers' work on the environment. Their filamentous structure, similar to the mycelium of fungi, is one of their most intriguing features (Olanrewaju and Babalola 2019). This unique morphology equips *Streptomyces* to colonize and effectively penetrate the soil matrix. The mycelial threads aid in nutrient absorption from the soil, contributing to the growth of diverse habitats (Chater et al. 2010). These filaments can also differentiate into aerial hyphae, producing spores that disperse through the air, facilitating colonization and survival across various habitats (Chater et al. 2010, Torres-Rodriguez et al. 2022).

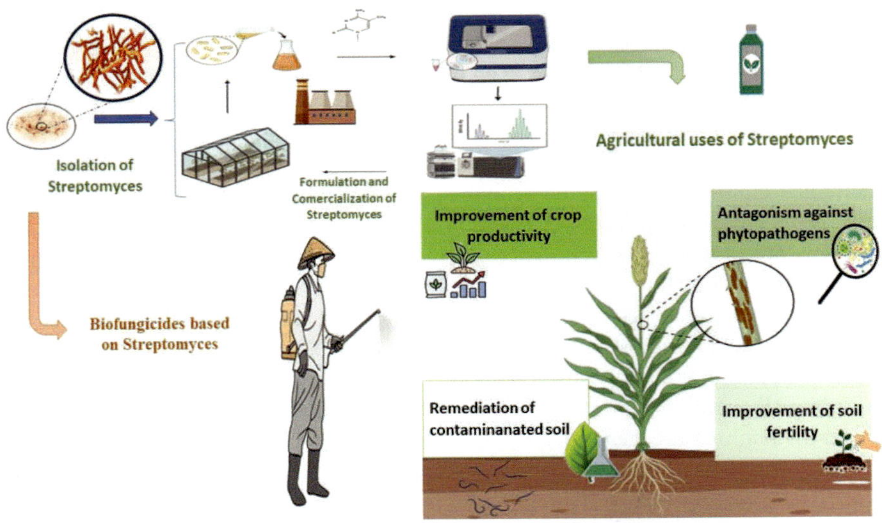

Figure 1: Exploring the Potential Uses of Streptomyces in Agriculture: A Schematic Overview.

Streptomyces, in addition to their soil-dwelling nature, play a substantial role in the nutrient cycle. They can produce numerous enzymes involved in the decomposition of complex organic substances like chitin, lignin, and cellulose (Yamunarani et al. 2019). Furthermore, these enzymes contribute to the degradation of organic substances, unlocking and releasing essential nutrients, such as nitrogen and phosphorus, to other organisms in the soil ecosystem (El-Hadi et al. 2020).

Streptomyces species are also very apt to establish mutualistic–antagonistic interactions with the roots of plants or other microorganisms (Takano et al. 2016, Krespach et al. 2023). In this way, they promote plant growth and development by producing phytohormones while, at the same time, protecting the roots from pathogenic fungi through antifungal compounds (Olanrewaju and Babalola 2019).

The history of *Streptomyces* dates back to 1943 when Waksman and Henrici made their groundbreaking discovery. This marked the eminent beginning of research associated with microbial biotechnology and pharmaceuticals, specifically the discovery of streptomycin, which had a profound effect on treating tuberculosis and other bacterial infections (Waksman and Henrici 1943, Khandelwal and Dubey 2020). Since then, *Streptomyces* has been known as one of the most productive genera in terms of secondary metabolite synthesis, contributing nearly 55% of the total antibiotics from natural sources. However, the overuse of *Streptomyces*-derived antibiotics has driven the development of antibiotic resistance in bacterial strains, posing a significant challenge in medicine (de Lima Procópio et al. 2012, Dow et al. 2023).

Streptomyces, particularly soil-dwelling strains, synthesize diverse bioactive substances, including antifungals, anticancer agents, and immunosuppressants. The antifungal agents from *Streptomyces*, especially their bioactive compounds, play a pivotal role in the development of biofungicides, offering a sustainable option in place of chemical fungicides (Zhang et al. 2015, Singh et al. 2017). *Streptomyces* compounds, such as streptochlorin and its derivatives, have shown promise in combating fungal plant pathogens, which suggests their potential as new compounds for agricultural use (Zhang et al. 2015, Shemedyuk et al. 2022). The synthesis of novel analogs of streptochlorin is a progressive strategy for enhancing the efficacy and spectrum of biofungicidal compounds (Jia et al. 2018). It further emphasizes the need to optimize fermentation conditions to produce these antifungal agents in maximum quantities. This optimization will have practical scaling-up significance for this production process when a critical, reliable supply of biofungicides must be achieved (Tan et al. 2020). For instance, marine-derived *Streptomyces* species have been explored for their unique antifungal compounds, which are effective against challenging pathogens like *Candida albicans* (Abou El-Enain et al. 2024).

The *Streptomyces* bioactive compounds have been discovered to be antifungals and also form a broad class of other bioactive molecules of an anticancer nature and those that exhibit activity in immunosuppression (Park et al. 2013, Sarmiento-Vizcaíno et al. 2018). These compounds, such as staurosporine, are potent biological compounds synthesized so far and have shown promise in kinase inhibition, an essential strategy in treating cancer diseases (Park et al. 2013).

2.2 Secondary Metabolites with Antifungal Properties

Streptomyces is renowned for the diversity of its bioactive products, such as antifungal, anticancer, and immunosuppressive agents. These make *Streptomyces* organisms necessary and even irreplaceable in many biotechnological applications (Jia et al. 2018). *Streptomyces* bioactive products include antifungal agents that participate in developing sustainable fungicides called biofungicides (Singh et al. 2017, Lacey and Rutledge 2022). In 2020, more than 70 new secondary metabolites were isolated from different *Streptomyces* species, demonstrating the enormous potential of this genus as a source of bioactive substances. These compounds, which encompass cyclic and linear peptides, polyketides, peptide derivatives, polyaromatics, macrocycles, and terpenoids, offer a wide range of structural diversity and biological activity (Lacey and Rutledge 2022).

Recent studies have brought to light the remarkable potential of specific compounds, notably the cyclic peptide secondary metabolites, which are the specialty of the prolific bacterium *Streptomyces yatensis* (Chen et al. 2021). With its unique ability to produce cyclic peptides, this bacterium has given rise to compounds such as yatencin A and yatencin B. These compounds have shown remarkable antifungal activities against root-rot pathogens of *Panax notoginseng*, which constitutes a significant breakthrough in the field. These compounds, classified as cyclic hexapeptides and pentapeptides, are not just any compounds. They are renowned disruptors, capable of breaking down cell membrane integrity and halting fungal growth. Their potent bioactivity positions them as strong candidates for agricultural biocontrol (García-Gutiérrez et al. 2024). They offer a promising prospect for the future of sustainable agriculture, ushering in a new era of effective biocontrol (Chen et al. 2021). The unique ring structure contributed by piperazic acid plays a pivotal role in the effectiveness of these peptides. This acid is crucial for the interaction of peptides with fungal cell components, typically by disrupting cell wall structures or inhibiting essential enzymes in fungal metabolic pathways. Incorporating piperazic acid into these peptides stabilizes their molecular structure and significantly boosts their ability to target and disrupt fungal cell functions (Lacey and Rutledge 2022).

Additionally, *streptomyces* sp. S–2, isolated from black soot, is renowned for its synthesis of a diverse array of secondary metabolites, including polyketides and non-ribosomal peptides, facilitated by non-ribosomal peptide synthetases (NRPS) and type I polyketide synthetases (PKS). The thorough genomic analysis of this strain has uncovered a vast collection of 55 distinct biosynthetic gene clusters, affirming its significant potential for novel antibiotic production (Siupka et al. 2020).

Streptomyces sp. S-2 is a fascinating strain that produces diverse bioactive compounds encompassing polyene and non-polyene structures. Among these, the polyenes candicidins are responsible for their potent antifungal activities, which disrupt fungal cell membranes and cause cell lysis and death. The strain also yields non-polyene compounds, such as lactophenol and surugamide A, which function through various mechanisms, including the inhibition of essential enzymes in fungal metabolic pathways. The diversity and complexity of these compounds present numerous opportunities for further investigation (Bobek et al. 2022).

The biosynthetic pathways utilized by *Streptomyces* sp. S-2 synthesize compounds through complex pathways involving multiple enzymatic steps. These steps ensure the formation of chemically diverse and biologically potent molecules. The NRPS and PKS systems, which are central to this process, play vital roles in the assembly of these molecules. This enzymatic assembly line is not only efficient but also highly versatile, allowing the organism to customize the production of a wide range of structural variants, each with unique biological properties (Siupka et al. 2020).

Another noteworthy example is *Streptomyces griseoruber* Acta 3662, which produces α-pyrone metabolites known as gombapyrones. These metabolites inhibit protein tyrosine phosphatase 1B and glycogen synthase kinase 3β, showcasing potential applications beyond antifungal activity, such as antitumor and antidiabetic therapies (Helaly et al. 2010). Indeed, gombapyrones stand out with their unique α-pyrone ring structure. This distinctive feature enables them to interact with and inhibit specific enzymes crucial in critical regulatory pathways. The inhibition of PTP1B and GSK-3β by gombapyrones suggests a fascinating mechanism where these metabolites modulate cellular processes (Helaly et al. 2009, 2010).

Streptomyces species are known for producing terpenoids and other aromatic compound derivatives with unique properties. They exhibit antifungal activities by generating reactive oxygen species within fungal cells or inhibiting specific fungal enzymes crucial for cell wall synthesis. The diverse range and high efficacy of these compounds highlight the remarkable chemical potential of *Streptomyces* in both ecological and medical contexts (Kirkpatrick 2009). Terpenoids and aromatic compounds from *Streptomyces* include a range of structures, from simple monoterpenes to complex polycyclic aromatic compounds. These metabolites interact with fungal cells in various ways, often disrupting cell wall integrity or interfering with key metabolic pathways (Lacey and Rutledge 2022).

3. Mechanisms of Action

3.1 Biocontrol Mechanisms against Phytopathogenic Fungi

As agricultural methodologies evolve to reduce the health and environmental dangers associated with chemical fungicides, the investigation into biological control agents has become more pronounced. Notably, *Streptomyces* is at the forefront of this field due to its significant biocontrol properties. Renowned for producing critical antibiotics, these microorganisms also play a pivotal role in agricultural realms, particularly in managing phytopathogenic fungi (Torres-Rodriguez et al. 2022). The expanding recognition of their diverse mechanisms of action, which include the disruption of fungal cell structure and the modulation of plant immune responses, underscores their potential in formulating sustainable crop protection strategies. This provides a foundation to explore the specific mechanisms through which *Streptomyces* manifests its antifungal activities, enhancing its utility as an effective biocontrol agent in agricultural settings (Figure 2).

Indeed, the biosynthesis of secondary metabolites in *Streptomyces* is crucial to their survival. These metabolites, produced through complex enzymatic pathways, result from a sophisticated conversion of primary metabolites into various bioactive compounds. These pathways, typically encoded by gene clusters, are crucial for controlling the synthesis of specific metabolites such as volatile organic compounds (VOCs), antifungal peptides, antibiotics, and other chemically diverse molecules (Torres-Rodriguez et al. 2022). In their evolutionary brilliance, *Streptomyces* utilize these metabolites to interact with their environment and inhibit the growth of competing microbial species, including fungi. These pathways produce vital metabolites such as rutamycins, quercetin, apigenin, and various organic acids (ferulic, vanillic, and gallic acids). The production of these compounds, which

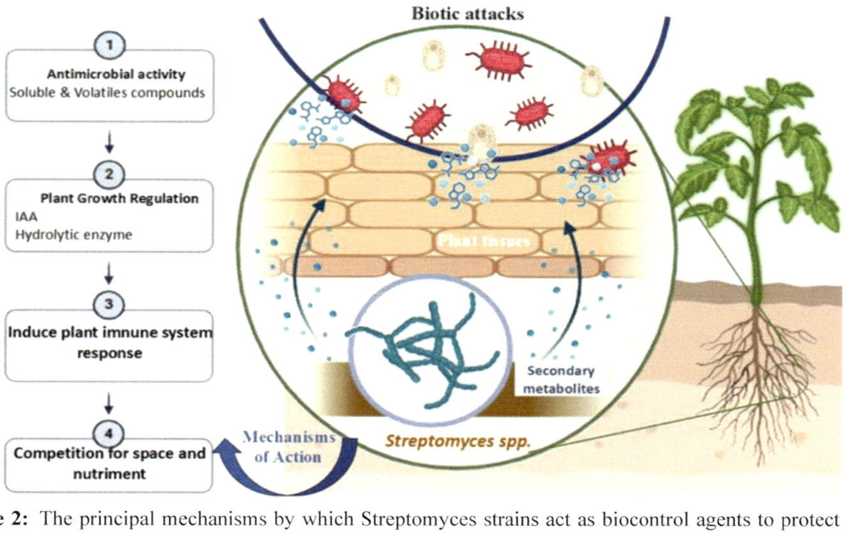

Figure 2: The principal mechanisms by which Streptomyces strains act as biocontrol agents to protect plants from diseases.

demonstrates the adaptability of *Streptomyces*, can be further enhanced under specific conditions that can be optimized in laboratory or industrial settings to increase yield and efficacy (Abodobara et al. 2020, Torres-Rodriguez et al. 2022).

The regulation of secondary metabolite synthesis in *Streptomyces* is a remarkably complex process that is tightly controlled by various genetic and environmental factors. Among these complex regulatory mechanisms, butyrolactone autoregulators emerge as pivotal signaling molecules that significantly influence the production of secondary metabolites (Thao et al. 2017). These autoregulators play a particularly crucial role in triggering the biosynthesis pathways that lead to the formation of various bioactive compounds, including antibiotics and antifungal agents (Kato et al. 2007, Kawabuchi et al. 2006). Butyrolactone autoregulators, such as the extensively studied A-factor in *Streptomyces griseus*, function as substantial molecular signals that initiate and regulate the synthesis of secondary metabolites. Binding to specific receptor proteins can activate or repress the transcription of genes involved in secondary metabolite biosynthesis (Kato et al. 2007, Thao et al. 2017). This interaction typically unfolds through a cascade of regulatory events, where the binding of a butyrolactone to its receptor alters the transcriptional activity of multiple genes across various biosynthetic clusters (Kato et al. 2007, Thao et al. 2017).

One of the remarkable features of *Streptomyces* species is their exceptional responsiveness to environmental changes. The synthesis and interaction of butyrolactone autoregulators facilitate this adaptability. These autoregulators play a crucial role in allowing *Streptomyces* to adjust their metabolic output based on nutrient availability and ecological competition (Mingyar et al. 2015). Furthermore, the synthesis of these autoregulators and their interaction with receptor proteins are intricately linked to the morphological development of *Streptomyces*, connecting the biosynthesis of secondary metabolites to the life cycle of the bacteria. This ensures that the energy-intensive process of metabolite production is only initiated when the organism is developmentally prepared and the environmental conditions are favorable (Mingyar et al. 2015, Xu and Yang 2019).

Quercetin and apigenin exert their antifungal effects primarily by disrupting essential cellular functions within fungi. These compounds inhibit key enzymatic pathways critical for cell wall synthesis, leading to compromised cell wall integrity and eventual cell lysis. This disruption not only inhibits growth but also prevents fungi from invading host plant tissues, thereby limiting the spread of infection (Jee et al. 2008, Magar and Sohng 2020). Rutamycins, on the other hand, target the protein synthesis machinery within fungal cells. By interfering with the translation process, these compounds reduce the production of vital proteins necessary for fungal growth and reproduction. This mode of action is particularly detrimental to the ability of fungi to recover and increase under adverse conditions, thereby reducing overall fungal proliferation in the environment (White et al. 1997, Hu Qiufen 2022).

Additionally, the competitive interaction of *Streptomyces* in the rhizosphere is not just a side effect but a crucial aspect of its role as a biocontrol agent against phytopathogenic fungi. This multifaceted competition for nutrients and space involves direct competition, changes in community dynamics, and the modulation of environmental conditions. These factors work together to suppress pathogen growth and colonization, underscoring the significance of *Streptomyces'* role in biocontrol (Torres-Rodriguez et al. 2022).

However, the activity of *Streptomyces* in the rhizosphere often leads to the secretion of organic acids and other compounds that can lower soil pH. This soil acidification, a crucial part of *Streptomyces'* biocontrol strategy, creates an environment that is less favorable for the growth of certain phytopathogenic fungi, which prefer neutral to slightly alkaline conditions (Sriragavi et al. 2023). A lower pH can make the soil environment hostile to certain phytopathogenic fungi. The secretion of enzymes by *Streptomyces* further assists in breaking down organic matter, which not only releases nutrients but can also lead to the generation of acidic by-products, further reducing the soil pH (Zhang et al. 2023). Beyond its metabolic activities, *Streptomyces* can also enhance populations of beneficial microbes in the rhizosphere. This enhancement arises from either improving the soil environment or directly suppressing pathogenic microbes, thus allowing beneficial microorganisms to thrive. Studies highlight the complex and intriguing interactions between soil amendments and microbial dynamics, including those influenced by *Streptomyces* (Viaene et al. 2016).

Furthermore, *Streptomyces* boosts plant immune responses by triggering a process called induced systemic resistance (ISR) (Viaene et al. 2016). Mainly, *Streptomyces* induces the accumulation of defense-related enzymes such as peroxidase, polyphenol oxidase, phenylalanine ammonia-lyase, and β-1,3 glucanase in plants. These enzymes bolster the plant cell walls, making it more challenging for pathogens to invade. For instance, *Streptomyces* strains have been found in a study to elevate these enzyme levels in rice plants, significantly reducing bacterial leaf streak disease and enhancing the plant's overall resistance (Hata et al. 2021). Also, specific *Streptomyces* isolates trigger innate immune responses in plants by improving the production of defense-related signaling molecules and up-regulating defense-related gene expression. For example, *Streptomyces cellulosae* isolate Actino 48 was found to increase peroxidase and chitinase enzyme levels and up-regulate defense-related gene expression in tomato plants, providing resistance against the tobacco mosaic virus (Abo-Zaid et al. 2020).

Moreover, some studies have shown that *Streptomyces* can influence broader defense mechanisms, including pathways related to salicylic acid, which are crucial for resistance against biotrophic pathogens. For instance, *Streptomyces* has been shown to induce systemic resistance in oak against powdery mildew by activating defense responses in pathways involving both salicylic acid and jasmonic acid/ethylene and enhancing the biosynthesis of phenylpropanoids (Kurth et al. 2014). *Streptomyces*, in addition to their role in systemic resistance, also play a pivotal function in activating defense responses. This is achieved through a process known as chemical priming, where the application of natural or synthetic compounds triggers faster and more robust plant defenses against pathogens. These responses are mediated by activating signaling pathways related to ISR and systemic acquired resistance (SAR) and using regulators like NPR1 and pathways involving salicylic acid signaling (Hönig et al. 2023).

3.3 Case Studies

The application of *Streptomyces* strains in agriculture is a significant part of global sustainability efforts. These strains are distinguished by their unique approach to crop disease management, relying on diverse mechanisms of action (Viaene et al. 2016). These include biochemical interactions, where the strains produce compounds that disrupt the development of pathogens, and molecular antagonism, where the strains directly attack and destroy pathogens. This dual defense equips plants with robust protection from a broad spectrum of fungal infections (Pang et al. 2022). Extensive research has further solidified the value of these biocontrol agents in modern agricultural practices.

A notable study by Montesdeoca-Flores et al. (2023) demonstrated that *Streptomyces netropsis* A52M exhibited significant antifungal activity against *Botrytis cinerea* on cherry tomatoes. The mechanism primarily involves the secretion of antifungal compounds that inhibit the synthesis of the cell wall of pathogens and disrupt cellular integrity. This strain likely produces polyketide compounds, characterized by strong, broad-spectrum antifungal effects which penetrate the membrane of fungi, leading to cell degradation. Furthermore, a comprehensive evaluation of *Streptomyces* spp., strains R18(6) and 6(4) demonstrated substantial antifungal activity against *Pyrenophora tritici-repentis*, a destructive pathogen of wheat. These strains produce glucanases and chitinases, which are important in degrading fungal cell walls. The enzymatic breakdown disrupts fungal growth and morphology, effectively reducing the disease index on wheat leaves with high biocontrol efficacy percentages ranging from 90% to 100%. Even more intriguing is the dual mechanism of action, where the production of small-molecule antibiotics further inhibits fungal growth, showcasing a complex interplay of factors that enhances biocontrol efficacy (Pereira et al. 2022).

Otherwise, Sánchez-García et al. (2022) identified and characterized three specific strains of *Streptomyces* for their antagonistic capabilities against phytopathogenic fungi. These strains exhibited biocontrol activity through the secretion of siderophores, which sequester iron from the environment, depriving fungal pathogens of the essential nutrients necessary for their proliferation. Additionally, these strains might utilize quorum-sensing inhibitors that disrupt the communication within fungal colonies, preventing them from spreading effectively across the plant tissues. Additionally, Akponah and Ubogu (2023) explored actinomycetes isolated from the rhizosphere of *Cymbopogon citratus*, where several isolates demonstrated significant fungicidal activity. The effectiveness of these *Streptomyces* isolates is attributed to their capacity to produce VOCs and lytic enzymes. VOCs can diffuse through the soil and inhibit fungal mycelial growth and spore germination, while the lytic enzymes degrade fungal structures directly.

4. Streptomyces-based Biopesticides: Formulation and Commercial Products

4.1 Commercial Products Based on Streptomyces

Recent advancements in agricultural methodologies, which shift from synthetic pesticides to organic farming, highlight the importance of utilizing microorganisms with similar functionalities (Chakraborty et al. 2023). In this context, many biopesticides based on *Streptomycetes* have been formulated for the management of plant diseases (Silva et al. 2022). Table 1 offers a synopsis of the current knowledge regarding various commercially available biofungicides based on *Streptomyces* and their target phytopathogenic fungi. The main challenge in developing commercial microorganism-based formulations is the inconsistency in biocontrol efficacy between laboratory tests and field conditions. Additionally, creating a commercial product involving microorganisms is a costly, lengthy, complex, and interactive process. The effectiveness of a bioagent depends on its formulation, which should contain the correct concentration of novel microorganisms along with a mix of inert ingredients to create a bioformulation suitable for commercial field application. This formulation should consistently demonstrate positive results, be reasonably priced, and be easy to handle (Torres-Rodriguez et al. 2022). The performance of these microbial agents is influenced by external variables such as precipitation, humidity, and temperature, challenges that chemical fungicides have successfully overcome (Farda et al. 2022). The challenges mentioned above pose obstacles in transitioning an efficient microbial biological agent from *in vitro* conditions to a commercially viable product for field use. Despite microorganisms being a current alternative to decrease reliance on chemical fungicides, the number of registered *Streptomyces* bioagents for commercial use against phytopathogenic fungi remains limited (Chaurasia et al. 2018).

Extensive efforts have led to the successful commercialization of nine actinomycetes-based products worldwide, all of which have been derived from various species within the genus *Streptomyces* (Copping and Duke 2007, Kabaluk et al. 2010). Currently, a total of ten active compounds are approved as commercial products, all sourced from different strains of *Streptomyces* (Aggarwal et al. 2016). These achievements underscore the efficacy of actinomycetes as prolific producers of bioactive compounds, including antibiotics and other industrially valuable substances. Furthermore, ongoing research endeavors in laboratories worldwide are focused on the development of new strains of actinomycetes, with the ultimate goal of expanding the repertoire of commercial products derived from these microorganisms (Chaurasia et al. 2018). The continuous exploration and exploitation of actinomycetes highlight their significant role in the pharmaceutical, agricultural, and biotechnological industries, with promising prospects for future innovation and commercialization.

Table 1: List of products based on *Streptomyces* spp. available and marketed worldwide.

Streptomyces	Biocontrol Mechanism	Target fungi/ Disease	Commercial Name	Country	References
***Streptomyces rimosus* DSMZ 12424**	Antibiotic effect	*Phytophthora* spp., *Botrytis* spp., *Pythium* spp., *Alternaria brassicola, Fusarium* spp., *Rhizoctonia solani, Phomopsis* spp.	Rhizovit®	Not explicitly indicated	(Berg et al. 2001)
***S. lydicus strain* WYEC108**	Hyperparasitism and antibiotic effect	*Sclerotinia* spp., *Laveillula spp., Fusarium* spp., *Phytophthora* spp., *Sphaeroteca* spp., *Rhizoctonia* spp., *Erisiphe spp., Pythium* spp.	Actino-Iron® Micro108® Actinovate®	USA, Canada, EU countries.	(Lichatowich 2007)
S. kasugaensis	Antibiotic effect	*R. solan*/large patch and sheath blight	Safegrow KIBC	South Korea	(Kabaluk et al. 2010)
***S. colombiensis* WYE20**	Antibiotic effect	Brown patch, grey mold, and powdery mildews	Mycocide KIBC	South Korea	(Kabaluk et al. 2010)
***S. hygroscopicus* 5008**	Antibiotic effect	*Rhizoctonia* spp.	Validacin, Mycin Hustler, Valimun, Valida Dantotsupadanvalida,	-	(Copping and Menn 2000)
S. cacaoi var. asoensis	Hyperparasitism and antibiotic effect	*Alternaria* spp., *Sclerotium* spp., *Helminthosporium* spp., *Sphaerotheca* spp., *Botrytis* spp., *Corynespora* spp., *Cochliobolus* spp.,/sheath blight, powdery mildews.	Polyoxorim (Polyoxin Z, Stopit and Endorse)	EU countries	(Copping and Duke 2007)
***S. strain griseoviridis* K61**	Hyperparasitism antibiotic effect and competition	*R. solani, Pythium* spp., *Fusarium* spp., *Ceratocystis radicicola, Phytophthora* spp., *Alternaria* spp.	Mycostop®	Canada and USA	(Suleman et al. 2002)

4.2 Formulations of BCAs

Formulations with a broad range of antimicrobial properties, which promote plant development, increase longevity, and efficiently control infections in the field, hold strong potential for commercialization and technological advancement. Research demonstrates that commercial biocontrol compounds that promote plant resistance to stress are easy to apply and raise yield. Still, the formulation process is the most significant obstacle to effective industrialization for agro-pharmaceutical companies. This involves using novel microorganisms as active ingredients in appropriate quantities, together with a variety of inert components, to produce field-ready commercial products (Pacios-Michelena et al. 2021). Furthermore, commercial formulations must adhere to all regulations concerning agrochemicals and microbial inoculants, while also meeting the demands of producers. These demands include consistent positive outcomes, ease of handling, and affordability (Elnahal et al. 2022). Taking these factors into account, microorganisms encounter a significant obstacle in preserving their activity during storage. This creates difficulties in ensuring both extended shelf life and stability under usual storage conditions for growers, as highlighted by Dhir (2017).

The formulation of the bioagent represents a multi-step procedure initiated immediately after selecting the most effective strain for the desired effect on a target pathogen. The primary goal of this process is to mix viable cells with adjuvants and carriers, resulting in formulations that offer both stability and established effectiveness. From a technical standpoint, achieving this requires a profound understanding of the biology and ecology of the bioagent, the pathosystem, and the specific application context (Bejarano and Puopolo 2020).

a. Production of microbial biocontrol agents

The concentration of microorganisms in the bioformulation is a critical determining factor for the effective colonization of plant tissues, so the number of viable cells available for crops needs to be sufficient to ensure successful plant biocontrol (Mitter et al. 2021). Among the major challenges in the development of *Streptomyces*-based biopesticides is attaining high culture densities on a large scale. This is attributed to their relatively slower multiplication rate compared to other types of bacteria (Sandoval-Powers et al. 2021). Furthermore, liquid cultures of *Streptomyces* tend to constitute aggregates and form mycelial pellets, posing challenges for product homogenization (Silva et al. 2022). The optimization of biotechnological processes related to the industrial production of *Streptomyces* spores can be performed through solid-state fermentation (SSF) and submerged fermentation (SF). The latter generally involves a more expensive process but enables improved control of parameters, streamlined industrialization, and a shorter culture period. Numerous investigations have been conducted to comprehend and optimize the culture variables that maximize spore production while mitigating the formation of aggregates and granules in this fermentation type (Glazebrook et al. 1990, Silva et al. 2022). The SSF generally involves a less expensive process but requires a longer growing period and more labor-intensive downstream processes. Similarly, within this fermentation process, certain investigations have explored enhancing spore production by *Streptomyces*

(Zhou et al. 2019). Similar challenges are encountered in the on-farm production of *Streptomyces*, particularly in the context of cultivating them through SSF or SF, often resulting in the formation of pellets and aggregates (Hamedi et al. 2017). The formations developed in submerged cultivation lead to reduced uniformity in the dispersion of cellular structures and can result in the obstruction of spray nozzles. Addressing these challenges might require the utilization of bioreactors and subsequent steps, potentially making the entire process economically impractical in on-farm production settings (Silva et al. 2022).

b. Additives used in microbial formulation

In addition to the live microorganisms, bioformulations incorporate inert compounds that improve the product's shelf life and ease of application in the field. Interaction of these ingredients is critical to ensure the bioactivity of the strain remains uncompromised (Bhardwaj et al. 2014). Besides conventional protective compounds such as Arabic gum, glycerol, and sucrose, which enhance the longevity of microorganisms, bioformulations can be modified by incorporating adjuvants to facilitate handling, mixing, efficiency, and application. For example, activator adjuvants enhance resistance to leaching, retention, spreading, penetration, and reduce the evaporation rate. These include oils and surfactants (Table 2). Non-ionic surfactants, such as sequenced copolymers and polyoxyethylenes, are among the most widely utilized in agriculture. Utility adjuvants modify the physical characteristics of the bioformulation, thereby mitigating challenges associated with handling and application. These comprise humectants like ethylene glycol, emulsifiers like ethylene oxide, spreading agents like alcohol ethoxylates, coupling agents like organosilanes, dispersants like celluloses, stabilizing agents like EDTA, co-solvents like glycol, thickeners like cellulose, buffering agents like citric acid, UV adsorbents like zinc oxide, stickers like polysaccharide gums, foaming and

Table 2: Adjuvants employed in microbial formulations.

Adjuvant	Type	Function
Hygroscopic agents	Humectants	Prevent premature drying of the deposit
Solvents	Coupling agents	Sustain the solubility of active ingredient
	Cosolvents	
Surfactant	Penetrants	Enhance dispersion, wetting, or spreading
	Spreaders	
	Wetting agents	
Oils	Crop oil	Enhance photostability, uptake
	Mineral oils	
Stabilizers	Anti-agglomeration agents	Ensure stability during the application process
	Dispersants	
	Emulsifiers	

defoaming agents like silica-based compounds, and colorants (Bejarano and Puopolo 2020). The efficacy of some biofungicides depends on providing abundant viable and active antagonistic microorganisms in the field, requiring high-quality inoculants. The carrier substrate is a critical component, essential for supporting a substantial microorganism population. Although carriers are inert ingredients without biocontrol activity, they can influence the formulation's efficacy and the shelf life of antagonistic microorganisms (Keswani et al. 2016). Improved survival of *Pseudomonas* species on minerals with small particle sizes, like montmorillonite, vermiculite, and zeolite, in comparison to minerals with larger particle sizes like kaolinite, pyrophyllite, and talc, was observed when stored at 20°C (Zhang et al. 2011). Previously, a comparison between attapulgus clay and diatomaceous earth was conducted to examine various physical characteristics such as water-holding ability and strength after autoclaving. Attapulgus clay granules expanded and lost their structure after autoclaving and exposure to water, whereas diatomaceous earth granules maintained their form and did not expand in water (Backman and Rodriguez-Kabana 1975). This characteristic allowed diatomaceous earth to absorb a molasses-based medium effectively for delivering *Trichoderma harzianum* to the field. In other instances, a mixture of wheat bran and perlite, as well as poplar bark compost, has been utilized as carriers with harmless *Fusarium* isolates combined with soil to induce protection from *Fusarium oxysporum* f. sp. *dianthi* (Garibaldi et al. 1985). Crushed tree bark has also proven to be a successful carrier for microbial formulations (Stack and Kenerley 1987). Peat commonly serves as a carrier for *Rhizobium* spp. and is also valuable for seed treatment and soil applications involving antagonistic microorganisms (Keswani et al. 2016).

Biofungicides are commonly introduced into plants using four distinct methods: seed treatment, which involves immersing seeds in a known concentration of microbial solution; soil application, where the active solution is incorporated into the soil or distributed around the root zone. Before planting, the roots can be treated with a biopesticide solution to support growth. The biopesticide can also be applied through foliar spraying, which employs spraying the solution on the leaves, allowing plants to receive treatment on infected leaves (Mishra et al. 2018).

Applying the inoculum to seeds supplies the microorganisms with nutrients derived from germination, enabling them to colonize roots promptly upon their formation. Conversely, introducing the inoculum directly into the soil or roots permits antagonistic bacteria to colonize roots that are already generated. Inoculating into the roots enables an immediate interaction between the antagonist and the plant, hastening the colonization process (Lopes and Gurgel 2021). Via foliar spraying, plants can undergo treatment consistently across the season. This can involve targeting specific growth stages to encourage plant development and improve resilience, or addressing precise climatic conditions to fight plant pathogens (Jakubiec-Krzesniak et al. 2018). Generally, formulations are conceived to reduce additional expenses over traditional synthetic pesticide applications (Abbey et al. 2019). In this context, biopesticides are frequently administered in the field using standard spraying methods on the plant and the soil (Sahayaraj 2014).

4.3 Constraints in the Production of Streptomyces-based Biofungicides

Sensitivity to abiotic factors, as well as nutritional requirements, constitutes problems when developing bioformulations based on *Streptomyces* (Saberi-Riseh and Roohallah 2021). Consequently, many formulations utilizing bioactive *Streptomyces* metabolites have been developed to protect plants (Pacios-Michelena et al. 2021). During the formulation processes of *Streptomyces*-based bioproducts, several additional challenges could be encountered. Effective formulations with bacterial mycelium/spores, whether in powder, liquid, or solid form, are required for farming applications, as fresh microbial cells are unsuitable for this purpose (Miyada et al. 2017). The optimization of spore production and the formulation of stable commercial products based on *Streptomyces* are key areas of research. Typically, this involves exploring encapsulation or drying technologies to safeguard spores from high water activity (Aw) and other abiotic factors that could affect their viability during shelf life. Several studies have investigated stabilization processes for *Streptomyces* in formulations, employing drying methods in conjunction with protectants (Chung et al. 2005, Saberi-Riseh and Roohallah 2021). Nevertheless, further advancements are required to enhance the overall stability of bioformulations.

The *in vitro* effectiveness of *Streptomyces* against various plant pathogens is frequently linked with positive outcomes, showcasing antagonistic activities that reinforce their potential application as components of biopesticides for plant disease control (Al Hamad et al. 2021). Regardless, the *in vitro* behavior does not consistently align with the results obtained when assessing the antagonistic activity *in vivo*. Likewise, several bacterial strains with limited *in vitro* impact demonstrate effective biocontrol activity when used in planta (Besset-Manzoni et al. 2019). This contrast in the two screening methods could be attributed to the influence of additional abiotic and biotic pressures, which either suppress or enhance the antagonistic effect of MBCAs. These pressures involve competition and other interactions between *Streptomyces* and foliar/soil microbiota, climatic variations (e.g., fluctuating humidity and temperature), challenges linked to the effective establishment in the plant tissues or rhizosphere, and fluctuations in nutrient availability (Jakubiec-Krzesniak et al. 2018). Therefore, the results obtained in controlled settings and sterile conditions require validation in field conditions to evaluate the actual effectiveness of the examined antagonistic microbe (Thambugala et al. 2020, Farhaoui et al. 2022).

5. Streptomyces in Biopesticides: Microbiota Interactions, Stress Responses and Root Colonization

Since many biopesticides are made up of living organisms, their effectiveness relies heavily on how these microbes interact with both environmental conditions and other organisms in the ecosystem. The soil contains a variety of organisms, and exploring the interaction of *Streptomyces* with the surrounding microbiota, as well as with other organisms and its relationship with the host plant, can assure the effectiveness of the product (Lopes et al. 2021). Additionally, *Streptomyces* may encounter substantial fluctuations in temperature, salinity, light exposure, humidity, nutrient availability, and soil pH (Egamberdieva et al. 2017). Stress resulting from these factors may interfere with the rhizosphere and its components, inducing changes in the activity of *Streptomyces* related to plant defense and growth stimulation (Etesami 2020). Many species can withstand these stresses by synthesizing osmoregulatory compounds, which can stop the plasmolysis of cell membranes and control homeostasis and enzymatic activity (Lopes et al. 2021). In response to high osmolarity conditions, strains of *Streptomyces* synthesize osmoprotectors such as trehalose, glycine, proline, and betaine—metabolites that trigger osmotic adaptation and maintain biological activity even in stressful situations (Ameur et al. 2011).

The root colonization ability is crucial for effective inoculation of a microorganism-based biopesticide (Doumbou et al. 2010). Generally, *Streptomyces* are among the effective root- and rhizosphere-colonizing bacteria (Merzaeva and Shirokikh 2006). This could be linked to the source of their isolation, related to the roots and conditions prevailing in the soil. Nevertheless, numerous bacteria exhibit high colonization ability in controlled condition tests but struggle to reproduce this ability when applied in the field (Bonaldi et al. 2015). This might be ascribed to the decreased production of active compounds during competition with other organisms in the soil microbiota, resulting in decreased rhizosphere competence and inefficient colonization of crop tissues (Compant et al. 2010). The ability to colonize the rhizosphere is also of crucial importance for biopesticides targeting soil pathogens like plant-pathogenic fungi (Walker et al. 2003). Colonization of crop tissues by the bioagent of the biopesticide could be necessary to combat many foliar diseases (Akbaba and Ozaktan 2018). This may be achieved through the use of endophytic *Streptomyces*, which release biomolecules, including lytic enzymes, that enhance their capacity to penetrate and persist in plant tissues (Liu et al. 2017). The selected inoculation technique is another factor influencing the colonization of the plant by the bioagent (Afzal et al. 2013). Generally, bacteria exhibit limited motility within the soil. Therefore, depending on the species and its bioactivity, applying it as close to the rhizosphere as possible seems essential to achieve satisfactory results (Lopes et al. 2021).

6. Synergistic Approaches

The collaboration between fungi and bacteria presents significant promise for enhancing crop cultivation through microbial assistance. Microorganisms like *Trichoderma* sp. and *Streptomyces* sp. can serve as biocontrol agents due to their ability to produce enzymes or components capable of altering cell structures, namely the enzyme chitinase (Fitriana et al. 2020). It possesses the capability to damage chitin, a component found in the cuticles or shells of insects. Furthermore, it has been demonstrated that the combination of metabolites from these microorganisms exhibits synergistic effects. In another study assessing the efficacy of a biopesticide containing a mixture of *Trichoderma* sp. and *Streptomyces* sp. in suppressing the insect *N. viridula*, signs of insect death were exhibited due to the decomposition of chitin in the thorax and abdomen (Agadhia et al. 2022). Previously, a combination of *Trichoderma* sp. and *Streptomyces* sp. has been shown to decrease the feeding activity of *Spodoptera litura* (Fitriana et al. 2020). Furthermore, these microorganisms may act as PGPR agents. Nduka et al. (2017) suggested that *Streptomyces* and *Trichoderma* are advantageous soil microorganisms that degrade organic debris, stimulate plant development, and function as biological agents. Notably, a considerable number of products based on *Trichoderma* are presently available on the market (Woo et al. 2023). The metabolome of this fungus is remarkably intricate and can exert direct effects on phytopathogens through antagonism. Moreover, it can indirectly influence the host plant's defense system (Vinale et al. 2012). *Trichoderma* has been observed to increase soil fertility by augmenting the nutrient uptake potential, such as potassium (K), phosphorus (P), and nitrogen (N), along with other micronutrients (Lee et al. 2015). Meanwhile, López-Mondéjar et al. (2018) highlighted the key function of *Streptomyces* as decomposers, contributing to the carbon cycle within ecosystems. *Streptomyces* is known to possess 10 enzymes utilized in the decomposition process of lignin and polysaccharides found in fungi and plants. Conversely, another study indicated that the utilization of *Streptomyces* sp. alone yielded more efficacy in deterring soybean pests like *Nezara viridula*, *Bemisia tabaci*, and *Aphid* spp., compared to employing a combination of *Trichoderma* sp. and *Streptomyces* sp. (Avrianto et al. 2022).

7. Conclusion and Future Prospects

Production of lytic enzymes, antibiotics, and various bioactive compounds highlights the considerable potential of *Streptomyces* as a valuable resource for microbial biopesticide development. Numerous strains of *Streptomyces* have shown a notable ability to produce extracts containing secondary metabolites with activity against phytopathogenic fungi. Despite extensive research on many of these bioactive metabolites, additional investigations are required to gain a comprehensive understanding of their chemical structures, diversity, molecular mechanisms of action, and their use in agriculture. Notwithstanding the significant antagonistic potential, only a limited number of biofungicidal products based on *Streptomyces* are currently available in the market. This scarcity can be attributed to various factors, including challenges related to large-scale bacterial production and formulation development. Likewise, a thorough grasp of the relationships between *Streptomyces,* host plants, and soil microbiota is imperative to identify the most effective bacterial inoculation technique and enhance overall efficacy. Thus, unlocking the vast biological treasure of *Streptomyces* can contribute to the development of bioformulations that enhance agriculture's productivity and sustainability, offering improved plant defense with microbial inoculants and bioformulations.

Acknowledgements

This study received financial support from the Phytopathology Unit at the Ecole Nationale d'Agriculture de Meknès, Morocco.

References

Abbey, Lord, Abbey, J., Leke-Aladekoba, A., Iheshiulo, E. M. A., and Ijenyo, M. (2019). Biopesticides and biofertilizers: Types, production, benefits, and utilization. *Byproducts from Agriculture and Fisheries: Adding Value for Food, Feed, Pharma and Fuels*, 479–500. https://doi.org/10.1002/9781119383956.ch20.

Abo-Zaid, G. A., Matar, S. M., and Abdelkhalek, A. (2020). Induction of plant resistance against tobacco mosaic virus using the biocontrol agent *streptomyces cellulosae* isolate actino 48. *Agronomy*, 10(11): https://doi.org/10.3390/agronomy10111620.

Abodobara, M., El-Falal, A., Mousa, M., and Shaaban, E. (2020). Optimization of cultural conditions for production of antifungal metabolites by *Streptomyces gougeroti*. *Scientific Journal for Damietta Faculty of Science*, 10(1): 29–36. https://doi.org/10.21608/sjdfs.2020.194997.

Abou El-Enain, I. M., Abed, N. N., Helal, E. E., Abdelkhalek, E. S., Suleiman, W., Safwat, N. A. et al. (2024). Eco-friendly Biosynthesis of Ag-NPs by *Streptomyces griseus* with *Anti-candida albicans* and Antitumor Activity. *Recent Advances in Anti-Infective Drug Discovery*, 19(1): 73–87. https://doi.org/10.2174/2772434418666230427165013.

Afzal, M., Khan, S., Iqbal, S., Mirza, M. S., and Khan, Q. M. (2013). Inoculation method affects colonization and activity of *Burkholderia phytofirmans* PsJN during phytoremediation of diesel-contaminated soil. *International Biodeterioration and Biodegradation*, 85(November): 331–336. https://doi.org/10.1016/j.ibiod.2013.08.022.

Agadhia, R. L., Suryaminarsih, P., Ramadhini, N., Mahendra, R., and Mujoko, T. (2022). Efficacy of biopesticide formula containing *Streptomyces* sp. and *Trichoderma* sp. against Southern Green Stink Bug (*Nezara viridula*) on Soybean (*Glycine max* L.). *Asian Research Journal of Agriculture*, 15(4): 218–226. https://doi.org/10.9734/arja/2022/v15i4372.

Aggarwal, N., Thind, S. K., and Sharma, S. (2016). Plant growth promoting actinobacteria: A new avenue for enhancing the productivity and soil fertility of grain legumes. 99–121, June 2016, 1–298. https://doi.org/10.1007/978-981-10-0707-1.

Akbaba, M., and Ozaktan, H. (2018). Biocontrol of angular leaf spot disease and colonization of cucumber (*Cucumis sativus* L.) by endophytic bacteria. *Egyptian Journal of Biological Pest Control*, 28(1): 1–10. https://doi.org/10.1186/s41938-017-0020-1.

Akponah, E., and Ubogu, M. (2023). Phytopathogenic Fungal Growth Inhibition by Actinomycetes Isolated from the Rhizosphere of *Cymbopogon citratus* (Lemongrass). *Acta Microbiologica Bulgarica*, 39(3): 308–316. https://doi.org/10.59393/amb23390312.

Al Hamad, B. M., Al Raish, S. M., Ramadan, G. A., Saeed, E. E., Alameri, S. S. A., Al Senaani, S. S. et al. (2021). Effectiveness of augmentative biological control of *Streptomyces griseorubens* uae2 depends on 1-aminocyclopropane-1-carboxylic acid deaminase activity against neoscytalidium dimidiatum. *Journal of Fungi*, 7(11). https://doi.org/10.3390/jof7110885.

Alblooshi, A. A., Purayil, G. P., Saeed, E. E., Ramadan, G. A., Tariq, S., Altaee, A. S. et al. (2022). Biocontrol potential of endophytic actinobacteria against fusarium solani, the causal agent of sudden decline syndrome on date palm in the UAE. *Journal of Fungi*, 8(1). https://doi.org/10.3390/jof8010008.

Ameur, H., Ghoul, M., and Selvin, J. (2011). The osmoprotective effect of some organic solutes on *Streptomyces* sp. mado2 and *Nocardiopsis* sp. mado3 growth. *Brazilian Journal of Microbiology*, 42(2): 543–553. https://doi.org/10.1590/S1517-83822011000200019.

Avrianto, N. I., Suryaminarsih, P., and Widajati, W. (2022). Keberadaan Serangga Tanaman Kedelai (*Glycines Max* L . Merril) Fase Pendahuluan Tanaman Kedelai (*Glycine max* (L.) Merill) merupakah salah satu komoditi tanaman pangan lebih ramah lingkungan. Beberapa jenis agensia hayati dengan penggunaan Streptomyc. *Seminar Nasional Agrote- Knologi Fakultas Pertanian UPN "Veteran" Jawa Timur 2021. NST Proceedings*, 2022: 31–39.

Backman, P. A., and Rodriguez-Kabana, R. (1975). A system for the growth and delivery of biological control agents to the soil. *Phytopathology*, 65(81): 3–21.

Bejarano, A., and Puopolo, G. (2020). *Bioformulation of Microbial Biocontrol Agents for a Sustainable Agriculture*. 275–293. https://doi.org/10.1007/978-3-030-53238-3_16.

Berg, G., Marten, P., Minkwitz, A., and Bruckner, S. (2001). Efficient biological control of fungal plant diseases by *Streptomyces* sp. DSMZ 12424/Effiziente biologische Bekämpfung pilzlicher Pflanzenkrankheiten durch *Streptomyces* sp. *DSMZ 12424. Zeitschrift für Pflanzenkrankheiten und Pflanzenschutz/Journal of Plan*.

Besset-Manzoni, Y., Joly, P., Brutel, A., Gerin, F., Soudière, O., Langin, T. et al. (2019). Does in vitro selection of biocontrol agents guarantee success in planta? A study case of wheat protection against Fusarium seedling blight by soil bacteria. *PLoS ONE*, 14(12): 1–18. https://doi.org/10.1371/journal.pone.0225655.

Bhardwaj, D., Ansari, M. W., Sahoo, R. K., and Tuteja, N. (2014). Biofertilizers function as key player in sustainable agriculture by improving soil fertility, plant tolerance and crop productivity. *Microbial Cell Factories*, 13(1): 1–10. https://doi.org/10.1186/1475-2859-13-66.

Bobek, J., Filipová, E., Bergman, N., Čihák, M., Petříček, M., Lara, A. C. et al. (2022). Polyenic antibiotics and other antifungal compounds produced by hemolytic *Streptomyces* species. *International Journal of Molecular Sciences*, 23(23). https://doi.org/10.3390/ijms232315045.

Bonaldi, M., Chen, X., Kunova, A., Pizzatti, C., Saracchi, M., and Cortesi, P. (2015). Colonization of lettuce rhizosphere and roots by tagged Streptomyces. *Frontiers in Microbiology*, 6(FEB): 1–10. https://doi.org/10.3389/fmicb.2015.00025.

Bonner, M. R., and Alavanja, M. C. R. (2017). Pesticides, human health, and food security. *Food and Energy Security*, 6(3): 89–93. https://doi.org/10.1002/fes3.112.

Boukaew, S., Chuenchit, S., and Petcharat, V. (2011). Evaluation of *Streptomyces* spp. for biological control of Sclerotium root and stem rot and Ralstonia wilt of chili pepper. *BioControl.*, 56: 365–374.

Chakraborty, N., Mitra, R., Pal, S., Ganguly, R., Acharya, K., Minkina, T. et al. (2023). Biopesticide Consumption in India: Insights into the Current Trends. *Agriculture (Switzerland)*, 13(3). https://doi.org/10.3390/agriculture13030557.

Chater, K. F., Biró, S., Lee, K. J., Palmer, T., and Schrempf, H. (2010). The complex extracellular biology of Streptomyces: Review Article. *FEMS Microbiology Reviews*, 34(2): 171–198. https://doi.org/10.1111/j.1574-6976.2009.00206.x.

Chaurasia, A., Meena, B. R., Tripathi, A. N., Pandey, K. K., Rai, A. B., and Singh, B. (2018). Actinomycetes: an unexplored microorganisms for plant growth promotion and biocontrol in vegetable crops. *World Journal of Microbiology and Biotechnology*, 34(9): 1–16. https://doi.org/10.1007/s11274-018-2517-5.

Chen, X., Zou, J. M., Li, L., Bao, K., Zhao, Y., Miao, C. P. et al. (2021). Cyclic peptide secondary metabolites with antifungal activity against root-rot pathogens of *Panax notoginseng* produced by *Streptomyces yatensis*. *Chemistry of Natural Compounds*, 57(6): 1181–1183. https://doi.org/10.1007/s10600-021-03582-w.

Chung, W. C., Huang, J. W., and Huang, H. C. (2005). Formulation of a soil biofungicide for control of damping-off of Chinese cabbage (*Brassica chinensis*) caused by *Rhizoctonia solani*. *Biological Control*, 32(2): 287–294.

Compant, S., Clément, C., and Sessitsch, A. (2010). Plant growth-promoting bacteria in the rhizo- and endosphere of plants: Their role, colonization, mechanisms involved and prospects for utilization. *Soil Biology and Biochemistry*, 42(5): 669–678. https://doi.org/10.1016/j.soilbio.2009.11.024.

Copping, L. G., and Menn, J. J. (2000). Biopesticides: a review of their action, applications and efficacy. *Pest Management Science: Formerly Pesticide Science*, 56(8): 651–676.

Copping, L. G., and Duke, S. O. (2007). Natural products that have been used commercially as crop protection agents. *Pest Management Science: Formerly Pesticide Science*, 63(6): 524–554.

de Lima Procópio, R. E., da Silva, I. R., Martins, M. K., de Azevedo, J. L., and de Araújo, J. M. (2012). Antibiotics produced by *Streptomyces*. In *Brazilian Journal of Infectious Diseases* (Vol. 16, Issue 5). https://doi.org/10.1016/j.bjid.2012.08.014.

Dhir, B. (2017). Biofertilizers and biopesticides: eco-friendly biological agents. *Advances in Environmental Biotechnology*, 167–188.

Doumbou, C. L., Salove, M. K. H., Beaulieu, D. L. C. C., Doumbou, C. L., and Salove, M. K. H. (2010). "Actinomycetes , promising tools to control plant diseases and to promote plant growth "Article de synthèse/Review article Actinomycetes, promising tools to control plant diseases and to promote plant growth. *Review Literature and Arts of the Americas*, 82(April), 85–102.

Dow, L., Gallart, M., Ramarajan, M., Law, S. R., and Thatcher, L. F. (2023). Streptomyces and their specialised metabolites for phytopathogen control – comparative in vitro and in planta metabolic approaches. In *Frontiers in Plant Science* (Vol. 14). https://doi.org/10.3389/fpls.2023.1151912.

Egamberdieva, D., Wirth, S., Jabborova, D., and Räsänen, L. A. (2017). Coordination between Bradyrhizobium and *Pseudomonas* alleviates salt stress in soybean through altering root system architecture. *Journal of Plant Interactions*, 12(1): 100–107. https://doi.org/10.1080/17429145.2017.1294212.

El-Hadi, A. A., Ahmed, H. M., Zaki, R. A., and Mohsen, A. M. (2020). Enhanced enzymatic activity of *Strepromyces griseoplanus* L-asparginase cia its incorporation in an oil-based nanocarrier. *Int. J. Appl. Pharm.*, 2020: 203–210.

El-Metwally, M., Abdel-Mogib, M., Elfedawy, M., Sohsah, G., Rezk, A., Moustafa, M. et al. (2021). Isolation, purification and structure elucidation of three new bioactive secondary metabolites from *Streptomyces lividans* AM. *Biocell*, 45(5): 1263–1271. https://doi.org/10.32604/BIOCELL.2021.013198.

El-shakh, A. S. A., Kakar, K. U., Wang, X., Almoneafy, A. A., Ojaghian, M. R., Li, B. et al. (2015). Controlling bacterial leaf blight of rice and enhancing the plant growth with endophytic and rhizobacterial *Bacillus* strains. *Toxicological and Environmental Chemistry*, 97(6): 766–785. https://doi.org/10.1080/02772248.2015.1066176.

Elnahal, A. S. M., El-Saadony, M. T., Saad, A. M., Desoky, E. S. M., El-Tahan, A. M., Rady, M. M. et al. (2022). The use of microbial inoculants for biological control, plant growth promotion, and sustainable agriculture: A review. In *European Journal of Plant Pathology* (Vol. 162, Issue 4). Springer Netherlands. https://doi.org/10.1007/s10658-021-02393-7.

Etesami, H. (2020). Plant–microbe interactions in plants and stress tolerance. In Plant life under changing environment (pp. 355–396). *Academic Press.*

Farda, B., Djebaili, R., Vaccarelli, I., Del Gallo, M., and Pellegrini, M. (2022). Actinomycetes from Caves: An overview of their diversity, biotechnological properties, and insights for their use in soil environments. *Microorganisms*, 10(2). https://doi.org/10.3390/microorganisms10020453.

Farhaoui, Abdelaaziz, Adadi, A., Tahiri, A., and Al., E. (2022). *Biocontrol* potential of plant growth-promoting rhizobacteria (PGPR) against *Sclerotiorum rolfsii* diseases on sugar beet (*Beta vulgaris* L.). *Physiological and Molecular Plant Pathology,* 119: 101829.

Farhaoui, A., Tahiri, A., Ezrari, S., Radouane, N., Khadiri, M., El, N. et al. (2022). *Identication, pathogenicity, fungicide sensitivity and biological control of Rhizoctonia solani associated with damping-off disease of sugar beet in Morocco*. 1–28. https://doi.org/10.21203/rs.3.rs-1517415/v1.

Farhaoui, A., Tahiri, A., Khadiri, M., El Alami, N., and Lahlali, R. (2023). Fungal Root Rots of Sugar Beets: A Review of Common Causal Agents and Management Strategies. *Gesunde Pflanzen,* 75(5): 1411–1440.

Fitriana, I. N., Suryaminarsih, P., and Mujoko, T. (2020). Potential of Multientomopa *Streptomyces* sp. and *Tripchoderma* sp. in Potato Extract Broth and Glucose Nitrate Broth Media on Pests (*Spodoptera litura*) Eating Behavior by *in Vitro* Test. 2019: 270–276. https://doi.org/10.11594/nstp.2019.0438.

García-Gutiérrez, C., Pérez-Victoria, I., Montero, I., Fernández-De la Hoz, J., Malmierca, M. G., Martín, J. et al. (2024). Unearthing a cryptic biosynthetic gene cluster for the piperazic acid-bearing depsipeptide diperamycin in the ant-dweller *Streptomyces* sp. CS113. *International Journal of Molecular Sciences*, 25(4). https://doi.org/10.3390/ijms25042347.

Glazebrook, M. A., Doull, J. L., Stuttard, C., and Vining, L. C. (1990). Sporulation of *Streptomyces venezuelae* in submerged cultures. *Journal of General Microbiology*, 136(3): 581–588. https://doi.org/10.1099/00221287-136-3-581.

Gradziel, T.M., Curtis, R., Socias i Company, R. (n.d.). *Almonds_ Botany, Production and Uses - book-Socias i Company, R., Gradziel, T.M., Eds.; CABI: Wallingford, UK, 2017;* pp. 70–86. https://books.google.co.ma/books?hl=fr&lr=&id=KxkxDwAAQBAJ&oi=fnd&pg=PA70&dq=Gradziel,+T.M.%3B+Curtis,+R.%3B+Socias+i+Company,+R.+Production+and+growing+regions.+In+Almonds.+ Botany,+Production+and+Uses%3B&ots=poOhreJW_K&sig=CGSvq4MKPgRo06G7RjYs36ARP1A &red.

Gullino, M. L., Garibaldi, A., and Romano, M. L. (1985). Identification and response to fungicides of *Colletotrichum gloeosporioides*, incitant of strawberry black rot in Italy. *Plant Disease*, 69(7).

Hamedi, J., Poorinmohammad, N., and Papiran, R. (2017). *Growth and life cycle of Actinobacteria. Biology and Biotechnology of Actinobacteria,* 29–50.

Hata, E. M., Yusof, M. T., and Zulperi, D. (2021). Induction of systemic resistance against bacterial leaf streak disease and growth promotion in rice plant by *Streptomyces shenzhenesis* TKSC3 and *Streptomyces* sp. SS8. *Plant Pathology Journal*, 37(2): 173–181. https://doi.org/10.5423/PPJ.OA.05.2020.0083.

Helaly, S., Schneider, K., Nachtigall, J., Vikineswary, S., Tan, G. Y. A., Zinecker, H. et al. (2010). ChemInform Abstract: Biosynthetic Capacities Actinomycetes. Part 52. Gombapyrones, new α-pyrone metabolites produced by *Streptomyces griseoruber* Acta 3662. *ChemInform*, 41(1). https://doi.org/10.1002/chin.201001216.

Helaly, S., Schneider, K., Nachtigall, J., Vikineswary, S., Tan, G. Y. A., Zinecker, H. et al. (2009). Gombapyrones, new α-pyrone metabolites produced by *Streptomyces griseoruber* Acta 3662. *Journal of Antibiotics*, 62(8): 445–452. https://doi.org/10.1038/ja.2009.70.

Hönig, M., Roeber, V. M., Schmülling, T., and Cortleven, A. (2023). Chemical priming of plant defense responses to pathogen attacks. In *Frontiers in Plant Science* (Vol. 14). https://doi.org/10.3389/fpls.2023.1146577.

Hu Qiufen, D. M. (2022). Study on the Secondary Metabolites from a Plant- Derived *Streptomyces* sp. D-154. *Biomedical Journal of Scientific & Technical Research*, 42(3). https://doi.org/10.26717/bjstr.2022.42.006748.

Huber, D. M., El-Nasshar, H., Moore, L. W., Mathre, D. E., and Wagner, J. E. (1989). Interaction between a peat carrier and bacterial seed treatments evaluated for biological control of the take-all diseases of wheat (*Triticum aestivum* L.). *Biology and Fertility of Soils,* 8: 166–171.

Jakubiec-Krzesniak, K., Rajnisz-Mateusiak, A., Guspiel, A., Ziemska, J., and Solecka, J. (2018). Secondary metabolites of actinomycetes and their antibacterial, antifungal and antiviral properties. *Polish Journal of Microbiology*, 67(3): 259–272. https://doi.org/10.21307/pjm-2018-048.

Jee, J. M., Chong, P. P., Ng, K. P., Ho, C. C., and Seow, H. F. (2008). Antifungal Properties of Secondary Metabolites from a *Streptomyces* species. *International Journal of Infectious Diseases*, 12: e285–e286. https://doi.org/10.1016/j.ijid.2008.05.766.

Jia, C. Y., Xu, L. Y., Yu, X., Ding, Y. B., Jin, B., Zhang, M. Z. et al. (2018). An efficient synthesis and antifungal evaluation of natural product streptochlorin and its analogues. *Fitoterapia*, 125: 106–110. https://doi.org/10.1016/j.fitote.2017.12.017.

Kabaluk, J. T., Svircev, A. M., Goettel, M. S., and Woo, S. G. (2010). The Use and Regulation of Microbial Pesticides in Representative Jurisdictions Worldwide. *IOBC Global*, 99.

Kato, J. Y., Funa, N., Watanabe, H., Ohnishi, Y., and Horinouchi, S. (2007). Biosynthesis of γ-butyrolactone autoregulators that switch on secondary metabolism and morphological development in *Streptomyces. Proceedings of the National Academy of Sciences of the United States of America*, 104(7): 2378–2383. https://doi.org/10.1073/pnas.0607472104.

Kawabuchi, M., Hara, Y., Nihira, T., and Yamada, Y. (2006). Production of butyrolactone autoregulators by Streptomyces coelicolor A3(2). *FEMS Microbiology Letters*, 157(1): 81–85. https://doi.org/10.1111/j.1574-6968.1997.tb12756.x.

Keinath, A. P., Fravel, D. R., and Papavizas, G. C. (1990). Evaluation of formulations of *Talaromyces flavus* for biocontrol of Verticillium wilt of potato. *Biol. Cult. Tests*, 7: 27.

Keswani, C., Bisen, K., Singh, V., Sarma, B. K., and Singh, H. B. (2016). Formulation technology of biocontrol agents: present status and future prospects. *Bioformulations: for Sustainable Agriculture,* 35–52.

Khan, M. Nasir, and F. M. (2014). Eutrophication: Challenges and Solutions. pp. 1–15. *In*: Abid, A. Ansari and Sarvajeet Singh Gill, (Eds.). Eutrophication: Causes, Consequences and Control. Dordrecht: Springer Netherlands. https://link.springer.com/10.1007/978-94-007-7814-6_1, accessed Februar.

Khandelwal, S., and Dubey, W. (2020). Overview of Mycobacterium: A Review. *European Journal of Molecular & Clinical Medicine*, 7(11): 6198–6213.

Kirkpatrick, P. (2009). Lead discovery: tailoring nuclear receptor activation. *Nature Reviews Drug Discovery*, 8(2): 107. https://doi.org/10.1038/nrd2807.

Krespach, M. K. C., Stroe, M. C., Netzker, T., Rosin, M., Zehner, L. M., Komor, A. J. et al. (2023). *Streptomyces polyketides* mediate bacteria–fungi interactions across soil environments. *Nature Microbiology*, 8(7): 1348–1361. https://doi.org/10.1038/s41564-023-01382-2.

Kurth, F., Mailänder, S., Bönn, M., Feldhahn, L., Herrmann, S., Groe, I. et al. (2014). Streptomyces-induced resistance against oak powdery mildew involves host plant responses in defense, photosynthesis, and secondary metabolism pathways. *Molecular Plant-Microbe Interactions*, 27(9): 891–900. https://doi.org/10.1094/MPMI-10-13-0296-R.

Lacey, H. J., and Rutledge, P. J. (2022). Recently Discovered Secondary Metabolites from Streptomyces Species. In *Molecules* (Vol. 27, Issue 3). https://doi.org/10.3390/molecules27030887.

Law, J. W. F., Tan, K. X., Wong, S. H., Mutalib, N. S. A., and Lee, L. H. (2018). Taxonomic and Characterization Methods of *Streptomyces*: A Review. In *Progress in Microbes and Molecular Biology* (Vol. 1, Issue 1). https://doi.org/10.36877/pmmb.a0000009.

Lee, J. H., Min, G. Y., Jeon, C. W., Choi, S. M., Shim, G. Y., and Kwak, Y. S. (2015). *Investigation of fungicide response of Streptomyces* spp. *isolated from rhizosphere in zoysiagrass. The Korean Journal of Pesticide Science*, 19(1): 54–63.

Lichatowich, T. (2007). The plant growth enhancing and biocontrol mechanisms of Streptomyces lydicus WYEC 108 and its use in nursery and greenhouse production. *In:* Riley, L. E., Dumroese, R. K., Landis, T. D. (eds.). *National Proceedings: Forest and Conservation Nursery Associations-2006. Procee.*

Liu, H., Carvalhais, L. C., Crawford, M., Singh, E., Dennis, P. G., Pieterse, C. M. J. et al. (2017). Inner plant values: Diversity, colonization and benefits from endophytic bacteria. *Frontiers in Microbiology*, 8(DEC), 1–17. https://doi.org/10.3389/fmicb.2017.02552.

Lopes, M. J. D. S., Dias-Filho, M. B., and Gurgel, E. S. C. (2021). Successful plant growth-promoting microbes: inoculation methods and abiotic factors. *Frontiers in Sustainable Food Systems*, 5: 606454.

López-Mondéjar, R., Brabcová, V., Štursová, M., Davidová, A., Jansa, J., Cajthaml, T. et al. (2018). Decomposer food web in a deciduous forest shows high share of generalist microorganisms and importance of microbial biomass recycling. *The ISME Journal*, 12(7): 1768–1778.

Magar, R. T., and Sohng, J. K. (2020). A review on structure, modifications and structure-activity relation of Quercetin and its derivatives. *Journal of Microbiology and Biotechnology*, 30(1): 11–20. https://doi.org/10.4014/jmb.1907.07003.

Malusà, E., Pinzari, F., and Canfora, L. (2016). Efficacy of biofertilizers: challenges to improve crop production. pp. 17–40. *In*: Dhananjaya Pratap Singh, Harikesh Bahadur Singh, and Ratna Prabha (Eds.). Microbial Inoculants in Sustainable Agricultural Productivity. New Delhi: Springer India. http://link.

Marín, L., Gutiérrez-del-Río, I., Villar, C. J., and Lombó, F. (2021). *De novo* biosynthesis of garbanzol and fustin in *Streptomyces albus* based on a potential flavanone 3-hydroxylase with 2-hydroxylase side activity. *Microbial Biotechnology*, 14(5): 2009–2024. https://doi.org/10.1111/1751-7915.13874.

Matsumoto, A., and Takahashi, Y. (2017). Endophytic actinomycetes: Promising source of novel bioactive compounds. *Journal of Antibiotics*, 70(5): 514–519. https://doi.org/10.1038/ja.2017.20.

Merzaeva, O. V., and Shirokikh, I. G. (2006). (2006). Colonization of plant rhizosphere by actinomycetes of different genera. *Microbiology*, 75: 226–230.

Mingyar, E., Feckova, L., Novakova, R., Bekeova, C., and Kormanec, J. (2015). A γ-butyrolactone autoregulator-receptor system involved in the regulation of auricin production in *Streptomyces aureofaciens* CCM 3239. *Applied Microbiology and Biotechnology*, 99(1): 309–325. https://doi.org/10.1007/s00253-014-6057-0.

Mishra, R. K., Bohra, A., Kamaal, N., Kumar, K., Gandhi, K., Sujayanand, G. K. et al. (2018). Utilization of biopesticides as sustainable solutions for management of pests in legume crops: Achievements and prospects. *Egyptian Journal of Biological Pest Control*, 28(1): 1–11. https://doi.org/10.1186/s41938-017-0004-1.

Mitter, E. K., Tosi, M., Obregón, D., Dunfield, K. E., and Germida, J. J. (2021). Rethinking crop nutrition in times of modern microbiology: innovative biofertilizer technologies. *Frontiers in Sustainable Food Systems*, 5: 606815.

Miyada, Z., Goudjal, Y., Sabaou, N., Mathieu, F., and Zitouni, A. (2017). Development of formulations based on *Streptomyces rochei* Strain PTL2 Spores for Biocontrol of *Rhizoctonia solani* Damping-off of Tomato Seedlings. *Biocontrol Science and Technology* 27(6): 723–738. https://www.tandfonline.com/doi/full/10.1080/09583157.2017.

Montesdeoca-Flores, D. T., Hernández-Bolaños, E., León-Barrios, M., Hernández-Amador, E., Díaz-González, S., Abreu-Acosta, N. et al. (2023). Antifungal Activity of *Streptomyces* spp. Extracts *In Vitro* and on Post-Harvest Tomato Fruits against Plant Pathogenic Fungi. *Horticulturae*, 9(12). https://doi.org/10.3390/horticulturae9121319

Nduka, B. A., Oduwaye, O. F., and Adewale, D. B. (2017). Potential of *Streptomyces* sp. and *Trichoderma* sp. as compost microbiota for coffee husk. *African Journal of Microbiology Research*, 11(14): 560–567.

Nicolopoulou-Stamati, P., Maipas, S., Kotampasi, C., Stamatis, P., and Hens, L. (2016). Chemical pesticides and human health: the urgent need for a new concept in agriculture. *Frontiers in Public Health*, 4(July): 1–8. https://doi.org/10.3389/fpubh.2016.00148.

Olanrewaju, O. S., and Babalola, O. O. (2019). Streptomyces: implications and interactions in plant growth promotion. In *Applied Microbiology and Biotechnology* (Vol. 103, Issue 3, pp. 1179–1188). https://doi.org/10.1007/s00253-018-09577-y.

Pacios-Michelena, S., Aguilar González, C. N., Alvarez-Perez, O. B., Rodriguez-Herrera, R., Chávez-González, M., Arredondo Valdés, R. et al. (2021). Application of *Streptomyces* antimicrobial compounds for the control of phytopathogens. *Frontiers in Sustainable Food Systems*, 5(September), 1–13. https://doi.org/10.3389/fsufs.2021.696518.

Pang, F., Solanki, M. K., and Wang, Z. (2022). Streptomyces can be an excellent plant growth manager. In *World Journal of Microbiology and Biotechnology* (Vol. 38, Issue 11). https://doi.org/10.1007/s11274-022-03380-8.

Park, B., Abdel-Azeem, A., Al-Sanea, M., Yoo, K., Tae, J., and Lee, S. (2013). *Staurosporine analogues* from microbial and synthetic sources and their biological activities. *Current Medicinal Chemistry*, 20(31): 3872–3902. https://doi.org/10.2174/09298673113209990176.

Penha, R. O., Vandenberghe, L. P., and Faulds, C., Soccol, V. T., Soccol, C. R. (2022). *Bacillus* lipopeptides as powerful pest control agents for a more sustainable and healthy agriculture: recent studies and innovations. *Planta*, 251: 1–15.

Pereira, P. M., Santana, F. M., and Van Der Sand, S. (2022). Evaluation of *Streptomyces* spp. strains as potential biocontrol agents for Pyrenophora tritici-repentis. *Biocontrol Science and Technology*, 32(9): 1095–1106. https://doi.org/10.1080/09583157.2022.2089974.

Raja, A., and Prabakarana, P. (2011). Actinomycetes and drug-an overview. In American Journal of Drug Discovery and Development (Vol. 1, Issue 2, pp. 75–84). https://doi.org/10.3923/ajdd.2011.75.84.

Saberi-Riseh, Roohallah, and M. M. (2021). A Novel Encapsulation of *Streptomyces fulvissimus* Uts22 by spray drying and its biocontrol efficiency against *Gaeumannomyces graminis*, the causal agent of take-all disease in Wheat. *Pest Management Science* 77(10): 4357–4364. https://onlinelibrary.wiley.c.

Sahayaraj, K. (2014). Nanotechnology and plant biopesticides: an overview. *Advances in plant Biopesticides,* pp. 279–293.

Sánchez-García, B. M., Ramírez-Pimentel, J. G., Rodríguez-Guerra, R., Guevara-Acevedo, L. P., Raya-Pérez, J. C., Covarrubias-Prieto, J. et al. (2022). Molecular identification and antagonistic potential of three strains of *Streptomyces* against phytopathogenic fungi. *Agrociencia*, 56(5): 1018–1043. https://doi.org/10.47163/agrociencia.v56i5.2793.

Sandoval-Powers, Fawwal, D. V., and Al., E. (2021). Králová, Stanislava, Megan *Streptomyces tardus* sp. Nov.: A slow-growing actinobacterium producing Candicidin, isolated from sediments of the trondheim fjord. Frontiers in Microbiology 12. Frontiers Media SA: 714233. https://www.frontiersin.org/articles/10.

Sarmiento-Vizcaíno, A., Espadas, J., Martín, J., Braña, A. F., Reyes, F., García, L. A. et al. (2018). Atmospheric precipitations, hailstone and rainwater, as a novel source of streptomyces producing bioactive natural products. *Frontiers in Microbiology*, 9(APR). https://doi.org/10.3389/fmicb.2018.00773.

Shcherbakov, A. V., Bragina, A. V., Kuzmina, E. Y., Berg, C., Muntyan, A. N., Makarova, N. M. et al. (2013). Endophytic bacteria of Sphagnum mosses as promising objects of agricultural microbiology. *Microbiology (Russian Federation)*, 82(3): 306–315. https://doi.org/10.1134/S0026261713030107.

Shemedyuk, N., Romashko, I., Butsiak, V., Dvylyuk, I., and Shved, O. (2022). The development of systemic plant stability and the prospects of using *Streptomyces* as biocontrol agents. *Chemistry, Technology and Application of Substances*, 5(1): 102–116. https://doi.org/10.23939/ctas2022.01.102.

Silva, Cruz, G. da, Isabella Takahashi Kitano, I. A. de F. R., and Lacava, P. T. (2022). The Potential Use of Actinomycetes as Microbial Inoculants and Biopesticides in Agriculture. Frontiers in Soil Science 2. Frontiers Media SA. https://www.frontiersin.org/articles/10.3389/fsoil.2022.833181/full, accessed February 14: 2024.

Singh, C., Singh Parmar, R., Jadon, P., and Kumar, A. (2017). Optimization of cultural conditions for production of antifungal bioactive metabolites by *Streptomyces* spp. Isolated from Soil. *International Journal of Current Microbiology and Applied Sciences*, 6(2): 386–396. https://doi.org/10.20546/ijcmas.2017.602.043.

Siupka, P., Piński, A., Babicka, D., and Piotrowska-Seget, Z. (2020). Genome mining revealed a high biosynthetic potential for antifungal *Streptomyces* sp. S-2 isolated from black soot. *International Journal of Molecular Sciences*, 21(7). https://doi.org/10.3390/ijms21072558.

Sriragavi, G., Sangeetha, M., Santhakumar, M., Lokesh, E., Nithyalakshmi, M., Saleel, C. A., and Balagurunathan, R. (2023). Exploring antibacterial properties of bioactive compounds isolated from *Streptomyces* sp. in Bamboo Rhizosphere Soil. *ACS Omega*, 8(39): 36333–36343. https://doi.org/10.1021/acsomega.3c04954.

Stack, J. P., C. M. Kenerley, and R. E. P. (1987). Influence of carbon and nitrogen sources, relative carbon and nitrogen concentrations, and soil moisture on the growth in nonsterile soil of soilborne fungal antagonists. *Canadian Journal of Microbiology* 33(7): 626–631. http://www.nrcresearchpress.com/doi.

Suleman, P., Al-Musallam, A., and Menezes, C. A. (2002). The effect of biofungicide Mycostop on *Ceratocystis radicicola*, the causal agent of black scorch on date palm. *BioControl.*, 47(2): 207–216. https://doi.org/10.1023/A:1014519726573.

Takano, H., Nishiyama, T., Amano, S. ichi, Beppu, T., Kobayashi, M., and Ueda, K. (2016). Streptomyces metabolites in divergent microbial interactions. *Journal of Industrial Microbiology and Biotechnology*, 43(2–3): 143–148. https://doi.org/10.1007/s10295-015-1680-z.

Tan, L. T. H., Lee, L. H., and Goh, B. H. (2020). Critical review of fermentation and extraction of anti-Vibrio compounds from *Streptomyces*. In *Progress in Microbes and Molecular Biology* (Vol. 3, Issue 1). https://doi.org/10.36877/pmmb.a0000051.

Thakur, Neelam, Simranjeet Kaur, Preety Tomar, Seema Thakur, and A. N. Y. (2020). Microbial Biopesticides: Current Status and Advancement for Sustainable Agriculture and Environment. In New and Future Developments in Microbial Biotechnology and Bioengineering pp. 243–282. Elsevier. https://www.sciencedirect.com/science/article/pii/B978.

Thambugala, K. M., Daranagama, D. A., Phillips, A. J. L., Kannangara, S. D., and Promputtha, I. (2020). Fungi vs. Fungi in Biocontrol: An Overview of Fungal Antagonists Applied Against Fungal Plant Pathogens. *Frontiers in Cellular and Infection Microbiology*, 10(November), 1–19. https://doi.org/10.3389/fcimb.2020.604923.

Thao, N. B., Kitani, S., Nitta, H., Tomioka, T., and Nihira, T. (2017). Discovering potential *Streptomyces* hormone producers by using disruptants of essential biosynthetic genes as indicator strains. *Journal of Antibiotics*, 70(10): 1004–1008. https://doi.org/10.1038/ja.2017.85.

Torres-Rodriguez, J. A., Reyes-Pérez, J. J., Quiñones-Aguilar, E. E., and Hernandez-Montiel, L. G. (2022). Actinomycete potential as biocontrol agent of phytopathogenic fungi: mechanisms, source, and applications. In *Plants* (Vol. 11, Issue 23). https://doi.org/10.3390/plants11233201.

Umesha, Sharanaiah, P. K. S., and Singh, R. P. (2018). Microbial Biotechnology and Sustainable Agriculture. In Biotechnology for Sustainable Agriculture. pp. 185–205. Elsevier. https://www.sciencedirect.com/science/article/pii/B9780128121603000064, accessed February 13: 2024.

Viaene, T., Langendries, S., Beirinckx, S., Maes, M., and Goormachtig, S. (2016). Streptomyces as a plant's best friend? *FEMS Microbiology Ecology*, 92(8). https://doi.org/10.1093/femsec/fiw119.

Vinale, F., Sivasithamparam, K., Ghisalberti, E. L., Ruocco, M., Woo, S., and Lorito, M. (2012). Trichoderma secondary metabolites that affect plant metabolism. *Natural Product Communications*, 7(11): 1545–1550. https://doi.org/10.1177/1934578x1200701133.

Waksman, S. A., and Henrici, A. T. (1943). The nomenclature and classification of the actinomycetes. *Journal of Bacteriology*, 46(4): 337–341. https://doi.org/10.1128/jb.46.4.337-341.1943.

Walker, T. S., Bais, H. P., Grotewold, E., and Vivanco, J. M. (2003). Root exudation and rhizosphere biology. *Plant Physiology* 132(1). *American Society of Plant Biologists:* 44–51. https://academic.oup.com/plphys/article-abstract/132/1/44/6111841, accessed February 16: 2024.

White, J. D., Ohba, Y., Porter, W. J., and Wang, S. (1997). Synthesis of the spiroketal segment (C19-C34) of the rutamycins, antifungal metabolites of *Streptomyces* species. *Tetrahedron Letters*, 38(18): 3167–3170. https://doi.org/10.1016/S0040-4039(97)00585-6.

Woo, S. L., Hermosa, R., Lorito, M., and Monte, E. (2023). Trichoderma: A multipurpose, plant-beneficial microorganism for eco-sustainable agriculture. *Nature Reviews Microbiology*, 21(5): 312–326.

Xu, G., and Yang, S. (2019). Regulatory and evolutionary roles of pseudo γ-butyrolactone receptors in antibiotic biosynthesis and resistance. In *Applied Microbiology and Biotechnology* (Vol. 103, Issues 23–24, pp. 9373–9378). https://doi.org/10.1007/s00253-019-10219-0.

Yamunarani, K., Sundaram, A. K., and Pandiyan, M. (2019). *Streptomycetes as a potential biocontrol agent. J. Entomol. Zool. Stud.*, 7: 637–644.

Zhang, M. Z., Chen, Q., Xie, C. H., Mulholland, N., Turner, S., Irwin, D. et al. (2015). Synthesis and antifungal activity of novel streptochlorin analogues. *European Journal of Medicinal Chemistry*, 92: 776–783. https://doi.org/10.1016/j.ejmech.2015.01.043.

Zhang, Y. feng, He, L. yan, Chen, Z. jin, Wang, Q. ya, Qian, M., and Sheng, X. fang. (2011). Characterization of ACC deaminase-producing endophytic bacteria isolated from copper-tolerant plants and their potential in promoting the growth and copper accumulation of *Brassica napus*. *Chemosphere*, 83(1): 57–62. https://doi.org/10.1016/j.chemosphere.2011.01.041.

Zhang, Y., Pan, L., Zhang, Y., Wang, K., Wang, L., Zhang, H. et al. (2023). Understanding the *Streptomyces albulus* response to low-pH stress at the interface of physiology and transcriptomics. *Applied Microbiology and Biotechnology*, 107(7–8): 2611–2626. https://doi.org/10.1007/s00253-023-12449-9.

Zhou, Si, Hu, X., Zhao, P., Niu, Y., and Liao, M. (2019). Optimization of sporulation conditions for *Streptomyces Galilaeus* AF1. Journal of Northwest A & F University-Natural Science Edition 47(1). Northwest A & F University (Natural Science Edition): 123–138. https://www.cabdirect.org/cabdirect/abstract/2019305.

10

Pseudomonas Based Biocontrol

Large-scale Production, Formulation, Quality Control and Delivery Methods

Anurag Yadav[1,*] and *Kusum Yadav*[2]

1. Introduction

Growing concerns over the reliance on chemical pesticides, particularly their associated detrimental impacts on the environment, the development of resistant pest species, and human health risks, are driving a worldwide search for 'greener' and more 'sustainable' pest management options (Lamichhane et al. 2016). In this context, *Pseudomonas*, a naturally growing soil bacterium inducing plant pathogen and pest control strategies (Miller et al. 2008, Oni et al. 2015), has reportedly received much attention among researchers and practitioners. Also, its capacity to produce antimicrobial compounds, compete for nutrients, and induce systemic resistance in plants makes it a candidate for ecologically responsible farming (Glick 2012). However, the scale-up of *Pseudomonas* for large-scale pest management is problematic due to inconsistent production, formulation problems, lack of quality control, and suboptimal delivery methods (Ab Rahman et al. 2018). This chapter addresses these obstacles, and strategies for large-scale biocontrol using *Pseudomonas* to promote a more sustainable future in pest management. From a broader perspective, chemical pesticides take a considerable environmental toll.

[1] Department of Microbiology, C.P. College of Agriculture, Sardarkrushinagar Dantiwada Agricultural University, Sardarkrushinagar, Dist. Banaskantha, Gujarat-385506, India.
[2] Department of Biochemistry, University of Lucknow, Lucknow, Uttar Pradesh-226007, India.
* Corresponding author: anuragyadav123@gmail.com

DOI: 10.1201/9781003603023-11

These compounds remain in soil and water in trace amounts, damage ecosystem health, and reduce biodiversity (Carvalho 2017). High pesticide usage over time has also led to the development of pest population resistance and/or evolved resistance to effective pesticides (Kim et al. 2017). In addition, there is concern about pesticide residues in food and direct human exposure. In view of such serious drawbacks, the agricultural sector is in search of alternative methods like *Pseudomonas* based application, whose benefits are effective and accompanied by fewer negative side effects.

A number of potent mechanisms against numerous pathogens and pests distinguish *Pseudomonas* as a biocontrol agent. They include the production of antimicrobial substances (pyocyanin and pyrrolnitrin), extremely aggressive competition in the rhizosphere, and stimulation of systemic resistance in plants (Durairaj et al. 2017). *Pseudomonas*' specificity, which targets harmful microbes without eliciting the kind of resistance chemical pesticides have a high propensity to cause (Liu et al. 2022), is also noteworthy. Of the four members of this genus, *P. fluorescens*, *P. putida*, *P. syringae*, and *P. aeruginosa* show great promise. They have been featured for their capabilities by researchers in many agricultural settings. For example, a number of cultured organisms in the *P. fluorescens* group produce antimicrobial compounds and are skilled at inducing systemic immunity in plants (Haas and Dëfago 2005, Höfte 2021). Similarly, other studies indicate the success of *P. putida*, including the eradication of *Salmonella* biofilms in broiler house water systems (Maes et al. 2020). Also, investigations of its success in controlling peanut stem rot by the bacterium *P. chlororaphis* Zm-1 point to the necessity of employing more advanced analytical techniques, for example, HPLC and UPLC-MS/MS, to clarify the antibacterial particles it creates (Liu et al. 2022).

Pseudomonas, although highly advantageous in various setups, has many hurdles to overcome on its way to wider use in commercial agriculture. This includes forming stable products that can survive field conditions, maintaining consistent quality for large-scale production, figuring out practical application methods, and conducting rigorous safety and potency tests in the field. Such barriers must be overcome if *Pseudomonas* is to be deployed as a mainstream tool in pest control. A multi-pronged strategy is needed to address such complexities. From the production side, there are more advanced fermentation methods, better encapsulation techniques, and cutting-edge carriers (alginate beads to chitosan capsules, among others) that are used to protect bacterial viability and ensure targeted delivery. It is important to establish proper protocols for training strains and selection/testing to ensure consistency and efficacy. From a deployment standpoint, innovative techniques like drone-based spraying and other forms of precision agriculture may also boost the reach and impact of *Pseudomonas* biocontrol agents (Nerek and Sokołowska 2022). This chapter was written against the backdrop of the objectives: to explore the potential of *Pseudomonas* as a sustainable biocontrol agent, to address the challenges of mass-producing *Pseudomonas* for large-scale agricultural use, to improve formulation and delivery methods for *Pseudomonas*-based biocontrol agents, to establish stringent quality control measures, and to promote the integration of *Pseudomonas* biocontrol into sustainable agricultural practices.

2. What Makes *Pseudomonas* an Effective Biocontrol Agent?

Pseudomonas species serve as effective biocontrol agents for sustainable agriculture through the use of their widespread groups *P. fluorescens*, *P. putida*, *P. syringae*, and *P. aeruginosa*. Through various helpful traits, these microorganisms support plant health functions and inhibit multiple kinds of pathogens. Members of the corrugata–protegens–chlororaphis cluster, including *P. chlororaphis* together with *P. protegens* and *P. corrugata*, demonstrate strong antagonistic behavior (Höfte and Altier 2010).

The production of antifungal metabolites including phenazines, 2,4–diacetylphloroglucinol, pyoluteorin, and pyrrolnitrin serves as a basic capability for their biocontrol success. The biocontrol compounds produced by microorganisms compromise fungal cell membranes and then interrupt vital cellular operations (Haas and Défago 2005, Mavrodi et al. 2006). *P. protegens* CHA0 produces the phenazine-1-carboxamide antifungal compound that reduces *Fusarium* wilt disease levels in tomato plants, resulting in significant infection reduction (Weller 2007). Pseudomonads promote disease inhibition because they manufacture siderophores, which extract iron from the rhizosphere where pathogenic organisms would normally acquire it, thus blocking their reproductive growth (Haas and Défago 2005). The disease-suppressive characteristics of *P. chlororaphis* strains become more powerful when they encounter certain amino acids that prompt increased hydrogen cyanide production (Kumar et al. 2017).

The bacteria activate plant defense mechanisms throughout the entire plant body following antipathogenic activities. Jasmonic acid along with ethylene hormones mediate the defense signaling pathway, which becomes activated through this process (Haas and Défago 2005). The rhizosphere colonization abilities and 2,4-diacetylphloroglucinol production of *P. fluorescens* Pf–5 led to effective damping-off disease control in cucumbers infected with *Pythium ultimum* (Haas and Défago 2005). Testing against potato diseases inflicted by *Rhizoctonia solani* revealed that *P. chlororaphis* O6 effectively diminished the severity of pathogen impact (Mavrodi et al. 2006). The plant's uptake of nutrients receives support from numerous pseudomonads, which increase plants' nutritional availability. The bacteria support nutrient acquisition by dissolving phosphorus and performing nitrogen fixation (Kumar et al. 2017). Plants receive increased vigor through plant-available iron, which bacteria contribute by secreting siderophores in the substrate. Organisms of the *Pseudomonas* species showcase environmental flexibility, which supports their profitable use in sustainable agricultural biocontrol programs (Alattas et al. 2024). These bacterial agents preserve ecological balance by avoiding the harmful effects of chemical pesticides, including damage to non-target organisms, resistance formation, and toxic residue (Alattas et al. 2024). Through their metabolic adaptability and genetic plasticity, *Pseudomonas* species can naturally overcome different types of soil environments. *Pseudomonas* cells rapidly establish themselves in the rhizosphere when they identify plant root exudates (Kumar et al. 2017). *P. fluorescens* shows strong attraction to tomato root-exuded amino acids, thus achieving effective root

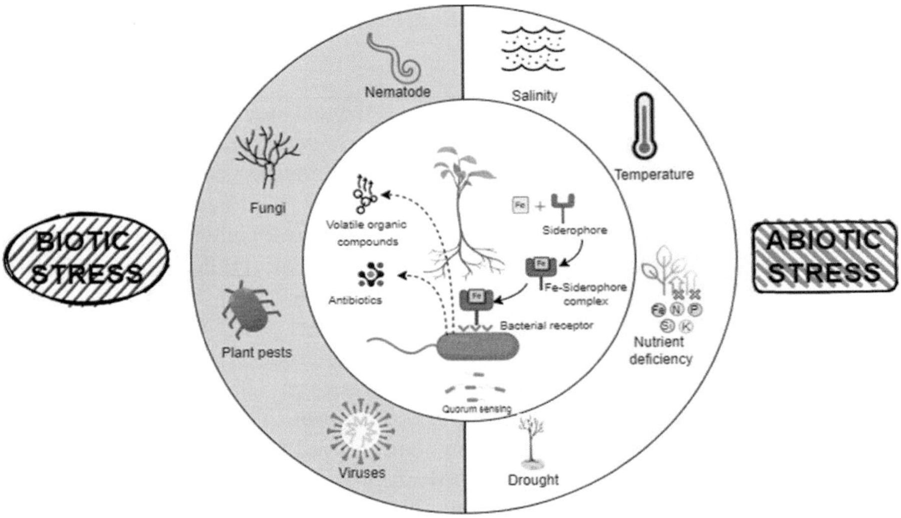

Figure 1: *Pseudomonas* as biocontrol agent.

colonization and prolonged protection of plants (Hofte and Altier 2010, Kumar et al. 2017). Pseudomonads achieve sustainable pest control through their dual capability to defend plants from pathogens and enhance nutrient absorption, which results in environmentally friendly pest management that increases yield productivity for sustainable agriculture (Höfte and Altier 2010, Kumar et al. 2017). The significance of *Pseudomonas* as a dependable biocontrol agent is shown in Figure 1.

3. Mass Production of *Pseudomonas* Based Inoculants

A systematic process exists for the mass production of *Pseudomonas* microbial inoculants, and their quality, effectiveness, as well as scalability, have been proven in agricultural applications. Isolation of *Pseudomonas* strains from environmental samples like the rhizosphere, soil, or plant tissues is a rich source of beneficial microbes and forms the first stage of the process. These strains are then selected based on their plant growth promotion (PGP) traits, such as phosphorus solubilization, nitrogen fixation, production of phytohormones, and their biocontrol potential, like the ability to produce antimicrobial compounds and to induce systemic resistance in plants (Glick 2012, Lugtenberg and Kamilova 2009). In addition, the strains are evaluated for their compatibility with certain agroclimatic conditions to guarantee that they will be adapted and effective against target regions (Díaz-Rodríguez et al. 2025).

Phenotypic, biochemical, and molecular approaches, such as 16S rRNA gene sequencing, are employed in a polyphasic taxonomic framework to confirm the identity of the selected *Pseudomonas* strains (Gomila et al. 2015). Strains are then identified and optimized; media composition and culture conditions are determined to maximize growth and productivity. Optimal carbon and nitrogen sources, pH, temperature, and aeration conditions are determined in this process (Patel et al. 2020).

Pseudomonas are cultured through fermentation techniques for mass production. High biomass yields are produced through submerged liquid fermentation, which is very common because of its scalability and efficiency. In other applications, where spore production or secondary metabolite synthesis is desired, solid-state fermentation may be used (Thomas et al. 2013). Downstream processing includes the harvesting of biomass, spores or secondary metabolites prior to formulation, stabilization, and packaging into biofertilizer or biocontrol products ready for field application (Agarwal et al. 2021). In order to obtain product stability and strength, the harvested microbial material is formulated further through addition of carriers (e.g., peat, clay, or talc) or additives (e.g., protectants or adhesives) to ensure viability and prolonged shelf life (Bashan et al. 2014). Viable cell count assessments, contamination checks, and PGP verification are all critical steps of the quality control to maintain product consistency and efficiency (Herrmann and Lesueur 2013). The performance of inoculants under controlled and field agricultural conditions is validated by using *Pseudomonas*-based inoculants in greenhouse and field trials. These trials assess the inoculant's capacity for enhancing crop growth, curbing pathogen activity, and improving yields, thereby ascertaining its practicality for commercial handling (Glick 2012, Vejan et al. 2016).

A systematic process to produce the mass culture of *Pseudomonas*-based inoculants that will induce plant growth and impart biocontrol in agriculture is outlined in Figure 2. First, *Pseudomonas* strains are isolated from environmental samples, and then selected strains are screened for plant growth-promoting and biocontrol ability. These tests involve metabolic and biocontrol assays along with compatibility tests with agroclimatic conditions like salt tolerance, temperature resilience, and their interaction with the existing native microbiota and agricultural input. Strains are confirmed for their identity through polyphasic taxonomic identification. This comprises steps of optimizing the media composition and growth conditions to maximize growth and productivity. Both solid-state fermentation and submerged liquid fermentation techniques are used for fermentation to achieve mass production.

The downstream processing includes the harvesting of biomass or spores and the extraction of metabolites after fermentation. The product is finally subjected to quality control, including viable cell counts and contamination checks. The inoculant is also formulated with carriers or additives to stabilize it and extend its shelf life. Greenhouse and field trials are conducted to confirm that the *Pseudomonas* inoculant is effective under real-world conditions. This comprehensive approach guarantees high-quality and effective production of microbial inoculants for agricultural use.

4. Fermenter Types

Proper balance of aeration and agitation is an integral factor in big-scale culturing. This need is met by a sterile air supply ranging from 3 to 10 liters per hour for every liter of medium (Maheshwari et al. 2022). Frequent sampling and analysis are necessary to keep track of growth dynamics. After a set incubation period, when cell density reaches around 10^9 cells/mL, the air supply is stopped, signifying that the culture is ready for harvesting.

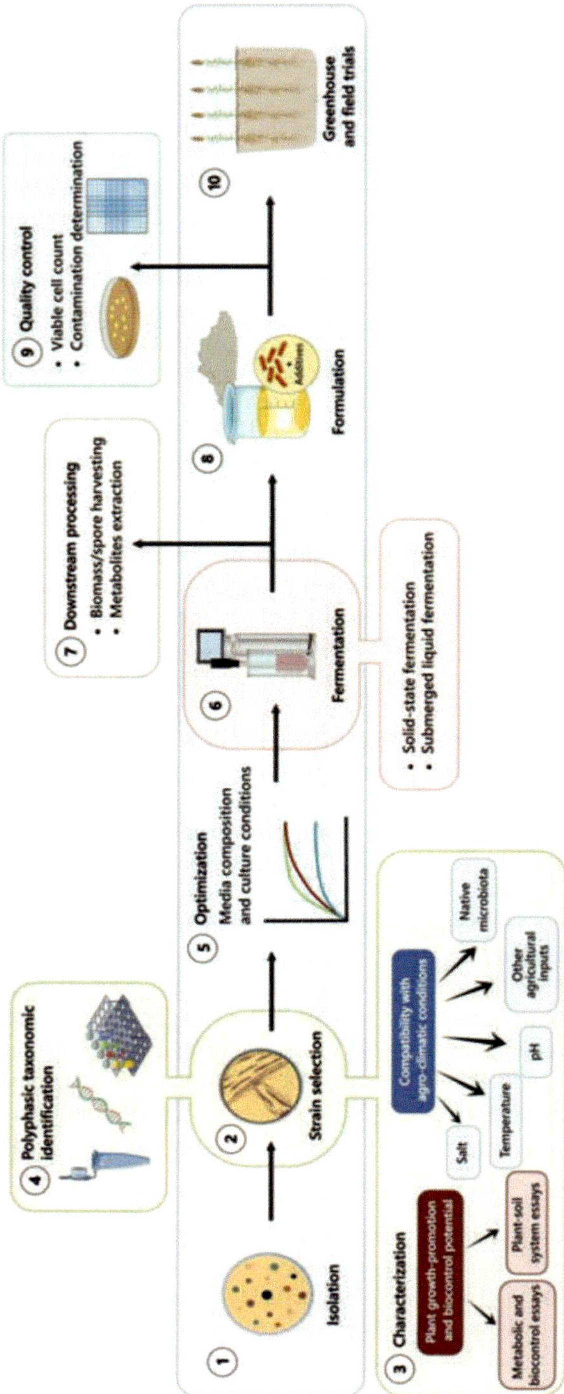

Figure 2: Schematic representation of the mass production process for *Pseudomonas* microbial inoculants. Taken from (Díaz-Rodríguez et al. 2025).

Table 1: Comparative analysis of various fermentation types employed for *Pseudomonas* mass production.

Fermentation Process	Efficacy	Viability	Cost Considerations	Suitability for Scale-up
Batch Fermentation	Moderate	High initially, decreases over time	Economical for small-medium scale	Good for experimental and smaller scale
Continuous Fermentation	High	Very high	Cost-effective for large-scale, long-term	Excellent for large-scale continuous production
Fed-Batch Fermentation	High	High throughout the process	Higher than batch fermentation	Suitable for large-scale, high yield production
Solid State Fermentation (SSF)	Good, substrate-dependent	High in suitable substrates	Variable, often lower due to cheaper substrates	Challenging, more suitable for niche applications
Submerged Fermentation (SmF)	High	High	Moderate to high due to aeration and agitation	Very good for industrial-scale production
Airlift Fermentation	Moderate to high	High	Generally lower energy requirements	Good, especially for shear-sensitive cultures
Membrane Fermentation	Can be high	High, with selective permeability	Higher due to membrane systems	Suitable, dependent on membrane maintenance at scale
Two-Stage Fermentation (SLF + SSF)	Combines high-efficiency growth with simple, low-cost finishing	Complex operational logistics, risk of contamination between stages	Moderate to high	Moderate, depends on specific stage design and automation

This process is of utmost importance for quality control. For the culture to be free of fungal or bacterial contamination to less than the log 10^{-6} dilution level, strict monitoring must precede further processing. Finally, regarding the shelf life of the broth after fermentation, another important factor to consider is that using the broth within one day allows it to maintain an appropriate cell count. The viability of these cells can be lost even under refrigeration if the culture is held for too long (Maheshwari et al. 2022). A comparison of the various fermentation approaches applied for the mass production of *Pseudomonas* is made in Table 1, and each fermenter is outlined in Table 2.

4.1 Stirred Tank Fermenter

Large-scale *Pseudomonas* cultivation is mostly carried out using stirred-tank reactors (STRs) owing to their ability to handle a variety of media, different oxygen requirements, and metabolic profile for different *Pseudomonas* species (Sarma et al. 2013). Excellent mixing due to mechanical agitation allows for the achievement of high cell densities by providing uniform nutrient availability. Furthermore, due to efficient oxygen transfer in STRs, they are suitable for aerobic *Pseudomonas* strains (Karimi et al. 2013). A second major benefit of STR consists in reliable heat

Table 2: Advantages and disadvantages of each fermenter type and its suitability for mass production of *Pseudomonas*.

Fermenter Type	Advantages	Disadvantages	Suitability for *Pseudomonas*
Stirred Tank Fermenter	• Excellent mixing and aeration • Good for shear-sensitive broths	• Complex design • Potential for foaming	High
Airlift Fermenter	• Good for shear-sensitive broths • Lower energy requirements	• Less efficient oxygen transfer than stirred tanks • Limited control of mixing patterns	Moderate
Bubble Column Fermenter	• Simple design • Good heat transfer	• Limited oxygen transfer rates • Potential for foaming	Moderate
Fluidized Bed Reactor	• High cell densities • Good for immobilized cells	• Complex design • Risk of cell washout	Moderate
Solid-State Fermenter	• Low energy consumption • Simple operation	• Slower growth rates • Difficult to control temperature and moisture	Moderate

dissipation and thus stabilization of the temperature, providing optimal conditions for bacterial growth (Müller et al. 2018). In addition, STRs are extremely suitable for use in *Pseudomonas* cultivation; in particular, they provide seamless operability in commercial bioreactors, from benchtop to industrial scales. These STR designs can interface with sophisticated temperature, pH, and dissolved oxygen monitoring and control technologies, which allow precise adjustment (Regonesi 2023). These systems enable automated feeding routines to predict the nutrient levels in fermentation for the duration (Regonesi 2023).

However, while the advantages of *Pseudomonas* cultivation are evident, some strains are susceptible to shear forces from impellers, leading to lower cell viability and productivity (Garcia-Ochoa and Gomez 2009, Schugerl 2001). To overcome this hurdle, operators tend to pick impeller designs that benefit from lower shear stress, for example, Rushton turbines or marine propellers, or by adjusting agitation speeds (Abo-Zaid et al. 2020, Doran 1995). The biosurfactants produced by some *Pseudomonas* strains can result in foam formation, and this is often accompanied by the need for antifoaming agents or mechanical breakers of foam (Tiso et al. 2024). As with any bioreactor, care must be taken in terms of rigorous sterilization and vigilant oversight to prevent contamination of an entire production batch (Gabelman 2022). Several design enhancements can further improve the performance of STRs for *Pseudomonas* growth. For instance, baffles can prevent vortexing and improve mixing as well as oxygen delivery (Regonesi 2023). Multiple impellers are often used in larger systems to increase oxygen transfer and mixing efficiency (Vrábel et al. 1999). Furthermore, fed-batch or continuous fermentation can be used to circumvent nutrient depletion and ensure robust *Pseudomonas* growth (Poontawee and Limtong 2020).

4.2 Airlift Fermenter

Stirred-tank reactors are predominant in large-scale *Pseudomonas* production, but airlift reactors are attractive for strains sensitive to mechanical shear. In airlift designs, upwardly moving air or gas bubbles circulate the culture medium, enabling hybrid mixing and oxygenation without impellers. For *Pseudomonas* strains, which might otherwise be adversely affected in terms of viability or productivity, this gentler environment is advantageous (Lecuyer et al. 2011). The bubble-driven flow is also effective in promoting efficient oxygen transfer, which is a very important aspect for metabolically active aerobic *Pseudomonas* species (Manal et al. 2005).

Yet, features of airlift fermenters need to be assessed with respect to limitations. For instance, these systems may be less effective with highly viscous broths where bubble-driven mixing alone could be insufficient (De Jesus et al. 2017). Additionally, foam buildup due to biosurfactant-producing *Pseudomonas* species requires the use of defoamers or foam-breaking mechanisms (Sarachat et al. 2010). Normally, at a large scale, one has to choose between airlift and stirred-tank reactors for *Pseudomonas* cultivation depending on a balance of various factors, including shear sensitivity and product output (e.g., maximal biomass or highest extraction of specific metabolites). Moreover, the viscosity of the culture medium and the strain's resilience are also key factors determining whether an airlift system is best for growth.

4.3 Bubble Column Fermenter

Pseudomonas can be readily and inexpensively grown at large scale in a bubble column reactor. Here, bubbles are formed at the bottom of a tower and travel upwards, mixing and delivering oxygen. Such a minimalistic configuration lowers both manufacturing and operational expenses, an excellent advantage for high-volume production. By supplying sufficient oxygen, *Pseudomonas* strains can grow robustly thanks to the continuous stream of bubbles (Zhang et al. 2017). However, bubble column reactors have less scope for control over mixing patterns compared to stirred-tank or airlift systems (Jakobsen et al. 2005). If the mixing conditions required by a certain *Pseudomonas* strain are very precise or intensive, this limitation may be crucial. Therefore, the simplicity and economic benefit of bubble columns should be weighed against their mixing quality. However, for strains where budget constraints are a major factor, the disadvantages of bubble columns may still be minor compared to their advantages.

4.4 Other Fermenter Types

Reactor designs other than stirred-tank, airlift, and bubble column reactors include packed-bed and fluidized-bed reactors as alternative large-scale *Pseudomonas* cultivation strategies. For the production of valuable metabolites with immobilized bacterial biomass, packed-bed reactors (PBRs) are commonly used, enabling easier downstream processing of extracellular products (Ayol et al. 2021). In these systems, *Pseudomonas* cells grow attached to a support material, nutrient-rich broth is continuously passed through the reactor, and cells are produced efficiently and continuously (Ayol et al. 2021).

Suspended fluidized-bed reactors (SFBRs) make use of a flowing medium that suspends the biomass and consequently increases mixing and mass transfer (Bello et al. 2017). This design is especially advantageous when *Pseudomonas* strains develop biofilms or adhere readily to surfaces. Further, these reactors may streamline continuous production of extracellular metabolites since there is constant flow and easy product separation.

Regardless, the oxygen and nutrient delivery requirements imposed upon immobilized biomass need to be sufficient in packed-bed and fluidized-bed systems when high-density cultures are used in bulk *Pseudomonas* production. Finally, the potential for clogging or channel formation must be considered, since these can obstruct flow and diminish overall productivity in large-scale operations. The basic types of fermenters in the production of *Pseudomonas* are shown in Figure 3.

5. Fermenter Scaling Up

Moving *Pseudomonas* biocontrol agents from laboratory production to industrial manufacturing requires successful fermenter scale-up. The phase produces uniform product quality alongside maximum output and economical production while satisfying the requirements for industrial biocontrol applications at scale. Further details about this process, along with the challenges it encounters, will appear in the upcoming subsections.

5.1 Challenges

The process of fermentative scale-up from laboratories to commercial factories requires the solution of various technical as well as biological hurdles. Table 3 outlines the limitations and solutions for the bulk production of *Pseudomonas* through fermentation.

5.1.1 Issues in Scaling Up Production While Maintaining Strain Efficacy

The use of *Pseudomonas* spp. for biocontrol is not simply a matter of mass production; it demands careful selection of the right strain, optimization of fermentation conditions, and an understanding of how the bacteria respond to environmental stress. Choosing a strain is almost like choosing the ideal teammate for an important task (Balthazar et al. 2022). To perform in the many ups and downs of large-scale manufacturing, the chosen *Pseudomonas* must excel at preventing unwanted microbes (Liu et al. 2022). As shown in Figure 4, in order to scale up the production of *Pseudomonas* spp., fermentation optimization, media formulation, and bioprocess control are integrated with high cell viability to ensure efficacy under large-scale conditions. Reliable performance of a strain depends on its resilience. To withstand processing and retain effectiveness, the bacteria may possess features such as cold shock proteins or efflux pumps. Careful fermentation optimization is achieved by balancing potency and adaptability; factors such as the choice of substrate or particular temperature and pH control (Pandey et al. 1999) are key in solid-state

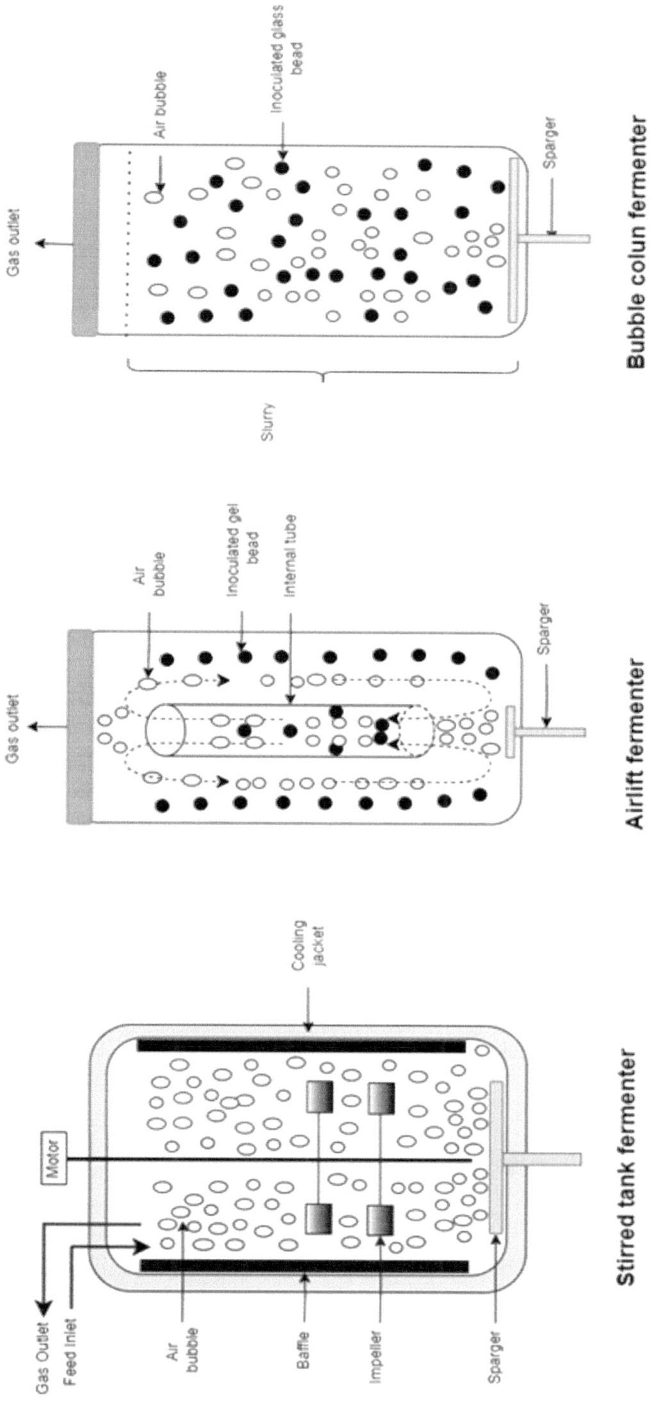

Figure 3: Three most common types of fermenter used in Pseudomonas mass production.

Table 3: Limitations and solutions for bulk production of *Pseudomonas* through fermentation.

Limitation	**Description**	**Proposed Solutions**
Oxygen Transfer	*Pseudomonas* cultures require adequate oxygenation for optimal growth. Inadequate oxygen transfer can limit cell density and product yield.	Increase agitation and aeration rates in bioreactors; use oxygen-enriched air; optimize bioreactor design for better gas exchange.
Contamination	Risk of contamination from other microorganisms can affect the desired product's purity and yield.	Implement stringent sterility controls; use closed systems; regular monitoring and rapid response to contamination.
Cost Considerations	Production can be costly due to expensive substrates, energy requirements, and downstream processing.	Use cheaper substrates like waste materials; optimize process efficiency; develop more cost-effective purification methods.
pH Stability	*Pseudomonas* cultures can be sensitive to pH changes, affecting growth and product formation.	Constant monitoring and adjustment of pH; use of buffer systems in the growth medium.
Genetic Stability	Instability of genetic modifications in *Pseudomonas* can lead to loss of product formation capability.	Use of advanced genetic tools for stable gene integration; regular monitoring of genetic stability.
Scale-up Challenges	Issues arise when scaling up from laboratory to industrial scale, affecting yield and consistency.	Thorough scale-up studies; optimization of process parameters at each scale; use of pilot-scale trials before full-scale production.
Nutrient Limitation	Nutrient limitations can restrict the growth and metabolic activity of *Pseudomonas*.	Optimize nutrient composition and concentrations in the growth medium; explore alternative sources.
Temperature Control	*Pseudomonas* may require specific temperature ranges for optimal growth and production.	Implement precise temperature control systems; optimize growth conditions based on temperature studies.
Metabolic By-products	Accumulation of metabolic by-products can inhibit growth and production in *Pseudomonas* cultures.	Develop strategies for by-product removal; modify metabolic pathways through genetic engineering.
Process Integration	Integrating the production process with other operations can be challenging.	Design integrated process flows; employ systems biology approaches for better process understanding.
Regulatory Compliance	Adherence to regulations and standards can be challenging in the production of bioproducts.	Stay updated with regulatory changes; ensure compliance through regular audits and quality checks.
Product Recovery and Purification	The recovery and purification of products from *Pseudomonas* cultures can be complex and costly.	Develop efficient downstream processing methods; use novel separation technologies.

or submerged fermentation. To maintain efficacy and viability during scale-up, several critical factors need to be considered. Sometimes, agricultural byproducts turn out to be good substrates, but temperature and pH must be controlled to allow both growth and production of key biocontrol compounds. Recent techniques such as statistical design approaches and AI-driven tools are already being developed to

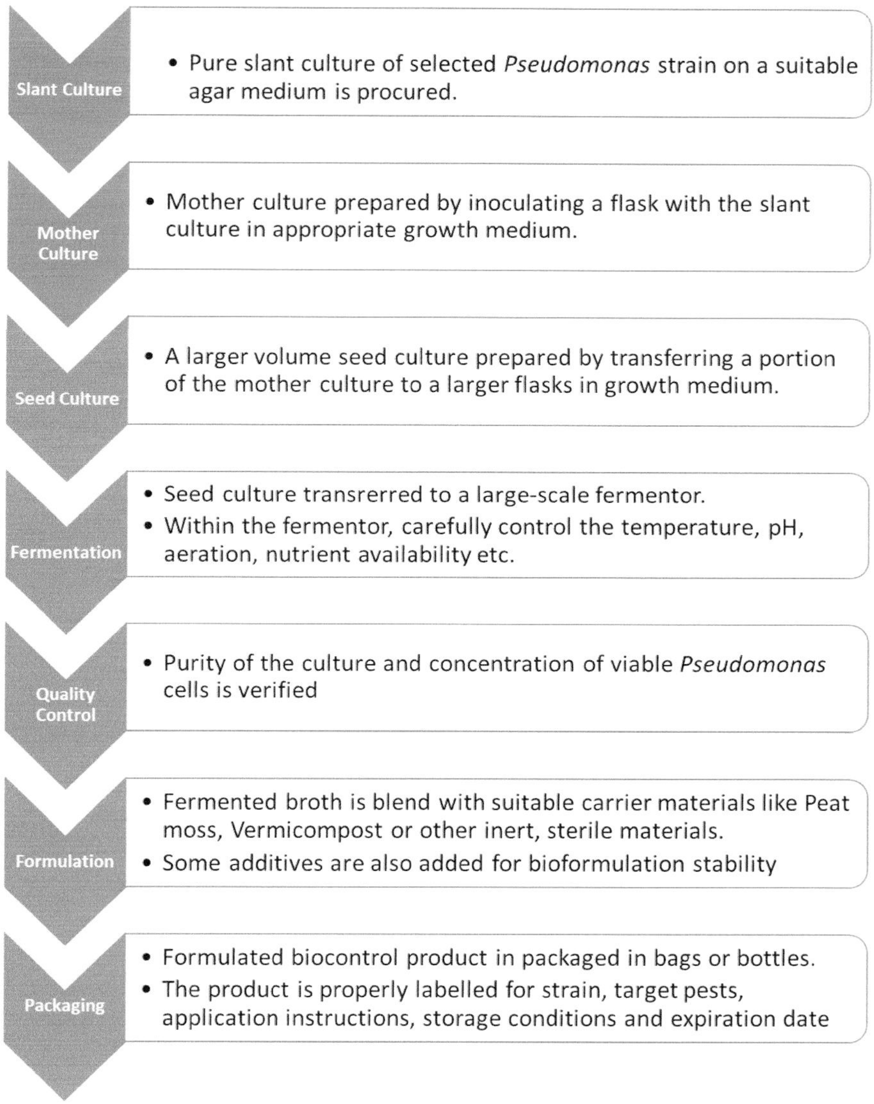

Figure 4: Process of bulk production of *Pseudomonas*.

establish ideal fermentation conditions, closely aligned with the strains' adaptability (Singh et al. 2016). As such, *Pseudomonas* also responds to environmental stress, such as shifting temperatures or fluctuating nutrients, with protective mechanisms that can affect biocontrol agent production (Humair et al. 2009). The solution lies in proactive strategies, such as using bacteria that can naturally cope with stress in stable fermentation environments. In addition, interest is growing in preconditioning bacteria to make them more resilient to harsh environments (Craig et al. 2021, Guan et al. 2017).

A further complication lies in regulatory standards, which also involve public safety. Therefore, authorities require strain characterization, safety evaluation, and risk assessment (Committee et al. 2020). Producers must have strong plans for addressing potential risks when working with genetically modified strains (Du et al. 2022). In fact, each stage, from selecting an appropriate bacterial strain to fermentation scale-up, brings researchers and stakeholders closer to developing a safer, effective, and thoroughly verified *Pseudomonas* product (Du et al. 2022).

5.1.2 High Costs and Inconsistent Quality Due to Small-Scale Fermentation

Transitioning to large-scale fermentation of *Pseudomonas* strains (e.g., *P. putida* KT2440 or *P. aeruginosa*) presents both obstacles and opportunities. *P. putida* KT2440's strict aerobic nature can complicate production, particularly when generating oxygen-sensitive compounds. This requirement drives up costs, as processes must ensure robust aeration, stirring, and other operational details. In addition, some researchers are testing anoxic cultivation methods or adding genes to the bacterial DNA, such as acetate kinase, which could help the bacteria survive low-oxygen conditions (Kampers et al. 2021). For *P. aeruginosa*, product yields are highly influenced by hydrodynamic variables (that is, how oxygen is supplied to and mixed in the broth). Thus, increasing mixing intensity and oxygen flow in experiments aimed at maximizing production of pyocyanin increases the level of secondary metabolites such as pyocyanin and rhamnolipids (Konopacki et al. 2022). Foaming is a huge headache when it comes to rhamnolipid production. These biosurfactants are extremely active, sometimes causing froth that interferes with nutrient and oxygen availability, and in the worst scenarios causing autolysis of cells. Foam formation conditions such as rhamnolipid concentration, biosurfactant makeup, pH, and bioreactor operations are all influenced (Gong et al. 2021). Additionally, some teams are testing various carbon and nitrogen options (i.e., glucose, glycerol, or soybean oil) for rhamnolipid synthesis under low-oxygen setups (Zhao et al. 2021). The genetic factors that permit or restrict growth under different conditions have been studied in *P. putida* KT2440 research (Kampers et al. 2021). These pathways need to be improved further to bolster productivity. Enzymes RhlA, RhlB, and RhlC are essential for rhamnolipid production in *P. aeruginosa*, and in particular RhlA makes the hydrophobic "backbone" (Chong and Li 2017). Therefore, attention has been drawn to cheaper ways to produce these biosurfactants, for example, with the use of agricultural waste for lignocellulose, lowering costs and increasing sustainability (Tan and Li 2018).

5.1.3 Environmental Changes

As fermentation scales up, the environment in which it occurs inside the bioreactor can deviate widely. It is much harder to maintain consistent conditions like temperature, pH, and nutrient levels at larger volumes (Crater and Lievense 2018). For example, *Pseudomonas* species that depend on a lot of oxygen often can't scale up because giving enough oxygen over enormous volumes is difficult (Alvarez-Ortega and Harwood 2007, Höfte 2021). Another concern is thermal stress, since larger volumes generate excess heat that, without proper regulation, can impair or kill cells (Bisutti and Stephan 2020). Delicate cultures may also be harmed by stirring, and even cross-scaling from bench scale to industry often introduces other factors that may not be purely linear (Brindley et al. 2011).

In addition, larger bioreactors consume more energy and generate more by-products, which means the environmental footprint of this process has to be managed (Takors 2012). The conversion can also be optimized using facilities that employ specialized bioreactor designs, advanced computational models, and metabolic engineering (Hewitt and Nienow 2007). Pilot-scale trials validate and refine processes under near-real-world conditions, enabling the identification and resolution of potential issues before full-scale rollout, thereby ensuring smoother large-scale operations (Hellsmark et al. 2016). Keeping the process sustainable is the ultimate goal of labs today, given that traditional fermentation relies heavily on non-renewable resources. Thus, world is shifting focus to alternative energies and waste feedstock to reduce resource use and make the process more sustainable (Aggarwal et al. 2023, Keasling et al. 2021).

5.1.4 Optimization of Feedstocks and Media

At industrial production scales, *Pseudomonas* bacteria use different raw materials known as feedstocks, which tend to shift from pure laboratory-grade substances toward agricultural waste products such as wheat bran and sugarcane bagasse (Ajala et al. 2021). Soil conditions, alongside local climate and harvest-time variations, affect product yields because of factors described as variability. By carefully controlling nutrient equilibrium together with the capacity to switch production between various crops and agricultural substrates like corn stover and potato peels, businesses can bypass supply chain issues (Höfte 2021). Combining statistical design methods such as Plackett-Burman and Response Surface Methodology with artificial intelligence algorithms such as genetic algorithms identifies optimal nutrient blends for microbial strains and targeted outcomes (Huang et al. 2019). Sustainable agriculture benefits from viewing feedstock selection and production conditions as part of a complete system that leads to increased microorganism productivity.

5.1.5 Oxygen Transfer and Mixing

The successful expansion of *Pseudomonas* production requires engineers to properly address the delivery of oxygen and nutrients within the bioreactor system. *Pseudomonas* flourishes under oxygen-rich conditions; therefore, sufficient oxygen transfer must be achieved (Nur Asshifa et al. 2017). The design of impellers, alongside their operating speeds, determines the mixability of oxygen with the culture. The selection between axial-flow impellers for vertical circulation and radial-flow types for lateral mixing enables engineers to fulfill high oxygen demands while maintaining operational efficiency. The selection of suitable aeration techniques, which include spargers and membrane-based aeration, and reliable foam control emerge as essential operational points (Abuchenari et al. 2020).

Hyper-aggressive mixing produces shear stress that harms cells, so it is essential to select agitation rates and impellers that gently treat the cultured cells. Tests for rheology enable scientists to optimize control parameters related to broth viscosity

(Sadino-Riquelme et al. 2022). Researchers use both computational fluid dynamics (CFD) models and real-time dissolved oxygen monitoring tools, and these instruments enable better efficiency in scale-up process developments through optimized oxygen delivery (Ihling et al. 2021). The production of *Pseudomonas* depends on effective oxygen management and efficient mixing systems because they enhance the bacterium's sustainability for environmentally friendly pest management.

5.1.6 Contamination and Sterility

The production of *Pseudomonas*-based biocontrol agents faces substantial challenges because of contamination issues (Höfte 2021). Large manufacturing systems contain more extensive surfaces, providing numerous entry points for possible contamination. A single trace of contamination has the power to halt production of an entire batch, wasting materials together with production time. The process of sterilization must be rigorous for all equipment, bioreactors, and media through autoclaving along with chemical disinfection or irradiation methods. Air filtration (using HEPA filters), maintaining positive air pressure, and proper aseptic protocols also help lower contamination risks. Frequent checks, like microbiological air and surface testing, ensure any intruders are caught early (Höfte 2021, Kaur 2015).

Scientists are looking into innovative strategies to bolster sterility on large scales. Some are developing genetically modified strains that can fend off contaminants, while others experiment with co-cultivation to outcompete unwanted microbes. Microfluidic platforms, which create tiny, highly controlled culture spaces, are also being studied for safer *Pseudomonas* production. All these steps aim to keep contamination in check, making biocontrol solutions more consistent and reliable for farmers.

5.1.7 Cost Considerations

There is always a fine line between producing enough *Pseudomonas* at scale and ensuring that costs do not skyrocket. To remain both effective and economically viable, the process must minimize losses. Small but crucial details, like maintaining tight temperature and pH ranges, can significantly raise yield and cut waste, helping keep the entire operation financially feasible (Anwar et al. 2019). Beyond the fermentation phase itself, paying attention to downstream processes and rigorous quality control can curb added expenses from reprocessing. Automation is another critical ally in managing costs. Automated systems reduce human error and increase reproducibility, ensuring consistent product quality. Applying real-time data analytics can further refine conditions moment by moment, optimizing for both output and cost. Single-use technologies, which forego the need for comprehensive cleaning, offer another cost advantage and make upscaling or downscaling much simpler.

5.1.8 Infrastructure Requirements

Scaling up *Pseudomonas* production demands specialized infrastructure far beyond the capacity of a typical laboratory (Eng et al. 2021). Different types of bioreactors anchor this effort, whether stirred-tank reactors for intense mixing and aeration or airlift reactors for shear-sensitive cultures (Tekere 2019). Membrane bioreactors can excel where product separation is a must. Controlling environmental factors, like temperature, pH, and oxygen, is vital for healthy bacterial growth. Automated heating or cooling, pH regulation, and well-designed aeration systems (e.g., spargers or membrane oxygenators) ensure the bacteria get what they need (Tekere 2019). Before the culture even begins, media need proper sterilization, meaning autoclaves and sophisticated feeding systems must be available. When it is time to harvest the cells or extract valuable by-products, additional equipment, like centrifuges and filtration units, steps in. Facilities also have to manage risks, particularly for genetically modified strains, and must handle waste properly to maintain environmental compliance. Automation and process control systems deliver real-time metrics, enabling quick adjustments for the best possible outcomes (Figure 5). In short, large-scale production of *Pseudomonas* calls for integrated infrastructures, combining specialized bioreactors, environmental monitoring, thorough sterilization, and mindful waste management, to ensure a cost-effective, safe, and high-quality supply of these vital biocontrol agents.

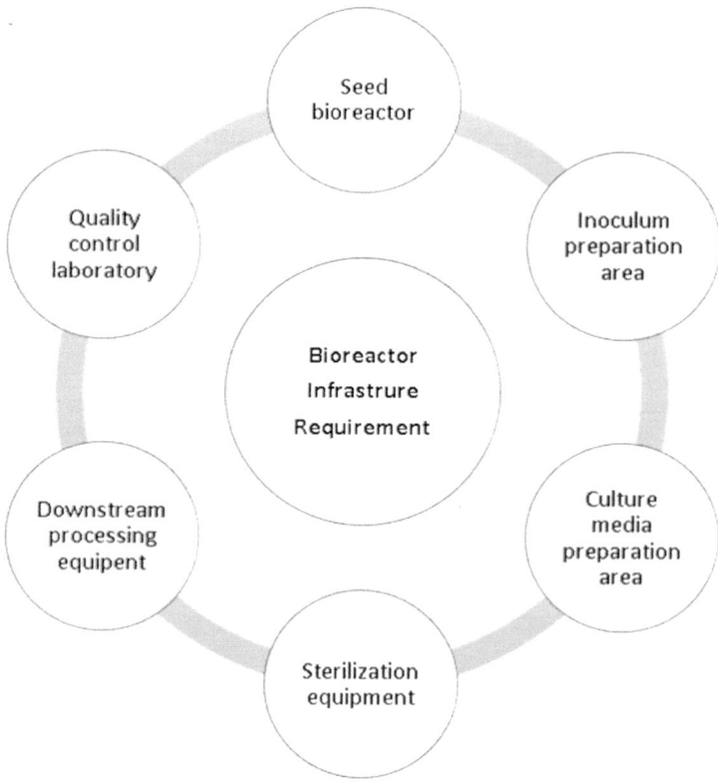

Figure 5: Essential infrastructure required for bulk production of *Pseudomonas*.

5.1.9 Data Analysis and Process Monitoring

The successful large-scale *Pseudomonas* fermentation requires the essential integration of solid infrastructure and rigorous data analysis with continuous process monitoring (Figure 6). Recent developments in computational biology, together with genomics, enable us to produce valuable compounds in an unprecedentedly precise manner. The advancement of bioprocess data management stands as a critical factor because sensors produce massive measurements of temperature, pH, and oxygen, but these readings need precise storage, preprocessing, and interpretation processes. Extensive genomic and transcriptomic studies help identify metabolic pathways of microorganisms while identifying zones of process limitation (Kampers et al. 2021). The raw data from bioprocessing become valuable insights through the application of machine learning technologies together with artificial intelligence methods. Through pattern recognition and deviation identification, the fermentation process becomes both more efficient and reliable (Bilbao et al. 2023). Metabolic modeling, together with simulation methods, permits virtual experimentations of various fermentation approaches, thus reducing the requirement for substantial physical experiments (Khaleghi et al. 2021). Real-time dashboards convert intricate data into easy-to-understand visual presentations that expedite knowledgeable decision-making (Feng et al. 2021). AI anomaly detection systems use standard fermentation patterns to learn normal conditions before detecting any deviations, so operators can solve problems swiftly (Newlands 2022, Yingngam et al. 2024). The inclusion of production data across all stages leads to a complete understanding of business processes, therefore triggering ongoing optimization and efficiency advancement for the entire manufacturing chain.

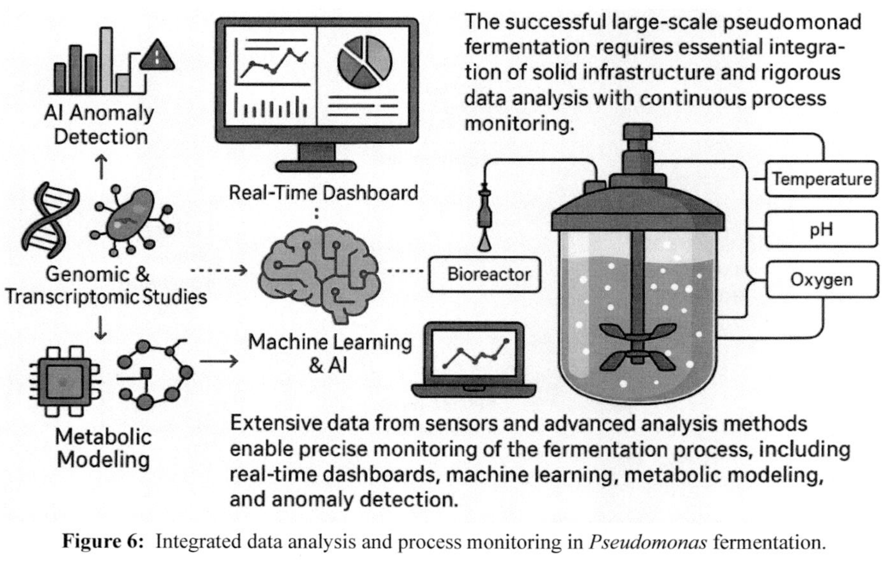

Figure 6: Integrated data analysis and process monitoring in *Pseudomonas* fermentation.

5.2 Solutions and Innovations

This section presents multi-dimensional advancements between advanced fermentation approaches and bioprocess design with alternative production platforms, which optimize the production scale and efficiency of *Pseudomonas*-based biocontrol agents. The combination of genetic improvements with real-time process tracking and co-culture methods proves to be a solution that establishes economical, sustainable production techniques that boost environmentally friendly agricultural practices.

5.2.1 Scalable Fermentation Technologies and Optimized Growth Media

Economical mass-scale production of biocontrol *Pseudomonas* strains like *P. fluorescens* requires fermentation process refinement to reach maximum output coupled with operational excellence. The research field works on improving cell propagation environments, protein manufacturing systems, and fermentation process management. The essential approach toward optimization requires precise adjustments to how proteins are expressed. The placement of proteins in the *P. fluorescens* periplasmic space enhances both the efficiency of downstream production and protein functional properties (Retallack et al. 2006, 2007). Multiple native secretion peptides, when combined, strengthen the efficiency of recombinant protein secretion (Retallack et al. 2006). Genome-scale metabolic modeling procedures with integrated thermodynamics function as an essential tool for comprehending *Pseudomonas* central carbon metabolism (Puchalka et al. 2008, Tokic et al. 2020). *P. putida*, i.e., KT2440's obligate aerobic limitations can now be prevented with knowledge obtained through analysis (Kampers et al. 2021). Academics now focus on developing cultivation methods suited for producing particular bioproducts. The production of poly(3-hydroxyoctanoic acid) (PHO) through *P. putida* GPo1 depends on adjusting conditions such as aeration, agitation, and temperature under nutrient-limited environments (Elbahloul and Steinbuchel 2009, Wang et al. 2007). Rhamnolipid manufacturing with *Pseudomonas* requires proper condition control and reduced foam production (Dabaghi et al. 2023, Gong et al. 2021).

The fermentation system development process becomes more refined through combined genetic and metabolic research alongside oxygen analysis studies (Kampers et al. 2021). The advancement of *Pseudomonas* biocontrol agent production benefits from technological progress and genetic engineering to achieve both effective pricing and large-scale manufacturing capability while providing a chemical pesticide alternative that respects environmental interests.

5.2.2 Integration of Bioprocess Engineering Principles for Improved Yield and Consistency

Implementation of bioprocess engineering approaches for *Pseudomonas* biocontrol agent manufacturing requires refined techniques based on genetic, molecular, and production process methods. Three species which demonstrate outstanding plant growth enhancement and nutrient acquisition abilities while fighting diseases are

P. fluorescens, *P. chlororaphis*, and *P. protegens*. Three effective combat mechanisms include nutrient-blocking methods, antibiotic compound synthesis, and systemic defense activation (Höfte 2021). Among the methods to enhance biocontrol capabilities stands genetic modification. Biocontrol of *Rhizoctonia solani* in controlled plant-growth environments was achieved by integrating the *chiA* gene, which directs chitinase production, into *P. fluorescens* (Majumdar 2023). The monitoring of key bioprocess variables in real-time through process optimization makes use of multiple sensor technologies such as optical sensors, electrochemical sensors, and biosensors. Real-time monitoring combined with control systems remains essential to produce dependable output from these advantageous bacterial strains.

Scientists can achieve superior reliability and better effectiveness of biocontrol agents through the integration of advanced bioprocess monitoring with *Pseudomonas* genetic expertise. Genetic engineering and bioprocess supervision continue to advance and will improve both the productivity and uniformity of *Pseudomonas*-based solutions (Majumdar 2023, Mark et al. 2006).

5.2.3 Exploring Alternative Production Platforms

Plant and *Pseudomonas* co-growth has become an important approach to produce biofertilizers while remediating contaminated environments. This method ensures beneficial interactions between specific plant species and *Pseudomonas* strains, primarily benefiting legumes when partnered with nitrogen-fixing *P. putida* species (Höfte 2021, Sanita Lima and Coutinho de Lucas 2021, Zhou et al. 2019). Plant-*Pseudomonas* compatibility validation stands as the essential initial process because it rewards both performance excellence and operational security (Sanita Lima and Coutinho de Lucas 2021).

Ring cultivation methods use sterile growth environments made with materials including sand, peat moss, and perlite to prevent contamination. The procedure introduces seeds or seedlings to environments where *Pseudomonas* culture receives inoculation. Plant growth systems receive *Pseudomonas* inoculations through two main methods: direct pre-plant inoculation of growth media or rhizosphere enrichment practices using sterile hydroponic systems. The key components for successful co-cultivation consist of maintaining nutritional equilibrium between microbial elements and comprehending their dynamic nature, specifically through metabolite exchanges and biofilm development (Kapoore et al. 2022, Selegato and Castro-Gamboa 2023). Bioflocculation serves as a method that simplifies the harvesting of plant material together with bacterial biomass when proper planning is executed. Researchers have introduced two new developments involving carrier-based *Pseudomonas* cell immobilization to control bacterial release and genetic strain modifications for better plant interactions (Hmelo et al. 2015, Shalev et al. 2022, Zheng et al. 2023). This co-culture fermentation system, based on plants and *Pseudomonas* microorganisms, allows for sustainable operations through plant-feeding sources, enhances operational scalability, and allows researchers to choose strains with superior performance. Improved crop cultivation through these innovations demonstrates the various applications of *Pseudomonas* within sustainable agricultural practices.

6. Formulation for Enhanced Stability and Delivery

6.1 Challenges

A detailed exposition of the preservation difficulties for cell viability and environmental stability, precise treatment mechanisms, and farm machine adaptation exists in the following subsections. Effective, sustainable *Pseudomonas*-based biocontrol formulation development requires solutions for strain selection as well as protective additives under optimized application conditions.

6.1.1 Maintaining Cell Viability During Storage and Transport

The viability of *Pseudomonas* cells during storage and transport operations is crucial for effective biocontrol solutions. Selecting strains with documented resistance to cold and desiccation marks a key first step in biocontrol storage (Vuolo et al. 2022), while *P. fluorescens* strains demonstrate optimal storage properties (Gade et al. 2014). A pure, high-density culture should be used from the beginning to minimize cell death during storage (Khan et al. 2023). Survival rates during freezing can be remarkably enhanced through the use of cryoprotectants including glycerol along with sucrose and DMSO (Criste et al. 2014). Silica gel is the recommended desiccant in desiccation-based systems because it protects cells (Bashan et al. 2014).

Storage protocols greatly influence viability. The storage period at 4°C is suitable for brief storage, but long-term preservation demands temperatures below –80°C (Ptaszek et al. 2023). Properly maintaining cell structure for desiccation-tolerant strains occurs through lyophilization (freeze-drying) procedures (Criste et al. 2014). Transport management requires airtight, moisture-proof containers to prevent exposure to oxygen and moisture changes (Bashan et al. 2014). The combination of insulated packaging with cold packs, along with fast transit through streamlined logistics systems, protects cultures by reducing stress. The mandatory function of quality control remains constant throughout the entire process. The value of pest control increases through optimal potency monitoring using colony-forming unit (CFU) counts before the agent is delivered to its final destination (Berninger et al. 2018).

6.1.2 Environmental Persistence (UV, desiccation)

When *Pseudomonas* biocontrol agents reach field applications, they must endure difficult environmental conditions, particularly UV radiation and desiccation. The identification of biocontrol strains occurs predominantly through selecting microorganisms that naturally withstand UV light and dry environments originating from harsh conditions (Bhattacharjee et al. 2023). The addition of protective UV agents such as melanin, natural pigments, or lignin derivatives provides external protection to cells (Vanhaelewyn et al. 2020). The moisture-bearing properties of hydrogels protect cells from desiccation stress (Mutlu et al. 2015) and desiccation protection capabilities of sugars and amino acids prevent drying-related osmotic damage (Saif et al. 2021). The timing of application plays a key role in success, as users should wait until evening hours or overcast skies to minimize UV exposure and target shaded areas.

6.1.3 Targeted Delivery and Adherence

Optimizing delivery and adherence of *Pseudomonas* strains relies on a clear understanding of their modes of action and environmental preferences. Foliar applications benefit from fine-mist nozzles and properly calibrated volumes to achieve uniform coverage (Devi et al. 2022). Alternative strategies, like seed treatments or root-level applications, support early colonization (Skrzypczak et al. 2021). Selecting weather conditions conducive to bacterial establishment and avoiding periods of intense sunlight or high temperatures fosters more effective colonization (Craig et al. 2021).

In formulations, non-toxic surfactants can lower surface tension (Silva et al. 2024). Specialized adjuvants that enhance adherence or help maintain moisture may also prove beneficial. Research indicates that certain biosurfactants, like syringafactin from *P. syringae*, increase bacterial survival on leaves by balancing fluctuations in humidity (Burch et al. 2014). Additionally, plant surface traits can influence bacterial attachment (Bogino et al. 2013). For strains that confer systemic immunity, precise application may be less critical, affording more flexibility in plant variety choices. Nonetheless, regular assessment of *Pseudomonas* populations and plant health is key; results guide adjustments in application strategies. Ensuring compatibility between the chosen strain, additives, and host plant, along with adherence to safety standards for microbial handling, supports successful deployment. When combined with integrated pest management (IPM) practices, *Pseudomonas* can form a robust part of an overall crop protection strategy.

6.1.4 Integration with Existing Farm Equipment

Large-scale *Pseudomonas* applications, especially *P. fluorescens*, depend heavily on the successful integration of bacterial timing and agricultural equipment with biocontrol agents and agrochemical practices. *Pseudomonas* delivery and survival rates depend on nozzle design (Alsved et al. 2018), UV radiation influence (Saha et al. 2013), and spray solution pH/temperature (Donnarumma et al. 2010), in addition to desiccation (Mputu Kanyinda et al. 2014). Fitting nozzles with low shear stress profiles preserves bacterial viability, and the addition of UV protectants enables improved bacterial functioning.

6.1.5 Synergies/antagonisms with Other Agrochemicals

The synergistic benefits obtained from combining *Pseudomonas* with *Trichoderma* and *Bacillus* biocontrol agents expand pathogen protection through their different mechanisms of action. *Pseudomonas* activates plant systemic resistance, which combines well with various biocontrol agents alongside fungicides (Jankiewicz and Koltonowicz 2012, Pieterse et al. 2014). The management of antagonistic behaviors requires extra attention because *Pseudomonas* produces antibiotics and competes for necessary resources, which may cause problems for helpful microbes. The selection of appropriate strains combined with strategic timing of applications reduces these issues. The timing of pesticide applications should be carefully considered because *Pseudomonas* displays variable susceptibility to different chemicals, which can be minimized through proper timing of exposure. Some *Pseudomonas* strains have shown combined effects with selected herbicides, creating potential applications in integrated pest management (de Oliveira et al. 2021).

6.2 Innovations

Agricultural practices benefit from technological developments that apply encapsulation for protection and UV defense through biodegradable carriers, leading to higher operational performance and greener environmental outcomes. Science has developed modern methods to improve durability, precise delivery capabilities, and reduced environmental effects of biopesticides and fertilizers, which has led to more sustainable agricultural systems. Further discussions about these innovations, together with their related effects, appear in subsequent sections.

6.2.1 Encapsulation Technologies for Improved Survival and Targeted Delivery

Progress in encapsulation science brings major impacts to agricultural operations by establishing precise delivery systems coupled with lower environmental effects. There has been significant development in agriculture through the creation of protective shells to enclose pesticides and fertilizers, which release only when triggered by certain environmental signals, including humidity and ultraviolet radiation. The encapsulation method both increases efficacy and minimizes the need for repeated applications while reducing leaching probabilities. Modern encapsulation methods enhance both biopesticide stability by protecting living microorganisms and bioactive compounds, allowing for precise, prolonged field applications. Research studies demonstrate that encapsulation technology can develop efficient farming systems while maintaining environmental responsibility (Li et al. 2019).

6.2.2 UV Protectants, Stabilizers, and Nutrient Supplements for Enhanced Field Persistence

The need to protect agricultural materials from environmental stresses, including UV radiation, has become essential in current agricultural production. The combination of UV protectants, stabilizers, and nutrient supplements extends durability, functional operation, and productivity yields of crops and agricultural materials (Kasirajan and Ngouajio 2012). Agricultural plastics receive protection from dangerous UV radiation through the implementation of UV protectants (Briassoulis et al. 2015). The high-quality agricultural plastic market uses Tinuvin® NOR 356 from BASF as a stabilizer, which its creators specifically designed for this purpose. The product from BASF protects against intense UV radiation, heat exposure, and elemental sulfur treatments, making it applicable to organic farming practices (BASF 2024).

Agricultural materials need stabilizers, as they preserve their structural quality even under difficult environmental conditions. The addition of polymer additives to greenhouse film materials strengthens their resilience against harsh weather conditions, chemical threats, and multiple forms of stressors (Kasirajan and Ngouajio 2012). Yet, some academic publications fail to discuss nutrient supplements directly, although extensive literature acknowledges their positive impacts on increasing crop vigor, yield, and quality enhancement (Fageria et al. 2010, Savvas et al. 2013). When

UV protectants, stabilizers, and nutrient supplements are combined in applications, they enable sustainable farming practices by promoting better crop yields and reducing resource expenses (Briassoulis et al. 2015).

6.2.3 Development of Carrier Materials (Biodegradable Polymers, Clays) for Slow-Release and Reduced Environmental Impact

The agricultural industry has adopted biodegradable polymers and clays as essential carrier materials to achieve slow-release delivery while lowering ecological risks. The chemical compounds of renewable origin, called biodegradable polymers, have gained environmental compatibility status because they originate from plant-based materials. Multiple applications exist for these materials, spanning from bioplastics to slow-release fertilizers, while showing a diverse variety of chemical and mechanical properties (Samir et al. 2022). Industrial applications utilize cellulose, starch, and pectin, which are obtained from agricultural waste during the production of environmentally sound products.

Advanced key nutrient carriers have been developed through clay-based nanocomposites, which represent major progress in the field. Scientists employ bentonite clay components in these composites to provide a method for controlled element distribution of phosphorus and nitrogen, thus raising fertilizer productivity. Clay-polymer composite behavior depends on the clay type selection, while neutralization control and the degree of chemical bonding influence its properties. The application of organo-clay composites specifically shows great promise for boosting phosphorus use efficiency and represents a better method for sustainable farming (Abuchenari et al. 2020). The advancements in carrier materials pave the way for sustainable and efficient agricultural systems by delivering inputs through optimized resource utilization alongside environmental protection mechanisms.

7. Stringent Quality Control Measures for Reliable Performance

7.1 Critical Quality Attributes of Pseudomonas Biocontrol Agents

The reliable function of *Pseudomonas* biocontrol agents depends on quality control programs that examine crucial aspects, including cell viability, virulence, genetic stability, purity, and bioactive metabolite production levels. The system maintains its safety and consistency through genomic screening combined with in-process testing and contamination controls, according to the subsections' explanations.

7.1.1 Cell Viability, Virulence, Genetic Stability, Purity, and Presence of Bioactive Metabolites

The agricultural sector is adopting *Pseudomonas* species as a leading biocontrol agent because of their high potential to control pests (Haas and Defago 2005, Lugtenberg and Kamilova 2009). Beneficial bacterial utilization requires special attention to particular critical quality attributes that determine their effectiveness. The safety and performance of *Pseudomonas* strains depend most strongly on five critical quality attributes: cell viability, virulence, genetic stability, purity, and bioactive metabolite presence.

Having an abundant and healthy community of *Pseudomonas* cells plays an essential role in successful rhizosphere colonization, pathogen suppression, and key metabolite generation within the system. Three main elements, namely storage environment, product production approaches, and substance dispensing routines, determine cell survival rates (Benhamou et al. 2000). Strains used for biocontrol programs must lack any health risks for human beings. The identification and removal of dangerous microorganisms from *Pseudomonas* populations become possible through screening programs combined with genomic analysis (Timmusk et al. 2005). Laboratory modifications of products can eliminate dangerous elements from organisms while maintaining functional properties (Weller et al. 2002). Mutational processes have the potential to damage essential metabolic pathways while introducing detrimental traits over time. Consistent outcomes depend on using strong production systems together with selecting naturally stable microorganisms that produce microbes.

The existence of competing microorganisms results in decreased solution effectiveness and introduces environmental safety issues. Multiple stages of production testing, together with proper sterilization methods, help ensure biocontrol agents remain free of contamination. *Pseudomonas* generates various antifungal as well as antibacterial metabolites that include antibiotics and siderophores (Cesa-Luna et al. 2020). Preparation of biocontrol strains can be optimized through the knowledge of when and how their metabolites are synthesized (Weller et al. 2002).

7.2 Standardization and Regulatory Aspects

Standardization processes, together with regulatory requirements, help *Pseudomonas* biocontrol agents fulfill robust safety, efficacy, and quality expectations. Strict strain characterization methods, alongside potency testing and contaminant detection procedures, help achieve global regulatory requirements and enable third-party certifications through adherence to Good Manufacturing Practices and rapid quality control testing. Subsequent points offer detailed information about these procedures, together with breakdowns of encountered implementation issues (Figure 7).

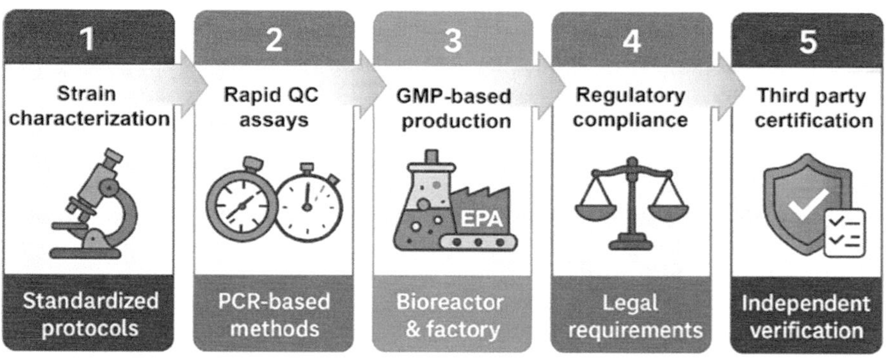

Figure 7: Standardization and regulatory aspects for *Pseudomonas* based biocontrol agents.

7.2.1 Standardized Protocols for Strain Characterization, Potency Testing, and Contaminant Detection

It is essential to conduct thorough identification and validate each distinct *Pseudomonas* strain because of its wide variety. Scientists rely on 16S rDNA sequencing to identify organisms but many researchers endorse additional powerful techniques due to their superior ability to minimize false identifications (Höfte 2021). Testing of the bacterial strains follows verification steps to ensure potency in counteracting plant pathogen infections through pathogen competition and compound production and plant defense activation (Hofte and Altier 2010). Contamination checks are similarly vital. Field research reliability requires constant product safety and purity verification from beginning to end. Genomic profiling reveals genes linked to antibiotic production and plant growth promotion, which helps predict how well each strain will perform (Mark et al. 2006). The requirement to meet environmental as well as safety standards stands without exceptions. A biocontrol product needs to demonstrate effectiveness alongside harmlessness in order to fulfill the principles of sustainable agriculture.

7.2.2 Rapid Quality Control Assays

The successful transfer of *Pseudomonas* biocontrol agents to farmers depends on immediate detection methods. Detection of target strains happens swiftly through the use of PCR-based methods and genomic approaches (Penazova et al. 2020). The fast detection methods provide significant value because *Pseudomonas* shows broad capabilities and many genetic variants (Höfte 2021, Zboralski et al. 2022). Determining the environmental survival patterns of these bacteria requires thorough comprehension. Through real-time PCR and next-generation sequencing, scientists can determine population sizes of the agents and understand their relationships with pathogens, as well as plants and soil microbes. Land-based investigations using accelerated testing methods help researchers screen newly developed strains with antifungal activities or metabolite production capabilities to select candidates for business applications.

7.2.3 Good Manufacturing Practices for Biocontrol Agents

Producers need Good Manufacturing Practices (GMP) for creating dependable *Pseudomonas* biocontrol agents that are safe (Bhartendu et al. 2024). The guidelines begin with using validated strains that belong to *P. fluorescens* or *P. fragi* groups along with other taxa from the genus (Höfte 2021). Standard operating procedures encompass the exact requirements for cultivating bacteria as well as the procedures needed to maintain consistency and purity during post-harvest (Lahlali et al. 2022, Teixidó et al. 2022). Laboratory-quality inspections verify that production bacteria maintain both their original genetic composition and capability to inhibit microbes. Specialized conditions needed for maintaining cell viability must be applied during final packaging, while complete records need to be maintained (Bhartendu et al. 2024). Product quality owes much of its assurance to GMP standards and the ability to fulfill all applicable legal criteria (Hoeschle-Zeledon et al. 2013).

7.2.4 Regulatory Compliance

Pseudomonas strains must meet various international biocontrol agent regulations to obtain commercial market entry. In the United States, for example, the Environmental Protection Agency (EPA) demands extensive evaluations for product safety, efficacy, and environmental footprint (EPA 2024, Leahy et al. 2014). A shared set of environmental regulations prevails between China and the European Union demonstrates rising international interest in sustainable agriculture (Silva et al. 2022).

Organic growers tend to prefer Organic Materials Review Institute (OMRI) certification in the U.S. and Ecocert certification in Europe, which adds value to market acceptance when used together (Beerling and Sahota 2013, Stemwedel 2009). Manufacturers should unite strong scientific trial data with regulatory expertise to succeed in market entry. The success of *P. putida* KT2440 depended on multiple factors, including its demonstrated abilities to fight fungal diseases and promote plant growth (Purtschert-Montenegro et al. 2022), which led to regulatory approval resulting in the commercial product "BlightStop."

7.2.5 Third-Party Certification

Biocontrol agents receive independent certification through assessments which verify their performance against plant pathogens as well as their safety standards (Glare et al. 2012). Third-party certification programs operate similarly to the Food and Drug Administration's (FDA) accredited third-party certification program by auditing both developmental operations and operational execution. Biocontrol products need to prove dependable performance together with strong pathogen control abilities and secure protection of non-target organisms (Sundh and Goettel 2012).

The certification process incorporates regular assessments that confirm the agent's safety characteristics remain unchanged. The biological control initiatives of U.S. Department of Agriculture (Coulson et al. 2000), together with the National Organic Program (Krasny 2002), demonstrate how proven biocontrol agents get implemented via integrated pest management (IPM). Advances in molecular biology, genomics, and allied disciplines are refining certification methods, making applications more precise and efficient. Using omics-enabled analyses, scientists dissect microbial species interactions (Lorito et al. 2010) and characterize bioactive compounds (Compant et al. 2005), thereby elucidating the functional basis of biocontrol systems (Kohl et al. 2019, Pal and Gardener 2006). Third-party oversight systems establish transparent, auditable standards to ensure verification and compliance. Third-party oversight systems establish trust while leading the pest management field to develop safer and intelligent solutions for pest control.

8. Efficient Delivery Methods for Field Efficacy

The ability to successfully deliver suitable microbes to pests remains equally important to discovering powerful biological control microorganisms in academic settings. Biocontrol agents like *Pseudomonas* must receive proper delivery to their target area because these living agents require protected transport for active function. Biological protection requires defending microbes from environmental factors, including sunlight exposure, temperature extremes, and changes in moisture, because these elements damage their survival capacity. Application must occur at the specific point when the targeted pest is most vulnerable. Precise placement of *Pseudomonas* at the pest's most affected areas leads to better control of infestation spread alongside increased efficiency. Direct delivery strategies increase both operational efficiency through better results and cost-effectiveness through effective resource utilization. Multiple delivery methods used for *Pseudomonas*-based biocontrol materials in field conditions are listed in Table 4.

Table 4: Methods of *Pseudomonas*-based biocontrol delivery in the field.

Delivery Method	Mechanism	Efficacy in Field Conditions	Compatibility with Crops	References
Precision Spraying	Targeted application to specific areas or crops	High, due to the targeted approach	Compatible with most crops; requires calibration	(Couillerot et al. 2009)
Seed Coatings	Coating seeds with biocontrol agents	Moderate, depending on the seed germination rate	High compatibility, especially with row crops	(Chin et al. 2022)
Soil Drenching	Applying agents directly to the soil	Varies with soil type and conditions	Broad compatibility, but efficacy varies	(Haas and Defago 2005)
Foliar Application	Spraying onto the leaves of plants	Good in early disease control stages	Suitable for leafy crops, less for root crops	(Fravel 2005)
Granular Formulation	Application of granules containing the agent	Effective in sustained release	Good for larger crops; varies with granule size	(Bloemberg and Lugtenberg 2001)
Irrigation Integration	Incorporation with irrigation systems	High, due to uniform distribution	Highly compatible, especially in irrigated farmland	(Perneel et al. 2007)
Aerial Broadcasting	Wide-area dispersal, often via drone or aircraft	Varies, dependent on coverage and wind conditions	Best for large, open fields or less accessible areas	(Thomashow and Weller 1996)
Root Dipping	Dipping roots in biocontrol solution before planting	High in root protection and early growth stages	Particularly effective for transplanted crops	(De Vleesschauwer and Höfte 2009)
Hydroponic Systems	Integration in water nutrient solutions in hydroponics	Highly controlled and efficient	Ideal for hydroponic systems, less for soil crops	(Pieterse et al. 2014)
Biofumigation	Incorporating dried biocontrol agent biomass into soil, releasing volatile compounds for pest control	Moderate-High: Broad-spectrum pest suppression, effective against soilborne pathogens, long-lasting residual activity	Moderate: May affect beneficial soil microbes, potential odor issues depending on agent strain	(Madhavi and Devi 2018)

8.1 Challenges

The conversion of *Pseudomonas*-based biocontrol as a laboratory concept into an effective field solution involves resolving multiple challenging obstacles. The main challenge involves maintaining bacterial survival when exposing them to environmental threats such as temperature variations, dehydration, and UV rays. Refined bioformulations with UV-protective additives and hydrogel water retention agents defend the bacteria from dangerous conditions (John et al. 2010). *Pseudomonas* must protect itself from typical soil microbes that fight for resources and at times attack the introduced bacterial strains (Stockwell et al. 2011). The selection of powerful *Pseudomonas* strains or their combination with helpful microbes provides better possibilities for survival.

Pseudomonas needs to achieve successful delivery to its desired target location. The deep penetration of *Pseudomonas* into roots as well as leaves for controlling diseases requires high-density populations for achieving substantial pest control outcomes. A combination of biosurfactants as adhesion enhancers and plant defense elicitors serving to boost plant resistance demonstrates both significant roles (Bais et al. 2004, Ongena and Jacques 2008).

On the practical side, scientists and manufacturers face the challenge of packaging these bacteria in a manner that preserves their potency during storage, transit, and field application. Low-cost yet protective carriers, such as clay or alginate, show potential, though fine-tuning these systems is key to maintaining long-term shelf life and reliable performance (Bashan et al. 2014).

8.1.1 Uniform Distribution and Adherence to Target Surfaces

In the quest to achieve maximum biocontrol impact, it is crucial to grasp how *Pseudomonas* operates (through nutrient competition, antimicrobial production, or boosting plant defenses) and identify the conditions in which it flourishes (Pieterse et al. 2014). At the outset, it helps to begin with a robust, high-density culture designed to keep cells viable (Scherwinski et al. 2008). For leaves, fine-mist sprays are ideal (Fu et al. 2020), while seed or root treatments can be particularly effective against soilborne threats (Reddy et al. 1999). Applications should be timed for *Pseudomonas* establishment because this strategy maximizes effectiveness (Krishna et al. 2022).

Surfactants serve as essential ingredients which help bacterial cells to distribute evenly across plant surfaces (D'aes et al. 2010). The utilization of adjuvants for enhancing bacterial cell compatibility and persistence represents a fundamental method to boost the process (Aguirre et al. 2023). Plant leaf characteristics determine how bacteria attach to them (Gupta et al. 2013, Orozco-Mosqueda et al. 2021). The determination of *Pseudomonas* densities through repeated testing together with pest or disease effects directs constant enhancements in the cultural media methods or delivery procedures as well as the addition of substances to bacterial solutions. The successful application of *Pseudomonas* treatment requires that both bacteria and plant hosts maintain a natural equilibrium with any additional components of the solution. *Pseudomonas* achieves greater effectiveness when combined with an already established Integrated Pest Management (IPM) plan (Kupferschmied et al. 2013).

8.1.2 Environmental Barriers

The extensive use of biocontrol agents faces barriers to their successful implementation at commercial scales because of environmental difficulties during pest and disease control operations. Two main categories of obstacles exist for biocontrol agents (BCA): their ecological and physiological characteristics, and approval-related requirements. The main area of concern focuses on BCA ecophysiology when used in field environments. Scientists create predictive models for research purposes regarding how distinct environmental elements and natural bacterial communities affect BCA populations and their capabilities to fight disease-causing pathogens. The goal behind this research is to achieve better and stable performance among BCAs during their field applications. DNA amplification together with antagonist-specific probes are expected to improve the tracking of BCA populations inside plants (Jaiswal et al. 2022).

Two main barriers stand in the way of BCA registration along with environmental safety requirements. The commercialization process for BCAs requires strict safety standards which become harder to achieve when the microorganism uses antibiosis as its mode of action and belongs to pathogenic strain groups. Companies face obstacles due to safety registration requirements coupled with high expenses for registration when the market demand for specific BCAs remains low (Barratt et al. 2018).

Soil ecology creates an advanced level of complexity when studying the control methods of plant-parasitic nematodes (PPNs). The translation from lab results to greenhouse or field accomplishments often fails since soil ecosystems comprise various interacting biological and environmental components (Abd-Elgawad and Askary 2020). A large number of BCAs applied in the field leads to only partial successful suppression of PPN populations (Antil et al. 2023). Researchers recommend environment-specific application techniques that combine biological and environmental considerations when trying to control PPN populations. The strategy of adjustable environmental management has emerged as a biocontrol approach called conservation biocontrol to decrease PPN effects (Stirling 2011).

8.1.3 Delivery Methods and Field Efficacy

Proper deployment methods serve as essential elements for improving BCA performance during practical applications. Research shows that water-dispersible granules (WDGs) can successfully fight diseases together with their ability to boost plant development. *Bacillus subtilis* WDG formulations in tomato cultivation reduced diseases while enabling superior growth parameter development compared to talc-based and alginate bead protocols (Kohl et al. 2019).

Basic chitinase enzymes produced by *Trichoderma harzianum* and *Bacillus subtilis* need to work for an effective biocontrol system. Research findings demonstrate that enzyme activity of chitinase and β-1,3-glucanase depends on incubation duration because pathogen suppression depends on these factors (Jangir et al. 2021). The improvement of formulation methods together with enzyme dynamic assessments allows researchers and growers to increase BCAs' success in agricultural applications.

8.1.4 Compatibility with Existing Spraying Equipment

Multiple factors determine precise pesticide and herbicide spraying in agricultural fields because it requires high efficiency alongside minimized off-target effects combined with present system compatibility. Modern technological developments target the improvement of application precision while maintaining increased safety for crops and environmental elements (Damalas and Koutroubas 2016).

The accuracy levels during spraying operations determine the success of this process. The delivery of chemicals depends on proper mixing of chemicals through pre-mixing agents, ongoing agitation techniques, and direct injectors (Nuyttens et al. 2007). It is equally vital to preserve the pre-established application rate. The system reaches operational stability through rate controllers, smooth speed control, and precise spray boom overlapping as specified by Kumar et al. (2023). The production of a uniform spray pattern depends critically on choosing the right nozzle angle and finding the best boom height in addition to achieving correct overlapping conditions. Droplet size determination represents a critical factor because it influences both the reduction of drift occurrences and application effectiveness (Nuyttens et al. 2007, Wolf and Peng 2011).

The precision of spraying tools depends on proper calibration methods which ensure efficient delivery. Sprayer calibration can be carried out successfully with the 1/128th method which provides simple and dependable results (Cromwell 1992). The exact chemical amount determination for treatment follows the label specifications and area measurements to achieve proper application rate recommendations (Planas et al. 2022). The application results heavily depend on uniform execution since misalignment and restrictions of nozzles as well as differing fan angles and variable nozzle positions can generate non-uniform coverage which produces streaking patterns alongside untreated areas and over-application zones (Doruchowski et al. 2012). The Smart Sprayer Analytical System (SSAS) has turned out to be a crucial innovation for the application requirements described. The SSAS enables professionals to optimize sprayer settings through air-assist velocity data collection, which increases precision while decreasing unintended toxicity outcomes (Bahlol et al. 2020).

8.2 Innovations

New methods to deliver biocontrol agents drive changes in agriculture by increasing accuracy and operational efficiency during production. Combining drones, sensor-integrated systems, electrostatic sprayers, seed coatings, and attractant-based delivery methods enables advanced precision spraying to be applied efficiently and reduces resource waste and environmental damage. This chapter presents extensive discussions about these modern strategies in subsequent sections.

8.2.1 Precision Spraying

Agriculture now demonstrates improved precision and efficiency because of recent developments concerning *Pseudomonas* species-based biocontrol agent delivery vehicles. Drone applications and electrostatic sprayers act as leading technology advancements in this progress. Modern technology is transforming pesticide delivery methods by focusing on precise methods that save resources while protecting the environment. Drones are one example of how unmanned aerial vehicles (UAVs) show the present technological revolution. Drone technology helps farmers direct their applications by treating restricted areas that traditional methods could not access. The systems collect current field information, which helps executives make better decisions while simultaneously minimizing waste (Montes-Colmenares et al. 2020). Advanced spraying technology advances through the combination of sensors with artificial intelligence systems. Advanced systems use a combination of image technology with plant-sensing devices while employing deep learning systems to distribute materials to precise locations (Ahmad et al. 2021). The system achieves precise measurements combined with lower chemical drift and optimal input resource usage (Li et al. 2023). Smart spraying technology delivers both minimal product waste and improved effectiveness in addition to strong efficiency (Giles and Billing 2015). The precision-based vision benefits from electrostatic sprayers as complementary methods that boost this specific delivery method. Electrostatic droplets obtain uniform coverage when the device charges droplets because plant surfaces carry opposing electrical charges (Scherm et al. 2007). The technique both lowers drift occurrence and strengthens the biocontrol agent's adhesion capabilities which leads to improved performance. The study conducted by Salcedo et al. (2020) demonstrates that electrostatic spraying provides substantial benefits for preserving control of diseases through its appointment as a valuable technique for *Pseudomonas*-based agent distribution to crops. Research findings about these novel distribution technologies indicate that their wider application will expand throughout the next couple of years (Höfte 2021).

8.2.2 Seed Coatings of Plant Growth-Promoting Inoculants

Modern agricultural practices benefit from *Pseudomonas*-based bioinoculants seeds treatment as an environmentally friendly method that supports disease control and plant advancement (O'Callaghan 2016). After seed surface coating, *Pseudomonas* spp. establishes a protective residential area on the seed surface and outcompetes damaging soil pathogens during seed germination (Weller 2007). Many *Pseudomonas* strains accomplish two purposes by protecting plants from pathogens while simultaneously boosting plant vitality. Plants require particular phytohormones and enzymes that *Pseudomonas* spp. release at defined periods to support both germination initiation and establishment of plant growth. The plant development process receives benefits from *Pseudomonas* sp. PIR3C and *Pseudomonas* sp. KS12, which supply indole-3-acetic acid (IAA) alongside phosphate solubilization and nitrogen fixation mechanisms (Glick 2012). Biocontrol properties of *Pseudomonas* sp. strain S3 are evident because it exhibits strong antagonistic behavior by directly sticking to *Rhizoctonia solani* mycelia until fungus growth stops (Lan et al. 2024).

Study findings establish that applying *Pseudomonas*-based inoculants to seeds enables plants to overcome both biological risks and environmental disturbances. The research revealed that seeds treated with *Pseudomonas* sp. strain S3 demonstrated improved growth and higher germination rates than standard tomato seedlings exposed to salt stress conditions (Pandey and Gupta 2020). The conducted research demonstrates that *Pseudomonas* offers plants great power for protecting against diseases while simultaneously enhancing their capacity to endure various environmental threats.

8.2.3 Attractant-Based Delivery Systems

The use of attractive delivery systems for biocontrol agents based on *Pseudomonas* spp. demonstrates potential as a sustainable agricultural approach. Special attractants serve as the core idea to bring target pests toward active areas of beneficial *Pseudomonas* spp., which boosts the chance of pest–biocontrol agent interaction. Scientific studies investigate pest attraction processes, which demonstrate how plant epicuticular lipids modify insect behavior in ways that make these chemicals effective attractants. Olfactory system research on insects demonstrates that some sensory pathways can effectively attract pests (Koner et al. 2022).

The advancement of precise technology methods creates transformative effects on pest and disease management approaches. Modern target spraying systems utilize enhanced Convolutional Neural Network models to achieve high-level weed detection accuracy (Tang et al. 2023). Field deployment of attractants and biocontrol agents requires exact precision to guarantee their delivery and reduce erroneous consequences in the environment. Experts suggest combining pest management strategies to both develop efficient attractants alongside a deep analysis of pest–disease–crop communication pathways, as well as pest protection mechanisms and regulatory pest control responses in crops (Tang et al. 2023). This comprehensive strategy is essential.

Bioproduct manufacturing through controlled cultivation facilities benefits from large-scale biocontrol agent production. The fermenter creates a flexible system that allows the production of substantial cell quantities. The inoculation port requires both alcohol treatment and flaming as part of efficient sterilization procedures, which stop contamination from occurring. The procedure continues with the addition of a log-phase culture that previously grew in a 5-L flask when the inoculation port has reached room temperature. The specified inoculum size range falls between 1–2%, and occasional modifications up to 5% can help achieve maximum growth under specific culture conditions.

9. Conclusion and Future Perspectives

The agricultural industry now emphasizes *Pseudomonas* bacteria as its fundamental choice for sustainable pest and disease management due to its microbial nature. Soil pathogens face strong competition from *Pseudomonas* bacteria because of their status as environmentally friendly microorganisms, which serve as a more sustainable solution than chemical control measures. Major scientific research on *Pseudomonas* biocontrol implementation has been established, but ongoing obstacles exist regarding massive production capabilities and increased understanding of microbial–plant relationships and effective delivery methods to crops.

The way *Pseudomonas* affects sustainable agricultural development holds an incredible amount of promise. Healthy crop fields have been projected to require significantly fewer chemical pesticide applications, thus leading to minimal environmental impact. The realization of this vision involves active participation by researchers, industry leaders, and regulators. Cooperative resource-sharing between groups makes it possible to develop *Pseudomonas* into its most effective biocontrol form while keeping farm implementation costs manageable. The key to progress includes making the public more knowledgeable about the subject matter and actively participating in it. The wider farming community will adopt *Pseudomonas* more easily through effective communication that promotes natural pest control benefits. Outreach programs become successful when they represent quantifiable improvements such as better soil conditions and purified water sources alongside enhanced agricultural productivity, which genuinely motivates farmers.

The evaluation of *Pseudomonas* for biocontrol use remains essential for upgrading agriculture toward sustainability. Research support must be steady, and innovation funding must remain constant in order for scientists and practitioners to achieve maximum use of bacterial capabilities. Such support enables our continued role to establish microbial solutions like *Pseudomonas* along with productivity maintenance and environmental protection objectives.

References

Ab Rahman, S. F. S., Singh, E., Pieterse, C. M., and Schenk, P. M. (2018). Emerging microbial biocontrol strategies for plant pathogens. *Plant Science*, 267: 102–111.

Abd-Elgawad, M. M. M., and Askary, T. H. (2020). Factors affecting success of biological agents used in controlling the plant-parasitic nematodes. *Egyptian Journal of Biological Pest Control*, 30(1): 17. https://doi.org/10.1186/s41938-020-00215-2.

Abo-Zaid, G. A., Soliman, N. A.-M., Abdullah, A. S., El-Sharouny, E. E., Matar, S. M., and Sabry, S. A.-F. (2020). Maximization of siderophores production from biocontrol agents, *Pseudomonas aeruginosa* f2 and *Pseudomonas fluorescens* JY3 using batch and exponential fed-batch fermentation. *Processes*, 8(4): 455.

Abuchenari, A., Hardani, K., Abazari, S., Naghdi, F., Ahmady Keleshteri, M., Jamavari, A. et al. (2020). Clay-reinforced nanocomposites for the slow release of chemical fertilizers and water retention. *Journal of Composites and Compounds*, 2(3): 85–91. https://doi.org/10.29252/jcc.2.2.4.

Agarwal, A., Jaiswal, N., Tripathi, A. D., and Paul, V. (2021). Downstream processing; applications and recent updates. *Bioprocessing for Biofuel Production: Strategies to Improve Process Parameters*, 29–55.

Aggarwal, N., Pham, H. L., Ranjan, B., Saini, M., Liang, Y., Hossain, G. S. et al. (2023). Microbial engineering strategies to utilize waste feedstock for sustainable bioproduction. *Nature Reviews Bioengineering*, 2(2): 155–174. https://doi.org/10.1038/s44222-023-00129-2.

Aguirre, K., Perez, L. M., and Montealegre, J. R. (2023). In vitro evaluation of drying supports and adhesive polymers as adjuvants for biocontrol of *Diplodia seriata* by *Trichoderma harzianum* and *Clonostachys rosea*. 227–237.

Ahmad, F., Khaliq, A., Qiu, B., Sultan, M., and Ma, J. (2021). Advancements of Spraying Technology in Agriculture. *In*: Fiaz, A., and Muhammad, S. (Eds.). *Technology in Agriculture* (pp. 33). IntechOpen. https://doi.org/10.5772/intechopen.98500.

Ajala, E. O., Ighalo, J. O., Ajala, M. A., Adeniyi, A. G., and Ayanshola, A. M. (2021). Sugarcane bagasse: a biomass sufficiently applied for improving global energy, environment and economic sustainability. *Bioresources and Bioprocessing*, 8(1): 87. https://doi.org/10.1186/s40643-021-00440-z.

Alattas, H., Glick, B. R., Murphy, D. V., and Scott, C. (2024). Harnessing *Pseudomonas* spp. for sustainable plant crop protection. *Frontiers in Microbiology*, 15: 1485197.

Alsved, M., Holm, S., Christiansen, S., Smidt, M., Rosati, B., Ling, M. et al. (2018). Effect of Aerosolization and Drying on the Viability of *Pseudomonas syringae* Cells [Original Research]. *Front Microbiol.*, 9: 3086. https://doi.org/10.3389/fmicb.2018.03086.

Alvarez-Ortega, C., and Harwood, C. S. (2007). Responses of *Pseudomonas aeruginosa* to low oxygen indicate that growth in the cystic fibrosis lung is by aerobic respiration. *Mol. Microbiol.*, 65(1): 153-165. https://doi.org/10.1111/j.1365-2958.2007.05772.x.

Antil, S., Kumar, R., Pathak, D. V., and Kumari, A. (2023). Recent advances in utilizing bacteria as biocontrol agents against plant parasitic nematodes emphasizing *Meloidogyne* spp. *Biological Control*, 183: 105244. https://doi.org/10.1016/j.biocontrol.2023.105244.

Anwar, F., Zafar, M., Anwar, Z., and Mehmood, R. T. (2019). Optimization of Fermentation and Production of Antimicrobial Agents as Secondary Metabolites from an Indigenous *Aspergillus niger*. *Punjab University Journal of Zoology*, 34(2): 165–173. https://doi.org/10.17582/journal.pujz/2019.34.2.165.173.

Ayol, A., Peixoto, L., Keskin, T., and Abubackar, H. N. (2021). Reactor designs and configurations for biological and Bioelectrochemical C1 Gas Conversion: A Review. *Int. J. Environ. Res. Public Health*, 18(21): 11683. https://doi.org/10.3390/ijerph182111683.

Bahlol, H. Y., Chandel, A. K., Hoheisel, G.-A., and Khot, L. R. (2020). Smart spray analytical system for orchard sprayer calibration: a-proof-of-concept and preliminary results. *Transactions of the ASABE*, 63(1): 29–35. https://doi.org/10.13031/trans.13196.

Bais, H. P., Fall, R., and Vivanco, J. M. (2004). Biocontrol of *Bacillus subtilis* against infection of *Arabidopsis* roots by *Pseudomonas syringae* is facilitated by biofilm formation and surfactin production. *Plant Physiol*, 134(1): 307–319. https://doi.org/10.1104/pp.103.028712.

Balthazar, C., St-Onge, R., Leger, G., Lamarre, S. G., Joly, D. L., and Filion, M. (2022). Pyoluteorin and 2,4-diacetylphloroglucinol are major contributors to *Pseudomonas protegens* Pf-5 biocontrol against *Botrytis cinerea* in cannabis [Original Research]. *Front Microbiol.*, 13: 945498. https://doi.org/10.3389/fmicb.2022.945498.

Barratt, B., Moran, V., Bigler, F., and Van Lenteren, J. (2018). The status of biological control and recommendations for improving uptake for the future. *BioControl.*, 63: 155–167.

BASF. (2024). Tinuvin® NOR® 356 (Light Stabilizer) - BASF- Technical Datasheet. https://polymer-additives.specialchem.com/product/a-basf-tinuvin-nor-356.

Bashan, Y., de-Bashan, L. E., Prabhu, S. R., and Hernandez, J.-P. (2014). Advances in plant growth-promoting bacterial inoculant technology: formulations and practical perspectives (1998–2013). *Plant and Soil*, 378(1): 1–33. https://doi.org/10.1007/s11104-013-1956-x.

Beerling, J., and Sahota, A. (2013). Green standards, certification and indices. *Sustainability: How the Cosmetics Industry is Greening Up*, 217–238.

Benhamou, N., Gagne, S., Le Quere, D., and Dehbi, L. (2000). Bacterial-mediated induced resistance in cucumber: Beneficial effect of the endophytic bacterium *Serratia plymuthica* on the protection against infection by *Pythium ultimum*. *Phytopathology*, 90(1): 45–56. https://doi.org/10.1094/PHYTO.2000.90.1.45.

Berninger, T., Gonzalez Lopez, O., Bejarano, A., Preininger, C., and Sessitsch, A. (2018). Maintenance and assessment of cell viability in formulation of non-sporulating bacterial inoculants. *Microb. Biotechnol.*, 11(2): 277–301. https://doi.org/10.1111/1751-7915.12880.

Bhartendu, P. S., Shukla, S., and Misra, G. (2024). 7 Good Practices Manufacturing in Quality. *Quality Control and Regulatory Aspects for Biologicals: Regulations and Best Practices*, 100.

Bhaskara Reddy, M. V., Arul, J., Angers, P., and Couture, L. (1999). Chitosan treatment of wheat seeds induces resistance to *Fusarium graminearum* and improves seed quality. *J. Agric. Food Chem.*, 47(3): 1208–1216. https://doi.org/10.1021/jf981225k.

Bhattacharjee, A., Dubey, S., and Sharma, S. (2023). Next-generation bioformulations for plant growth promotion and stress mitigation: A promising approach for sustainable agriculture. *Journal of Plant Growth Regulation*, 42(10): 6741–6759.

Bilbao, A., Munoz, N., Kim, J., Orton, D. J., Gao, Y., Poorey, K. et al. (2023). PeakDecoder enables machine learning-based metabolite annotation and accurate profiling in multidimensional mass spectrometry measurements. *Nat. Commun.*, 14(1): 2461. https://doi.org/10.1038/s41467-023-37031-9.

Bisutti, I. L., and Stephan, D. (2020). Influence of fermentation temperature and duration on survival and biocontrol efficacy of *Pseudomonas fluorescens* Pf153 freeze-dried cells. *J. Appl. Microbiol.*, 128(1): 232–241. https://doi.org/10.1111/jam.14458.

Bloemberg, G. V., and Lugtenberg, B. J. (2001). Molecular basis of plant growth promotion and biocontrol by rhizobacteria. *Curr. Opin. Plant Biol.*, 4(4): 343–350. https://doi.org/10.1016/s1369-5266(00)00183-7.

Bogino, P. C., Oliva Mde, L., Sorroche, F. G., and Giordano, W. (2013). The role of bacterial biofilms and surface components in plant-bacterial associations. *Int. J. Mol. Sci.*, 14(8): 15838–15859. https://doi.org/10.3390/ijms140815838.

Briassoulis, D., Babou, E., Hiskakis, M., and Kyrikou, I. (2015). Analysis of long-term degradation behaviour of polyethylene mulching films with pro-oxidants under real cultivation and soil burial conditions. *Environ. Sci. Pollut. Res. Int.*, 22(4): 2584–2598. https://doi.org/10.1007/s11356-014-3464-9.

Brindley, D., Moorthy, K., Lee, J. H., Mason, C., Kim, H. W., and Wall, I. (2011). Bioprocess forces and their impact on cell behavior: implications for bone regeneration therapy. *J. Tissue Eng.*, *2011*, 620247. https://doi.org/10.4061/2011/620247.

Burch, A. Y., Zeisler, V., Yokota, K., Schreiber, L., and Lindow, S. E. (2014). The hygroscopic biosurfactant syringafactin produced by *Pseudomonas syringae* enhances fitness on leaf surfaces during fluctuating humidity. *Environmental Microbiology*, 16(7): 2086–2098.

Carvalho, F. (2017). Pesticides, environment, and food safety. Food and Energy Security 6: 48–60. In.

Cesa-Luna, C., Baez, A., Quintero-Hernández, V., Cruz-Enríquez, J. D. L., Castañeda-Antonio, M. D., and Muñoz-Rojas, J. (2020). The importance of antimicrobial compounds produced by beneficial bacteria on the biocontrol of phytopathogens. *Acta Biológica Colombiana*, 25(1): 140–154.

Chin, J. M., Lim, Y. Y., and Ting, A. S. Y. (2022). Biopriming *Pseudomonas fluorescens* to vegetable seeds with biopolymers to promote coating efficacy, seed germination and disease suppression. *Journal of the Saudi Society of Agricultural Sciences*, 21(8): 493–505.

Chong, H., and Li, Q. (2017). Microbial production of rhamnolipids: opportunities, challenges and strategies. *Microb. Cell Fact*, 16(1): 137. https://doi.org/10.1186/s12934-017-0753-2.

Committee, E. S., More, S., Bampidis, V., Benford, D., Bragard, C., Halldorsson, T. et al. (2020). Evaluation of existing guidelines for their adequacy for the microbial characterisation and environmental risk assessment of microorganisms obtained through synthetic biology. *EFSA Journal*, 18(10): e06263.

Compant, S., Duffy, B., Nowak, J., Clement, C., and Barka, E. A. (2005). Use of plant growth-promoting bacteria for biocontrol of plant diseases: principles, mechanisms of action, and future prospects. *Appl. Environ. Microbiol.*, 71(9): 4951–4959. https://doi.org/10.1128/AEM.71.9.4951-4959.2005.

Couillerot, O., Prigent-Combaret, C., Caballero-Mellado, J., and Moënne-Loccoz, Y. (2009). *Pseudomonas fluorescens* and closely-related fluorescent pseudomonads as biocontrol agents of soil-borne phytopathogens. *Letters in Applied Microbiology*, 48(5): 505–512.

Coulson, J. R., Vail, P., Dix, M., Nordlund, D., and Kauffman, W. (2000). 110 years of biological control research and development in the United States Department of Agriculture. *USDA ARS Admin. Rep*, 1.

Craig, K., Johnson, B. R., and Grunden, A. (2021). Leveraging *Pseudomonas* stress response mechanisms for industrial applications. *Front Microbiol.*, 12: 660134. https://doi.org/10.3389/fmicb.2021.660134.

Crater, J. S., and Lievense, J. C. (2018). Scale-up of industrial microbial processes. *FEMS Microbiol. Lett.*, 365(13): fny138. https://doi.org/10.1093/femsle/fny138.

Criste, A., Giuburuncă, M., Negrea, O., Dan, S., and Zăhan, M. (2014). Research concerning use of long-term preservation techniques for microorganisms. *Scientific Papers Animal Science and Biotechnologies*, 47(2): 73–73.

Cromwell, R. P. (1992). *The 1/128th of an Acre Sprayer Calibration Method* (Vol. EDIS). University of Florida Cooperative Extension Service, Institute of Food and Agriculture Sciences.

D'aes, J., De Maeyer, K., Pauwelyn, E., and Höfte, M. (2010). Biosurfactants in plant–*Pseudomonas* interactions and their importance to biocontrol. *Environmental microbiology reports*, 2(3): 359–372.

Dabaghi, S., Ataei, S. A., and Taheri, A. (2023). Production of rhamnolipid biosurfactants in solid-state fermentation: process optimization and characterization studies. *BMC Biotechnol*, 23(1): 2. https://doi.org/10.1186/s12896-022-00772-4.

Damalas, C. A., and Koutroubas, S. D. (2016). Farmers' Exposure to Pesticides: Toxicity Types and Ways of Prevention. *Toxics*, 4(1): 1. https://doi.org/10.3390/toxics4010001.

De Jesus, S. S., Neto, J. M., and Maciel Filho, R. (2017). Hydrodynamics and mass transfer in bubble column, conventional airlift, stirred airlift and stirred tank bioreactors, using viscous fluid: A comparative study. *Biochemical Engineering Journal*, 118: 70–81.

de Oliveira, E. P., Rovida, A., Martins, J. G., Pileggi, S. A. V., Schemczssen-Graeff, Z., and Pileggi, M. (2021). Tolerance of Pseudomonas strain to the 2,4–D herbicide through a peroxidase system. *Plos One*, 16(12): e0257263. https://doi.org/10.1371/journal.pone.0257263.

De Vleesschauwer, D., and Höfte, M. (2009). Rhizobacteria-induced systemic resistance. *Advances in Botanical Research*, 51: 223–281.

Devi, L., Sheikh, S., Kumari, N., and Kumar, R. (2022). Role of Biofertilizers in Organic Agriculture. *A Monthly Peer Reviewed Magazine for Agriculture and Allied Sciences*, 1: 42–45.

Díaz-Rodríguez, A. M., Parra Cota, F. I., Cira Chávez, L. A., García Ortega, L. F., Estrada Alvarado, M. I., Santoyo, G. et al. (2025). Microbial Inoculants in Sustainable Agriculture: Advancements, Challenges, and Future Directions. *Plants*, 14(2): 191. https://www.mdpi.com/2223-7747/14/2/191.

Donnarumma, G., Buommino, E., Fusco, A., Paoletti, I., Auricchio, L., and Tufano, M. A. (2010). Effect of temperature on the shift of *Pseudomonas fluorescens* from an environmental microorganism to a potential human pathogen. *Int. J. Immunopathol. Pharmacol.*, 23(1): 227–234. https://doi.org/10.1177/039463201002300120.

Doran, P. M. (1995). *Bioprocess engineering principles*. Elsevier.

Doruchowski, G., Hołownicki, R., Godyń, A., and Świechowski, W. (2012). Sprayer calibration training-concept and performance. Fourth European Workshop on standardised prodedure for the inspection of sprayers – SPISE 4, March 27-29, 2012, Lana (South Tyrol).

Du, Y. H., Wang, M. Y., Yang, L. H., Tong, L. L., Guo, D. S., and Ji, X. J. (2022). Optimization and Scale-Up of Fermentation Processes Driven by Models. *Bioengineering (Basel)*, 9(9): 473. https://doi.org/10.3390/bioengineering9090473.

Durairaj, K., Velmurugan, P., Park, J. H., Chang, W. S., Park, Y. J., Senthilkumar, P. et al. (2017). Potential for plant biocontrol activity of isolated *Pseudomonas aeruginosa* and *Bacillus stratosphericus* strains against bacterial pathogens acting through both induced plant resistance and direct antagonism. *FEMS Microbiol. Lett.*, 364(23), fnx225. https://doi.org/10.1093/femsle/fnx225.

Elbahloul, Y., and Steinbuchel, A. (2009). Large-scale production of poly(3-hydroxyoctanoic acid) by *Pseudomonas putida* GPo1 and a simplified downstream process. *Appl. Environ. Microbiol.*, 75(3): 643–651. https://doi.org/10.1128/AEM.01869-08.

Eng, T., Banerjee, D., Lau, A. K., Bowden, E., Herbert, R. A., Trinh, J. et al. (2021). Engineering *Pseudomonas putida* for efficient aromatic conversion to bioproduct using high throughput screening in a bioreactor. *Metab Eng*, 66: 229–238. https://doi.org/10.1016/j.ymben.2021.04.015.

EPA. (2024). *Biopesticides*. EPA. Retrieved January 12, 2024 from https://www.epa.gov/pesticides/biopesticides.

Fageria, N. K., Baligar, V. C., and Jones, C. A. (2010). *Growth and Mineral Nutrition of Field Crops*. CRC Press. https://books.google.co.in/books?id=AP3Mcm0NdJMC.

Feng, Y., Tian, X., Chen, Y., Wang, Z., Xia, J., Qian, J. et al. (2021). Real-time and on-line monitoring of ethanol fermentation process by viable cell sensor and electronic nose. *Bioresources and Bioprocessing*, 8(1): 37. https://doi.org/10.1186/s40643-021-00391-5.

Fravel, D. R. (2005). Commercialization and implementation of biocontrol. *Annu. Rev. Phytopathol.*, 43: 337–359. https://doi.org/10.1146/annurev.phyto.43.032904.092924.

Fu, H. Z., Marian, M., Enomoto, T., Hieno, A., Ina, H., Suga, H. et al. (2020). Biocontrol of tomato bacterial wilt by foliar spray application of a novel strain of endophytic *Bacillus* sp. *Microbes Environ*, 35(4): ME20078. https://doi.org/10.1264/jsme2.ME20078.

Gabelman, A. (2022). Fermentation and downstream processing: Part 1. pp. 13–68. *In: Integration and Optimization of Unit Operations*. Elsevier.

Gade, R., Chaithanya, B., and Khurade, K. (2014). A comparative study of different carriers for shelf life of *Pseudomonas fluorescens*. *The Bioscan.*, 9(1): 287–290.

Garcia-Ochoa, F., and Gomez, E. (2009). Bioreactor scale-up and oxygen transfer rate in microbial processes: an overview. *Biotechnol. Adv.*, 27(2): 153–176. https://doi.org/10.1016/j.biotechadv.2008.10.006.

Giles, D., and Billing, R. (2015). Deployment and performance of a UAV for crop spraying. *Chemical Engineering Transactions*, 44: 307–312.

Glare, T., Caradus, J., Gelernter, W., Jackson, T., Keyhani, N., Kohl, J. et al. (2012). Have biopesticides come of age? *Trends Biotechnol.*, 30(5): 250–258. https://doi.org/10.1016/j.tibtech.2012.01.003.

Glick, B. R. (2012). Plant growth-promoting bacteria: mechanisms and applications. *Scientifica (Cairo)*, 2012: 963401. https://doi.org/10.6064/2012/963401.

Gomila, M., Peña, A., Mulet, M., Lalucat, J., and García-Valdés, E. (2015). Phylogenomics and systematics in Pseudomonas. *Frontiers in Microbiology*, 6: 214.

Gong, Z., Yang, G., Che, C., Liu, J., Si, M., and He, Q. (2021). Foaming of rhamnolipids fermentation: impact factors and fermentation strategies. *Microb. Cell Fact*, 20(1): 77. https://doi.org/10.1186/s12934-021-01516-3.

Guan, N., Li, J., Shin, H. D., Du, G., Chen, J., and Liu, L. (2017). Microbial response to environmental stresses: from fundamental mechanisms to practical applications. *Appl. Microbiol. Biotechnol.*, 101(10): 3991–4008. https://doi.org/10.1007/s00253-017-8264-y.

Gupta, R., Bisaria, V. S., and Sharma, S. (2013). Bioinoculants: More than just plant growth promoting agents. *Endocytobiosis & Cell Research*, 24: 8.

Haas, D., and Defago, G. (2005). Biological control of soil-borne pathogens by fluorescent pseudomonads. *Nat. Rev. Microbiol.*, *3*(4): 307–319. https://doi.org/10.1038/nrmicro1129.

Hellsmark, H., Frishammar, J., Söderholm, P., and Ylinenpää, H. (2016). The role of pilot and demonstration plants in technology development and innovation policy. *Research Policy*, 45(9): 1743–1761.

Herrmann, L., and Lesueur, D. (2013). Challenges of formulation and quality of biofertilizers for successful inoculation. *Appl. Microbiol. Biotechnol.*, 97(20): 8859–8873. https://doi.org/10.1007/s00253-013-5228-8.

Hewitt, C. J., and Nienow, A. W. (2007). The scale-up of microbial batch and fed-batch fermentation processes. *Adv. Appl. Microbiol.*, 62: 105–135. https://doi.org/10.1016/S0065-2164(07)62005-X.

Hmelo, L. R., Borlee, B. R., Almblad, H., Love, M. E., Randall, T. E., Tseng, B. S. et al. (2015). Precision-engineering the *Pseudomonas aeruginosa* genome with two-step allelic exchange. *Nat. Protoc.*, 10(11): 1820–1841. https://doi.org/10.1038/nprot.2015.115.

Hoeschle-Zeledon, I., Neuenschwander, P., and Kumar, P. (2013). Regulatory challenges for biological control. *Systemwide Program on Integrated Pest Management, Ibadan, Nigeria: IITA*.

Höfte, M. (2021). The use of Pseudomonas spp. as bacterial biocontrol agents to control plant diseases. pp. 301–374. *In: Microbial Bioprotectants for Plant Disease Management*. https://doi.org/10.19103/as.2021.0093.11.

Hofte, M., and Altier, N. (2010). Fluorescent pseudomonads as biocontrol agents for sustainable agricultural systems. *Res. Microbiol.*, 161(6): 464–471. https://doi.org/10.1016/j.resmic.2010.04.007.

Huang, J., Zhang, G., Zheng, L., Lin, Z., Wu, Q., and Pan, Y. (2019). Plackett-Burman design and response surface optimization of conditions for culturing *Saccharomyces cerevisiae* in *Agaricus bisporus* industrial wastewater. *Acta Sci. Pol. Technol. Aliment.*, 18(1): 65–74. https://doi.org/10.17306/J.AFS.0620.

Humair, B., Gonzalez, N., Mossialos, D., Reimmann, C., and Haas, D. (2009). Temperature-responsive sensing regulates biocontrol factor expression in *Pseudomonas fluorescens* CHA0. *ISME J.*, 3(8): 955–965. https://doi.org/10.1038/ismej.2009.42.

Ihling, N., Munkler, L. P., Berg, C., Reichenbacher, B., Wirth, J., Lang, D. et al. (2021). Time-resolved monitoring of the oxygen transfer rate of chinese hamster ovary cells provides insights into culture behavior in shake flasks. *Front. Bioeng. Biotechnol.*, 9: 725498. https://doi.org/10.3389/fbioe.2021.725498.

Jaiswal, D. K., Gawande, S. J., Soumia, P. S., Krishna, R., Vaishnav, A., and Ade, A. B. (2022). Biocontrol strategies: an eco-smart tool for integrated pest and diseases management. *BMC Microbiol.*, 22(1): 324. https://doi.org/10.1186/s12866-022-02744-2.

Jakobsen, H. A., Lindborg, H., and Dorao, C. A. (2005). Modeling of bubble column reactors: progress and limitations. *Industrial & Engineering Chemistry Research*, 44(14): 5107–5151.

Jangir, M., Sharma, S., and Sharma, S. (2021). Development of next-generation formulation against Fusarium oxysporum and unraveling bioactive antifungal metabolites of biocontrol agents. *Sci. Rep.*, 11(1): 22895. https://doi.org/10.1038/s41598-021-02284-1.

Jankiewicz, U., and Koltonowicz, M. (2012). The involvement of *Pseudomonas* bacteria in induced systemic resistance in plants (review). *Prikl. Biokhim. Mikrobiol.*, 48(3): 276–281. https://www.ncbi.nlm.nih.gov/pubmed/22834297.

John, R. P., Tyagi, R. D., Prévost, D., Brar, S. K., Pouleur, S., and Surampalli, R. Y. (2010). Mycoparasitic *Trichoderma viride* as a biocontrol agent against *Fusarium oxysporum* f. sp. adzuki and *Pythium arrhenomanes* and as a growth promoter of soybean. *Crop Protection*, 29(12): 1452–1459. https://doi.org/10.1016/j.cropro.2010.08.004.

Kampers, L. F. C., Koehorst, J. J., van Heck, R. J. A., Suarez-Diez, M., Stams, A. J. M., and Schaap, P. J. (2021). A metabolic and physiological design study of *Pseudomonas putida* KT2440 capable of anaerobic respiration. *BMC Microbiol.*, 21(1): 9. https://doi.org/10.1186/s12866-020-02058-1.

Kapoore, R. V., Padmaperuma, G., Maneein, S., and Vaidyanathan, S. (2022). Co-culturing microbial consortia: approaches for applications in biomanufacturing and bioprocessing. *Crit. Rev. Biotechnol.*, 42(1): 46–72. https://doi.org/10.1080/07388551.2021.1921691.

Karimi, A., Golbabaei, F., Mehrnia, M. R., Neghab, M., Mohammad, K., Nikpey, A. et al. (2013). Oxygen mass transfer in a stirred tank bioreactor using different impeller configurations for environmental purposes. *Iranian J. Environ. Health Sci. Eng.*, 10(1): 6. https://doi.org/10.1186/1735-2746-10-6.

Kasirajan, S., and Ngouajio, M. (2012). Polyethylene and biodegradable mulches for agricultural applications: a review. *Agronomy for Sustainable Development*, 32(2): 501–529. https://doi.org/10.1007/s13593-011-0068-3.

Kaur, G. (2015). Strategies for Large-scale Production of Polyhydroxyalkanoates. *Chemical and Biochemical Engineering Quarterly*, 29(2): 157–172. https://doi.org/10.15255/cabeq.2014.2255.

Keasling, J., Garcia Martin, H., Lee, T. S., Mukhopadhyay, A., Singer, S. W., and Sundstrom, E. (2021). Microbial production of advanced biofuels. *Nat. Rev. Microbiol.*, 19(11): 701–715. https://doi.org/10.1038/s41579-021-00577-w.

Khaleghi, M. K., Savizi, I. S. P., Lewis, N. E., and Shojaosadati, S. A. (2021). Synergisms of machine learning and constraint-based modeling of metabolism for analysis and optimization of fermentation parameters. *Biotechnol. J.*, 16(11): e2100212. https://doi.org/10.1002/biot.202100212.

Khan, A., Singh, A. V., Gautam, S. S., Agarwal, A., Punetha, A., Upadhayay, V. K. et al. (2023). Microbial bioformulation: a microbial assisted biostimulating fertilization technique for sustainable agriculture. *Front Plant Sci.*, *14*: 1270039. https://doi.org/10.3389/fpls.2023.1270039.

Kim, K.-H., Kabir, E., and Jahan, S. A. (2017). Exposure to pesticides and the associated human health effects. *Science of The Total Environment*, 575: 525–535.

Kohl, J., Kolnaar, R., and Ravensberg, W. J. (2019). Mode of Action of Microbial Biological Control Agents Against Plant Diseases: Relevance Beyond Efficacy [Review]. *Front Plant Sci.*, 10: 845. https://doi.org/10.3389/fpls.2019.00845.

Koner, A., Das, S., Mobarak, S. H., and Barik, A. (2022). Short-range attraction and oviposition stimulant of a biocontrol agent, *Galerucella placida* Baly (Coleoptera: Chrysomelidae) toward weed leaf surface waxes. *Bull. Entomol. Res*, *112*(2): 204–218. https://doi.org/10.1017/S0007485321000730.

Konopacki, M., Jablonska, J., Dubrowska, K., Augustyniak, A., Grygorcewicz, B., Glizniewicz, M. et al. (2022). The influence of hydrodynamic conditions in a laboratory-scale bioreactor on *Pseudomonas aeruginosa* metabolite production. *Microorganisms*, 11(1): 88. https://doi.org/10.3390/microorganisms11010088.

Krasny, L. T. (2002). USDA's National Organic Program: The Final Rule. *FDLI Update*, 23.

Krishna, P. S., Woodcock, S. D., Pfeilmeier, S., Bornemann, S., Zipfel, C., and Malone, J. G. (2022). *Pseudomonas syringae* addresses distinct environmental challenges during plant infection through the coordinated deployment of polysaccharides. *J. Exp. Bot.*, 73(7): 2206–2221. https://doi.org/10.1093/jxb/erab550.

Kumar, A., Verma, H., Singh, V. K., Singh, P. P., Singh, S. K., Ansari, W. A. et al. (2017). Role of Pseudomonas sp. in Sustainable Agriculture and Disease Management. In *Agriculturally Important Microbes for Sustainable Agriculture* (pp. 195–215). https://doi.org/10.1007/978-981-10-5343-6_7.

Kumar, M., Mehta, C. R., Bhargav, V. K., Tripathi, M. K., Agrawal, K. N., and Babu, V. B. (2023). Effect of Spray Operating Parameters on Viability of Bacillus thuringiensis Based Biopesticide under Laboratory Condition. *Agricultural Research*, 12(2): 189–196.

Kupferschmied, P., Maurhofer, M., and Keel, C. (2013). Promise for plant pest control: root-associated pseudomonads with insecticidal activities. *Frontiers in Plant Science*, 4: 287.

Lahlali, R., Ezrari, S., Radouane, N., Kenfaoui, J., Esmaeel, Q., El Hamss, H. et al. (2022). Biological Control of Plant Pathogens: A Global Perspective. *Microorganisms*, 10(3): 596. https://doi.org/10.3390/microorganisms10030596.

Lamichhane, J. R., Dachbrodt-Saaydeh, S., Kudsk, P., and Messéan, A. (2016). Toward a reduced reliance on conventional pesticides in European agriculture. *Plant Disease*, 100(1): 10–24.

Lan, Q., Liu, Y., Mu, R., Wang, X., Zhou, Q., Islam, R. et al. (2024). Biological control effect of antagonistic bacteria on potato black scurf disease caused by *Rhizoctonia solani*. *Agronomy*, *14*(2): 351. https://www.mdpi.com/2073-4395/14/2/351.

Leahy, J., Mendelsohn, M., Kough, J., Jones, R., and Berckes, N. (2014). Biopesticide oversight and registration at the US Environmental Protection Agency. *Biopesticides: State of the Art and Future Opportunities*, 3–18.

Lecuyer, S., Rusconi, R., Shen, Y., Forsyth, A., Vlamakis, H., Kolter, R. et al. (2011). Shear stress increases the residence time of adhesion of *Pseudomonas aeruginosa*. *Biophys J*, 100(2): 341–350. https://doi.org/10.1016/j.bpj.2010.11.078.

Li, S., Li, J., Yu, S., Wang, P., Liu, H., and Yang, X. (2023). Anti-Drift Technology Progress of Plant Protection Applied to Orchards: A Review. *Agronomy*, 13(11): 2679.

Li, T., Teng, D., Mao, R., Hao, Y., Wang, X., and Wang, J. (2019). Recent progress in preparation and agricultural application of microcapsules. *J. Biomed. Mater. Res. A*, 107(10): 2371–2385. https://doi.org/10.1002/jbm.a.36739.

Liu, F., Yang, S., Xu, F., Zhang, Z., Lu, Y., Zhang, J. et al. (2022). Characteristics of biological control and mechanisms of *Pseudomonas chlororaphis* ZM-1 against peanut stem rot. *BMC Microbiol.*, 22(1): 9. https://doi.org/10.1186/s12866-021-02420-x.

Lorito, M., Woo, S. L., Harman, G. E., and Monte, E. (2010). Translational research on Trichoderma: from 'omics to the field. *Annu. Rev. Phytopathol.*, *48*: 395–417. https://doi.org/10.1146/annurev-phyto-073009-114314.

Lugtenberg, B., and Kamilova, F. (2009). Plant-Growth-Promoting Rhizobacteria. *Annual Review of Microbiology*, (Volume 63: 2009): 541–556. https://doi.org/https://doi.org/10.1146/annurev.micro.62.081307.162918.

Madhavi, G. B., and Devi, G. U. (2018). Effect of combined application of biofumigant, *Trichoderma harzianum* and *Pseudomonas fluorescens* on *Rhizoctonia solani* f. sp. sasakii. *Indian Phytopathology*, *71*: 257–263.

Maes, S., De Reu, K., Van Weyenberg, S., Lories, B., Heyndrickx, M., and Steenackers, H. (2020). *Pseudomonas putida* as a potential biocontrol agent against Salmonella Java biofilm formation in the drinking water system of broiler houses. *BMC Microbiol.*, 20(1): 373. https://doi.org/10.1186/s12866-020-02046-5.

Maheshwari, P., Gayathry, G., Sankar, P. M., Sangeetha, P., and Anandaraj, P. (2022). Mass Multiplication and Production Cost Analysis of Phosphate Solubilizing Microorganisms. pp. 287–302. *In: Industrial Microbiology Based Entrepreneurship: Making Money from Microbes*. Springer.

Majumdar, A. (2023). Molecular techniques for the improvement of microbial biocontrol agents against plant pathogens. *Egyptian Journal of Biological Pest Control*, 33(1): 103. https://doi.org/10.1186/s41938-023-00746-4.

Manal, M. A., El-Naggar, S., El-Aasar, A., and Barakat Khlood, I. (2005). Bioremediation of crystal violet using air bubble bioreactor packed with *Pseudomonas aeruginosa*. *Water Res*, 39(20): 5045–5054. https://doi.org/10.1016/j.watres.2004.08.001.

Mark, G., Morrissey, J. P., Higgins, P., and O'Gara, F. (2006). Molecular-based strategies to exploit *Pseudomonas biocontrol* strains for environmental biotechnology applications. *FEMS Microbiol Ecol*, 56(2): 167–177. https://doi.org/10.1111/j.1574-6941.2006.00056.x.

Mavrodi, D. V., Blankenfeldt, W., and Thomashow, L. S. (2006). Phenazine compounds in fluorescent *Pseudomonas* spp. biosynthesis and regulation. *Annu. Rev. Phytopathol.*, 44(1). 417–445.

Miller, S., Mark, G., Franks, A., and O'Gara, F. (2008). Pseudomonas–plant interactions. *Pseudomonas: Model Organism, Pathogen, Cell Factory*, 353–376.

Montes-Colmenares, O., Castro-Brindis, R., Uribe-Gómez, M., and Colmenares, C. (2020). Biopesticides and Biofertilizers Spraying Using UAV For Horticultural Agroforestry Systems. 24: 00115–00117.

Mputu Kanyinda, J.-n., Pierart, C., Destain, J., Noki, P., and Thonart, P. (2014). Impact of drying on *Pseudomonas fluorescens* viability. A review. *Biotechnology, Agronomy and Society and Environment*, 18: 134–141.

Müller, M., Husemann, U., Greller, G., Meusel, W., and Kraume, M. (2018). Heat transfer characteristics of a stirred single-use bioreactor. *Biochemical Engineering Journal*, 140: 168–177. https://doi.org/10.1016/j.bej.2018.09.022.

Mutlu, B. R., Hirschey, K., Wackett, L. P., and Aksan, A. (2015). Long-term preservation of silica gel-encapsulated bacterial biocatalysts by desiccation. *Journal of Sol-Gel Science and Technology*, 74: 823–833.

Nerek, E., and Sokołowska, B. (2022). *Pseudomonas* spp. in biological plant protection and growth promotion. *AIMS Environmental Science*, 9(4): 493–504. https://doi.org/10.3934/environsci.2022029.

Newlands, N. K. (2022). Artificial Intelligence and Big Data Analytics in Vineyards: A Review. *Grapes and Wine*, 8: 65.

Nur Asshifa, M. N., Zambry, N. S., Salwa, M. S., and Yahya, A. R. M. (2017). The influence of agitation on oil substrate dispersion and oxygen transfer in *Pseudomonas aeruginosa* USM-AR2 fermentation producing rhamnolipid in a stirred tank bioreactor. *3 Biotech.*, 7(3): 189. https://doi.org/10.1007/s13205-017-0828-0.

Nuyttens, D., Baetens, K., De Schampheleire, M., and Sonck, B. (2007). Effect of nozzle type, size and pressure on spray droplet characteristics. *Biosystems Engineering*, 97(3): 333–345. https://doi.org/10.1016/j.biosystemseng.2007.03.001.

O'Callaghan, M. (2016). Microbial inoculation of seed for improved crop performance: issues and opportunities. *Appl. Microbiol. Biotechnol.*, 100(13): 5729–5746. https://doi.org/10.1007/s00253-016-7590-9.

Ongena, M., and Jacques, P. (2008). Bacillus lipopeptides: versatile weapons for plant disease biocontrol. *Trends Microbiol*, 16(3): 115–125. https://doi.org/10.1016/j.tim.2007.12.009.

Oni, F. E., Kieu Phuong, N., and Höfte, M. (2015). Recent advances in *Pseudomonas* biocontrol. *Bacteria-Plant Interactions: Advanced Research and Future Trends*, 167–198.

Orozco-Mosqueda, M. d. C., Flores, A., Rojas-Sánchez, B., Urtis-Flores, C. A., Morales-Cedeño, L. R., Valencia-Marin, M. F. et al. (2021). Plant growth-promoting bacteria as bioinoculants: Attributes and challenges for sustainable crop improvement. *Agronomy*, 11(6): 1167.

Pal, K. K., and Gardener, B. M. (2006). Biological control of plant pathogens.

Pandey, A., Selvakumar, P., Soccol, C. R., and Nigam, P. (1999). Solid state fermentation for the production of industrial enzymes. *Current Science*, 149–162.

Pandey, S., and Gupta, S. (2020). Evaluation of Pseudomonas sp. for its multifarious plant growth promoting potential and its ability to alleviate biotic and abiotic stress in tomato (*Solanum lycopersicum*) plants. *Scientific Reports*, 10(1): 20951. https://doi.org/10.1038/s41598-020-77850-0.

Patel, G., Khobragade, T. P., Avaghade, S. R., Patil, M. D., Nile, S. H., Kai, G. et al. (2020). Optimization of media and culture conditions for the production of tacrolimus by *Streptomyces tsukubaensis* in shake flask and fermenter level. *Biocatalysis and Agricultural Biotechnology*, 29: 101803.

Penazova, E., Dvorak, M., Ragasova, L., Kiss, T., Pecenka, J., Cechova, J. et al. (2020). Multiplex real-time PCR for the detection of *Clavibacter michiganensis* subsp. michiganensis, *Pseudomonas syringae* pv. tomato and pathogenic *Xanthomonas* species on tomato plants. *Plos One*, 15(1): e0227559. https://doi.org/10.1371/journal.pone.0227559.

Perneel, M., Heyrman, J., Adiobo, A., De Maeyer, K., Raaijmakers, J. M., De Vos, P. et al. (2007). Characterization of CMR5c and CMR12a, novel fluorescent *Pseudomonas* strains from the cocoyam rhizosphere with biocontrol activity. *J. Appl. Microbiol.*, 103(4): 1007–1020. https://doi.org/10.1111/j.1365-2672.2007.03345.x.

Pieterse, C. M., Zamioudis, C., Berendsen, R. L., Weller, D. M., Van Wees, S. C., and Bakker, P. A. (2014). Induced systemic resistance by beneficial microbes. *Annu. Rev. Phytopathol.*, 52: 347–375. https://doi.org/10.1146/annurev-phyto-082712-102340.

Planas, S., Roman, C., Sanz, R., and Rosell-Polo, J. R. (2022). Bases for pesticide dose expression and adjustment in 3D crops and comparison of decision support systems. *Sc.i Total Environ.*, 806(Pt 1), 150357. https://doi.org/10.1016/j.scitotenv.2021.150357.

Poontawee, R., and Limtong, S. (2020). Feeding Strategies of Two-Stage Fed-Batch Cultivation Processes for Microbial Lipid Production from Sugarcane Top Hydrolysate and Crude Glycerol by the Oleaginous Red Yeast *Rhodosporidiobolus fluvialis*. *Microorganisms*, 8(2): 151. https://doi.org/10.3390/microorganisms8020151.

Ptaszek, M., Canfora, L., Pugliese, M., Pinzari, F., Gilardi, G., Trzcinski, P. et al. (2023). Microbial-Based Products to Control Soil-Borne Pathogens: Methods to Improve Efficacy and to Assess Impacts on Microbiome. *Microorganisms*, 11(1): 224. https://doi.org/10.3390/microorganisms11010224.

Puchalka, J., Oberhardt, M. A., Godinho, M., Bielecka, A., Regenhardt, D., Timmis, K. N. et al. (2008). Genome-scale reconstruction and analysis of the *Pseudomonas putida* KT2440 metabolic network facilitates applications in biotechnology. *PLoS Comput. Biol.*, 4(10): e1000210. https://doi.org/10.1371/journal.pcbi.1000210.

Purtschert-Montenegro, G., Carcamo-Oyarce, G., Pinto-Carbo, M., Agnoli, K., Bailly, A., and Eberl, L. (2022). *Pseudomonas putida* mediates bacterial killing, biofilm invasion and biocontrol with a type IVB secretion system. *Nat. Microbiol.*, 7(10): 1547–1557. https://doi.org/10.1038/s41564-022-01209-6.

Regonesi, G. (2023). *Bioreactors : A Complete Review*.

Retallack, D., Schneider, J. C., Chew, L., Ramseier, T., Allen, J., Patkar, A. et al. (2006). *Pseudomonas fluorescens* – a robust expression platform for pharmaceutical protein production. *Microb. Cell Fact.*, 5(1): S28. https://doi.org/10.1186/1475-2859-5-S1-S28.

Retallack, D. M., Schneider, J. C., Mitchell, J., Chew, L., and Liu, H. (2007). Transport of heterologous proteins to the periplasmic space of *Pseudomonas fluorescens* using a variety of native signal sequences. *Biotechnol. Lett.*, 29(10): 1483–1491. https://doi.org/10.1007/s10529-007-9415-5.

Sadino-Riquelme, M. C., Rivas, J., Jeison, D., Donoso-Bravo, A., and Hayes, R. E. (2022). Investigating a Stirred Bioreactor: Impact of Evolving Fermentation Broth Pseudoplastic Rheology on Mixing Mechanisms. *Fermentation*, 8(3): 102. https://www.mdpi.com/2311-5637/8/3/102.

Saha, R., Donofrio, R. S., and Bagley, S. T. (2013). Determination of the effectiveness of UV radiation as a means of disinfection of metalworking fluids. *Annals of Microbiology*, 64(2): 831–838. https://doi.org/10.1007/s13213-013-0722-x.

Saif, S., Abid, Z., Ashiq, M. F., Altaf, M., and Ashraf, R. S. (2021). Biofertilizer formulations. *Biofertilizers: Study and Impact*, 211–256.

Salcedo, R., Llop, J., Campos, J., Costas, M., Gallart, M., Ortega, P. et al. (2020). Evaluation of leaf deposit quality between electrostatic and conventional multi-row sprayers in a trellised vineyard. *Crop Protection*, 127: 104964. https://doi.org/https://doi.org/10.1016/j.cropro.2019.104964.

Samir, A., Ashour, F. H., Hakim, A. A. A., and Bassyouni, M. (2022). Recent advances in biodegradable polymers for sustainable applications. *npj Materials Degradation*, 6(1): 68. https://doi.org/10.1038/s41529-022-00277-7.

Sanita Lima, M., and Coutinho de Lucas, R. (2021). Co-cultivation, Co-culture, Mixed Culture, and Microbial Consortium of Fungi: An Understudied Strategy for Biomass Conversion [Opinion]. *Front Microbiol.*, 12: 837685. https://doi.org/10.3389/fmicb.2021.837685.

Sarachat, T., Pornsunthorntawee, O., Chavadej, S., and Rujiravanit, R. (2010). Purification and concentration of a rhamnolipid biosurfactant produced by *Pseudomonas aeruginosa* SP4 using foam fractionation. *Bioresour. Technol.*, 101(1): 324–330. https://doi.org/10.1016/j.biortech.2009.08.012.

Sarma, M. V. R. K., Gautam, A., Kumar, L., Saharan, K., Kapoor, A., Shrivastava, N. et al. (2013). Bioprocess strategies for mass multiplication of and metabolite synthesis by plant growth promoting pseudomonads for agronomical applications. *Process Biochemistry*, 48(9): 1418–1424. https://doi.org/10.1016/j.procbio.2013.04.017.

Savvas, D., Gianquinto, G., Tuzel, Y., and Gruda, N. (2013). Soilless culture. In (pp. 303-354).

Scherm, H., Savelle, A., and Law, S. (2007). Effect of electrostatic spray parameters on the viability of two bacterial biocontrol agents and their deposition on blueberry flower stigmas. *Biocontrol Science and Technology*, 17(3): 285–293.

Scherwinski, K., Grosch, R., and Berg, G. (2008). Effect of bacterial antagonists on lettuce: active biocontrol of *Rhizoctonia solani* and negligible, short-term effects on nontarget microorganisms. *FEMS Microbiol. Ecol.*, 64(1): 106–116. https://doi.org/10.1111/j.1574-6941.2007.00421.x.

Schugerl, K. (2001). Progress in monitoring, modeling and control of bioprocesses during the last 20 years. *J. Biotechnol.*, 85(2): 149–173. https://doi.org/10.1016/s0168-1656(00)00361-8.

Selegato, D. M., and Castro-Gamboa, I. (2023). Enhancing chemical and biological diversity by co-cultivation [Review]. *Front Microbiol.*, 14: 1117559. https://doi.org/10.3389/fmicb.2023.1117559.

Shalev, O., Karasov, T. L., Lundberg, D. S., Ashkenazy, H., Pramoj Na Ayutthaya, P., and Weigel, D. (2022). Commensal *Pseudomonas* strains facilitate protective response against pathogens in the host plant. *Nat Ecol Evol*, 6(4): 383–396. https://doi.org/10.1038/s41559-022-01673-7.

Silva, M. d. G. C., Medeiros, A. O., Converti, A., Almeida, F. C. G., and Sarubbo, L. A. (2024). Biosurfactants: Promising Biomolecules for Agricultural Applications. *Sustainability*, 16(1): 449. https://www.mdpi.com/2071-1050/16/1/449.

Silva, V., Yang, X., Fleskens, L., Ritsema, C. J., and Geissen, V. (2022). Environmental and human health at risk–Scenarios to achieve the Farm to Fork 50% pesticide reduction goals. *Environment International*, *165*: 107296.

Singh, V., Haque, S., Niwas, R., Srivastava, A., Pasupuleti, M., and Tripathi, C. K. (2016). Strategies for Fermentation Medium Optimization: An In-Depth Review [Review]. *Front Microbiol.*, 7: 2087. https://doi.org/10.3389/fmicb.2016.02087.

Skrzypczak, D., Jarzembowski, L., Izydorczyk, G., Mikula, K., Hoppe, V., Mielko, K. A. et al. (2021). Hydrogel Alginate Seed Coating as an Innovative Method for Delivering Nutrients at the Early Stages of Plant Growth. *Polymers (Basel)*, 13(23): 4233. https://doi.org/10.3390/polym13234233.

Stemwedel, D. A. (2009). A Fundamental Flaw in the National Organic Program: The Case for National Regulation of Organic Input Materials. *San Joaquin Agric. L. Rev.*, 19: 199.

Stirling, G. (2011). Biological Control of Plant-Parasitic Nematodes: An Ecological Perspective, a Review of Progress and Opportunities for Further Research. pp. 1–38. *In:* Davies, K., and Spiegel, Y. (Eds.). *Biological Control of Plant-Parasitic Nematodes: Building Coherence Between Microbial Ecology and Molecular Mechanisms*. https://doi.org/10.1007/978-1-4020-9648-8_1.

Stockwell, V. O., Johnson, K. B., Sugar, D., and Loper, J. E. (2011). Mechanistically compatible mixtures of bacterial antagonists improve biological control of fire blight of pear. *Phytopathology*, 101(1): 113–123. https://doi.org/10.1094/PHYTO-03-10-0098.

Sundh, I., and Goettel, M. S. (2012). Regulating biocontrol agents: a historical perspective and a critical examination comparing microbial and macrobial agents. *BioControl*, 58(5): 575–593. https://doi.org/10.1007/s10526-012-9498-3.

Takors, R. (2012). Scale-up of microbial processes: impacts, tools and open questions. *J. Biotechnol.*, 160(1–2): 3–9. https://doi.org/10.1016/j.jbiotec.2011.12.010.

Tan, Y. N., and Li, Q. (2018). Microbial production of rhamnolipids using sugars as carbon sources. *Microb. Cell Fact*, 17(1): 89. https://doi.org/10.1186/s12934-018-0938-3.

Tang, Y., Chen, C., Leite, A. C., and Xiong, Y. (2023). Editorial: Precision control technology and application in agricultural pest and disease control [Editorial]. *Front Plant Sci.*, 14: 1163839. https://doi.org/10.3389/fpls.2023.1163839.

Teixidó, N., Usall, J., and Torres, R. (2022). Insight into a successful development of biocontrol agents: Production, formulation, packaging, and shelf life as key aspects. *Horticulturae*, 8(4): 305.

Tekere, M. (2019). Microbial bioremediation and different bioreactors designs applied. pp. 1–19. *In:* Eduardo Jacob, L., and Leila Queiroz, Z. (Eds.). *Biotechnology and Bioengineering*. IntechOpen. https://doi.org/10.5772/intechopen.83661.

Thomas, L., Larroche, C., and Pandey, A. (2013). Current developments in solid-state fermentation. *Biochemical Engineering Journal*, 81: 146–161. https://doi.org/https://doi.org/10.1016/j.bej.2013.10.013.

Thomashow, L. S., and Weller, D. M. (1996). Current concepts in the use of introduced bacteria for biological disease control: mechanisms and antifungal metabolites. *Plant-microbe interactions*, 187–235.

Timmusk, S., Grantcharova, N., and Wagner, E. G. (2005). *Paenibacillus polymyxa* invades plant roots and forms biofilms. *Appl. Environ. Microbiol.*, 71(11): 7292–7300. https://doi.org/10.1128/AEM.71.11.7292-7300.2005.

Tiso, T., Demling, P., Karmainski, T., Oraby, A., Eiken, J., Liu, L. et al. (2024). Foam control in biotechnological processes—challenges and opportunities. *Discover Chemical Engineering*, 4(1): 2. https://doi.org/10.1007/s43938-023-00039-0.

Tokic, M., Hatzimanikatis, V., and Miskovic, L. (2020). Large-scale kinetic metabolic models of *Pseudomonas putida* KT2440 for consistent design of metabolic engineering strategies. *Biotechnol Biofuels*, 13(1): 33. https://doi.org/10.1186/s13068-020-1665-7.

Vanhaelewyn, L., Van Der Straeten, D., De Coninck, B., and Vandenbussche, F. (2020). Ultraviolet Radiation From a Plant Perspective: The Plant-Microorganism Context [Review]. *Front Plant Sci*, *11*: 597642. https://doi.org/10.3389/fpls.2020.597642.

Vejan, P., Abdullah, R., Khadiran, T., Ismail, S., and Nasrulhaq Boyce, A. (2016). Role of Plant Growth Promoting Rhizobacteria in Agricultural Sustainability—A Review. *Molecules*, 21(5): 573. https://www.mdpi.com/1420-3049/21/5/573.

Vrábel, P., Van der Lans, R. G. J. M., Cui, Y. Q., and Luyben, K. C. A. M. (1999). Compartment Model Approach. *Chemical Engineering Research and Design*, 77(4): 291–302. https://doi.org/10.1205/026387699526223.

Vuolo, F., Novello, G., Bona, E., Gorrasi, S., and Gamalero, E. (2022). Impact of Plant-Beneficial Bacterial Inocula on the Resident Bacteriome: Current Knowledge and Future Perspectives. *Microorganisms*, 10(12): 2462. https://doi.org/10.3390/microorganisms10122462.

Wang, L., Armbruster, W., and Jendrossek, D. (2007). Production of medium-chain-length hydroxyalkanoic acids from *Pseudomonas putida* in pH stat. *Appl. Microbiol. Biotechnol.*, 75(5): 1047–1053. https://doi.org/10.1007/s00253-007-0920-1.

Weller, D. M. (2007). *Pseudomonas* biocontrol agents of soilborne pathogens: looking back over 30 years. *Phytopathology*, 97(2): 250–256. https://doi.org/10.1094/PHYTO-97-2-0250.

Weller, D. M., Raaijmakers, J. M., Gardener, B. B., and Thomashow, L. S. (2002). Microbial populations responsible for specific soil suppressiveness to plant pathogens. *Annu. Rev. Phytopathol.*, 40(1): 309–348. https://doi.org/10.1146/annurev.phyto.40.030402.110010.

Wolf, T. M., and Peng, G. (2011). Improving spray deposition on vertical structures: the role of nozzle angle, boom height, travel speed, and spray quality. *Pest Technology*, 5(1): 67–72.

Yingngam, B., Navabhatra, A., and Sillapapibool, P. (2024). AI-Driven Decision-Making Applications in Pharmaceutical Sciences. pp. 1–63. *In: Using Traditional Design Methods to Enhance AI-Driven Decision Making*. IGI Global.

Zboralski, A., Biessy, A., Ciotola, M., Cadieux, M., Albert, D., Blom, J. et al (2022). Harnessing the genomic diversity of *Pseudomonas* strains against lettuce bacterial pathogens. *Front Microbiol.*, 13: 1038888. https://doi.org/10.3389/fmicb.2022.1038888.

Zhang, T., We, C., Ren, Y., Feng, C., and Wu, H. (2017). Advances in airlift reactors: modified design and optimization of operation conditions. *Reviews in Chemical Engineering*, 33(2): 163–182.

Zhao, F., Wu, Y., Wang, Q., Zheng, M., and Cui, Q. (2021). Glycerol or crude glycerol as substrates make *Pseudomonas aeruginosa* achieve anaerobic production of rhamnolipids. *Microb. Cell Fact*, 20(1): 185. https://doi.org/10.1186/s12934-021-01676-2.

Zheng, W., Xia, Y., Wang, X., Gao, S., Zhou, D., Ravichandran, V. et al. (2023). Precise genome engineering in *Pseudomonas* using phage-encoded homologous recombination and the Cascade-Cas3 system. *Nat. Protoc.*, 18(9): 2642–2670. https://doi.org/10.1038/s41596-023-00856-1.

Zhou, Z., Liu, Y., Zanaroli, G., Wang, Z., Xu, P., and Tang, H. (2019). Enhancing bioremediation potential of *Pseudomonas putida* by developing its acid stress tolerance with glutamate decarboxylase dependent system and global regulator of extreme radiation resistance [Original Research]. *Front Microbiol.*, 10: 2033. https://doi.org/10.3389/fmicb.2019.02033.

11

Trichoderma Biocontrol

Large-scale Production, Formulation Techniques, Stringent Quality Control Measures, and Efficient Delivery Methods

Saidul Islam,[1,3,*] *Arghya Banerjee*,[2] *Bholanath Mondal*,[3] *Malay Kumar Samanta*,[1] *Kamel Abd-Elsalam*,[4,*] *Maksud Hasan Shah*,[5] *Sk Naim Aktar*[6] and *Mahatab Ali*[7]

1. Introduction

Different species of the filamentous fungi belonging to *Trichoderma* spp. are very beneficial in agriculture, most particularly as plant growth promoters and protectants. These fungi can be prized for their capacity to inhibit plant diseases, facilitate nutrient acquisition and benefit the soil in which growth is occurring. They are effective, versatile and that is why they are integrated into the conception of effective

[1] Nadia Krishi Vigyan Kendra, Bidhan Chandra Krishi Viswavidyalaya (BCKV), Gayeshpur, Nadia, West Bengal, India

[2] Plant Pathology Laboratory, School of Agriculture and Allied Sciences, The Neotia University, Jhinger Pole, Diamond Harbour Rd, Sarisha, West Bengal, India

[3] Department of Plant Pathology, Palli-Siksha Bhavana, Institute of Agriculture, Visva-Bharati University, Sriniketan, West Bengal, India

[4] Plant Pathology Research Institute, Agricultural Research Center, 12619, Giza, Egypt

[5] Department of Agronomy, Bidhan Chandra Krishi Viswavidyalaya (BCKV), Mohanpur, Nadia, West Bengal, India

[6] Department of Agronomy, Palli-Siksha Bhavana, Institute of Agriculture, Visva-Bharati University, Sriniketan, West Bengal, India

[7] Department of Genetics and Plant Breeding, Bidhan Chandra Krishi Viswavidyalaya (BCKV), Mohanpur, Nadia, West Bengal, India

* Corresponding authors: islamsaidulr23@gmail.com; kamelabdelsalam@yandex.com

DOI: 10.1201/9781003603023-12

Table 1: Background of Trichoderma.

Years	Milestones of *Trichoderma*	Source
1794	The term *Trichoderma* was officially introduced.	Persoon 1794
1865	It has been observed that *Trichoderma viride* (*Hypocrearufa*) has a sexual stage.	Tulasne and Tulasne 1863
1932	*Trichoderma lignorum* (*Hypocrea virens*) was first recognized for its mycoparasitic and biocontrol capabilities.	Weindling 1932
1934	The first antimicrobial compound, such as gliotoxin, was invented.	Weindling 1934
1957	It was found that *T. viride*, also called *T. reesei*, was affected by light.	Gutter 1957
1972	*T. harzianum* was shown to have biocontrol activity against *Sclerotium rolfsii* in field trials.	Wells 1972
1979	Through UV-induced mutagenesis, the *T. reesei* carbon catabolite mutant RUT C30 was produced.	Montenecourt and Eveleigh 1979
1983	Cloning the first *Trichoderma* species such as *T. reesei* was accomplished successfully.	Shoemaker et al. 1983
1984	*Trichoderma*'s first worldwide workshop was held.	-
1985	The first thorough analysis of *Trichoderma*'s biology, ecology, and potential for biocontrol was published by Papavizas.	Papavizas1985
1986	It has been claimed that *Trichoderma* can stimulate the growth of roots.	Chang et al. 1986
1987	The genetic transformation of *T. reesei* was accomplished successfully.	Penttilä et al. 1987
1989	Binab T, for example, was the first commercial *Trichoderma* product to be registered.	Harkki et al. 1989
1992	Evidence was presented showing *Trichoderma* species cloning lectin-coated fibers.	Inbar and Chet 1992
1993	The cloning of the prb1 gene, linked to mycoparasitism and cell wall-induced expression, was accomplished.	Geremia et al. 1993
1997	*Trichoderma* species were shown to enhance plant immunity (ISR) through application.	Bigirimana et al. 1997
1998	Factors that activate genes involved in mycoparasitism were identified.	Cortes et al. 1998
1999	*Trichoderma* was shown to colonize plant roots internally.	Yedidiaet al. 1999
2002	The functions of G proteins and signaling pathways in biocontrol processes and conidia development were clarified.	Rocha-Ramírez et al. 2002
2003	It was determined which mitogen-activated protein kinase (MAPK) in *T. virens*is in charge of adversely controlling conidiation.	Mendoza et al. 2003, Mukherjee et al. 2003
2004	In *T. atroviride*, photoreceptors (Brl1 and Brl2) were found.	Casas-Flores et al. 2004

Table 1 contd. ...

...Table 1 contd.

Years	Milestones of *Trichoderma*	Source
2005	The function of *Trichoderma* MAPK in inducing systemic resistance (ISR) was established.	Viterbo et al. 2005
2006	The first true elicitor protein (Sm1/Epl1) of *Hypocrea atroviridis* was purified using glucose.	Djonović et al. 2006
2006	Regulators of cellulases and xylanases (e.g., XYR1) in *T. reesei* were identified.	Stricker et al. 2006
2008	The genome of a *Trichoderma* species (e.g., *T. reesei*) was sequenced for the first time.	Martinez et al. 2008
2009	The role of endophytic *Trichoderma* in improving plant stress tolerance was discovered.	Baeet al. 2009
2010	Analyzing biocontrol agents (e.g., *T. virens*), other *Trichoderma* species, and their mutual combinations, the VELVET protein (Vel1) was recognized as an essential factor. Peptide pheromone precursor genes in *Trichoderma* were found.	Mukherjee and Kenerley2010
2011	A genome comparison of three *Trichoderma* species was performed.	Kubicek et al. 2011
2012	Complete genome of *T. asperellim, T. longibrachiati, T. harzianum* and *T. citrinoviride* were sequenced using next-generation sequencing and high-throughput sequencing.	Schuster et al. 2012
2022	There were five new *Trichoderma* species found and documented.	Zhang et al. 2022
2014–2022	In 40 countries, 144 strains of *Trichoderma* were registered.	Woo 2023

measures in maintaining sustainable agriculture (Harman et al. 2004, Shoresh et al. 2010). Both solid-state and submerged fermentation techniques are employed for the commercial production of *Trichoderma* (Table 1). SSF is commonly employed due to its ease and low cost; the substrates employed are agricultural residuals (Pandey et al. 2021). New trends in SSF have included substrate modification, pretreatment, and fermentation parameters to enhance the overall spores of *S. bolili* and spore germination (Martins et al. 2023). Additionally, operating under such an environment, SmF allows the set environment for growth that enhances the development of high-density spore production. Ongoing developments in bioreactor structures and control process systems are making SmF more effective and scalable (Rao et al. 2022, Zhao et al. 2023). The formulation of products containing *Trichoderma*, which are used to control plant diseases, must be prepared. The most common formulations include powder, liquid, and granules. Pelletized forms, which might be based on carriers like talc or clay, are utilized as soil and seed applications. Recent studies were directed toward obtaining a stable and effective treatment through the help of new carriers and stabilizers (Goswami et al. 2022). Concentrated solutions for use by spraying or by soil soak are enhanced through improvements in stabilizers and emulsifiers to boost spore viability (Sharma et al. 2023). Solid forms are fine particles, with the use of controlled-release systems providing an extension of activity duration and a decreased rate of application (Huang et al. 2023). Tests are performed quite a lot for quality control of *Trichoderma* products, including the determination of spore viability, purity, and effectiveness. Respective concentrations of viable cells in products are often evaluated by assessing viability via electron and molecular microscopy (Singh et al. 2022). As a means to identify impurities and confirm product purity, PCR-based techniques and molecular diagnostic approaches must be employed (Jaiswal et al. 2023). Efficiency testing in the form of field and controlled environment tests is vital to the confirmation of *Trichoderma* products in different settings (Agarwal et al. 2023). There are numerous methods to apply *Trichoderma*, including use as a seed treatment, a foliar application, and soil application. Hence, soil treatment strategies are being continuously designed to ensure enhanced *Trichoderma* colonization and performance in the rhizosphere (Mehta et al. 2022). Due to improvements in sprayer technology and formulation stability, foliar sprays have been reformulated to ensure better coverage and adherence (Li et al. 2023). Innovative coating approaches could be developed with the seed treatment of newly developed *Trichoderma* formulations for seedlings, which could lead to improved early establishment (Patel et al. 2023).

Research focused on controlled-release and microencapsulation techniques for improving the stability and shelf life of *Trichoderma* products is being conducted (Yadav et al. 2023, Verma et al. 2023). All these current studies on formulation, production, quality control, and delivery mechanisms indicate the role of *Trichoderma* spp. in sustainable agriculture. All these enhancements add up to the efficacy and dependability of biocontrol approaches, leading to stronger agricultural systems and healthier crops.

Objective: This review focuses on the mechanisms of action, large-scale production optimization, and the potential to improve sustainable agriculture practices with respect to the production, formulation, and application methods of *Trichoderma* species as biocontrol agents and plant growth promoters.

2. Mechanisms Employed by *Trichoderma*as Biological Control Agent

Trichoderma species have emerged as efficient biological control agents due to their versatile interactions with host plants and their pathogens. The most widely understood mechanism is the direct attack of *Trichoderma* on the pathogens, in which they release certain hydrolytic enzymes including chitinases, glucanases, and proteases that aid in degrading the fungal cell wall and releasing nutrients for *Trichoderma*. This process is prone to certain limitations, including pH, temperature, and moisture of the environment. *Trichoderma* also elicits ISR, allowing the plants to activate their defense mechanisms for future pathogenic events. Such resistance is dependent on the induction of several proteins and genes that regulate salicylic acid and jasmonic acid activation pathways in plants (Van Wees et al. 2000). These proteins and genes also confer certain metabolic activities that elevate stress endurance in plants.

Several studies have pointed out the role of ISR in protection against bacteria and viruses, thus making it reliable. In addition, *Trichoderma* creates a challenge for the soil-inhabiting pathogens by competing for the nutrient resources, including essential minerals and elements. The advantages of this approach are visible in soils with limited nutrients. Besides this, it also benefits the plant by directly inhibiting the growth of pathogenic microorganisms (Benítez et al. 2004). *Trichoderma* also secretes certain metabolites responsible for repressing pathogen growth and enhancing the growth of beneficial microbes simultaneously. These fungi also develop mycorrhizal associations with plant roots; as a result, the plant has a higher surface area for absorption of water and nutrients due to a higher degree of root branching and elongation. Various plant growth regulators (PGRs) such as auxins and gibberellins are provided by *Trichoderma* that are responsible for stimulating plant growth and defense against pathogens (Contreras-Cornejo et al. 2016). Due to this, plants having mycorrhizal associations with *Trichoderma* are vigorous and show some degree of tolerance toward abiotic stresses. Moreover, healthy roots ensure assimilation of iron, zinc, and other micronutrients which are obligatory for various metabolic processes in plants. *Trichoderma*, like all other fungi, releases secondary metabolites which help it in degradation of substrates. These metabolites are different from the metabolites of pathogenic species and they provide antifungal properties to *Trichoderma*, e.g., gliotoxins, trichodermins, and viridicatins. The antifungal roles are achieved by disruption of cellular mechanisms of pathogenic microorganisms (Frisvad et al. 2011). *Trichoderma* has also been reported to suppress viral infection by the aid of its secondary metabolites that create a challenging environment for viruses. These properties enable *Trichoderma* to provide protection against a wide range of pathogens. The mycorrhizal association also serves as a barrier against pathogens and prevents their entry into host plants and has been seen to inhibit bacterial infection by preventing its entry into plants (Marx and Krupa 1978). Plants with these mycorrhizal associations have enhanced nutrient uptake because of improved vigor of the rooting system. Moreover, *Trichoderma* also encourages certain microbes that are necessary for soil health. A diverse microbiome in soil can thus aid in reduction of diseases arising due to soil-inhabiting pathogens (Larkin 2008). *Trichoderma* also increases moisture retention and aeration in soil by degradation of organic matter in the soil. Plant health is directly dependent on soil quality; thus, *Trichoderma* improves plant health by making soil conditions favorable for the plants. *Trichoderma* plays mutualistic roles with other microbes in the rhizosphere, and thus a rugged community arises in soil that protects plants from competing pathogens. Poveda and Echeverría (2015) stated that these ecological interactions enhance the resilience of the plant-soil ecosystem. *Trichoderma* secretes certain organic acids that lower soil pH, thus increasing available nutrients in soil and simultaneously inducing challenges for pathogen survival (Meyer et al. 2020). This mechanism of pH alteration enables *Trichoderma* to provide resistance to plants and improve plant health. Some signal molecules, i.e., volatile organic compounds (VOCs), are also produced by *Trichoderma* that trigger defense in plants and bacteria. These chemicals also induce group defense mechanisms by initiating ISR in nearby plants (Intana et al. 2021). *Trichoderma* also helps in maintaining ecological balance by attracting beneficial microbes in its vicinity. *Trichoderma* enhances soil health by degrading toxic materials present in the soil, including pollutants and mycotoxins. This not only enhances soil health but decreases phytotoxic effects of pathogens on crop plants (Abdel-Fattah 2021). *Trichoderma* application is in line with the slogan of sustainable agriculture and plays a pivotal role in IPM. Thus, it ensures long-term sustainability and uplifts crop yield without damaging the environment (Abdul-Halim et al. 2023).

2.1 Mycoparasitism

Another antagonistic approach employed by *Trichoderma* for controlling plant diseases is mycoparasitism. As BCAs, they directly penetrate the fungal tissues using their hyphae and release certain metabolic compounds that degrade the fungal tissues, thus enabling *Trichoderma* to absorb nutrients from them (Prasun et al. 2022). It has been experimentally shown that *T. koningii* is effective against various species of *Aspergillus* responsible for post-harvest losses in tomato (Oyesola et al. 2020). This shows the versatility of *Trichoderma* in control of pathogens as well as saprophytes causing severe post-harvest losses. Moreover, *Trichoderma* has the ability to grow rapidly in comparison to other fungi, making it suitable to create competition for the pathogenic microorganisms (Guzmán-Guzmán et al. 2023, Tyśkiewiczet al. 2022, Asad 2022, Yao et al. 2023). Studies were conducted on various *Trichoderma* species (*T. virens*, *T. atroviride*, and *T. reesei*). These fungi participate in root interactions since the fungal pathogen activates the rhizosphere and provides nutrient root exudates (Prasun et al. 2022). Its functions in terms of its ecological context gradually changed from exclusively mycoparasitic to more plant-oriented activity (Hough et al. 2023). Three aspects of mycoparasitism in *Trichoderma* are the use of biochemical signals to develop ureVIC structures and enzymes. It is identified that *Trichoderma* can perceive oligochitins in the target's cell wall and initiate chemotropic growth. It encircles its own hyphae around that of the pathogen and secretes, beyond the interface, chitinases and β-1,3-glucanases to disintegrate the cell wall of the pathogen. Through this process, *Trichoderma* acquires nutrients through pores developed from this process (Sood et al. 2020, Macías-Rodríguez et al. 2020, Guzmán-Guzmán et al. 2019).

2.2 Mycovirus-Mediated Cross-Protection (MMCP)

MMCP is the best-understood mechanism through which biocontrol fungi guard plants from the attacks of other pathogenic fungi (Pandit et al. 2022). In addition to mycoparasitism, antibiosis, induced systemic resistance (ISR), and competition for nutrients and space, MMCP is one of the five major strategies employed by beneficial fungi in controlling plant pathogens. In the course of horizontal mycovirus transmission for strains including *Cryphonectria parasitica* hypovirus 1 (CHV1), virulence in the recipient strain is generally reduced. Reduction of pathogenicity leads to cross-immunity with other related fungal diseases (Ghorbanpour et al. 2018, Hough et al. 2023). Upon the plant's perception of MAMPs or PAMPs specific to fungi, the plant initiates MTI. Likewise, as soon as a plant detects fungal effectors, it launches ETI. Stimulation of defense mechanisms is preceded by intracellular signaling, essentially by activation of mitogen-activated protein kinase (MAPK) pathways (Jaroszuk-Ścisełet al. 2019, Oyesolaet al. 2024).

As soon as a plant detects MAMPs or PAMPs characteristic offungi, it starts MTI. Similarly, once a plant recognizes fungal effectors, it deploys ETI. Subversion of defense mechanisms must occur in spatial conjunction with intracellular signaling, fundamentally involving the stimulation of mitogen-activated protein kinase (MAPK) pathways (Jaroszuk-Ścisełet al. 2019, Oyesolaet al. 2024).

2.3 Production of Cell Wall Degrading Compounds (CWDCs)

At the final phase of mycoparasitism, the synthesis of Cell Wall Degrading Compounds (CWDCs) is vital (Ghorbanpour et al. 2018). The major structural polymers of the fungal cell wall include mannan, chitosan, 1,6-glucans, chitin, and others such as proteins (Sood et al. 2020). Chitinases and proteases, which degrade common cell constituents, are released in large amounts by *Trichoderma* (Ribeiro et al. 2019). For instance, glucanases, which weaken the structural cell walls of pathogenic fungi, and chitinases, which break the glycosidic bonds that hold carbon molecules in chitin together (Loc et al. 2020). Evaluation of genomic sequences indicates that many of the genes encoding these enzymes are unique to *Trichoderma* (Goughenour et al. 2021). For instance, putative cross-talk between *Trichoderma harzianum* and *Fusarium solani* occurs, leading to upregulation of a β-1,3-glucanase gene (Ghorbanpour et al. 2018). The absence of the Tv bgn-3 gene in *Trichoderma virens*, which creates β-1,6-glucanase, limits the ability of the fungus to control *Pythium ultimum* (Sood et al. 2020). Increased chitinase and glucanase activities have been reported when *Trichoderma* is grown on lyophilized cells of the pathogenic *Fusarium culmorum* as the sole carbon source. Therefore, more hydrolytic enzymes are secreted during fungal antagonism (Vicente et al. 2022, Sood et al. 2020). Proteolytic enzymes also contribute to mycoparasitism by breaking peptide bonds within proteins (Volkov et al. 2020). For instance, the Prb1 protease of *Trichoderma* enhances the resistance of cotton seedlings to fungal diseases (Yao et al. 2023, Gueye et al. 2020,Olasehindeet al. 2016). In addition, an aspartic protease produced by *T. harzianum* reduces the impact of fruit graymold and preserves plant growth and seed germination. Proteases can also regulate the production of other hydrolytic enzymes that participate in mycoparasitism (De Zotti et al. 2020, Deng et al. 2018).

2.4 Antibiotic and Other Antifungal Agents Production

Production of antimicrobial substances to inhibit plant diseases is one of the known mechanisms through which biocontrol agents develop antibiosis. Many process-interfering compounds, including enzymes such as pectinase, xylanase, cellulase, lipase, amylase, glucanase, and protease; volatile organic compounds; and antibiotics such as trichodermin, gliotoxin, viridin, and peptaibols are produced by *Trichoderma* species (Volkov et al. 2020). To date, over three hundred seventy secondary metabolites from *Trichoderma* have been identified and reported, and they have been proven to be potential antagonists of pathogens (Ghorbanpour et al. 2018, Gajera et al. 2020). Several *Trichoderma* species also produce VOCs such as peptaibols and polyketides (Xiao et al. 2023). In the "Peptaibiotics Database," the majority of the substances are peptide antibiotics with molecular weights in the 500–2200 Da range, which constitutes over 80% of the total. These substances have medium-acylated N-termini, contain amino alcohols at terminal ends, and include a large proportion of protein amino acids, isovaline, and α-aminoisobutyric acid (Fernández and Pablo 2022, Tamandegani et al. 2020, Marik et al. 2019). Thus, *T. longibrachiatum* produces peptaibols such as trichobrachins, suzukacillins, trichoaureocins, and longibrachins. These peptaibols are produced by NRPSs in different species such as *T. citrinoviride*, *T. koningi*, and *T. pseudokoningi* (Vicente et al. 2022, Siddiquee 2017). Also, PKs, a large group of chemicals produced by bacteria, plants, and fungi, are synthesized by *Trichoderma*. PKSs use building blocks such as acetyl-CoA and malonyl-CoA to build polyketides (Zeilinger et al. 2016, Cardoza et al. 2022). However, there are few scientific works that focus on the genes of *Trichoderma* harboring PK biosynthesis and their metabolic pathways. More recently, Mary and Russell (2021), Sobowale (2019), and Deepa et al. (2021) have postulated the genes involved in the biosynthesis of PK and their metabolic pathways. Many PKSs in *Trichoderma* have been linked to the production of fumonisin, and *T. reesei*, *T. virens*, and *T. atroviride* all possess orthologs (Atanasova et al. 2013, Chen et al. 2021). Pyrones, anthraquinones, terpenoids, and epipolythiodioxopiperazines (ETPs) are also among the secondary metabolites produced by *Trichoderma* (Shenouda et al. 2022). For example, *T. asperellum* possesses secondary metabolites, such as 6-phenyl-α-pyrone (6PAP), that act against *Fusarium oxysporum*. Other bioactive constitutive compounds include the anthraquinone pigments pachybasin and emodin, which are produced by *T. harzianum*, *T. aureoviride*, and *T. viride*. These substances possess mycoparasitic capacity and demonstrate high antifungal effectiveness (Tyśkiewiczet al. 2022, Sehimet al. 2023, Yusuf et al. 2019). The immunosuppressive effects found among ETPs are due in part to compounds such as the antifungal harzianic acid of *T. harzianum* and gliotoxin, a metabolite responsible for its immunosuppressive action (Kubiak et al. 2022, Scharf et al. 2016, Kumar et al. 2023, Yao et al. 2023).

2.5 Competition for Vital Nutrients and Space

Trichoderma uses highly efficient strategies to invade plants by competing with pathogenic microorganisms for space and nutrients (Zaid et al. 2022, Halifuet al. 2019). It invades environment niches including plant tissues, rhizospheres, and phyllospheres, thus growing at a rate faster than other organisms for the available resources (Oszust et al. 2019, Martinez et al. 2023). This serves as aprimary carbon source,supporting high levels of sugar and sucrose within *Trichoderma* and conferring an advantage in nutrient uptake over other fungal genera (Pandit et al. 2022, Rauf et al. 2024). This competitive advantage is reinforced by its ability to secrete organic acids, such as gluconic, citric, and fumaric acids, thereby lowering the pH of the soil and making phosphates, iron, manganese, and magnesium available to plants, which can also better protect against pathogens (Volkov et al. 2020,Vinale et al. , 2006). Siderophores are low–molecular-weight chelating agents that have a high affinity for iron. They play a major role in *Trichoderma*'s capacity to compete in environments with low iron levels (Zhao et al. 2020, Haseeb et al. 2023). Many mycopathogenic fungi require iron ions for their growth, but there is evidence that *Trichoderma* siderophores capture and convert these ions into a more soluble Fe^{3+} form used by microorganisms and plants (Robinson and Parkin 2021, Chowdappa et al. 2020, Zhang et al. 2021, Pecoraro et al. 2021). *Trichoderma* has the ability to hamper the mycelial growth of soil-borne pathogens and strip ferric ions from its surroundings (Dutta et al. 2022, Lahlaliet al. 2022). For example, studies carried out in the laboratory showed that *T. koningii* can effectively inhibit the growth of *Aspergillus niger, A. flavus, and A. nidulans*. Also, pathogens regulated by *T. harzianum* include *Alternaria alternata, Alternaria infectoria*, and *Fusarium acuminatum* (Oyesolaet al. 2021, Patkowska et al. 2020). Moreover, *Trichoderma* is commonly applied in plant protection and improvement of soil fertility; its ability to take up metal ions makes it a better biocontrol agent (Rauf et al. 2024). Enzyme-mediated nutrient competition and environmental modification neutralize various phytopathogens such as Colletotrichum, Botrytis, Verticillium, and Phytophthora. This results in substantial decreases in the pathogenicity of Macrophominaphaseolina, Fusarium graminearum, and Botrytis cinerea (Saravanakumar and Wang 2019, López-Coria et al. 2023).

2.6 Induction of Plant Resistance

MAMPs or PAMPs are microbial signals that are sensed by specific PRRs that plants have evolved to use (Ghorbanpour et al. 2018, Abdul et al. 2020) (Figure 1). At the time of invasion, they are useful for sensing pathogens and mounting defenses against them. To enhance disease tolerance, *Trichoderma*, which is a rhizospheric fungus, interacts with plant roots and induces both localized and systemic immune responses (Sood et al. 2020, Zeilinger et al. 2016). Several inducer chemicals, or molecules that trigger defense responses, contribute positively to resistance. These

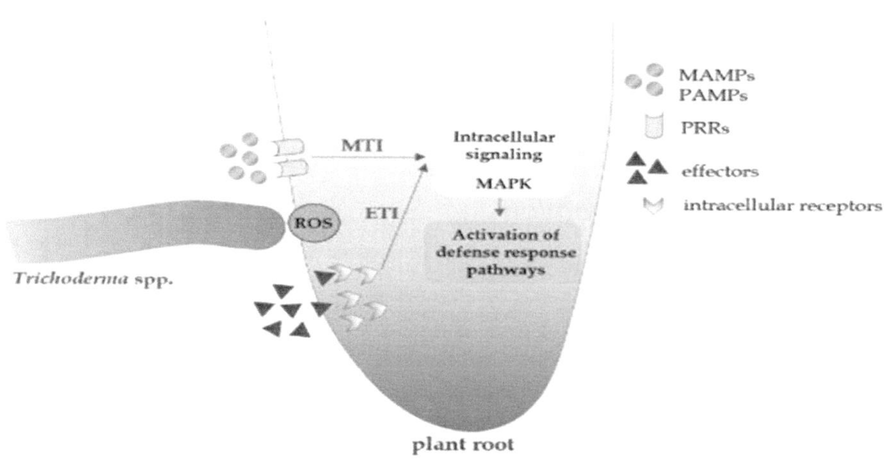

Figure 1: Through recognizing fungal effectors and microbe associated molecular patterns (MAMPs), plants respond defensively to fungi

include peptides from plant tissues and other proteins or peptides from microbial cells. Elicitors fall into two categories: non–race-specific elicitors, which generate broader responses in both host and pathogen, and race-specific elicitors that trigger gene-for-gene defense reactions in particular host cultivars (Zehra et al. 2021, Köhl et al. 2019). The physical, chemical, and biological reactions elicitors induce in plants include reactive oxygen species (ROS) and changes in ion flux. These reactions lead to production of immune compounds such as phytoalexins, enzymes, and phytohormones, which restrict infections (Aljbory and Chen 2018, Juan et al. 2021). *Trichoderma*, for example, leads to an increase in immunological activity in a range of monocots and dicots (Harman et al. 2021). Recognition of MAMPs/PAMPs by intracellular receptors (IRs) and transmembrane pattern recognition receptors (PRRs) is a vital form of defense in plants (Jones and Dangl 2006) (Figure 1). Plants trigger MTI or ETI when they recognize certain molecular patterns (Jaroszuk-Ściseł et al. 2019). Though localized, the hypersensitive response (HR) is generally more restricted than the long-lasting and systemic form of defense known as effector-triggered immunity (ETI), which may lead to localized cell death (De et al. 2015). Since mitogen-activated protein kinases are involved in relaying the signal from these receptors to plant cells, immunity is triggered throughout the plant (De et al. 2015). *Trichoderma* increases defense against both biotrophic and necrotrophic pathogens and certain abiotic stressors by triggering SAR against biotrophic pathogens, ISR against necrotrophic pathogens, and general induced resistance (IR) (Jagodziket al. 2018). STM is shown to be governed by the hormones ethylene (ET) and jasmonic acid (JA), while SAR is associated with salicylic acid (SA) as a signaling molecule and induction of pathogenesis-related proteins. Also, aminobutyric acid triggers activation of IR defenses via abscisic acid signaling (Boccardo et al. 2019). *Trichoderma* enhances levels of various metabolites and enzymes that contribute to the plant's immunobiological defense (Pandit et al. 2022). For example, the effectiveness of *T. harzianum* and *T. asperellum* in managing Colletotrichum truncatum on Capsicum annuum revealed antifungal activity (Hartman et al. 2016, Kim et al. 2023). At the same time, *T. longibrachiatum* enhances cucumber plants' tolerance to *B. cinerea* through stimulation of specific signal transduction cascades (Yadav et al. 2023). Hence, *Trichoderma* is considered an effective biocontrol agent because of the multiple mechanisms that are crucial in plant disease management (Yuan et al. 2019).

2.7 Plant Growth Promotion

Trichoderma and other BCAs increase plant growth and disease resistance (Emmanuel and Babalola 2020, Ayub et al. 2023). Plant development is significantly improved by the endophytic interactions of these fungi with plant roots (Contreras-Cornejo et al. 2024). Additionally, they possess mechanisms that boost plant defenses, produce siderophores, produce growth-regulating compounds and secondary metabolites, and improve plant development (Al-Ani 2018) (Figure 2). *Trichoderma* can facilitate nutrient uptake, promote efficient use of internal resources, enhance interactions

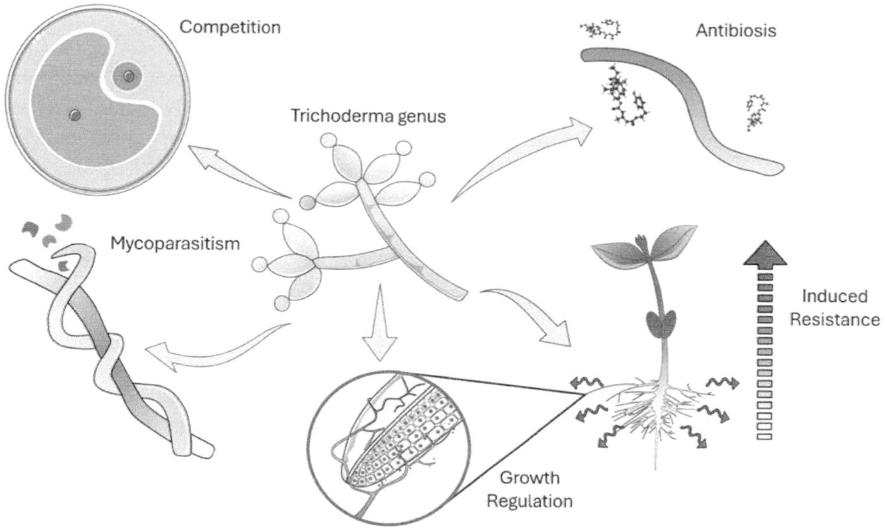

Figure 2: This diagram illustrates some of the ways used by Trichoderma in disease management for plants. Competition for resources and space is one of these strategies, as is the synthesis of compounds that halt the spread of bacterial diseases; these compounds may also serve as medications in addition to providing plant protection, trigger activating defense mechanism in plants, and encouraging plant growth. The following sections are reprinted from Vindas-Reyes et al. (2024).

with the plant environment (including plant-to-plant signaling via phytohormones), and elicit adaptive responses to abiotic stimuli, thereby promoting plant growth (Hermosa et al. 2013).

2.7.1 Mineral Solubilization

Trichoderma and other microorganisms living in plant roots contribute to growth by dissolving essential minerals for the plants that otherwise cannot be acquired (Altomare et al. 1999). Soil and nutrient solubilization and immobilization alter the ease with which plants access necessary nutrients. Amoo et al. (2023) and Alshammari et al. (2024) have observed that the solubility of soil minerals through *Trichoderma* species makes essential nutrients available to the plant. For example, *T. harzianum* solubilizes rock phosphate, manganese oxide, zinc, and iron oxide while being central to the biocontrol mechanism (Li et al. 2015). In other studies, it mobilizes phytate and micro-fertilizers in tomato plants and probably raises biomass yield (Tandon et al. 2019, Bononi et al. 2020, Boat et al. 2022, Song et al. 2023).

2.7.2 Nitrogen Fixation

Trichoderma assists plants in taking up nitrogen to higher levels. *Trichoderma* enhances plant growth and disease tolerance by improving photosynthetic carbon assimilation and nitrogen acquisition and metabolism (Rascio and Rocca 2012, Shirvani and Yahaghi 2021, Mei et al. 2022).

2.7.3 Regulation of Phytohormone

Trichoderma play a role in phytohormone turnover for the defense and growth of plants (Contreras-Cornejo et al. 2024). To modulate several hormones, it synthesizes IAA, GA, ABA, SA, and CK among others (Stewart and Hill 2013). Also, ethylene stress is minimized because the plant-associated fungus *Trichoderma* possesses ACC deaminase—the enzyme that modulates ethylene levels (Rauf et al. 2021, Jha et al. 2021). *Trichoderma* enhances biomass yield, particularly under abiotic stress such as water stress, and relieves stress by stabilizing phytohormone concentrations, offering multipurpose resistance against many diseases, and improving overall plant health (Isabel and Rubén 2022 Bahadur and Dutta 2022).

2.7.4 Soil pH Alteration

Another aspect is that *Trichoderma* alters the pH of the soil to better suit plant growth. In producing these beneficial by-products, it forms organic acids and solubilizes minerals, effects that may reduce pH to limit pathogen growth while enhancing nutrient availability (Esparza-Reynoso et al. 2019). These conditions also induce the secretion of antimicrobial molecules. In addition, competition for resources in the form of mycoparasitism of other soil-borne plant pathogens enhances plant protection and helps maintain soil microecology rather than relying on extensive chemical fungicides (Abbas et al. 2022). When harnessed, the abilities of *Trichoderma* can lead to sustainable disease-management strategies that enable the development of environmentally friendly, productive, and stable agrosystems (Alfiky and Weisskopf 2020).

3. *Trichoderma* Production on a Large Scale: Techniques and Optimization

To enhance microbial cell stability for conventional storage, transportation, and application of biocontrol agents, microbial or microbe-derived active components are formulated with an appropriate carrier material and additives. For the microorganisms to be effective, the right formulation is essential (Bashan et al. 2014). In field applications, the formulation plays a crucial role in preserving performance. To improve distribution, extend shelf life, and lower costs, formulation primarily entails combining the bioagents with carriers and additives. For instance, the solid-state fermentation process is used extensively in the industrial production of *Trichoderma*, employing inexpensive coarse cereal grains including ragi, millet, and sorghum (Jeyarajan 2006). These grains are inoculated with *Trichoderma* for ten to fifteen days after they have been sterilized and moistened; during this time, the fungus produces the dark green spores that coat the grains. After that, the grains are either immediately applied to soil or used to enrich farmyard manure, or they are processed into a powder for seed coating. The method's disadvantages include a lengthy production period and contamination hazards, although it is suitable for small-scale production. Granules, dust, or wettable powders are examples of dry products; emulsions, oils, and aqueous suspensions are examples of liquid products; and microencapsulated products are other formulation types. Microbial inoculant effectiveness depends mainly on the formulation type and the delivery system, which should support a long shelf life with active antagonistic or PGP agents (Gopalakrishnan et al. 2016). However, they also have to be inexpensive to produce in order to keep handling and application costs low once formulated into microbial preparations. The formulation should maintain the viability of the PGP microbes and stability in field environments while being able to deliver the microbes to the target site. More information on the different formulation types, additives, and PGP/antagonistic microbes in crops has been elaborated by Nakkeeran et al. (2005) and Bashan et al. (2014). The principal function of a formulation is to establish the appropriate microenvironment surrounding introduced PGP microorganisms in the soil, under which their rate of decline is reduced. Although BCAs can be produced in different ways, most commercial products are available in liquid, granular, or powder form (Keswani et al. 2016). There are several stages in the formulation process, including selecting the best BCA for the target pathogen, incorporating live microbes, adjuvants, and carriers in order to produce a reliable product with a defined shelf life and assurance that it will perform optimally as envisaged. This includes understanding the target pathogen system, the environment into which the BCA is applied, the ecology of the BCA, and the application context, which involves the growth stage of the agri-food system, application techniques, and irrigation methods (Bejarano and Puopolo 2020).

3.1 Formulation

The formulation process often involves blending active fungal spores with other inert materials, such as diluents and surfactants, to improve the physical characteristics of the formulation. A good formulation should possess the following important characteristics: (1) easy application; (2) thermal stability at 5–35°C; and (3) room-temperature stability for more than two years from the time of packaging (Junaid et al. 2013). The long-term survival of the propagules is probably the most crucial problem in commercializing biocontrol agents. Storage temperature and the formulation materials used may greatly affect the biocontrol product. For example, *T. virens* spores sustained 82% viability in talc-based formulations at 5°C for six months and could be maintained under the same conditions at room temperature for three months (Chaube et al. 2003). As mentioned earlier, several formulations (see Figure 3) have been used for pilot production of *Trichoderma* (see Table 2), and these formulations can be broadly classified based on their ability to improve the stability of *Trichoderma* cultures.

Table 2: Trichoderma biomass production with their efficacy.

Substrate	Composition/Components	Efficacy	References
Cereals (wheat, moong, maize)	Grains (200 g) + sugar (1%) + *Trichoderma harzianum*	12.96%	Nabi et al. 2017
Vermicompost fortified with *Trichoderma*	Vermicompost + cereals + pulses	10.01% wilt incidence in chili	
Wheat seed-based formulation	Ground grain + sugar solution (1%) + *Trichoderma harzianum*	38×10^7 CFU/g	Mohiddin et al. 2017
Formulation enhanced with cow dung	*Trichoderma* formulation combined with decomposed cow dung	37.5×10^7 CFU/g	
Formulation based on talc	*Trichoderma* culture + CMC + talcum powder	37×10^7 CFU/g	
Grain of sorghum	Grain that has been partially crushed, a solution of sugar (1%) and purified water	6.1×10^4 CFU/g	Ananda et al. 2011
Formulation based on wood pellets	Conidial suspensions of *T. atroviride* combined with beech, fir, and chestnuts, distilled water, and soy flour	Tenfold increase in growth	Chammem et al. 2021
Formulation based on vermiculite and wheat bran	100 g of vermiculite + 33 g of wheat bran + 20 g of fermented *Trichoderma* biomass +175 mL of 0.05N HCl	-	Lewis 1989
Formulation based on vermiculite and bentonite	20 g of oats + 50 mL of bentonite + vermiculite + *T. harzianum* + 60 mL of water	Maintained CFU after 8 weeks	Martínez-Medina et al. 2009
Formulation based on coffee husks	Cow dung + poultry manure + coffee fruit skin + a suspension of *T. harzianum*	9×10^{11} to 3×10^{12} CFU/g substrate	Sawant and Sawant 1996
Formulation based on oil	Coconut oil + distilled water + glycerol (1%) + PVP (1%) + Tween 20 (1%) + $ZnSO_4$ (0.5%)	More than 180 days	Sreenayana and Nakkeeran 2019
A formulation based on pesta granules	100 g of wheat flour with 52 mL of sterile water and fermentation biomass	Viable for a long time	Connick et al. 1991
Based on banana waste	Rock phosphate plus banana trash (5–6 cm chunks)	Six months	Bhai 2020, Thangavelu et al. 2004
Encapsulation of *Trichoderma* sodium alginate	*Trichoderma* suspension plus 0.6 percent sodium alginate solution plus 1.5% $CaCl_2$	Viable for more than six years at room temp	Abd-Elsalam et al. 2010
Formulation of wheat flour and kaolin	80 g of wheat flour, 20 g of kaolin, and 52 mL of fermenter biomass	Few months	Ramanujam et al. 2010

Table 2 contd. ...

...Table 2 contd.

Substrate	Composition/Components	Efficacy	References
Formulation T2-liquid (NIPHM medium)	250 mL of *Trichoderma* filtrate, 750 mL of water, and 3% glycerol	-	Komala et al. 2019
Compost activator based on *Trichoderma*	50 grams of soil, 5 grams of rice straw, and 500 mg of *Trichoderma* biomass	5.3×10^{10} CFU/g	Organo et al. 2022
Formulation based on graphite and silica	*B. subtilis* + *T. harzianum* + graphite (5%) + silica nanoparticles (1%)	35–54% efficacy (in vitro)	Marcelo et al. 2021
Formulation based on chitosan & PEG	*T. harzianum* spores + glacial acetic acid (0.1%) + chitosan-PEG	More than six months	Prasad et al. 2022
Formulation based on rice powder	*T. viride* + talc powder + dextrose + sterilized rice powder	10×10^{9} CFU/g up to six months at room temp	Rini et al. 2018
Formulation based on dextrin	500 g of *T. harzianum* filtrate + 500 mL of paraffin oil + 0.2% CMC + 0.1% chitosan	26.10% efficacy; 4.33×10^{7} CFU/g (6 months)	Rai and Dewari 2016
Liquid formulation based on oil	Paraffin oil with *T. asperellum* together	28.67×10^{8} CFU/mL for 30 days	Patilet al. 2021 (b)
Formulation based on glycerol	Glycerol (3%) (v/v) + MYE medium + *T. harzianum* + talc powder	Shelf life extended to 7–12 months	Sriram et al. 2011
T. harzianum paste formulation	10% starch + 20% copper sulfate + *T. harzianum*	11.6×10^{10} CFU/g for 120 days	Balakrishnan et al. 2017

Figure 3: The formulation process for Trichoderma involves several key steps: (a) collecting rhizospheric soil, (b) isolating the fungus using a selective medium, (c) obtaining a pure culture, (d) conducting liquid fermentation, (e) performing soil-based fermentation, (f) preparing talcum powder, (g) mixing Trichoderma with talcum powder, (h) packaging the formulation, (i) incorporating the formulation into farmyard manure (FYM), (j) priming seeds with the treatment, (k) applying to seedlings, and (l) applying in pots (Kumar et al. 2022).

3.2 Liquid Formulations

Liquid formulations for biological control agents (BCAs) can vary in composition, typically consisting of microbial cultures or suspensions mixed with water, oils (either mineral or organic), polymers, or a combination of these components. Suspension concentrates for microbial inoculants are commonly created by dispersing solid active ingredients, such as free or immobilized microbial cells, into water or an aqueous solution using standard methods (Gopalakrishnan et al. 2016). Liquid formulations are widely used in biopesticides. For example, *Pseudomonas fluorescens* was formulated in coconut water with the help of glycerol or polyvinylpyrrolidone (PVP) to prepare SC (Anith et al. 2017). Mbarga et al. (2014) described an oil dispersion of *T. asperellum* conidia that included soybean oil, an emulsifier, a structural agent, and glucose. Peeran et al. (2014) also stabilized an oil-in-water emulsion of P. fluorescens using water, glycerol, polyethylene glycol, Tween 20, coconut oil, rice bran oil, and castor oil. Herrera et al. (2020) formulated *T. asperellum* TV190 in liquid using emulsified mineral and vegetable oils; viability rates were 37–43% for mineral oils and 56–63% for vegetable oils, compared with 8–12% for the control. In a more recent study, Patil et al. (2021) (a) evaluated the efficacy of an oil-based liquid formulation of *T. asperellum* to manage Fusarium wilt of chickpea using paraffin, soybean, and groundnut oils as carriers. The count of *T. asperellum* in paraffin oil was 28.67 × 10^8 cfu/mL at 30 days, decreasing to 18.00 × 10^8 cfu/mL at 180 days. Other low-cost growth media, such as molasses from fermented sugarcane or brewer's yeast, have been used to formulate liquids and to manage pH, temperature, and nutrients to minimize contamination (Papavizas 1985, Whipps 1997).

3.3 Solid Formulations

Trichoderma inoculum in this study was prepared by solid-state fermentation using agri-waste such as rice bran, corn cob meal, sawdust, wheat straw, and sugarcane bagasse, or their mixtures. For example, solid and liquid formulations are often dried to achieve a product that will be stable for a much longer period in storage (Jin et al. 1992). Solid formulations require considerable space for processes including grinding, drying, storage, inoculation, and preparation. Spray drying is a low-cost method of producing dried microbial particles for large-scale production (Morgan et al. 2006). Laboratory solid fermentation is frequently used to produce large quantities of *Trichoderma*. The fungus is first cultivated on Petri plates, after which the spores and other propagules are collected and processed. Low-cost agricultural leftovers and by-products can also be used for large-scale production. For example, many large-scale producers use cereal grains like sorghum, bajra, and wheat as substrates. After soaking in water for 12 hours and draining any excess, the grains are sterilized in heat-resistant polypropylene bags. Following sterilization, *Trichoderma* culture is added to the grains, which are then incubated for 10–15 days to produce dark green spores. These processed grains can be ground into a fine powder or used directly to enrich farmyard manure for soil application and seed treatment (Cumagun 2014).

3.4 Talc-Based Formulation

Trichoderma viride formulations based on talc have been created in India, namely at the Agricultural University of Tamil Nadu, for use in seed, soil, and seedlings (Jeyarajan et al. 1994). *Trichoderma* is cultivated in a liquid medium, combined with talcum powder in a 1:2 ratio, and dried in the shade until the moisture content reaches 8%. These formulations, which have been shown to be effective in controlling certain soil-borne illnesses, are stored for up to three or four months. The rate of seed treatment is 4–5 g/kg. Most non-profit organizations in India prepare large quantities of talc-based formulations, while 50% of the area of the country will be covered and the 5000 MT/year requirement will be met every year, as told by Jeyarajan(2006). The ability of a talc-based formulation of *T. viride* to control Fusarium wilt of chickpea and its potential for seed and soil application was demonstrated by Pradhan et al. (2022). Similarly, Sundaramoorthy and Balabaskar (2013) formulated *T. harzianum* with talc and evaluated its potential against tomato wilt caused by *Fusarium oxysporum*. The findings revealed a 15.33% decline in wilt incidence when plants were grown under greenhouse conditions. The seed treatment rate is 4–5 g/kg. Many nongovernmental companies in India manufacture large quantities of these talc-based formulations, reportedly enough to cover nearly half of the country's agricultural land; the annual requirement was estimated at about 2,000 quints (5,000 metric tons) by Jeyarajan (2006). On the same topic, scientists, including Pradhan et al. (2022), established that a talc-based formulation of *T. viride* can effectively control Fusarium wilt of chickpea, besides the prospects of seed and soil formulation (Pradhan et al. 2022). Similarly, Sundaramoorthy and Balabaskar (2013) developed *T. harzianum* based talc and tested against Fusarium oxysporum-caused tomato wilt disease and reduce the incidence of wilt to a maximum of 15.33% in greenhouse conditions.

3.5 Vermiculite-Wheat Bran-Based Formulation

Trichoderma is grown for 240 hours in molasses–yeast media to develop a formulation containing vermiculite–wheat bran. An organic composition of 3.3 g wheat bran and 100 g vermiculite is autoclaved for 3 days at 70°C. After sterilization, 17.5 mL of 0.05 N HCl and 14 mL of liquid nutrient media are added to the substrate. In some cases, the substrate is left to dry in the shade, after which it is packed (Lewis 1991).

3.6 Water in Oil/Mineral Emulsions

The formulation of *Trichoderma* as a type of liquid preparation uses a liquid medium with the an appropriate oil carrier, including mineral, vegetable, or coconut oil. Typically, formulation process can be initiated when *Trichoderma* is inoculated in broth first for proper stability. These oil carriers have several benefits: they protect the cargo from negative environmental factors such as UV light and high temperatures, enhance attachment to seeds and plant surfaces, control the release of *Trichoderma*, and increase their efficacy against pathogens. Surfactants in non-aqueous solvents (diesel or mineral oil) can also be used to remove bacterial contamination.

These compositions create stable emulsions when combined with water. The presence of oil-soluble emulsifying agents enables emulsifiable concentrates to form homogeneous emulsions upon dilution. The use of *Trichoderma*-based formulations as biocontrol agents is growing, and because of their longer shelf life, oil-based formulations work especially well for foliar applications in arid environments. Batta (2007) presented a *T. harzianum* emulsion formulation that significantly decreased postharvest fruit rot caused by *Botrytis cinerea*. To protect peanuts from root rot caused by soil-borne pathogens, PDBC created an emulsified formulation of *T. harzianum* in India that could be stored for up to eight months.

3.7 Sodium Alginate Encapsulation of *Trichoderma*

Trichoderma formulations that use sodium alginate encapsulate spores in an alginate matrix, a naturally occurring polymer derived from seaweed. These formulations are made by combining *Trichoderma* spores with a 0.6% sodium alginate solution at a 1:4 ratio. During encapsulation, the mixture is dropped into a 1.5% calcium chloride ($CaCl_2$) solution and left to settle for half an hour, yielding relatively homogeneous beads measuring about 3 mm in diameter. After that, beads are filtered through sterile muslin cloth and aseptically air-dried depending on bead type. The obtained *Trichoderma* beads encapsulated with sodium alginate are kept at room temperature in plastic containers containing sterile distilled water (Bhai 2020).

3.8 Formulation Based on Press Mud

One of the best media for the growth of *Trichoderma* is press mud, a sugarcane-industry waste. An actively growing 9-day-old culture of *Trichoderma* isolated from potato broth forms the basis of the process. This culture is mixed with 120 kg of moist press mud. The resulting mixture is left in the shade and spread out on burlap to prevent drying while receiving adequate ventilation. After 25 days, the seed culture is prepared for field application by adding 8 tons of press mud; it then requires fermentation for another eight days at low temperature (Byakodiand Babu 2021).

3.9 Formulation Based on Coffee Husk

Trichoderma has been cultivated on a large scale using coffee husk, a waste product from the Karnataka coffee sector. Sawant and Sawant (1996) reported that application of such formulations was effective in preventing *Phytophthora* foot rot in black pepper. Likewise, Rosane et al. (2008) identified that *Trichoderma* can be cultured on inexpensive carriers including rice, corn, and wheat bran, especially the latter, which boosts spore yield (Cavalcante et al.2008.

3.10 Formulations Based on Banana Waste

A new substrate derived from the production of solid formulations with *Trichoderma* is banana trash. A technology that can incorporate banana waste, urea, rock phosphate, and a group of bio-control agents, including Bacillus polymyxa, Pleurotus sajor-caju and *Trichoderma* viride was developed by Balasubramanian et al. (2008). Chopped banana peels, pseudo-stems, and cores between 5 and 8 cm in length are layered and then add biocontrol chemicals, urea, and rock phosphate to the trench. These layers break down over the course of 45 days, following which they are cultivated for use in agriculture.

3.11 Dry Flowable Formulation

Dry flowable or water-dispersible granule formulations are environmentally-friendly and have easy application characteristics (Knowles 2008). Their powder form limits the uptake of moisture from the air, and hence the formulation remains stable for a longer period. Encapsulation is very crucial in the initial steps of making these formulations because it enhances the survival and performance of the product (Jin and Custis 2011). Encapsulation can be done using dry techniques or modified alginate microbeads (Dos Santos et al. 2015). Another approach is the production of a spray-dried flowable formulation containing polyacrylic acid, citric acid, and sodium bicarbonate, with polyvinyl alcohol as a binding agent and lecithin as a wetting agent (Oancea et al. 2016).

3.12 Saw Dust Formulation

Sawdust is recommended as a carrier material because it is an abundant natural product and an excellent absorber of organic material and moisture (Arora et al. 2001, Moses 2014). According to Arora et al. (2008), sawdust showed the highest activity as a substrate for microorganisms among the various carriers examined. Method of preparation: in heat-resistant polybags, one kilogram each of sawdust, soil, and molasses are combined. After the bags are securely sealed, they are sterilized for 15 minutes at 121°C under a pressure of 15 kg/cm^2. Chloramphenicol is added to *Trichoderma harzianum* formulations at a dose of 10 mg/kg. Using a sterile needle and syringe, 5 mL of a homogenized liquid culture of the biocontrol agent is inoculated into the bags following autoclaving. A small hole is made in the polybag for inoculation and afterward sealed with cello tape. The inoculated bags are incubated in a BOD incubator at 28 ± 2°C for 10–15 days with occasional gentle shaking to ensure uniform colonization of the biocontrol agent (Khan et al. 2011).

3.13 Major characteristics of a Trichoderma formulation

The following crucial qualities are necessary for a *Trichoderma* formulation to be successful (Jeyarajan and Nakkeeran 2000):

i. **High Rhizosphere Competence:** The formulation should be able to effectively compete with other microorganisms which live in the rhizosphere soil.

ii. **Strong Saprophytic Ability:** It must be capable of showing high saprophytic indices to grow on organic substrates.

iii. **Disease Management and Plant Growth Promotion:** It involves disease control of plant diseases as well as healthy growth of plants.

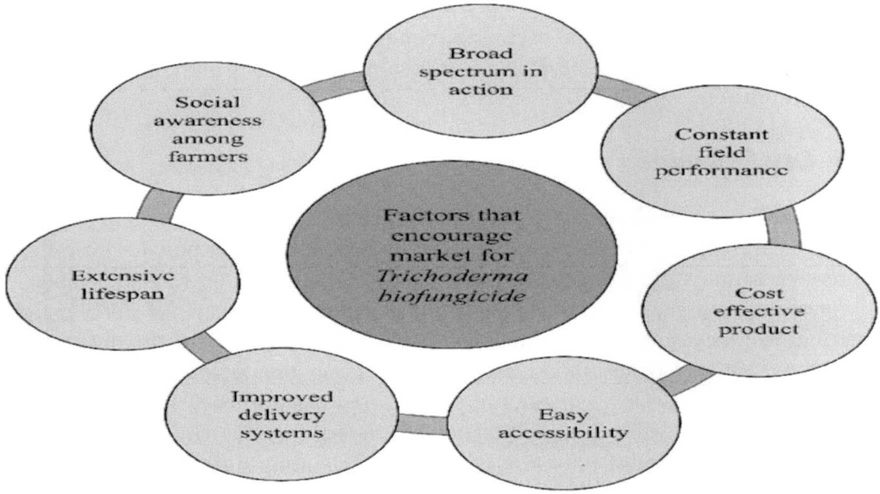

Figure 4: The following are the factors that have been depicted in the subsequent figures as influencing the market for *Trichoderma* bio-fungicide.

iv. **Mass Multiplication and Safety:** It needs to be effective for the production in large quantity and harmless for humans and atmosphere.

v. **Effective Action and Compatibility:** In addition to being resilient against stressors such as desiccation heat and UV radiation, it should offer improved efficacy that is compatible with other biocontrol agents and biofertilizers (Figure 4).

All these attributes guarantee that formulation of *Trichoderma*is able to meet the various agricultural conditions in the field.

4. Quality Control Parameters

Before recommending a *Trichoderma* formulation as a final product, specific quality parameters must be met to ensure its successful application:

i. **Colony Forming Units (CFUs):** The specimen formulation should be at least 2×10^6 Colony Forming Units per gram (solid formulation) or milliliter (liquid formulation).

ii. **Shelf Life:** Additionally, the product should have a four to six month shelf life.

iii. **Particle Size:** Talcum powder used in the formulation should also have particle size of not more than 500 microns.

iv. **Packaging:** The last product should constitute standard white polythene bags for packaging.

v. **Moisture Content:** The maximum allowable moisture content should be 20%.

vi. **Contamination Levels:** Contamination is allowed to a lowest level of 1*10^4 colony forming units per millilitre (liquid formulation) or gram (solid formulation).

vii. **Pathogen Elimination:** Additionally, food formulation mandates that food be free of harmful pathogens including Vibrio, Shigella, and Salmonella.

There are certain restrictions on evaluating the full potential of biocontrol agents due to the Central Insecticides Board (CIB) quality control standards for biopesticides. A central authority to supervise quality assurance is expected to be established by the Department of Agriculture. This body would assess the propagules' efficacy against plant diseases in addition to their viability in the formulation. To guarantee that farmers obtain dependable, high-quality products, these characteristics must be regularly assessed.

5. Methods of Application of *Trichoderma*

The effectiveness of *Trichoderma* in disease control depends on its distribution and establishment at the treatment site. Application methods vary from seed biopriming and seedling root soaking to foliar spraying, furrow or soil application, and covering plants with treated cloth. These application techniques ensure *Trichoderma* is introduced correctly for improved disease control.

5.1 Seed Treatment

One efficient, easy, and cost-effective strategy for biocontrol agent delivery to combat soil- and seed-borne diseases is seed coating with *Trichoderma* species (O'Callaghan 2016). In this way, only a small quantity of *Trichoderma* inoculum is placed on the seed coat immediately before sowing. It has also been used to treat seeds of different sizes as depicted by Rocha and colleagues (2019), Jogani and John, 2014. Following treatment, the spores germinate on the seed surface and then spread to the roots of the growing plantlets and their surrounding environment (rhizosphere) (Tewari and Mukhopadhyay 2001, Paravar et al. 2023). This approach works if *Trichoderma* can rapidly develop in the rhizosphere and also possess biocontrol characteristics. Papavizas (1985) showed that the biocontrol agent can help to control pathogens that cause seed rot and seedling diseases if it establishes in the soil surrounding the treatment site. Lamichhane et al. (2022) reported that the use of seed bioproducts led to increased crop productivity (21.2%), disease suppression (55.1%), plant mass (53.5%), increased emergence of seedlings (91.5%), and better seed germination (76%). This study also demonstrated the efficacy of a *T. viride* talc-based formulation (1% w/w) for controlling *F. oxysporum* f. sp. *ciceri*, which causes chickpea wilt, as well as improving physico-chemical qualities of the crop (Pradhan et al. 2022). Two *Trichoderma* strains, *T. harzianum* and *T. asperellum* (2 × 10^12 CFU per 100 kg of seed) used for seed priming enhanced seedling emergence, root and shoot elongation, and dry shoot biomass in a controlled greenhouse environment (Couto et al. 2021). In a study involving *T. koningii*, *T. viride*, and *T. asperellum* applied to peanut seeds, all three greatly enhanced both germination and vegetative growth rates (Sudana and

Raka 2020). Records indicate that *Trichoderma* seed treatment enhances yield and seedling vigor in several vegetables as well as cereal crops, including maize, tomato, and cucumber (Hossain and Akter 2020, Kim et al. 2020).

5.2 Seed Biopriming

An important type of seed treatment, called "seed biopriming," involves coating the seeds with beneficial microbes before sowing (Sarkar et al. 2021). Here, a mixture of *Trichoderma* formulations (6–10 g/kg of seed) is applied to seeds, which are then exposed to heat and moisture before sowing. This enhances germination of *Trichoderma* spores, which form a sheath around the seed coat, thereby protecting it from environmental stress factors and enabling rapid and uniform germination. Therefore, this technique enhances yield quality and improves plant stand (Swain et al. 2018, Rajput et al. 2019). Singh et al. (2016) reported that seeds treated with *T. asperellum* BHUT8 grew better than untreated seeds, as indicated by greater root and shoot length, higher dry and fresh weight, and more leaves. Testing of those *Trichoderma* strains in biopriming of rice seeds helped to increase crop yield, chlorophyll index, and germination rate (Swain et al. 2021). As highlighted by Degani and Dor (2021), maize seeds treated with priming techniques with *T. longibrachiatum* and *T. asperelloides* improved growth and manage the soil-borne disease by *Magnaporthiopsis maydis*. Additionally, when preventing and controlling important rice diseases such as *R. solani* and *S. oryzae*, seed priming with *T. hebeiensis* and *T. erinaceum* reduced the germination time, improved seedling growth and increased yield (Jambhulkar et al. 2022).

5.3 Root Dipping

Singh and Zaidi (2002) opined that root dipping is one of the effective methods of applying biocontrol agents such as *Trichoderma* on rice, brinjal, chili, tomato, and capsicum with a view to enhancing seedling vigor and mitigating the impact of disease. In this method, plant roots are immersed in a *Trichoderma* spore solution or a spore-laden slurry before planting. Chili, cucumber, tomato, and rice are some of the crops for which this treatment has been used to effectively manage the disease (Yang et al. 2022, Ferreira and Musumeci 2021).

5.4 Foliar Spraying/Wound Dressing

Temperature, humidity, and radiation are exogenous factors that affect the phyllosphere, or the surface of plant leaves, and their functionality. This change, in turn, influences the ability of biocontrol agents to survive and succeed. Several researchers have also noted that foliar spray of *T. viride* controls linseed blight under greenhouse conditions. In foliar application, others have used it and found efficacy for foliar application from research of Panwar et al. (2014), Karthikeyan et al. (2008) and shelf stability from Gade et al. (2009). To be used as a foliar disease-management measure, the stress tolerance of biocontrol agents must be evaluated (Köhl et al. 2019). When *Trichoderma asperellum* and *Trichoderma harzianum* were used as part of a *Trichoderma* consortium for phyllosphere treatment of blue Passiflora, the consortium prompted synergistic reactions whereby leaf size increased, chloroplasts were larger, and antioxidant activity in leaves was enhanced. Additionally, studies have shown that foliar application of *Trichoderma* to crops such as tomatoes, cucumbers, and pepper enhances plant health and contributes to the suppression of foliar diseases (Uddin et al. 2015, Dursun et al. 2010, Syam et al. 2021). This experiment revealed that T. asperellum foliar sprays enhanced plant height, leaf area, fruit weight, and total soluble solids compared to the control (Yao et al. 2023, Sehimet al. 2023).

5.5 Seed Material Treatment

Used on seeds, sugarcane setts, turmeric and ginger rhizomes, banana suckers, and potato tubers: mix 8–10 g of *Trichoderma* formulation with a liter of water and allow it to stand for ½ an hour. It has been revealed that *Trichoderma* species offer a means to manage Panama disease in bananas caused by *Fusarium oxysporum* f. sp. *cubense*. This has ensured that the incidence and severity of the disease have been reduced drastically (Cannon et al. 2022). Furthermore, evidence has shown that the use of *Trichoderma* before storing potato tubers greatly reduces incidence of dry rot (Tiwari et al. 2020, Xue et al. 2023).

5.6 Soil Application

Applying *Trichoderma* to soils is a powerful way to promote plant growth, improve soil health, and manage soil-borne diseases. *Trichoderma* preparation can be made by mixing 0.5–1 L of liquid *Trichoderma* culture or 1–2 kg of talcum-based suspension with 25–50 kg FYM and burying the mixture in sacks or sugarcane blades, where it is kept in the shade for two to three weeks. Depending on the method of distribution, it may be deposited; it is recommended to soak and remix every three to four days, preferably starting 15 days prior to planting. Two and a half kilograms per hectare of *T. viride* decreased disease incidence and increased crop yields, as supported by Jeyalakshmi et al. (2013). At an optimum concentration of 50 mL/L as a soil suspension, *T. harzianum* reduced root-rot disease incidence in grapevines by 80% (El-Mohamedy et al. 2010). Other evidence reveals that *T. harzianum* increases the population density of beneficial microbes and thus improves the soil microbial community, crop yield, and soil fertility (Liu et al. 2022, Xiao et al. 2023). Moreover, *T. harzianum* enhanced soil microbial community structure and microbial activity, and it reduced *Fusarium* wilt disease in cucumbers, as indicated by Wang et al. (2022). In combination with mycorrhizal consortia such as *T. harzianum* and arbuscular mycorrhizae, *Trichoderma* also enhances soil quality, root colonization, and growth of maize plants, particularly in low-zinc soils. This combined use enhanced organic matter, enzyme activity, and microbial-rich properties, and consequently enhanced the growth and yield of maize.

5.7 Cutting/Seedling Root Dip Application

Trichoderma treatment is safe for the environment because it enhances seed germination while discouraging pathogen germination. Basically, 1 liter of water is stirred with 20–25 g of *Trichoderma* sp. in powder form, or 5–10 mL in liquid form, and allowed to stand for 30 minutes. Seedling roots and cuttings are immersed in the resulting solution for nearly half an hour before transplantation. It has been used to reduce incidence of disease transmission through the soil. A study on olive cuttings treated with *T. harzianum* showed that the treatment significantly enhanced chlorophyll content, rooting percentage, and root and shoot length. Moreover, the percentage of fungal infections observed in cuttings treated with *Trichoderma* was significantly lower than in the control (Yedidia et al. 2001).

5.8 Nursery Bed Treatment

A method of reducing the number of seedlings with soil diseases is to use *Trichoderma* in nursery beds. By this process, mix 500 g of a *Trichoderma* formulation with 10–20 kg of vermicast or well-decomposed FYM. One acre should be sprayed at night. Moreover, for *Trichoderma* to grow well in such soils, the soil moisture at the time of spraying should be sufficient. Hence, a study shows that incorporation of *Trichoderma*-enriched compost in nursery beds enhances germination and yield of tomato seeds (Kumhar et al. 2022). Furthermore, this treatment was found to enhance enzyme activity and soil fertility, which are fundamental to improved soil health. To reduce the incidence of soil-borne infections and improve seedling growth of crops such as chili, tomato, and brinjal, the use of *Trichoderma*-modified nursery beds was another area of focus of the study. Also, the application of *Trichoderma* enhanced nutrient content and soil enzyme activity that leads to better soil health (Yu et al. 2023). Moreover, there was enhanced seedling vigor and nutrient uptake, higher chlorophyll content, and reduced damping-off disease incidence in tomato seedlings biocontrolled by *Trichoderma* treatment of nursery beds (Pascual et al. 2018). This, in turn, was beneficial for the health of the plant as a whole.

5.9 Soil Drenching

By this procedure, 200 liters of fresh water and 1–2 kg of *Trichoderma* formulations are used and applied directly to the soil close to the base of plants. This method involves the use of *Trichoderma* in the soil; it acts as a defense against soil pathogens that cause damping-off. Therefore, during fertilizer application the soil should be at its recommended moisture level. In pepper, tomato, and maize, *Trichoderma* applied by soil drenching or by seed soaking has been found to enhance plant growth and yield (Van Jaarsveld et al. 2021, Lanzuiseet al. 2022).

5.10 Other Applications of Trichoderma

Trichoderma functions in a host of capacities because of its multifaceted applicability across various industrial sectors. For bioremediation, degradation of agricultural residues, wine and beer production, food preservation, textile dyeing, the paper and pulp industries, and bioethanol production from agricultural residues, this fungus has vast potential. It degrades part of the plant cell wall, allowing the various lytic enzymes of *Trichoderma*, including cellulase, hemicellulase, and pectinase, to be useful in the brewing and animal feed sectors. This process enhances the digestibility and nutritional quality of feed, which has a positive effect on animal growth and milk yield (Galante et al. 1998). Moreover, *Trichoderma* has been found to alter fiber characteristics of pulp and to decrease lignin content (Galante et al. 1998). Its high versatility makes *Trichoderma* applicable across a wide range of industries. In bioremediation, biodegradation of agricultural waste, winemaking, brewing, as a food digestive agent, textile processing, the paper and pulp industry, and in conversion of farm waste to bioethanol, this fungus is a useful agent. Due to the microorganism's ability to partially hydrolyze plant cell walls, *Trichoderma*'s lytic enzymes, including cellulase, hemicellulase, and pectinase, could help the brewing and animal feed industries. This procedure enhances food and nutrient digestibility and utilization, hence benefiting animal growth and milk yield (Galante et al. 1998). Fiber alteration and lignin degradation in pulp by *Trichoderma* may also be applied by the paper industry (Galante et al. 1998).

6. Important Issues and Future Directions of Biopesticides

Biopesticides, especially biofungicides, are significant because they pose low hazards, have targeted activity, cause less environmental pollution, and carry a lower likelihood of pathogen resistance. The recent increase in biopesticide products is due to consumer demand for disease-free, safe food products and market demand for such products. However, the global market for biopesticides is relatively small, especially in some nations, including India. Greater effort and improvements in technology must be pursued to enhance the efficiency of commercial biopesticides and to overcome current challenges and limitations (Sachdev et al. 2016). Most of these are requirements for PE adoption in conventional farming practices: better targeted delivery systems, lower costs, certified field performance, greater shelf life, ease of acquisition, and increased social awareness among farmers of biopesticides. It also leads to hit-and-trial development and calls for simplification of registration and regulatory processes for their integration (Mishra et al. 2015, Chakraborty et al. 2023) (Figure 3).

6.1 Biological Constraints in the Development of Biopesticides Derived from Trichoderma

The development of *Trichoderma*-based biopesticides faces several constraints:

i. **Lack of Screening Methods:** Lack of adequate protocols hinder the identification of promising *Trichoderma* strains.
ii. **Limited Understanding of Ecology**: At present, there is comparatively very limited information available regarding the ecological aspects of *Trichoderma* and its antagonism with other plant pathogens.

iii. **Technological Limitations:** This implies that the improvement of fermentation technologies and effective high volume production systems are still in their infancy.

iv. **Variability in Performance:** The performance of products is inconsistent, and they often have a very short shelf life.

v. **Lack of Patent Protection:** Absence of strong patent laws discourages investment in research and development of biopesticides.

vi. **High Registration Costs:** The expenses associated with the registration of biopesticides are typically very high.

vii. **Need for Training and Awareness:** Enhanced training and educational materials for farmers and industry personnel are necessary.

viii. **Lack of a Multidisciplinary Approach:** Currently, development of biopesticides is fragmented and lacks a clearly defined, multidisciplinary framework in terms of scope and process.

ix. **Technological Barriers:** Certain technological considerations need to be addressed to enhance product functionality.

x. **Market Entry and Size:** The main problem involves getting new products to market, and the limited number of customers may be a significant barrier (Topolovec-Pintari 2019).

Short shelf life of *Trichoderma* biopesticides is also a major challenge in their development and marketability (Kumar et al. 2014).

7. Conclusion

Trichoderma species have become prominent biocontrol agents for sustainable agriculture and employ multiple approaches to benefit plants and combat pathogens. These include mycoparasitism (parasitizing destructive fungi), resource competition for nutrients and space that can exclude pathogenic forms, and production of inhibitory compounds and enzymes that suppress numerous plant pathogens. Furthermore, when effectively applied, *Trichoderma* induces systemic resistance in plants, strengthening their natural immunity and allowing them to withstand both biotic and abiotic stresses. In addition to pathogen suppression, *Trichoderma* is beneficial for plant growth and soil health by altering nutrients such as phosphorus, making them more soluble. It enhances nutrient availability, promoting yield, and also improves nitrogen availability. Positive interactions with soil microbes enhance soil fertility and contribute to a healthier environment for agricultural practices. Some advanced production processes, such as solid-state and submerged fermentations, have eased the marketing of *Trichoderma*-based bioformulations. In IPM tactics, specific formulations based on crop type and environmental factors have been effective in reducing the use of chemical fertilizers and pesticides. To cater to diverse farming systems, *Trichoderma* can be formulated into liquid suspensions, powders, and/or granules. Issues such as deviations in field performance, differences in efficacy across strains, and the impact of environmental conditions on the effectiveness of applied methods pose challenges.

There has been emphasis on how to overcome these limitations, which include strain improvement, genetic modification, and better formulation techniques. Further, the high acceptance and success maybe explained by enhanced regulatory support and farmers' awareness of the practice. In other words, *Trichoderma* is part of the shift in agriculture toward natural and organic methods of production. They advocate for international goal-oriented activities in food production and environmental conservation by providing a natural, cheap and environmentally friendly alternative to agrochemicals. With more research, innovation,and comparative education, *Trichoderma* has the potential to revolutionize agriculture as a practice and to contribute to improved crops, soils and ecosystem all over the world.

References

Abdel-Fattah, G. M. (2021). Biodegradation of environmental pollutants by *Trichoderma* species: A review. *Applied Microbiology and Biotechnology,* 105(2): 569–585. https://doi.org/10.1007/s00253-020-10843-0.

Abd-Elsalam, K. A., Yassin, M. A., Moslem, M. A., Bahkali, A. H., de Wit, P. J., McKenzie, E. H. et al. (2010). Culture collections, the new herbaria for fungal pathogens. *Fungal Diversity,* 45: 21–32. https://doi.org/10.1007/s13225-010-0053-1.

Abdul-Halim, A. M., Shivanand, P., Krishnamoorthy, S., and Taha, H. (2023). A review on the biological properties of *Trichoderma* spp. as a prospective biocontrol agent and biofertilizer. *Journal of Applied Biology & Biotechnology*, 11(5): 34–46.

Abdul Malik, N. A., Kumar, I. S., and Nadarajah, K. (2020). Elicitor and receptor molecules: Orchestrators of plant defense and immunity. *International Journal of Molecular Sciences,* 21(3). https://doi.org/10.3390/ijms21030963.

Al-Ani, L. K. T. (2018). *Trichoderma*: beneficial role in sustainable agriculture by plant disease management. *Plant Microbiome: Stress Response*, 105–126.

Alfiky, A., and Weisskopf, L. (2020). Deciphering *Trichoderma*–plant–pathogen interactions for better development of biocontrol applications. *Journal of Fungi,* 7(1). https://doi.org/10.3390/jof7010061

Aljbory, Z., & Chen, M. S. (2018). Indirect plant defense against insect herbivores: A review. *Journal of Insect Science,* 25: 2–23.

Alshammari, W., Bairum, R., Sulieman, A. M., Alshammari, N., and Elamin, H. (2024). *In vitro* and *in vivo* study of antagonistic and biocontrol of *Trichoderma harzianum* strains against wood decay pathogens. *Polish Journal of Environmental Studies,* 33(1): 515–521. https://doi.org/10.15244/pjoes/172043.

Altomare, C., Norvell, W. A., Björkman, T., and Harman, G. E. (1999). Solubilization of phosphates and micronutrients by the plant-growth-promoting and biocontrol fungus *Trichoderma harzianum Rifai* 1295-22. *Applied and Environmental Microbiology,* 65(7): 2926–2933. https://doi.org/10.1128/aem.65.7.2926-2933.1999.

Amira, R. D., Roshanida, A., Rosli, M., Zahrah, M. S. F., Anuar, J. M., and Adha, C. N. (2011). Bioconversion of empty fruit bunches (EFB) and palm oil mill effluent (POME) into compost using *Trichoderma virens*. *African Journal of Biotechnology,* 10: 18775–18780.

Amoo, A. E., Olanrewaju, O. S., Babalola, O. O., Ajilogba, C. F., Chukwuneme, C. F., Ojuederie, O. B. et al. (2023). The functionality of plant-microbe interactions in disease suppression. *Journal of King Saud University – Science,* 35(8): 102893. https://doi.org/10.1016/j.jksus.2023.102893.

Ananda, N., Vadlani, P. V., and Prasad, P. V. V. (2011). Evaluation of drought and heat-stressed grain sorghum (*Sorghum bicolor*) for ethanol production. *Industrial Crops and Products,* 33: 779–782.

Anith, K., Vaishakhi, A., Viswanathan, A., Varkey, S., and Aswini, S. (2017). Population dynamics and efficiency of coconut water-based liquid formulation of *Pseudomonas fluorescens* AMB-8. *Journal of Tropical Agriculture,* 54: 184.

Arora, N. K., Kumar, V., and Maheshwari, D. K. (2001). Constraints, development, and future of the inoculants with special reference to *Rhizobial inoculants*. pp. 241–245. *In:* Maheshwari, D. K., and Dubey, R. C. (Eds.). *Innovative Approaches in Microbiology*. Bishen Singh Mahendra Pal Singh.

Arora, N., Khare, E., Naraian, R., and Maheshwari, D. (2008). Sawdust as a superior carrier for production of multipurpose bioinoculant using plant growth-promoting rhizobial and pseudomonad strains and their impact on the productivity of *Trifolium repens*. *Current Science,* 95: 90–94.

Asad, S. A. (2022). Mechanisms of action and biocontrol potential of *Trichoderma* against fungal plant diseases – A review. *Ecological Complexity,* 49: 100978. https://doi.org/10.1016/j.ecocom.2021.100978.

Atanasova, L., Knox, B. P., Kubicek, C. P., Druzhinina, I. S., and Baker, S. E. (2013). The polyketide synthase gene *pks4* of *Trichodermareesei* provides pigmentation and stress resistance. *Eukaryotic Cell,* 12: 1499–1508.

Ayub, J., Tahir, A., Iqbal, U., Ayub, H., Hyder, M. Z., Kiyani, A. et al. (2023). Effective biological control of lentil (*Lens culinaris*) *Fusarium* wilt and plant growth promotion through native rhizobacteria. *Physiological and Molecular Plant Pathology,* 129: 102203. https://doi.org/10.1016/j.pmpp.2023.102203.

Bae, H., Sicher, R. C., Kim, M. S., Kim, S. H., Strem, M. D., Melnick, R. L. et al. (2009). The beneficial endophyte *Trichoderma hamatum* isolate DIS 219b promotes growth and delays the onset of the drought response in *Theobroma cacao*. *Journal of Experimental Botany,* 60: 3279–3295.

Bahadur, A., and Dutta, P. (2022). *Trichoderma* spp: Their impact in crops diseases management. *IntechOpen*. https://doi.org/10.5772/intechopen.101846.

Balakrishnan, S., Parthasarathy, S., Kamalakannan, A., Kuppusamy, K., and Gopalakrishnan, C. (2017). A novel method to develop paste formulation of *Trichoderma harzianum*. *International Journal of Current Microbiology and Applied Sciences,* 6: 3286–3293.

Balasubramanian, C., Udaysoorian, P., Prabhu, C., and Kumar, G. (2008). Enriched compost for yield and quality enhancement in sugarcane. *Journal of Ecobiology,* 22: 173–176.

Bashan, Y., de-Bashan, L. E., Prabhu, S., and Hernandez, J.-P. (2014). Advances in plant growth-promoting bacterial inoculant technology: Formulations and practical perspectives (1998–2013). *Plant and Soil,* 378(1): 1–33. https://doi.org/10.1007/s11104-013-1952-x.

Batta, Y. (2007). Control of postharvest diseases of fruit with an invert emulsion formulation of *Trichoderma harzianum Rifai*. *Postharvest Biology and Technology,* 43(2): 143–150. https://doi.org/10.1016/j.postharvbio.2006.08.012.

Bejarano, A., and Puopolo, G. (2020). Bioformulation of microbial biocontrol agents for sustainable agriculture. pp. 275–293. *In: How Research Can Stimulate the Development of Commercial Biological Control Against Plant Diseases*. Springer, Berlin/Heidelberg, Germany.

Benítez, T., Rincón, A. M., Limón, M. C., and Codón, A. C. (2004). Biocontrol mechanisms of *Trichoderma* strains. *International Microbiology,* 7(4): 249–260. https://doi.org/10.1007/s10123-004-0151-5.

Bhai, R. (2020). Preservation and long-term storage of *Trichoderma* spp. by sodium alginate encapsulation. *Journal of Plant Crops,* 48(1): 36–44.

Bigirimana, J., De Meyer, G., Poppe, J., Elad, Y., and Höfte, M. (1997). Induction of systemic resistance on bean (*Phaseolus vulgaris*) by *Trichoderma harzianum*. *Mededelingen van de FaculteitLandbouwwetenschappen Universiteit Gent,* 62(3b): 1001–1007.

Boat Bedine, M. A., Iacomi, B., Tchameni, S. N., Sameza, M. L., and Fekam, F. B. (2022). Harnessing the phosphate-solubilising ability of *Trichoderma* strains to improve plant growth, phosphorus uptake, and photosynthetic pigment contents in common bean (*Phaseolus vulgaris*). *Biocatalysis and Agricultural Biotechnology,* 45: 102510. https://doi.org/10.1016/j.bcab.2022.102510.

Boccardo, N. A., Segretin, M. E., Hernandez, I., Mirkin, F. G., Chacón, O., López, Y. et al. (2019). Expression of pathogenesis-related proteins in transplastomic tobacco plants confers resistance to filamentous pathogens under field trials. *Scientific Reports,* 9(1): 2791. https://doi.org/10.1038/s41598-019-39690-w.

Bononi, L., Chiaramonte, J. B., Pansa, C. C., Moitinho, M. A., and Melo, I. S. (2020). Phosphorus-solubilizing *Trichoderma* spp. from Amazon soils improve soybean plant growth. *Scientific Reports,* 10(1): 1–13. https://doi.org/10.1038/s41598-020-59793-8.

Byakodi, A., and Babu, B. S. (2021). Experimental studies on the effects of selected species of fungi on aerobic composting of sugar pressmud. *Environmental Research, Engineering and Management,* 77(3): 74–85. https://doi.org/10.5755/j01.erem.77.3.26511.

Cannon, S., Kay, W., Kilaru, S., Schuster, M., Gurr, S. J., and Steinberg, G. (2022). Multi-site fungicides suppress banana Panama disease, caused by *Fusarium oxysporum* f. sp. cubense Tropical Race 4. *PLoS Pathogens,* 18(8): e1010860. https://doi.org/10.1371/journal.ppat.1010860.

Cardoza, R. E., McCormick, S. P., and Izquierdo-Bueno, I. (2022). Identification of polyketide synthase genes required for aspinolide biosynthesis in *Trichoderma arundinaceum*. *Applied Microbiology and Biotechnology,* 106(17): 7153–7171. https://doi.org/10.1007/s00253-022-12182-9.

Casas-Flores, S., Rios-Momberg, M., Bibbins, M., Ponce-Noyola, P., and Herrera-Estrella, A. (2004). BLR-1 and BLR-2, key regulatory elements of photo-conidiation and mycelial growth in *Trichoderma atroviride*. *Microbiology,* 150(11): 3561–3569.

Cavalcante, R. S., Lima, H. L. S., Pinto, G. A. S., Gava, C. A. T., and Rodrigues, S. (2008). Effect of moisture on *Trichoderma* conidia production on corn and wheat bran by solid-state fermentation. *Food and Bioprocess Technology,* 1(1): 100–104. https://doi.org/10.1007/s11947-007-0003-z.

Chakraborty, N., Mitra, R., Pal, S., Ganguly, R., Acharya, K., Minkina, T. et al. (2023). Biopesticide consumption in India: Insights into the current trends. *Agriculture,* 13(3): 557. https://doi.org/10.3390/agriculture13030557.

Chammem, H., Nesler, A., and Pertot, I. (2021). Wood pellets as carriers of conidia of *Trichoderma atroviride* SC1 for soil application. *Fungal Biology,* 125(12): 989–998. https://doi.org/10.1016/j.funbio.2021.09.007.

Chang, Y. C., Chang, Y. C., and Baker, R. (1986). Increased growth of plants in the presence of the biological control agent *Trichoderma harzianum*. *Plant Disease,* 70(2): 145–148.

Chaube, H. S., Mishra, D. S., and Varshney, S. (2003). Biocontrol of plant pathogens by fungal antagonists: A historical background, present status, and future prospects. *Annual Review of Plant Pathology,* 2: 1–42.

Dos Santos, G., Locatelli, G., Coêlho, D., Botelho, P., De Amorim, M., De Vasconcelos, T. et al. (2015). Factorial design, preparation, and characterization of new beads formed from alginate, polyphosphate, and glycerol gelling solution for microorganism microencapsulation. *Journal of Sol-Gel Science and Technology,* 75(2): 345–352.

Dursun, A., Ekinci, M., and Dönmez, M. F. (2010). Effects of foliar application of plant growth-promoting bacterium on chemical contents, yield, and growth of tomato (*Lycopersicon esculentum* L.) and cucumber (*Cucumis sativus* L.). *Pakistan Journal of Botany,* 42: 3349–3356.

Dutta, P., Deb, L., and Pandey, A. K. (2022). *Trichoderma*—From lab bench to field application: Looking back over 50 years. *Frontiers in Agronomy,* 4: 932839. https://doi.org/10.3389/fagro.2022.932839.

El-Mohamedy, R. S., Ziedan, E., and Abdalla, A. (2010). Biological soil treatment with *Trichoderma harzianum* to control root rot disease of grapevine (*Vitis vinifera* L.) in newly reclaimed lands in Nobaria province. *Archives of Phytopathology and Plant Protection,* 43(1): 73–87.

Emmanuel, O. C., and Babalola, O. O. (2020). Productivity and quality of horticultural crops through co-inoculation of arbuscular mycorrhizal fungi and plant growth-promoting bacteria. *Microbiological Research,* 239: 126569. https://doi.org/10.1016/j.micres.2020.126569.

Esparza-Reynoso, A., Mubeen, M., Zheng, H., Sohail, M. A., Shakeel, Q., Solanki, M. K. et al. (2022). *Trichoderma* spp. genes involved in the biocontrol activity against *Rhizoctonia solani*. *Frontiers in Microbiology,* 13. https://doi.org/10.3389/fmicb.2022.884469.

Esparza-Reynoso, S., Pelagio-Flores, R., and López-Bucio, J. (2019). Mechanism of plant immunity triggered by *Trichoderma*. pp. 57–73. *In: New and future developments in microbial biotechnology and bioengineering*. https://doi.org/10.1016/B978-0-12-819453-9.00003-9.

Fernández, J., and Pablo, J. (2022). Peptaibol production and characterization from *Trichoderma asperellum* and their action as biofungicide. *Journal of Fungi,* 8(10), 1037. https://doi.org/10.3390/jof8101037.

Ferreira, F. V., and Musumeci, M. A. (2021). *Trichoderma* as a biological control agent: Scope and prospects to improve efficacy. *World Journal of Microbiology and Biotechnology,* 37: 90.

Frisvad, J. C., Nielsen, K. F., and Samson, R. A. (2011). Fungi producing mycotoxins and other secondary metabolites. *Fungal Diversity,* 50(1): 1–2. https://doi.org/10.1007/s13225-010-0024-6.

Gade, R., Wardhe, S., and Armarker, S. (2009). Shelf life study of *Trichoderma* spp. in different carrier materials. *Journal of Maharashtra Agricultural Universities,* 34: 181–182.

Gajera, H., Hirpara, D. G., Savaliya, D. D., and Golakiya, B. (2020). Extracellular metabolomics of *Trichoderma*biocontroller for antifungal action to restrain *Rhizoctonia solani* Kuhn in cotton. *Physiological and Molecular Plant Pathology,* 112: 101547. https://doi.org/10.1016/j.pmpp.2020.101547.

Galante, Y., De Conti, A., and Monteverdi, R. (1998). *Trichoderma* and *Gliocladium*. In Trichoderma & Gliocladium: Enzymes, biological control and commercial applications (pp. 1–15). Taylor & Francis.

Geremia, R. A., Goldman, G. H., Jacobs, D., Ardets, W., Vila, S. B., Van Montagu, M. et al. (1993). Molecular characterization of the proteinase-encoding gene, *prb1*, related to mycoparasitism by *Trichoderma harzianum*. *Molecular Microbiology,* 8: 603–613.

Ghorbanpour, M., Omidvari, M., Abbaszadeh-Dahaji, P., Omidvar, R., and Kariman, K. (2018). Mechanisms underlying the protective effects of beneficial fungi against plant diseases. *Biological Control,* 117: 147–157. https://doi.org/10.1016/j.biocontrol.2017.11.006.

Ghosh, S. K., Bera, T., and Chakrabarty, A. M. (2020). Microbial siderophore—A boon to agricultural sciences. *Biological Control,* 144: 104214. https://doi.org/10.1016/j.biocontrol.2020.104214.

Gopalakrishnan, S., Sathya, A., Vijayabharathi, R., and Srinivas, V. (2016). Formulations of plant growth-promoting microbes for field applications. *Microbial Inoculants for Sustainable Agriculture Production: Functional Applications,* 2: 239–251.

Goughenour, K. D., Whalin, J., Slot, J. C., and Rappleye, C. A. (2021). Diversification of fungal chitinases and their functional differentiation in *Histoplasma capsulatum*. *Molecular Biology and Evolution,* 38(4): 1339–1355. https://doi.org/10.1093/molbev/msaa293.

Gueye, N., Kumar, G. K., Ndiaye, M., Sall, S. Y. D., Ndiaye, M. A. F., Diop, T. A. et al. (2020). Factors affecting the chitinase activity of *Trichoderma asperellum* isolated from agricultural field soils. *Journal of Applied Biology and Biotechnology,* 8(2): 41–44. https://doi.org/10.7324/JABB.2020.80207.

Gutter, Y. (1957). Effect of light on sporulation of *Trichoderma viride*. *Bulletin of the Research Council of Israel Section D,* 5: 273–286.

Guzmán-Guzmán, P., Kumar, A., de Los Santos-Villalobos, S., Parra-Cota, F. I., Orozco-Mosqueda, M. D. C., Fadiji, A. E. et al. (2023). *Trichoderma* species: Our best fungal allies in the biocontrol of plant diseases – A review. *Plants,* 12(3): 432. https://doi.org/10.3390/plants12030432.

Guzmán-Guzmán, P., María, D. P., Vianey, O., and Alfredo, H. (2019). *Trichoderma* species: Versatile plant symbionts. *Phytopathology,* 109(1): 6–16.

Haddadin, M. S., Haddadin, J., Arabiyat, O. I., and Hattar, B. (2009). Biological conversion of olive pomace into compost by using *Trichoderma harzianum* and *Phanerochaetechrysosporium*. *Bioresource Technology,* 100: 4773–4782.

Halifu, S., Deng, X., Song, X., and Song, R. (2019). Effects of two *Trichoderma* strains on plant growth, rhizosphere soil nutrients, and fungal community of *Pinus sylvestris* var. *mongolica* annual seedlings. *Forests,* 10(9): 758. https://doi.org/10.3390/f10090758.

Harkki, A., Uusitalo, J., Bailey, M., Penttilä, M., and Knowles, J. K. (1989). A novel fungal expression system: Secretion of active calf chymosin from the filamentous fungus *Trichodermareesei*. *Biotechnology,* 7: 596–603.

Harman, G. E., Howell, C. R., Viterbo, A., Chet, I., and Lorito, M. (2021). *Trichoderma* species—Opportunistic, avirulent plant symbionts. *Nature Reviews Microbiology,* 2: 43–56.

Hartman, G. L., Pawlowski, M. L., Chang, H.-X., and Hill, C. B. (2016). Successful technologies and approaches used to develop and manage resistance against crop diseases and pests. pp. 43–66. *In:* Madramootoo, C. (Ed.). *Emerging Technologies for Promoting Food Security*. Woodhead Publishing.

Haseeb, M., Shafiq, M., Noman, F., Abbas, S. A. A. A., and Ali, N. (2023). Structure and potential applications of bacterial siderophores. pp. 159–175. *In: Bacterial secondary metabolites*. https://doi.org/10.1016/B978-0-323-95251-4.00007-7.

Hermosa, R., Rubio, M. B., Cardoza, R. E., Nicolás, C., Monte, E., and Gutiérrez, S. (2013). The contribution of *Trichoderma* to balancing the costs of plant growth and defense. *Int. Microbiol.*, 16(2): 69–80.

Herrera, W., Valbuena, O., and Pavone-Maniscalco, D. (2020). Formulation of *Trichoderma asperellum* TV190 for biological control of *Rhizoctonia solani* on corn seedlings. *Egyptian Journal of Biological Pest Control,* 30: 1–8.

Hossain, S., and Akter, F. (2020). Effects of *Trichoderma*-enriched biofertilizer and farmyard manure on the growth and yield of brinjal (*Solanum melongena* L.). *Dhaka University Journal of Biological Sciences,* 29: 1–8.

Hough, B., Steenkamp, E., Wingfield, B., and Read, D. (2023). Fungal viruses unveiled: A comprehensive review of mycoviruses. *Viruses,* 15(5): 1202. https://doi.org/10.3390/v15051202.

Inbar, J., and Chet, I. (1992). Biomimics of fungal cell-cell recognition by use of lectin-coated nylon fibers. *Journal of Bacteriology,* 174: 1055–1059.

Intana, W., Yusakul, G., Wanchanthuek, P., Wongkaew, S., and Sunpapao, A. (2021). Volatile organic compounds released by *Trichoderma asperellum* T76-14 suppress postharvest fruit rot in muskmelon caused by *Fusarium incarnatum. Plants,* 10(4): 629.

Isabel, M., and Rubén, S. (2022). Plant defensive responses triggered by *Trichoderma* spp. as tools to face stressful conditions. *Horticulturae,* 8(12): 1181. https://doi.org/10.3390/horticulturae8121181.

Jagodzik, P., Tajdel-Zielińska, M., Cieśla, A., Marczak, M., and Ludwikow, A. (2018). Mitogen-activated protein kinase cascades in plant hormone signalling. *Frontiers in Plant Science,* 9: 1387.

Jambhulkar, P. P., Raja, M., Singh, B., Katoch, S., Kumar, S., and Sharma, P. (2022). Potential native *Trichoderma* strains against *Fusarium verticillioides* causing post-flowering stalk rot in winter maize. *Crop Protection,* 152: 105838.

Jaroszuk-Ściseł, J., Tyśkiewicz, R., Nowak, A., Ozimek, E., Majewska, M., Hanaka, A. et al. (2019). Phytohormones (auxin, gibberellin) and ACC deaminase *in vitro* synthesized by the mycoparasitic *Trichoderma DEMTkZ3A0* strain and changes in the level of auxin and plant resistance markers in wheat seedlings inoculated with this strain conidia. *International Journal of Molecular Sciences,* 20(19). https://doi.org/10.3390/ijms20194923.

Jeyalakshmi, C., Rettinassababady, C., and Nema, S. (2013). Integrated management of sesame diseases. *Journal of Biopesticides,* 6: 68.

Jeyarajan, R. (2006). Prospects of indigenous mass production and formulation of *Trichoderma*. pp. 74–80. *In*: Proceedings of the Group Meeting on Antagonistic Organisms in Plant Disease Management. Project Directorate of Biological Control, Indian Council of Agricultural Research.

Jeyarajan, R., and Nakkeeran, S. (2000). Exploitation of microorganisms and viruses as biocontrol agents for crop disease management. *In*: Dwivedi, B. K. (Ed.). *Biocontrol Potential and its Exploitation in Sustainable Agriculture: Crop Diseases, Weeds, and Nematodes* (pp. 95–116). Springer.

Jeyarajan, R., Ramakrishnan, G., Dinakaran, D., and Sridar, R. (1994). Development of products *Trichoderma viride* and *Bacillus subtilis* for biocontrol of root rot diseases. In *Biotechnology in India* (pp. 25–36). Bioved Research Society.

Jha, C. K., Sharma, P., Shukla, A., Parmar, P., Patel, R., Goswami, D. et al. (2021). Microbial enzyme, 1-aminocyclopropane-1-carboxylic acid (ACC) deaminase: An elixir for plants under stress. *Physiological and Molecular Plant Pathology,* 115: 101664. https://doi.org/10.1016/j.pmpp.2021.101664.

Jin, X., and Custis, D. (2011). Microencapsulating aerial conidia of *Trichoderma harzianum* through spray drying at elevated temperatures. *Biological Control,* 56(2): 202–208.

Jin, X., Hayes, C. K., and Harman, G. E. (1992). Principles in the development of biological control systems employing *Trichoderma* species against soil-borne plant pathogenic fungi. In *Frontiers in Industrial Mycology* (pp. 174–195). Springer, Berlin/Heidelberg, Germany.

Jogani, V., and John, P. (2014). Evaluation of different application methods of *Trichoderma harzianum* (Rifai) against *Fusarium* wilt of tomato. *Crop Research,* 48: 76–79.

Jones, J., and Dangl, J. (2006). The plant immune system. *Nature,* 444(7117): 323–329. https://doi.org/10.1038/nature05286.

Juan, C. A., Plou, F. J., and Pérez-Lebeña, E. (2021). The chemistry of reactive oxygen species (ROS) revisited: Outlining their role in biological macromolecules (DNA, lipids, and proteins) and induced pathologies. *International Journal of Molecular Sciences,* 22(9): 4642. https://doi.org/10.3390/ijms22094642.

Junaid, J. M., Dar, N. A., Bhat, T. A., Bhat, A. H., and Bhat, M. A. (2013). Commercial biocontrol agents and their mechanism of action in the management of plant pathogens. *International Journal of Modern Plant & Animal Science,* 1(12): 39–57.

Kaczmarek, J., and Jedryczka, M. (2011). Characterization of two coexisting pathogen populations of *Leptosphaeria* spp., the cause of stem canker of brassicas. *Acta Agrobotanica,* 64(2): 1–8.

Karthikeyan, M., Radhika, K., Bhaskaran, R., Mathiyazhagan, S., Sandosskumar, R., Velazhahan, R. et al. (2008). Biological control of onion leaf blight disease by bulb and foliar application of powder formulation of antagonist mixture. *Archives of Phytopathology and Plant Protection,* 41(5): 407–417.

Kaur, R., Kalia, A., Lore, J. S., Kaur, A., Yadav, I., Sharma, P. et al. (2021). *Trichoderma* sp. endochitinase and -1,3-glucanase impede *Rhizoctonia solani* growth independently, and their combined use does not enhance impediment. *Plant Pathology,* 70(6): 1388–1396.

Keswani, C., Bisen, K., Singh, V., Sarma, B. K., and Singh, H. B. (2016). Formulation technology of biocontrol agents: Present status and future prospects. pp. 35–52. *In: Bioformulations: For Sustainable Agriculture* . Springer, New Delhi, India.

Khan, M. R., Majid, S., Mohidin, F. A., and Khan, N. (2011). A new bioprocess to produce low-cost powder formulations of biocontrol bacteria and fungi to control fusarial wilt and root-knot nematode of pulses. *Biological Control,* 59(2): 130–140.

Kim, K., Heo, Y. M., Jang, S., Lee, H., Kwon, S.-L., Park, M. S. et al. (2020). Diversity of *Trichoderma* spp. in marine environments and their biological potential for sustainable industrial applications. *Sustainability,* 12(10): 4327. https://doi.org/10.3390/su12104327.

Kim, S. H., Lee, Y., Balaraju, K., and Jeon, Y. (2023). Evaluation of *Trichodermaatroviride* and *Trichoderma longi brachiatum* as biocontrol agents in controlling red pepper anthracnose in Korea. *Frontiers in Plant Science,* 14: 1201875. https://doi.org/10.3389/fpls.2023.1201875.

Knowles, A. (2008). Recent developments of safer formulations of agrochemicals. *Environmentalist,* 28(1): 35–44.

Köhl, J., Kolnaar, R., and Ravensberg, W. J. (2019). Mode of action of microbial biological control agents against plant diseases: Relevance beyond efficacy. *Frontiers in Plant Science,* 10: 845. https://doi.org/10.3389/fpls.2019.00845.

Komala, G., Madhavi, G. B., and Nath, R. (2019). Shelf life studies of different formulations of *Trichoderma harzianum. Plant Cell Biotechnology and Molecular Biology,* 20(1): 1100–1105.

Kubiak, A., Pilarska, A. A., and Niewiadomska, A. (2022). Fungi of the *Trichoderma* genus: Future perspectives of benefits in sustainable agriculture. *Applied Sciences,* 13(11): 6434. https://doi.org/10.3390/app13116434.

Kubicek, C. P., Herrera-Estrella, A., Seidl-Seiboth, V., Martinez, D. A., Druzhinina, I. S., Thon, M. et al. (2011). Comparative genome sequence analysis underscores mycoparasitism as the ancestral lifestyle of *Trichoderma. Genome Biology,* 12(5): R40.

Kullnig-Gradinger, C. M., Zeilinger, S., and Schwarz, P. (2002). Mycoparasitism of *Trichoderma*: A potential tool for biocontrol. *Journal of Biotechnology,* 98(1): 41–49. https://doi.org/10.1016/S0168-1656(02)00014-1.

Kumar, S., Thakur, M., and Rani, A. (2014). *Trichoderma*: Mass production, formulation, quality control, delivery, and its scope in commercialization in India for the management of plant diseases. *African Journal of Agricultural Research,* 9(38): 3838–3852.

Kumar, V., Koul, B., Taak, P., Yadav, D., and Song, M. (2023). Journey of *Trichoderma* from pilot scale to mass production: A review. *Agriculture,* 13(10): 2022. https://doi.org/10.3390/agriculture13102022.

Kumhar, K. C., Kumar, K., Arora, I., Bhatia, A. K., Batra, V. K., and Raj, H. (2022). Management of tomato damping-off disease in the nursery using *Trichoderma asperellum. International Journal of Economic Plants,* 9: 145–151.

Lahlali, R., Ezrari, S., Radouane, N., Kenfaoui, J., Esmaeel, Q., Hamss, H. E. et al. (2022). Biological control of plant pathogens: A global perspective. *Microorganisms,* 10(3). https://doi.org/10.3390/microorganisms10030596.

Lamichhane, J. R., Corrales, D. C., and Soltani, E. (2022). Biological seed treatments promote crop establishment and yield: A global meta-analysis. *Agronomy for Sustainable Development,* 42: 45.

Lanzuise, S., Manganiello, G., Guastaferro, V. M., Vincenzo, C., Vitaglione, P., Ferracane, R. et al (2022). Combined biostimulant applications of *Trichoderma* spp. with fatty acid mixtures improve biocontrol activity, horticultural crop yield, and nutritional quality. *Agronomy,* 12: 275.

Larkin, R. P. (2008). Influence of compost amendments on disease development and microbial communities in the rhizosphere of cucumber. *Soil Biology & Biochemistry,* 40(5): 1346–1356. https://doi.org/10.1016/j.soilbio.2007.12.017.

Lewis, J. (1991). Formulation and delivery systems of biocontrol agents with emphasis on fungi. pp. 279–287. *In: The Rhizosphere and Plant Growth: Proceedings of the Beltsville Symposia in Agricultural Research, Beltsville, MD, USA, 8–11 May 1989*. Springer: Berlin/Heidelberg, Germany.

Li, R.-X., Cai, F., Pang, G., Shen, Q.-R., Li, R., and Chen, W. (2015). Solubilisation of phosphate and micronutrients by *Trichoderma harzianum* and its relationship with the promotion of tomato plant growth. *PLOS ONE,* 10(6): e0130081. https://doi.org/10.1371/journal.pone.0130081.

Liu, L., Xu, Y., Cao, H., Fan, Y., Du, K., Bu, X. et al. (2022). Effects of *Trichoderma harzianum* biofertilizer on growth, yield, and quality of *Bupleurum chinense*. *Plant Direct,* 6: e461.

Loc, N. H., Huy, N. D., Quang, H. T., Lan, T. T., and Ha, T. T. T. (2020). Characterization and antifungal activity of extracellular chitinase from a biocontrol fungus, *Trichoderma asperellum* PQ34. *Mycology,* 11: 38–48.

López-Coria, M., Guzmán-Chávez, F., Carvente-García, R., Muñoz-Chapul, D., Sánchez-Sánchez, T., Arciniega-Ruíz, J. M. et al. (2023). Maize plant expresses SWEET transporters differently when interacting with *Trichoderma asperellum* and *Fusarium verticillioides*, two fungi with different lifestyles. *Frontiers in Plant Science,* 14. https://doi.org/10.3389/fpls.2023.1253741.

Macías-Rodríguez, L., Contreras-Cornejo, H. A., Adame-Garnica, S. G., Del-Val, E., and Larsen, J. (2020). The interactions of *Trichoderma* at multiple trophic levels: Inter-kingdom communication. *Microbiological Research,* 240: 126552. https://doi.org/10.1016/j.micres.2020.126552.

Marcelo, L. R., de Gois, J. S., da Silva, A. A., and Cesar, D. V. (2021). Synthesis of iron-based magnetic nanocomposites and applications in adsorption processes for water treatment: A review. *Environmental Chemistry Letters,* 19: 1229–1274.

Marik, T., Tyagi, C., Balázs, D., Urbán, P., Szepesi, Á., Bakacsy, L. et al. (2019). Structural diversity and bioactivities of peptaibol compounds from the *Longibrachiatum clade* of the filamentous fungal genus *Trichoderma*. *Frontiers in Microbiology,* 10: 1434. https://doi.org/10.3389/fmicb.2019.01434.

Martinez, D., Berka, R. M., Henrissat, B., Saloheimo, M., Arvas, M., Baker, S. E. et al. (2008). Genome sequencing and analysis of the biomass-degrading fungus *Trichodermareesei* (syn. *Hypocreajecorina*). *Nature Biotechnology,* 26: 553–560.

Martinez, Y., Ribera, J., and Schwarze, F. W. M. R. (2023). Biotechnological development of *Trichoderma*-based formulations for biological control. *Applied Microbiology and Biotechnology,* 107: 5595–5612. https://doi.org/10.1007/s00253-023-12687-x.

Martínez-Medina, A., Roldán, A., and Pascual, J. A. (2009). Performance of a *Trichoderma harzianum* bentonite–vermiculite formulation against Fusarium wilt in seedling nursery melon plants. *HortScience,* 44: 2025–2027.

Martins, M., Melo, I., and Oliveira, A. (2023). Solid-state fermentation of *Trichoderma*: Advances and future directions. *Biotechnology Advances,* 54: 107778.

Marx, D. H., and Krupa, S. V. (1978). Ectomycorrhizae and the role of fungi in plant disease. *Plant Disease Reporter,* 62(1): 5–9. https://doi.org/10.1094/PD-62-5.

Mary, L. S., and Russell, J. C. (2021). Molecular methods unravel the biosynthetic potential of *Trichoderma* species. *RSC Advances,* 11(6): 3622–3635. https://doi.org/10.1039/d0ra09627j.

Mbarga, J. B., Begoude, B., Ambang, Z., Meboma, M., Kuaté, J., Schiffers, B. et al. (2014). A new oil-based formulation of *Trichoderma asperellum* for the biological control of cacao black pod disease caused by *Phytophthora megakarya*. *Biological Control,* 77: 15–22.

Mei, H. M., Ruan, Y. N., Zhang, J. X., Cui, J. X., Yan, K., Dong, X. Y. et al. (2022). Effects of *Trichoderma* on nitrogen absorption and use efficiency in *Lycium chinense* roots under saline stress. *Ying yong sheng tai xue bao = The Journal of Applied Ecology,* 33(9): 2539–2546. https://doi.org/10.13287/j.1001-9332.202209.015.

Mendoza-Mendoza, A., Pozo, M. J., Grzegorski, D., Martínez, P., García, J. M., Olmedo-Monfil, V. et al. (2003). Enhanced biocontrol activity of *Trichoderma* through inactivation of a mitogen-activated protein kinase. *Proceedings of the National Academy of Sciences of the United States of America,* 100: 15965–15970.

Meyer, S. T. (2020). The impact of *Trichoderma* on soil pH and nutrient availability. *Applied Soil Ecology,* 149: 103546. https://doi.org/10.1016/j.apsoil.2019.103546.

Mishra, J., Tewari, S., and Singh, S. (2015). Biopesticides: Where we stand. pp. 135–154. *In:* Arora, N. K. (Ed.). *Plant Microbes Symbiosis*. Springer.

Mohiddin, F., Bashir, I., Padder, S. A., and Hamid, B. (2017). Evaluation of different substrates for mass multiplication of *Trichoderma* species. *Journal of Pharmacognosy and Phytochemistry,* 6: 563–569.

Montenecourt, B. S., and Eveleigh, D. E. (1979). Selective screening methods for the isolation of high-yielding cellulase mutants of *Trichodermareesei*. *Advances in Chemistry,* 181: 289–301.

Morgan, C. A., Herman, N., White, P., and Vesey, G. (2006). Preservation of microorganisms by drying: A review. *Journal of Microbiological Methods,* 66: 183–193.

Moses, K. (2014). Assessment of sawdust as carrier material for fungal inoculum intended for faster composting. *International Journal of Current Microbiology and Applied Sciences,* 3: 608–613.

Mukherjee, P. K., and Kenerley, C. M. (2010). Regulation of morphogenesis and biocontrol properties in *Trichoderma virens* by a VELVET protein, Vel1. *Applied and Environmental Microbiology,* 76: 2345–2352.

Mukherjee, P. K., Latha, J., Hadar, R., and Horwitz, B. A. (2003). TmkA, a mitogen-activated protein kinase of *Trichoderma virens*, is involved in biocontrol properties and repression of conidiation in the dark. *Eukaryotic Cell, 2*: 446–455.

Mukhopadhyay, A. N., Shrestha, S. M., and Mukherjee, P. K. (1992). Biological seed treatment for control of soil-borne plant pathogens. *FAO Plant Protection Bulletin,* 40: 21–30.

Nabi, S. U., Raja, W. H., Sharma, A., and Malik, G. (2017). Evaluation of different substrates for development of *Trichoderma harzianum* based stock cultures and their utilization in management of chilli wilt disease. *Chemistry Science Review and Letters,* 6: 2229–2235.

Nakkeeran, S., Fernando, W. D., and Siddiqui, Z. A. (2005). Plant growth-promoting rhizobacteria formulations and its scope in commercialization for the management of pests and diseases. pp. 257–296. *In: PGPR: Biocontrol and Biofertilization*. Springer.

O'Callaghan, M. (2016). Microbial inoculation of seed for improved crop performance: Issues and opportunities. *Applied Microbiology and Biotechnology,* 100: 5729–5746.

Oancea, F., Raut, I., Sesan, T. E., and Cornea, P. C. (2016). Dry flowable formulation of biostimulants *Trichoderma* strains. *Agriculture and Agricultural Science Procedia,* 10: 494–502.

Olasehinde, G. I., Okolie, Z. V., Oniha, M. I., Adekeye, B. T., and Ajayi, A. A. (2016). In vitro antibacterial and antifungal activities of *Chrysophyllum albidum* and *Diospyros monbuttensis* leaves. *Journal of Pharmacognosy and Phytotherapy,* 8(1): 1–7.

Organo, N. D., Granada, S. M. J. M., Pineda, H. G. S., Sandro, J. M., Nguyen, V. H., and Gummert, M. (2022). Assessing the potential of a *Trichoderma*-based compost activator to hasten the decomposition of incorporated rice straw. *Scientific Reports,* 12: 448.

Oszust, K., Cybulska, J., and Frąc, M. (2019). How do *Trichoderma* genus fungi win a nutritional competition battle against soft fruit pathogens? A report on niche overlap nutritional potentiates. *International Journal of Molecular Sciences,* 21(12): 4235. https://doi.org/10.3390/ijms21124235.

Oyesola, O. L., Aworunse, O. S., Oniha, M. I., Obiazikwor, O. H., Bello, O., Atolagbe, O. M. et al. (2021). Impact and management of diseases of *Solanum tuberosum*. IntechOpen. https://doi.org/10.5772/intechopen.98899.

Oyesola, O. L., Sobowale, A. A., and Obembe, O. O. (2020). Effectiveness of *Trichodermakoningii* extract on *Aspergillus* species isolated from rotting tomato (*Solanum lycopersicum* Mill). *Tropical Journal of Products and Research,* 4(11): 961–965. https://doi.org/10.26538/tjnpr/v4i11.19.

Oyesola, O. L., Tonjock, R. K., Bello, A. O., Taiwo, O. S., and Obembe, O. O. (2024). *Trichoderma*: A review of its mechanisms of action in plant disease control. https://doi.org/10.20944/preprints202405.1378.v1.

Pandit, M. A., Kumar, J., Gulati, S., Bhandari, N., Mehta, P., Katyal, R. et al. (2022). Major biological control strategies for plant pathogens. *Pathogens,* 11(2): Article 273. https://doi.org/10.3390/pathogens11020273.

Panwar, V., Aggarwal, A., Singh, G., Verma, A., Sharma, I., and Saharan, M. (2014). Efficacy of foliar spray of *Trichoderma* isolates against *Fusarium graminearum* causing head blight of wheat. *Journal of Wheat Research,* 6: 59–63.

Papavizas, G. (1985). *Trichoderma* and *Gliocladium*: Biology, ecology, and potential for biocontrol. *Annual Review of Phytopathology,* 23: 23–54.

Paravar, A., Piri, R., Balouchi, H., and Ma, Y. (2023). Microbial seed coating: An attractive tool for sustainable agriculture. *Biotechnology Reports,* 37, Article e00781.

Pascual, J. A., Ceglie, F., Tuzel, Y., Koller, M., Koren, A., Hitchings, R. et al. (2018). Organic substrate for transplant production in organic nurseries: A review. *Agronomy for Sustainable Development, 38*: 1–23.

Patil, P. D., Patil, S. P., Kelkar, R. K., Patil, N. P., Pise, P. V., and Nadar, S. S. (2021) (a). Enzyme-assisted supercritical fluid extraction: An integral approach to extract bioactive compounds. *Trends in Food Science & Technology,* 116: 357–369.

Patil, S. S., Guldegar, D. D., Potdukhe, S. R., and Khobragade, H. M. (2021) (b). Investigation of liquid formulation of *Trichodermaasperellum* against Fusarium wilt of chickpea. *Biology Forum - International Journal,* 13: 571–576.

Patkowska, E., Mielniczuk, E., and Jamiołkowska, A. (2020). The influence of *Trichoderma harzianum* Rifai T-22 and other biostimulants on rhizosphere beneficial microorganisms of carrot. *Agronomy,* 10(11): 1637. https://doi.org/10.3390/agronomy10111637.

Pecoraro, L., Wang, X., Shah, D., Song, X., Kumar, V., Shakoor, A. et al. (2021). Biosynthesis pathways, transport mechanisms, and biotechnological applications of fungal siderophores. *Journal of Fungi,* 8(1). https://doi.org/10.3390/jof8010021.

Peeran, M. F., Krishnan, N., Thangamani, P. R., Gandhi, K., Thiruvengadam, R., and Kuppusamy, P. (2014). Development and evaluation of water-in-oil formulation of *Pseudomonas fluorescens* (FP7) against *Colletotrichum musae*, incitant of anthracnose disease in banana. *European Journal of Plant Pathology,* 138: 167–180. https://doi.org/10.1007/s10658-013-0392-5.

Penttilä, M., Nevalainen, H., Rättö, M., Salminen, E., and Knowles, J. (1987). A versatile transformation system for the cellulolytic filamentous fungus *Trichodermareesei*. *Gene,* 61: 155–164.

Pereira-Dias, L., Oliveira-Pinto, P. R., Fernandes, J. O., Regalado, L., Mendes, R., Teixeira, C. et al. (2023). Peptaibiotics: Harnessing the potential of microbial secondary metabolites for mitigation of plant pathogens. *Biotechnology Advances,* 68: Article 108223. https://doi.org/10.1016/j.biotechadv.2023.108223.

Persoon, C. H. (1794). *Deposit MethodicaFungorum. Romer's Neues Magazin für Botanik,* 1: 81–128.

Poveda, J., and Echeverría, J. (2015). The role of beneficial microorganisms in plant health. *The Scientific World Journal*, Article 576508. https://doi.org/10.1155/2015/576508.

Pradhan, P. C., Mukhopadhyay, A., Kumar, R., Kundu, A., Patanjali, N., Dutta, A. et al. (2022). Performance appraisal of *Trichoderma viride* based novel tablet and powder formulations for management of Fusarium wilt disease in chickpea. *Frontiers in Plant Science,* 13: Article 990392.

Prasad, R. D., Poorna Chandrika, K. S., Desai, S., Greeshma, K., and Vijaykumar, S. (2022). Development, production, and storage of *Trichoderma* formulations for agricultural applications. In *Advances in Trichoderma Biology for Agricultural Applications* (pp. 371–385). Springer.

Prasun, K., Mukherjee, A. M., Susanne, Z., and Benjamin, A. H. (2022). Mycoparasitism as a mechanism of *Trichoderma*-mediated suppression of plant diseases. *Fungal Biology Reviews,* 39: 15–33. https://doi.org/10.1016/j.fbr.2021.11.004.

Rai, D., and Tewari, A. (2016). Shelf-life studies of different formulations based on *Trichoderma harzianum* (Th14). *Annals of Biological Research,* 7: 1–5.

Rajput, R. S., Singh, P., Singh, J., Vaishnav, A., Ray, S., and Singh, H. (2019). *Trichoderma* mediated seed biopriming augments antioxidant and phenylpropanoid activities in tomato plant against *Sclerotium rolfsii*. *Journal of Pharmacognosy and Phytochemistry,* 8: 2641–2647.

Ramanujam, B., Prasad, R., Sriram, S., and Rangeswaran, R. (2010). Mass production, formulation, quality control, and delivery of *Trichoderma* for plant disease management. *Journal of Plant Protection Science,* 2: 1–8.

Rascio, N., and La Rocca, N. (2012). Biological nitrogen fixation. In *Encyclopedia of Ecology* (2nd ed., pp. 264–279). https://doi.org/10.1016/B978-0-444-63768-0.00685-5.

Rauf, A., Subhani, M. N., Siddique, M., Shahid, H., Chattha, M. B., Alrefaei, A. F. et al. (2024). Cultivating a greener future: Exploiting *Trichoderma* derived secondary metabolites for *Fusarium wilt* management in peas. *Heliyon,* 10(7): Article e29031. https://doi.org/10.1016/j.heliyon.2024.e29031.

Rauf, M., Awais, M., Ud-Din, A., Ali, K., Gul, H., Rahman, M. M. et al. (2021). Molecular mechanisms of the 1-aminocyclopropane-1-carboxylic acid (ACC) deaminase-producing *Trichoderma asperellum* MAP1 in enhancing wheat tolerance to waterlogging stress. *Frontiers in Plant Science,* 11: 614971. https://doi.org/10.3389/fpls.2020.614971.

Ribeiro, M. S., Voltan, A. R., Carraro, C. B., Steindorff, A. S., Goldman, G. H., Silva, R. N. et al. (2019). Endo-β-1,3-glucanase (GH16 family) from *Trichoderma harzianum* participates in cell wall biogenesis but is not essential for antagonism against plant pathogens. *Biomolecules,* 9(12). https://doi.org/10.3390/biom9120781.

Rini, C., Ramya, J., Jayakumar, G., and Shajan, V. (2018). Low-cost carrier material for mass production of *Trichoderma* inoculants. *Journal of Tropical Agriculture,* 56: 63–68.

Robinson, J. R., Isikhuemhen, O. S., and Anike, F. N. (2021). Fungal–metal interactions: A review of toxicity and homeostasis. *Journal of Fungi,* 7(3). https://doi.org/10.3390/jof7030225.

Rocha, I., Ma, Y., Souza-Alonso, P., Vosátka, M., Freitas, H., and Oliveira, R. S. (2019). Seed coating: A tool for delivering beneficial microbes to agricultural crops. *Frontiers in Plant Science,* 10: 1357.

Rocha-Ramírez, V., Omero, C., Chet, I., Horwitz, B. A., and Herrera-Estrella, A. (2002). *Trichoderma atroviride* G-protein-subunit gene tga1 is involved in mycoparasitic coiling and conidiation. *Eukaryotic Cell,* 1: 594–605.

Sachdev, S., and Singh, R. P. (2016). Current challenges, constraints, and future strategies for development of successful market for biopesticides. *Climatic Change and Environmental Sustainability,* 4: 129–136.

Saravanakumar, K., and Wang, M. (2019). Isolation and molecular identification of *Trichoderma* species from wetland soil and their antagonistic activity against phytopathogens. *Physiological and Molecular Plant Pathology,* 109: 101458. https://doi.org/10.1016/j.pmpp.2020.101458.

Sarkar, D., Singh, S., Parihar, M., and Rakshit, A. (2021). Seed bio-priming with microbial inoculants: A tailored approach towards improved crop performance, nutritional security, and agricultural sustainability for smallholder farmers. *Current Research in Environmental Sustainability,* 3: 100093. https://doi.org/10.1016/j.crsust.2021.100093.

Sawant, I. S., and Sawant, S. (1996). A simple method for achieving high cfu of *Trichoderma harzianum* on organic wastes for field applications. *Indian Phytopathology,* 49: 185–187.

Scharf, D. H., Brakhage, A. A., and Mukherjee, P. K. (2016). Gliotoxin–bane or boon? *Environmental Microbiology,* 18(4): 1096–1109. https://doi.org/10.1111/1462-2920.13080.

Schuster, A., Bruno, K. S., Collett, J. R., Baker, S. E., Seiboth, B., Kubicek, C. P. et al. (2012). A versatile toolkit for high throughput functional genomics with *Trichodermareesei*. *Biotechnology for Biofuels,* 5: 1–10.

Sehim, A. E., Hewedy, O. A., Altammar, K. A., Alhumaidi, M. S., and Abd Elghaffar, R. Y. (2023). *Trichoderma asperellum* empowers tomato plants and suppresses *Fusarium oxysporum* through priming responses. *Frontiers in Microbiology,* 14: 1140378. https://doi.org/10.3389/fmicb.2023.1140378.

Sesan, T. E., Oancea, A. O., Ştefan, L. M., Mănoiu, V. S., Ghiurea, M., Răut, I. et al. (2020). Effects of foliar treatment with a *Trichoderma* plant biostimulant consortium on *Passiflora caerulea* L. yield and quality. *Microorganisms,* 8: 123.

Shenouda, M. L., Ambilika, M., Skellam, E., and Cox, R. J. (2022). Heterologous expression of secondary metabolite genes in *Trichodermareesei* for waste valorization. *Journal of Fungi,* 8(4): 355. https://doi.org/10.3390/jof8040355.

Shirvani, M., and Yahaghi, Z. (2021). Role of Pb-solubilizing and plant growth-promoting bacteria in Pb uptake by plants. In *Advances in Microbe-Assisted Phytoremediation of Polluted Sites* (pp. 231–270). https://doi.org/10.1016/B978-0-12-823443-3.00002-8

Shoemaker, S., Schweickart, V., Ladner, M., Gelfand, D., Kwok, S., Myambo, K. et al. Molecular cloning of exocellohydrolase derived from *Trichodermareesei* strain L27. *Biotechnology,* 1: 691–696.

Shoresh, M., Harman, G. E., and Mastouri, F. (2010). Induced systemic resistance and plant responses to fungal biocontrol agents. *Annual Review of Phytopathology,* 48: 21–43.

Siddiquee, S. (2017). Fungal volatile organic compounds: Emphasis on their plant growth-promoting. pp. 313–333. *In: Volatiles and Food Security*. Springer.

Singh, U. S., and Zaidi, N. W. (2002). Current status of formulation and delivery of fungal and bacterial antagonists for disease management in India. In *Microbial Biopesticide Formulations and Application* (p. 269). Project Directorate of Biological Control.

Singh, V., Upadhyay, R. S., Sarma, B., and Singh, H. (2016). Seed bio-priming with *Trichoderma asperellum* effectively modulates plant growth promotion in pea. *International Journal of Agricultural and Environmental Biotechnology,* 9(4): 361–365.

Sobowale, A. (2019). Probable effects of dual inoculation of maize (*Zea mays*) stem with *Fusarium verticillioides* and certain *Trichoderma* species on fumonisin content of maize seeds. *American Journal of Plant Sciences,* 10: 752–759. https://doi.org/10.4236/ajps.2019.105055.

Song, M., Wang, X., Xu, H., Zhou, X., and Mu, C. (2023). Effect of *Trichoderma viride* on insoluble phosphorus absorption ability and growth of *Melilotus officinalis*. *Scientific Reports,* 13(1): 1–11. https://doi.org/10.1038/s41598-023-39501-y.

Sood, M., Kapoor, D., Kumar, V., Sheteiwy, M. S., Ramakrishnan, M., Landi, M. et al. (2020). *Trichoderma*: The "secrets" of a multitalented biocontrol agent. *Plants,* 9: 762.

Sreenayana, B., and Nakkeeran, S. (2019). Effect of oil-based formulation of *Trichoderma* spp. on growth parameters of cucumber seedlings. *International Journal of Current Microbiology and Applied Sciences,* 8: 200–209.

Sriram, S., Roopa, K. P., and Savitha, M. J. (2011). Extended shelf-life of liquid fermentation-derived talc formulations of *Trichoderma harzianum* with the addition of glycerol in the production medium. *Crop Protection,* 30: 1334–1339.

Stewart, R., and Hill, R. (2013). Applications of *Trichoderma* in plant growth promotion. In *Biotechnology and Biology of Trichoderma* (pp. 415-428). https://doi.org/10.1016/B978-0-444-59576-8.00031-X.

Stricker, A. R., Grosstessner-Hain, K., Würleitner, E., and Mach, R. L. (2006). Xyr1 (xylanase regulator 1) regulates both the hydrolytic enzyme system and D-xylose metabolism in *Hypocreajecorina*. *Eukaryotic Cell,* 5: 2128–2137.

Sudana, I. M., and Raka, I. G. N. (2020). Seed treatment using *Trichoderma* spp. formulated in bioslurry and vermicompost to induce resistance of peanut (*Arachis hypogaea*) diseases. *International Journal of Current Microbiology and Applied Sciences,* 9: 1253–1265.

Sundaramoorthy, S., and Balabaskar, P. (2013). Biocontrol efficacy of *Trichoderma* spp. against wilt of tomato caused by *Fusarium oxysporum* f. sp. *lycopersici*. *Journal of Applied Biology and Biotechnology,* 1: 036–040.

Swain, H., Adak, T., Mukherjee, A. K., Mukherjee, P. K., Bhattacharyya, P., Behera, S. et al. (2018). Novel *Trichoderma* strains isolated from tree barks as potential biocontrol agents and biofertilizers for direct seeded rice. *Microbiological Research,* 214: 83–90.

Swain, H., Adak, T., Mukherjee, A. K., Sarangi, S., Samal, P., Khandual, A. et al. (2021). Seed biopriming with *Trichoderma* strains isolated from tree bark improves plant growth, antioxidative defense system in rice and enhances straw degradation capacity. *Frontiers in Microbiology,* 12: 633881.

Syam, N., Hidrawati, Sabahannur, S., and Nurdin, A. (2021). Effects of *Trichoderma* and foliar fertilizer on the vegetative growth of black pepper (*Piper nigrum*) seedlings. *International Journal of Agronomy,* 2021: 9953239.

Tamandegani, P. R., Marik, T., Zafari, D., Balázs, D., Vágvölgyi, C., Szekeres, A. et al. (2020). Changes in peptaibol production of *Trichoderma* species during in vitro antagonistic interactions with fungal plant pathogens. *Biomolecules,* 10(5). https://doi.org/10.3390/biom10050730.

Tandon, A., Fatima, T., Shukla, D., Tripathi, P., Srivastava, S., and Singh, P. C. (2019). Phosphate solubilisation by *Trichodermakoningiopsis* (NBRI-PR5) under abiotic stress conditions. *Journal of King Saud University - Science,* 32(1): 791–798. https://doi.org/10.1016/j.jksus.2019.02.001.

Tewari, A., and Mukhopadhyay, A. (2001). Testing of different formulations of *Gliocladium virens* against chickpea wilt-complex. *Indian Phytopathologist,* 54: 67–71.

Thangavelu, R., Palaniswami, A., and Velazhahan, R. (2004). Mass production of *Trichoderma harzianum* for managing fusarium wilt of banana. *Agricultural Ecosystems and Environment,* 103: 259–263.

Tiwari, R. K., Kumar, R., Sharma, S., Sagar, V., Aggarwal, R., Naga, K. C. et al. (2020). Potato dry rot disease: Current status, pathogenomics, and management. *3 Biotech,* 10: 503.

Topolovec-Pintarić, S. (2019). *Trichoderma*: Invisible partner for visible impact on agriculture. In *Trichoderma The Most Widely Used Fungicide* (p. 15). IntechOpen.

Tulasne, L. R., and Tulasne, C. (1863). *Selecta Fungorum Carpologia* 2. i–xix, 319 pp., 34 tabs. Paris.

Tyśkiewicz, R., Nowak, A., Ozimek, E., and Jaroszuk-Ściseł, J. (2022). *Trichoderma*: The current status of its application in agriculture for the biocontrol of fungal phytopathogens and stimulation of plant growth. *International Journal of Molecular Sciences,* 23(4): 2329. https://doi.org/10.3390/ijms23042329.

Uddin, A., Hussain, M., Rahman, S., Ahmad, H., and Roni, M. (2015). Effect of *Trichoderma* concentrations on growth and yield of tomato. *Bangladesh Research Publication Journal,* 11: 228–232.

Van Jaarsveld, W. J., Halleen, F., Bester, M. C., Pierron, R. J., Stempien, E., and Mostert, L. (2021). Investigation of *Trichoderma* species colonization of nursery grapevines for improved management of black foot disease. *Pest Management Science,* 77: 397–405.

Van Wees, S. C. M., Van der Ent, S., and Pieterse, C. M. J. (2000). Induced resistance in plants. *Plant Physiology,* 124(2): 511–519. https://doi.org/10.1104/pp.124.2.511.

Vázquez, M. B., Barrera, V., and Bianchinotti, V. (2015). Molecular identification of three isolates of *Trichoderma harzianum* isolated from agricultural soils in Argentina, and their abilities to detoxify in vitro metsulfuron methyl. *Botany,* 93: 793–800.

Vicente, I., Baroncelli, R., Hermosa, R., Monte, E., Vannacci, G., and Sarrocco, S. (2022). Role and genetic basis of specialised secondary metabolites in *Trichoderma* ecophysiology. *Fungal Biology Reviews,* 39: 83–99. https://doi.org/10.1016/j.fbr.2021.12.004.

Vinale, F., Marra, R., Scala, F., Ghisalberti, E. L., Lorito, M., and Sivasithamparam, K. (2006). Major secondary metabolites produced by two commercial *Trichoderma* strains active against different phytopathogens. *Letters in Applied Microbiology,* 43: 143–148.

Vindas-Reyes, E., Chacón-Cerdas, R., and Rivera-Méndez, W. (2024). *Trichoderma* Production and Encapsulation Methods for Agricultural Applications. *AgriEngineering*, 6(3): 2366–2384.

Viterbo, A., Harel, M., Horwitz, B. A., Chet, I., and Mukherjee, P. K. (2005). *Trichoderma* mitogen-activated protein kinase signaling is involved in induction of plant systemic resistance. *Applied and Environmental Microbiology,* 71: 6241–6246.

Volkov, P., Rubtsova, E., Rozhkova, A., Sinitsyna, O., Zorov, I., Kondratyeva, E. et al. (2020). Properties of recombinant endo-β-1,6-glucanase from *Trichoderma harzianum* and its application in the pustulan hydrolysis. *Carbohydrate Research,* 499: 108211. https://doi.org/10.1016/j.carres.2020.108211.

Wang, R., Liu, C., Jiang, X., Tan, Z., Li, H., Xu, S. et al. (2022). The newly identified *Trichoderma harzianum* partitivirus (ThPV2) does not diminish spore production and biocontrol activity of its host. *Viruses,* 14: 1532.

Weindling, R. (1932). *Trichodermalignorum* as a parasite of other soil fungi. *Phytopathology,* 22: 837–845.

Weindling, R. (1934). Studies on a lethal principle effective in the parasitic action of *Trichodermalignorum* on *Rhizoctonia solani* and other soil fungi. *Phytopathology,* 24: 1153–1179.

Wells, H. D. (1972). Efficacy of *Trichoderma harzianum* as a biocontrol for *Sclerotium rolfsii. Phytopathology,* 62: 442.

Whipps, J. M. (1997). Developments in the biological control of soil-borne plant pathogens. *Advances in Botanical Research,* 26: 1–134.

Woo, S. L., Hermosa, R., Lorito, M., and Monte, E. (2023). *Trichoderma*: A multipurpose, plant-beneficial microorganism for eco-sustainable agriculture. *Nature Reviews Microbiology,* 21: 312–326.

Xiao, Z., Zhao, Q., Li, W., Gao, L., and Liu, G. (2023). Strain improvement of *Trichoderma harzianum* for enhanced biocontrol capacity: Strategies and prospects. *Frontiers in Microbiology,* 14: 1146210.

Xue, H., Liu, Q., and Yang, Z. (2023). Pathogenicity, mycotoxin production, and control of potato dry rot caused by *Fusarium* spp.: A review. *Journal of Fungi,* 9: 843.

Yadav, M., Divyanshu, K., Dubey, M. K., Rai, A., Kumar, S., Tripathi, Y. N. et al. (2023). Plant growth promotion and differential expression of defense genes in chilli pepper against *Colletotrichum truncatum* induced by *Trichoderma asperellum* and *T. harzianum. BMC Microbiology,* 23. https://doi.org/10.1186/s12866-023-02789-x.

Yang, Z., Zhi, P., and Chang, C. (2022). Priming seeds for the future: Plant immune memory and application in crop protection. *Frontiers in Plant Science,* 13: 961840.

Yao, X., Guo, H., Zhang, K., Zhao, M., Ruan, J., and Chen, J. (2023). *Trichoderma* and its role in biological control of plant fungal and nematode disease. *Frontiers in Microbiology,* 14: 1160551. https://doi.org/10.3389/fmicb.2023.1160551.

Yedidia, I., Benhamou, N., and Chet, I. (1999). Induction of defense responses in cucumber plants (*Cucumis sativus* L.) by the biocontrol agent *Trichoderma harzianum. Applied and Environmental Microbiology,* 65: 1061–1070.

Yedidia, I., Srivastva, A. K., Kapulnik, Y., and Chet, I. (2001). Effect of *Trichoderma harzianum* on microelement concentrations and increased growth of cucumber plants. *Plant and Soil,* 235: 235–242.

Yu, C., Jiang, X., Xu, H., and Ding, G. (2023). *Trichoderma longi brachiatum* inoculation improves drought resistance and growth of *Pinus massoniana* seedlings through regulating physiological responses and soil microbial community. *Journal of Fungi,* 9: 694. https://doi.org/10.3390/jof9030694.

Yuan, M., Huang, Y., Ge, W., Jia, Z., Song, S., Zhang, L. et al. (2019). Involvement of jasmonic acid, ethylene, and salicylic acid signaling pathways behind the systemic resistance induced by *Trichoderma longi brachiatum* H9 in cucumber. *BMC Genomics,* 20: 144.

Yusuf, M. A., Singh, B. N., Sudheer, S., Kharwar, R. N., Siddiqui, S. M. A., Fernandes Fraceto, L. et al. (2019). Chrysophanol: A natural anthraquinone with multifaceted biotherapeutic potential. *Biomolecules,* 9(2): 68. https://doi.org/10.3390/biom9020068.

Zafra, G., Moreno-Montaño, A., Absalón, Á. E., and Cortés-Espinosa, D. V. (2015). Degradation of polycyclic aromatic hydrocarbons in soil by a tolerant strain of *Trichoderma asperellum. Environmental Science and Pollution Research,* 22: 1034–1042.

Zaid, R., Koren, R., Kligun, E., Gupta, R., Leibman-Markus, M., Mukherjee, P. K. et al. (2022). Gliotoxin, an immunosuppressive fungal metabolite, primes plant immunity: Evidence from *Trichoderma virens*-tomato interaction. *Microbiology,* 13(4). https://doi.org/10.1128/mbio.00389-22.

Zehra, A., Raytekar, N. A., Meena, M., and Swapnil, P. (2021). Efficiency of microbial bio-agents as elicitors in plant defense mechanism under biotic stress: A review. *Current Research in Microbial Sciences, 2*. https://doi.org/10.1016/j.crmicr.2021.100054.

Zeilinger, S., Gruber, S., Bansal, R., and Mukherjee, P. K. (2016). Secondary metabolism in *Trichoderma*—Chemistry meets genomics. *Fungal Biology Reviews,* 30: 74–90.

Zhang, G.-Z., Yang, H.-T., Zhang, X.-J., Zhou, F.-Y., Wu, X.-Q., Xie, X.-Y. et al. (2022). Five new species of *Trichoderma* from moist soils in China. *MycoKeys,* 87: 133.

Zhang, J., Zhang, P., Zeng, G., Wu, G., Qi, L., Chen, G. et al. (2021). Transcriptional differences guided discovery and genetic identification of coprogen and dimerumic acid siderophores in *Metarhizium robertsii*. *Frontiers in Microbiology,* 12: 783609. https://doi.org/10.3389/fmicb.2021.783609.

Zhao, L., Wang, Y., and Kong, S. (2020). Effects of *Trichoderma asperellum* and its siderophores on endogenous auxin in *Arabidopsis thaliana* under iron-deficiency stress. *International Microbiology,* 23: 501–509.

Zhao, L., Li, Y., Yu, M., Peng, Y., and Ran, F. (2023). Electrolyte-wettability issues and challenges of electrode materials in electrochemical energy storage, energy conversion, and beyond. *Advanced Science*, 10(17): 2300283.

Zhao, X., Zhou, L., Riaz Rajoka, M. S., Yan, L., Jiang, C., Shao, D. et al. (2018). Fungal silver nanoparticles: Synthesis, application, and challenges. *Critical Reviews in Biotechnology,* 38: 817–835.

Index

A

Application methods 314, 331

B

Bacillus sp. 216, 218, 220
bioactive compounds 69, 73–75, 79, 80, 86, 89, 90, 98, 289, 293
Biocontrol agents 3
Biofungicide agent 14, 17, 25
Biofungicide Formulation 8, 118–120, 122–126, 129, 130, 132, 133, 135, 138, 141–143
Biofungicides 1, 3, 4, 6–9, 35, 43, 44–51, 53, 63–70, 73–77, 79, 100–104, 175–179, 181–184, 186–201
Biological control agents 209–210, 225
Bionanofungicides 24, 26
biosynthetic pathways 73
Botanical 13, 15, 17, 18

C

Carrier compatibility 159, 160
Carrier Material 168, 190, 218, 222, 290
Carrier selection 149, 155, 158–160
Chromatography 66, 79, 86, 97, 98
Commercial Development 51
commercialization 209, 222, 223
Culture Maintenance 34, 42, 53

D

Disease management 3, 6, 8, 9
Dried formulation 167, 168

E

Encapsulation 157, 158, 160, 162, 164–166

F

formulation 63–66, 68, 74, 98, 100, 102, 103, 267, 268, 271, 276, 287, 288, 294–296
Formulation of biofungicides 153, 158
Formulation techniques 311, 337

I

improvement 222, 224–226
Integrated pest management 26

M

Mechanisms 243, 246–250, 258, 314–317, 320
Microbial 13–15, 18, 20, 22, 25, 26
Microbial strains 178, 191, 195, 197

N

Nanobiofungicide 25
Nanoemulsion 156–158, 161–163, 165, 166, 168, 169
Nanotechnology 187, 197–199

P

phytopathogenic fungi 40, 65, 175, 176, 221, 242, 247–251, 258
Purification 78, 79, 94–96

R

Resistance development 6, 9

S

Scale-up Production 175, 189
Screening 210, 213, 214, 216, 222, 226, 227
Screening of biocontrol agent 151
secondary metabolite 271, 280
Solid-State Fermentation 35, 44
Streptomyces 157, 242–254, 256–258
Submerged Fermentation 35, 45
Sustainable agriculture 245, 313, 314, 316, 336
Synthetic fungicides 5–8

T

Trichoderma spp. 155, 162, 163, 168, 255, 257, 258, 311–337

About the Editors

Kamel A. Abd-Elsalam, Ph.D., is currently a Research Professor and Deputy Director for Research at the Plant Pathology Research Institute, Agricultural Research Center, Giza, Egypt. Dr. Kamel's research interests include developing and improving plant biosecurity diagnostic tools, understanding fungal pathogen genomes, and creating eco-friendly hybrid nanomaterials for controlling toxigenic fungi and plant diseases, as well as applications in agroecosystems. Kamel earned his Ph.D. in Molecular Plant Pathology from Christian Albrechts University of Kiel (Germany) and Suez Canal University (Egypt). In 2008, he was awarded a postdoctoral fellowship from the same institution. Dr. Kamel served as a Visiting Associate Professor at Mae Fah Luang University in Thailand, the Institute of Microbiology at TUM in Germany, the Laboratory of Phytopathology at Wageningen University in the Netherlands, and the Plant Protection Department at Sassari University in Italy.

He has published 35 books related to nanobiotechnology applications in agriculture and plant protection with major publishing houses such as Springer, Taylor & Francis, and Elsevier. Since 2019, he has served as the Editor-in-Chief of the Elsevier book series *Nanobiotechnology for Plant Protection* and as the Series Editor for *Applications of Genome Modified Plants and Microbes in Food and Agriculture*. Additionally, he has been an active member of the Elsevier Advisory Panel since 2020, providing feedback and suggestions to improve Elsevier's products and services. He has published more than 302 scientific research articles in international and regional specialized journals with high impact factors and has an H-index of 51, an i10-index of 163, and over 10,134 citations. He served as a Guest Editor for the *Journal of Fungi, Plants, and Microorganisms*, and as a Review Editor for *Frontiers in Genomic Assay Technology, Agrochemicals, and Plant Nano Biology*. He was ranked in the top 2% of the most influential scientists in the world in nanobiotechnology for 2021, 2022, 2023, and 2024 by Elsevier and Stanford University. In 2014, he was awarded the Federation of Arab Scientific Research Councils Prize for excellent scientific research in biotechnology (fungal genomics), ranking first. Additionally, he received the Egyptian State Award for Agricultural Scientific Excellence from ASRT, Egypt, in 2022 with a silver medal.

Dr. Mirza Abid Mehmood is a Lecturer in Plant Pathology at the Institute of Plant Protection, Muhammad Nawaz Shareef University of Agriculture, Multan, Punjab, Pakistan. He completed his Ph.D. in Molecular Plant Pathology from Huazhong Agricultural University, Wuhan, P.R. China. He is currently a Postdoctoral Research Fellow in the School of Ecology and Environmental Science, Yunnan University, Kunming, China. Dr. Mirza's research interests include the use of eco-friendly approaches to manage biotic and abiotic stress, soil microbial ecology, phytobiomes, metagenomics, metatranscriptomics, metaproteomics, and metabolomics. He has published 7 book chapters with Elsevier and Springer Nature. He has published 35 research and review articles in international peer-reviewed journals, including *Current Opinion in Green and Sustainable Chemistry, Microbiology Spectrum, Plants, Horticulturae, Journal of King Saud University-Science, Stress Biology, International Journal of Tropical Insect Science, Saudi Journal of Biological Sciences*, and the *Journal of Environmental Management*. Dr. Mirza is a reviewer for journals including the *Journal of Environmental Management, Phytopathology Research, Frontiers in Microbiology, Journal of Soil Science and Plant Nutrition* and *SN Applied Sciences, Soil Systems,* and *Agronomy*.

For Product Safety Concerns and Information please contact our EU representative GPSR@taylorandfrancis.com Taylor & Francis Verlag GmbH, Kaufingerstraße 24, 80331 München, Germany